GEOCHEMISTRY, GROUNDWATER AND POLLUTION

GEOCHEMISTRY, GROUNDWATER AND POLLUTION

C.A.J. APPELO
Faculty of Earth Sciences, Free University, Amsterdam

D. POSTMA
Department of Geology & Geotechnical Engineering, Technical University of Denmark, Lyngby

A.A. BALKEMA / ROTTERDAM / BROOKFIELD / 1999

CIP-DATA KONINKLIJKE BIBLIOTHEEK, DEN HAAG

Appelo, C.A.J.

Geochemistry, groundwater and pollution/C.A.J. Appelo,
D. Postma. – Rotterdam [etc.]: Balkema. – Ill.
With index, ref.
ISBN 90 5410 105 9 bound
ISBN 90 5410 106 7 pbk.
Subject headings: geochemistry/groundwater.

First print: 1993
Second corrected print: 1994
Third corrected print: 1996
Fourth corrected print: 1999

Published by
A.A. Balkema, P.O. Box 1675, 3000 BR Rotterdam, Netherlands
Fax: +31.10.4135947; E-mail: balkema@balkema.nl; Internet site: http://www.balkema.nl

A.A. Balkema Publishers, Old Post Road, Brookfield, VT 05036-9704, USA
Fax: 802.276.3837; E-mail: info@ashgate.com

ISBN 90 5410 105 9 hardbound edition
ISBN 90 5410 106 7 student paper edition

Preface

There exists a number of practical reasons for studying groundwater geochemistry. Groundwater is used for drinking water, irrigating lands, or general industrial purposes. The quality of groundwater determines if it is suitable for consumption, whether it may adversely affect soil properties, or precipitate mineral scales. The problems may be caused by the concentration levels of only one or two solutes and the processes controlling these levels must be understood to make optimal use of the groundwater resource. Will concentrations increase or remain stable during exploitation of the aquifer? What happens when air enters the saturated zone due to the drawdown of piezometric levels by production from the aquifer? What are the changes in water composition due to upconing of saltwater in an over-exploited reservoir?

In recent years also pollution of aquifers has become an important issue. Sources of pollution may be either diffuse or localized. Groundwater acidification by acid rain and nitrate pollution by the use of fertilizers both result from diffuse sources of pollution. Examples of point sources are landfills and chemical waste deposits from which organic substances and heavy metals may leach. The important question is in both cases whether pollutants are transported as conservative components, or to what extent they will be retarded by geochemical processes within the aquifer.

This book attempts to offer a theoretical framework and to present practical applications to address these questions. The first eight chapters treat the main chemical processes that affect groundwater composition. These chapters are suitable for a first course in groundwater geochemistry. Examples are provided of how natural groundwaters have obtained their composition, and the effects of pollution are discussed. Under ideal field conditions a single chemical reaction may adequately describe the evolution of groundwater compositions, but in general, the interplay of several processes must be considered. The important, and difficult task is to trace and verify which of the processes have dominant control in a natural, and often complex, situation. Here, the laws of chemistry and physics offer the tools to be used in the investigation. Their application may offer protection against the violation of principles that have proved their validity in much broader disciplines than ours.

The remaining chapters present the systematics of transport in aquifers and how to couple transport to chemical reactions. It treats the basics of writing computer programs to calculate the combined effects of transport and chemical reactions. Applications of hydrogeochemical transport modeling of complex systems are demonstrated and geochemical approaches to enhance aquifer clean-up schemes are evaluated. This second part of the book is more appropriate for courses at an advanced level although we have tried to treat all the material in such a way that first year courses in earth sciences and chemistry are sufficient background.

We thank colleagues and students for commenting and reviewing various parts of the book. Among these, we particularly would like to mention N.M. de Rooy, P. Engesgaard, J. Griffioen, C.J. Hemker, L. Revallier, I. Simmers, C.G.E.M. van Beek, P. van Rossum, P. Smit, C. Grøn, C. Boesen, S. Laursen and B.K. Hansen. Furthermore we are grateful to Vibeke Knudsen and Henri Sion for skillful draftwork, to Susanne Kaarøe for competent editing and Linda Brodersen for finding numerous errors in calculus.

The Technical University of Denmark provided a grant to Tony Appelo for a two month stay in Lyngby, while the Free University financed a sojourn of Dieke Postma in Amsterdam; both visits were essential for this book. Additional funding has been provided by the Groundwater Research Center at the Technical University of Denmark.

February, 1993
Amsterdam, Tony Appelo
Lyngby, Dieke Postma

Table of contents

LIST OF EXAMPLES   XII

NOTATION   XIV

CHAPTER 1: INTRODUCTION TO GROUNDWATER GEOCHEMISTRY   1
1.1 Groundwater quality   1
1.2 Units of analysis   5
    1.2.1 Standards for drinking water   9
1.3 Sampling of groundwater   10
    1.3.1 Depth integrated or depth specific sampling   10
    1.3.2 Procedures for sampling of groundwater   11
1.4 Chemical analysis of groundwater   14
    1.4.1 Field analysis and sample conservation   14
    1.4.2 Accuracy of chemical analysis   16
Problems   19
References   20

CHAPTER 2: FROM RAINWATER TO GROUNDWATER   22
2.1 Rainwater   22
    2.1.1 Sources and transport of atmospheric pollutants   27
    2.1.2 Dry deposition   28
2.2 From rainwater to groundwater   30
2.3 Overall controls on water quality   32
Problems   41
References   42

CHAPTER 3: SOLUTIONS, MINERALS AND EQUILIBRIA   45
3.1 Solubility of minerals   45
3.2 Corrections for solubility calculations   49
    3.2.1 Concentration and activity   49
    3.2.2 Complexes   51
    3.2.3 Combined complexes and activity corrections   52
    3.2.4 Calculation of saturation states   54
    3.2.5 Computer programs for equilibrium calculations   56

3.3 Mass action constants and thermodynamics   59
   3.3.1 The calculation of mass action constants   59
   3.3.2 Calculation of mass action constants at different temperature   61
3.4 Solid solution   62
   3.4.1 Basic theory   62
3.5 Kinetics of geochemical processes   66
   3.5.1 Kinetics and equilibrium   66
   3.5.2 Rate laws and chemical reactions   68
   3.5.3 Crystal formation   71
   3.5.4 Dissolution and crystal growth   72
   3.5.5 Mechanisms of dissolution and crystallization   74
Problems   81
References   83

CHAPTER 4: CARBONATES AND CARBON DIOXIDE   86
4.1 Carbonate minerals   87
4.2 Carbonate equilibria   90
   4.2.1 CO_2 species in water   90
   4.2.2 CO_2 in soil water   96
   4.2.3 Open system dissolution of calcite   99
   4.2.4 Closed system dissolution of calcite   102
   4.2.5 Dolomite dissolution   103
4.3 Field conditions for calcite/dolomite dissolution   105
4.4 Kinetics of carbonate reactions   119
   4.4.1 Dissolution   121
   4.4.2 Precipitation   127
   4.4.3 Inhibitors   129
4.5 Solid solution: Mg-calcites   130
4.6 Pleistocene carbonate aquifers   134
Problems   137
References   138

CHAPTER 5: ION EXCHANGE AND SORPTION   142
5.1 Cation exchange in salt/fresh water intrusions   143
5.2 Adsorbents in soils and aquifers   148
   5.2.1 Clay minerals   150
   5.2.2 Charge from surface reactions   153
5.3 Exchange equations   155
   5.3.1 Values for exchange coefficients   159
   5.3.2 Calculation of exchanger composition   160
   5.3.3 Distribution coefficients   162
5.4 Chromatography of cation exchange   165
   5.4.1 Field examples of refreshening   165
   5.4.2 Quality patterns with salinization   173
5.5 Specific adsorption   175
   5.5.1 Surface complexation and electrostatic models   177
   5.5.2 Specific adsorption of heavy metals   181

5.6 The Gouy-Chapman theory of the double layer   184
    5.6.1 Thickness of the double layer   186
    5.6.2 Practical aspects of double layer theory   187
    5.6.3 Potentials in a double layer   190
5.7 Irrigation water quality   191
5.8 Final remarks   195
Problems   195
References   198

CHAPTER 6: SILICATE WEATHERING   202
6.1 Weathering processes   202
6.2 The stability of weathering products   207
6.3 Stability of primary silicates   210
6.4 The mass balance approach to weathering   215
6.5 Kinetics of silicate weathering   220
6.6 Groundwater acidification   223
    6.6.1 Buffering processes in aquifers   226
    6.6.2 Field weathering rates   230
Problems   235
References   236

CHAPTER 7: REDOX PROCESSES   239
7.1 Basic theory   239
    7.1.1 The significance of redox measurements   243
    7.1.2 Redox reactions and the pe concept   245
7.2 Redox diagrams   246
    7.2.1 Stability of water   247
    7.2.2 The stability of dissolved species and gases   248
    7.2.3 The stability of minerals   252
7.3 Sequences of redox reactions and redox zoning   257
7.4 Oxygen consumption in aquifers   261
    7.4.1 Pyrite oxidation   263
    7.4.2 Kinetics of pyrite oxidation   265
7.5 Nitrate in groundwater   271
    7.5.1 Processes of nitrate reduction   271
    7.5.2 Nitrate reduction by organic matter   272
    7.5.3 Nitrate reduction by pyrite oxidation   275
    7.5.4 Nitrate reduction by Fe(II)   278
7.6 Iron in groundwater   279
    7.6.1 Sources of iron in groundwater   279
    7.6.2 Solubility controls of iron in groundwater   282
7.7 Sulfate reduction and iron sulfide formation   285
    7.7.1 The formation of iron sulfides   287
7.8 Methane in groundwater   288
Problems   289
References   290

CHAPTER 8: SALT WATER, MIXING AND DISCHARGE/QUALITY
              RELATIONS   296
8.1 Salt water   296
        8.1.1 Evaporative concentration   297
        8.1.2 Subsurface brines   303
8.2 Mixing of different water types   305
        8.2.1 pH change in mixed waters and solubility effects   306
        8.2.2 The common ion effect in mixed waters   309
        8.2.3 Mixing of reduced and oxygenated water   311
8.3 Changing water quality with changes in discharge   314
        8.3.1 Runoff models   314
        8.3.2 Mixing models   315
        8.3.3 Lag effects   319
        8.3.4 Water quality routing   320
Problems   323
References   324

CHAPTER 9: SOLUTE TRANSPORT IN AQUIFERS   327
9.1 Groundwater flow   327
        9.1.1 Flow velocity in the unsaturated zone   329
        9.1.2 Flowlines and flow velocities in aquifers   331
        9.1.3 Effects of non-homogeneity   335
        9.1.4 The aquifer, an ideal mixed reservoir   336
        9.1.5 Interpretation of tracer transport: tritium profiles 337
9.2 Retardation of chemicals   339
        9.2.1 Distribution coefficients for organic micropollutants 341
        9.2.2 Distribution coefficients for trace metals   344
        9.2.3 Sorption isotherms and their effect on retardation   347
9.3 Dispersion and diffusion   349
        9.3.1 Diffusion as a random process   352
        9.3.2 Column breakthrough-curves   356
        9.3.3 Values for dispersion coefficients; dispersivity   359
        9.3.4 Macrodispersivity   361
        9.3.5 Diffusion influences on transport   367
9.4 Combining the processes   369
9.5 Numerical modeling of transport   373
        9.5.1 Only diffusion   374
        9.5.2 Advection and diffusion/dispersion   378
        9.5.3 Advection, diffusion/dispersion and non-linear reactions 383
        9.5.4 Permeability variations in flow models   387
Problems   390
References   392

CHAPTER 10: HYDROGEOCHEMICAL TRANSPORT MODELING   396
10.1 Geochemical models, what can they do?   397
        10.1.1 Structure of the models and governing equations   398

10.1.2 Available models 401
10.1.3 Cation exchange in the geochemical models 403
10.2 Guide for the models PHREEQE/PHREEQM 408
10.2.1 The geochemical model PHREEQE 409
10.2.2 The hydrogeochemical transport model PHREEQM 420
10.3 Examples of geochemical transport modeling 427
References 446

CHAPTER 11: ASPECTS OF FLUSHING AND AQUIFER CLEAN-UP 449
11.1 Flushing factors from desorption isotherms 449
11.1.1 Desorption isotherms from a single elution curve 454
11.1.2 $C(x)$ and $C(V)$ profiles and sharp fronts 454
11.1.3 Acceleration of aquifer clean-up 456
11.2 Ion exchange 458
11.2.1 Concentration effects in heterovalent exchange 462
11.2.2 Multicomponent exchange (sharp fronts) 464
11.3 Processes that affect sorption 466
11.3.1 Hysteresis 467
11.3.2 Physical non-equilibrium 468
11.3.3 Chemical disequilibrium 471
11.3.4 Modeling the processes 472
11.4 Conclusions and a further outlook 475
Problems 476
References 476

APPENDIX A: ACTIVITIES, STANDARD STATES AND THE LAW OF
          MASS ACTION 480

APPENDIX B: STABILITY CONSTANTS 485

APPENDIX C: ANSWERS TO PROBLEMS 503

INDEX 521

List of examples

1.1 Relations between water quality and rock weathering 3
1.2 Effect of iron oxidation on analytical results 15
1.3 Estimating the reliability of water analyses 17
1.4 Analytical inaccuracies due to precipitation in the sample bottle 18

3.1 Gypsum addition to high fluor groundwater 47
3.2 Solubility of gypsum 53
3.3 Calculation of solubiliy products from Gibbs free energy data 60
3.4 Temperature dependency of the solubility product 61
3.5 Cadmium partitioning in calcite 65
3.6 Oxidation of Fe(II) 69
3.7 Dissolution of hydroxyapatite; transport or surface reaction controlled? 75

4.1 Calculate TIC at pH = 7 and 10 when the CO_2 pressure is 0.01 atm 92
4.2 Denudation rates in limestone area 101
4.3 Dissolution of dolomite in the Italian Dolomites 104

5.1 Recalculate CEC (meq/100 g soil) to concentration (meq/l pore water) 149
5.2 Structural charge of smectite 153
5.3 Exchange coefficients for different conventions as function of normality of solution 158
5.4 Cation exchange complex in equilibrium with coastal dune groundwater 161
5.5 Estimate the distribution coefficient of Zn^{2+} 163
5.6 Distribution coefficient for Sr^{2+} in multicomponent solution 163
5.7 Exchange coefficient of Cd^{2+} vs. Na^+ on montmorillonite 164
5.8 Flushing of an exchange complex 166
5.9 Water composition after a salinity front 168
5.10 Calculation of the exchangeable sodium ratio (ESR) 192
5.11 Calculation of SAR adjusted for calcite precipitation 194

6.1 Mass balance and the water chemistry of the Sierra Nevada (USA) 216
6.2 Mass balance calculation with the code BALANCE 218
6.3 Acid groundwater formation and gibbsite buffering 227

7.1 Redox speciation with the Nernst equation 242
7.2 Calculate E^0 from ΔG_r^0 243
7.3 Redox speciation using the pe concept 245
7.4 Mass action constant from thermodynamic data 246

7.5 Oxidation of the atmosphere's N_2 content to nitrate 252
7.6 Redox balance for nitrate reduction by organic matter 273

8.1 Concentration of a $CaSO_4$ solution 298
8.2 pH calculation in mixed carbonate water 306
8.3 Seepage of acid groundwater in a river with neutral pH 309
8.4 Mixing of gypsum-water with high F^- containing water 310
8.5 Contributions of rock types to total flow 311
8.6 Calculate discharge along a stream from *EC*-measurements 323

9.1 Calculation of infiltration depths and flow velocities in an aquifer 334
9.2 Flushing of NO_3^- from an aquifer 337
9.3 Retardation of lindane and PCB 344
9.4 Retardation of Zn^{2+} 345
9.5 Retardation calculated from Freundlich isotherm data 349
9.6 Dispersion coefficient from a single shot input 355
9.7 Front dispersion in a column 357
9.8 Pollutant spreading during transport in an aquifer 365
9.9 Longitudinal dispersivity in the Borden aquifer 365
9.10 Calculation of aquifer pollution by waste site leachate 371
9.11 Diffusive flux through a clay barrier 373
9.12 A Pascal code to model Cl^- diffusion from seawater into fresh water sediment 376
9.13 Numerical model of linear retardation of γ-HCH in a laboratory column 380
9.14 Effect of the Freundlich exponent on breakthrough curves from a column 384

10.1 Calculation of Ca/Al cation exchange 405
10.2 The redox state (THSP) of PHREEQE species 413
10.3 Examples of SPECIES definition in PHREEQE 414
10.4 Examples of LOOK MIN mineral definitions 415
10.5 Estimate pH from *TIC* and electroneutrality 416
10.6 Calculation of the exact pe for water in equilibrium with atmospheric oxygen 416
10.7 NO_3^- reduction in groundwater from agricultural land by organic matter 418
10.8 Examples of SUMS of species 419
10.9 Calcite precipitation from groundwater that is heated for use in aquifer thermal energy storage (ATES) 419
10.10 A few exchangeable species defined for PHREEQM 426
10.11 Evolution of water in a limestone/dolostone aquifer 427
10.12 Evapotranspiration of acid rainwater 430
10.13 Flushing of Na^+ and K^+ from a column with $CaCl_2$ solution 431
10.14 Flushing a cation exchange complex with $SrCl_2$ solution 435
10.15 Recovery of fresh water injected in a brackish water aquifer 437
10.16 Aquifer thermal energy storage (ATES) at St Paul, USA 439
10.17 Reduction of MnO_2 by $FeSO_4$ solution in a column 442
10.18 Acidification of groundwater 443

11.1 Analytical modeling of column elution 452
11.2 Elution of TCE from an aquifer 453
11.3 Flushing of K^+ from a column 461
11.4 Flushing of Cd^{2+} from a sediment with NaCl solution 462
11.5 Retardation of a K^+, Rb^+, Cs^+ mixture 464

Notation

Greek characters

α dispersivity (m)
            subscripts L and T denote longitudinal and transversal dispersivity
$\alpha_{0, 1, ...}$ parameters in Margules equation (Chapter 3, (–))
α_i equivalent fraction of cation i in water (–)
β_i equivalent fraction of i on exchanger (–)
β_i^M molar fraction of i on exchanger (–)
γ_i activity-coefficient of species i in water (–)
Γ_c correlation coefficient
ε porosity (a fraction), or
            dielectric constant in Chapter 5 (F/m)
ε_W water filled porosity (–)
θ tortuosity, i.e. ratio of actual path over straight line path (–)
$\kappa\varepsilon$ specific capacitance (F/m^2)
$1/\kappa$ Debije length (m)
λ_i activity coefficient of i in solid solution (–)
ν stoichiometric coefficient in reaction (–)
ρ charge density (C/m^3)
ρ_b bulk density of solid (with air filled pores) (g/cm^3)
σ charge density at solid surface (eq/m^2)
σ^2 variance in a set of parameters; of the diffusion curve (units depend on parameter)
Φ_m mobile fraction of pore water (–)
Ψ potential in a double layer (V)
Ψ_0 potential at a solid surface (V)
Ω saturation state, IAP/K (–)

Capital letters

A temperature dependent coefficient in Debije-Hückel theory, or
            surface area in Darcy's Law (Chapter 9, (m^2)), or
            surface area of crystals (Chapter 3, (m^2)), or
            quality parameter (Chapter 8, units depend on parameter)
A_0 total cation concentration in solution (eq/l)
A_s specific surface (m^2/kg)
C solute concentration (mol/l, mol/kg, mg/l, etc.)
CEC cation exchange capacity (meq/100 g)
D diffusion coefficient (m^2/s), or
            distribution coefficient (–), or
            dry deposition (Chapter 2, (μg/m^2 .yr)), or
            thickness of aquifer (Chapter 9, (m))

D_{app}	apparent diffusion coefficient, in crystal dissolution (m^2/s)
D_f	diffusion coefficient in free water (m^2/s)
D_e	effective diffusion coefficient in a porous medium (m^2/s)
D_L	hydrodynamic dispersion coefficient used in 1D formulations of the advective-dispersion equation (m^2/s), subscripts L and T with dispersion coefficients and dispersivities indicate longitudinal, resp. transversal tensor
DDL	diffuse double layer
E	electrical field (V/m), or
	redox potential (V)
EC	electrical conductivity of water sample (μS/cm)
Eh	redox potential relative to the standard state H_2/H^+ reaction (V)
E^0	standard (redox) potential (all species are in their standard state) (V)
ESR	exchangeable sodium ratio (–)
F	Faraday constant (96485 C/mol or J/Volt.g eq)
F_{Cl}	fractionation factor with respect to Cl^- (–)
ΔG_r^0	Gibbs free energy (enthalpy) of a reaction at standard state conditions (25°C, 1 atm, unit activity of species) (J/mol or kcal/mol)
ΔG_f^0	Gibbs free energy (enthalpy) of formation from elements (J/mol or kcal/mol)
ΔH_r^0	reaction heat or enthalpy at 25°C, 1 atm (J/mol or kcal/mol)
ΔH_f^0	enthalpy of formation of a species from elements (J/mol or kcal/mol)
I	ionic strength (mol/l)
IAP	ion activity product (–)
J	flux of crystallization or dissolution (mol/s)
K	mass action constant (–), or
	solubility product (–)
K_a	acid dissociation constant (–)
K_d	distribution coefficient [(mg sorbed/l porewater)/(mg solute/l porewater)] (–)
K_d'	distribution coefficient [(mg sorbed/g solid)/mg solute/l porewater)] (l/kg)
K_{int}	intrinsic constant (chemical term only) (–)
K_{oc}	distribution coefficient for chemical amongst 100% organic carbon and water (l/kg)
K_{ow}	idem, amongst octanol and water (l/kg)
$K_{i \backslash j}$	coefficient for exchange reaction, the subscript ions indicate the solute ions according to their appearance in the mass action equation: i is removed from solution, j goes in solution, Gaines and Thomas convention (–)
$K_{i \backslash j}^G$	idem, Gapon convention (–)
L	length of column or flowtube (m)
M	molality (mol/kg H_2O)
N	normality (eq/l)
N_a	Avogadro's number ($6 \cdot 10^{23}$/mol)
P	precipitation surplus (m/yr), or
	gas pressure (atm)
Pe	Peclet number (–)
P_i	gas pressure of i (atm)
Q	discharge (m^3/s)
R	retardation factor (–), or
	gas constant (8.314 J/mol.deg), or
	rate of dissolution ($mmol/cm^2.s$)
R_{fw}	rate of dissolution ($mmol/cm^2.s$)
R_b	rate of precipitation ($mmol/cm^2.s$)
S	sink term for chemical through adsorption (mol/l.s)
ΔS_r^0	reaction entropy at 25°C, 1 atm (J/mol.deg)
SAR	sodium adsorption ratio (–)
SI	saturation index, log (IAP/K) (–)

T	temperature in K ($= °C + 273.15$)
TDS	total dissolved solids (mg/l, mmol/l, etc.)
TIC	total inorganic carbon (mg/l, mmol/l, etc.)
TOC	total organic carbon (mg/l, mmol/l, etc.)
U	potential energy in Boltzmann equation (J/atom)
V_0	pore volume of a column (m^3)
V_m	molar volume of crystals (m^3/mol)
V^s	sharp front flushing factor (–)
V^*	flushing factor (–)
-X	soil exchanger
X_i	molar fraction of i in solid solution

Lower case letters

d	depth in aquifer (in Section 9.1), or
	particle diameter (in the Peclet number)
f_{oc}	fraction of organic carbon in soil or sediment (–)
h	hydraulic head or potential for water flow (m)
k	Boltzmann constant ($13.8 \cdot 10^{-24}$ J/K), or
	permeability or hydraulic conductivity (m/day), or
	rate constant in kinetic rate law (units depend on rate law)
k_a	distribution coefficient with adsorption (–)
k_d	distribution coefficient with desorption (–)
m_i	molal concentration of i (mol/kg H$_2$O, mmol/kg H$_2$O)
m/m_0	mass fraction
$meq_{I\text{-}X}$	exchangeable cation I (meq/100 g)
q	(on solid) sorbed concentration, expressed per liter porewater (mol/l, mg/l, etc.)
q_e	charge of the electron ($1.6 \cdot 10^{-19}$ C)
r	crystal radius (m)
s	(on solid) sorbed concentration (mol/kg, mg/g, etc.); $q = (\rho_b/\varepsilon) \cdot s$
t_R	residence time (s)
v	porewater flow velocity (m/s)
v_D	Darcy flow velocity, $v_D = v \cdot \varepsilon$ (m/s)
v_{C_i}	flow velocity of a given concentration of i (m/s)
z_i	charge number of i

CHAPTER 1

Introduction to groundwater geochemistry

Groundwater geochemistry is an interdisciplinary science concerned with the chemistry of water in the subsurface environment. The chemical composition of groundwater is the combined result of the composition of water that enters the groundwater reservoir and reactions with minerals present in the rock that may modify the water composition. Apart from natural processes as controlling factors on the groundwater quality, in recent years the effect of pollution, such as nitrate from fertilizers and acid rain, also influences the groundwater chemistry. Due to the long residence time of groundwater in the invisible subsurface environment, the effects of pollution may first become apparent tens to hundreds of years afterwards. It is clear that a proper understanding of the processes occurring in aquifers is required in order to predict what the effects of present day human activities will be on such a time scale.

The interest of society in groundwater geochemistry is mainly to ensure good quality drinking water. Although drinking water can be manufactured in a chemical plant, as for example in desalinization plants, this is a very costly affair. Preservation of good ground-water resources therefore has a high priority for environmental authorities.

This chapter gives a general introduction to groundwater geochemistry: some examples of different water qualities are presented and it also introduces basic subjects such as units of analysis, water sampling, and accuracy of analysis. Some knowledge of these subjects is required to set the stage for a more quantitative geochemical approach.

1.1 GROUNDWATER QUALITY

The classical use of water analyses in hydrology is to produce information concerning the water quality. Maps are usually constructed which display the regional distribution of water quality. Such maps are very useful for water resources purposes and are a sound motive for Surveys to routinely perform chemical analyses of well waters (at a nominal cost of at least US$50).

Groundwater geochemistry also has a potential use for tracing the origins and the history of water. Water compositions change through reactions with the environment, and water quality may yield information about the environments through which the water has circulated. There is a great need in present-day hydrology to obtain better knowledge concerning residence times, flowpaths and aquifer characteristics, a need which often arises

1

from man-made pollution problems. The water chemistry can provide the required information since chemical reactions are time and space dependent.

The number of possible chemical reactions which may occur in a groundwater reservoir is tremendous, and the interpretation of observed water quality differences requires a good chemical background. A textbook which provides such a chemical background is 'Aquatic Chemistry' by Stumm and Morgan (1981). Other books focus more on the hydrological interpretation of water quality. Hem (1985) presents many examples of water compositions and concentrations of individual elements in different terrains. Drever (1988) provides insight into the development of water chemistry interpretations by geochemists. Domenico and Schwartz (1990) and Lloyd and Heathcote (1985) have done the same from a hydrogeological point of view, using a modern approach to the interpretation of groundwater quality.

Traditionally, water qualities are interpreted from a static point of view, i.e. assuming stationary conditions both at the site of infiltration and along the flowline, without reversals of flow patterns in time. Such static conditions are seldom valid for hydrological systems in which the flow of water so clearly indicates a state of perpetual change (Heraclitus, cf. Biswas, 1970). The complexity implied by permanently changing fluxes can, fortunately, be elucidated by the solid principles and process descriptions from physics and chemistry. The essential point in the present text is to demonstrate how water obtains its composition and to describe the processes which modify its composition. With a good understanding of the processes, the observed water qualities can be interpretated hydrologically in terms of the origin and flowpath of water. Calculation and modeling is very important here, to lift conjectures from the marsh of possibilities and vague terms.

An example of the hydrological use of chemical parameters is to calculate the precipitation surplus from the concentration of a conservative (inert) parameter in rain and groundwater. Chloride is such a conservative parameter and is often used to obtain an approximate evapotranspiration factor from rainwater to groundwater (Eriksson, 1960; Schoeller, 1960). Table 1.1 compares the composition of groundwater in the Veluwe (the Netherlands) with rainwater. It shows that the Cl^- concentration in precipitation in the area

Table 1.1. Composition of rainwater in 1938 (Leeflang, 1938) and 1978-1980 (Deelen Airport) and groundwater in the Netherlands Veluwe area. Concentrations in mmol/l.

	Rainwater		Groundwater		Concentration factor
	1938	Average 1978-1980	range (n = 30)	Average	
Depth (m.b.s.)			1.5 -192	46	
Na^+	{0.07}	0.070	0.16 -0.36	0.30	4.2
K^+		0.004	0.018-0.039	0.025	6.2
Mg^{2+}	0.03	0.009	0.02 -0.20	0.089	{(calcite dissolves)}
Ca^{2+}	0.043	0.015	0.01 -1.5	0.49	
NH_4^+	0.03	0.12	< 0.005	0.005	
Cl^-	0.09	0.078	0.19 -0.45	0.30	3.8
HCO_3^-	0.05	0	0 -3.6	1.04	(calcite dissolves)
SO_4^{2-}	0.045	0.074	0.01 -0.61	0.15	2-3 (?)
NO_3^-	0.002	0.064	0.002-0.41	0.063	?

remained unchanged during the last fifty years. The ratio of the Cl^- concentration in groundwater over the concentration in rain is 3.8. Thus, with an average precipitation of 840 mm/year, the annual recharge of groundwater in the area is estimated to be 220 mm. This classical use of hydrochemistry, simple and effective, has in recent times become popular again for estimating the groundwater recharge in arid areas where soil moisture flow is very difficult to measure (Edmunds et al., 1988). We should note, however, that dry deposition may increase the amount of atmospheric Cl^- input, and reliable estimates of dry deposition are also difficult to obtain (see Chapter 2.1.2).

The subsoil of the Veluwe area consists of Pleistocene sands of fluvioglacial origin. The sediments contain up to 95% quartz with few reactive minerals. Therefore, the concentration of several other elements also behave more or less as conservatively, with concentrations that are determined almost exclusively by concentrations in the rain multiplied by an evaporation factor. An exception are enhanced concentrations of Ca^{2+}, Mg^{2+} and HCO_3^- which are affected by calcite dissolution. Otherwise, the sands from the Veluwe show an exceptionally low reactivity. The average sandstone on earth contains about 79% SiO_2 (Wedepohl, 1969). The quartz content is even lower than indicated by this percentage, because SiO_2 is partly incorporated in other minerals in sandstones. Such minerals are in fact more reactive than quartz, and their dissolution creates different water qualities. Some examples of relations between water chemistry and geology are presented in Example 1.1.

EXAMPLE 1.1: *Relations between water quality and rock weathering*
A laboratory has mixed up the sample bottles containing water from different geological units in a field area (Table 1.2).
The samples come from springs draining limestone, dolomite, granite-gneiss, serpentinite and schists, and a sample of rainwater must be present as well. Choose the right geological unit for each sample, and determine the precipitation surplus.

It is difficult to obtain an immediate overview over different water compositions from tabulated data. A useful technique is therefore to display the data graphically, for example in a Stiff diagram (Hem, 1985). Stiff-diagrams are constructed by recalculating the analysis into milliequivalents per liter and then plotting anions and cations on three axes as shown in Figure 1.1. The advantages of Stiff-diagrams are dual. Firstly, different water types yield different shapes which are recognized at first glance. Secondly, the absolute concentrations are visualized by the width of the figure.

Figure 1.1 shows that the samples are very different. It seems logical to take the least mineralized Sample (1) as rainwater. Using the averaged Cl^- concentrations in the other samples, an evaporation factor of $0.40/0.28 = 1.43$ is calculated.

Table 1.2. Examples of water quality from the Ahrn-Valley, N.-Italy. Concentrations in mmol/l.

	Sample no.					
	1	2	3	4	5	6
Na^+	0.13	0.18	0.20	0.19	0.50	0.30
K^+	0.02	0.01	0.05	0.01	0.07	0.20
Ca^{2+}	0.09	0.15	0.89	1.30	0.80	0.28
Mg^{2+}	0.08	0.30	0.95	0.11	0.12	0.12
Cl^-	0.28	0.38	0.40	0.41	0.40	0.50
HCO_3^-	0.001	0.47	3.39	2.34	2.14	< 0.001
SO_4^{2-}	0.10	0.12	0.14	0.14	0.07	0.45
pH	5.3	6.83	7.27	7.17	8.37	4.70

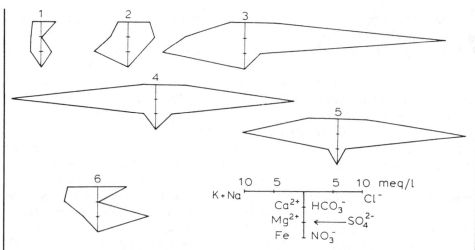

Figure 1.1. Stiff-diagrams of the analyses in Table 1.2.

Multiplying all other concentrations in rain water with this factor, and subtracting the results from the analyses, gives Table 1.3. Concentrations which are less then 0.05 mmol/l are omitted. The remaining concentrations should be the result from the interaction with different rocks. Limestone and dolomite are often monomineralic. Limestone consists of pure $CaCO_3$, and dolomite of $CaMg(CO_3)_2$. A proper choice would be that Sample (4) comes from a limestone aquifer, since only Ca^{2+} and HCO_3^- have increased. It is known that dolomite dissolves congruently, i.e. the quantities of components that appear in solution are proportional to those in the dissolving mineral. In Sample (3), equal increases of Ca^{2+} and Mg^{2+} are observed, so that this water could be derived from dolomite. These basic relationships are simple for carbonate rocks, but become more complicated for silicate rocks.

Serpentinite rock contains high proportions of talc or serpentine, and the rock is rich in magnesium. Sample (2) shows the largest increase in Mg^{2+} and is therefore attributed to the serpentine.

The remaining Samples (5) and (6) must originate from the granite or schist. A schist may contain some pyrite (FeS_2) which easily oxidizes to SO_4^{2-} and produces acid. Since the pH and HCO_3^- concentrations are lowest in Sample (6), it is probably derived from the schist. Cl^- also increases in this sample, perhaps because salt is present as a remnant of connate water (water included in the sediment during burial).

Sample (5) must then come from the granite. The major reactive minerals in granite are plagioclase

Table 1.3. The contribution of rock weathering to the water compositions in Table 1.2 (mmol/l).

| | Sample no. | | | | |
	2	3	4	5	6
Na^+				0.31	0.11
K^+				0.04	0.17
Ca^{2+}		0.76	1.17	0.67	0.15
Mg^{2+}	0.19	0.84			
Cl^-					0.10
HCO_3^-	0.47	3.39	2.34	2.14	
SO_4^{2-}					0.31

and mica. The ratio of Ca^{2+}/Na^+ is somewhat larger than would be normal for a granite-plagioclase. However, the dissolution of silicate minerals such as plagioclase is not an easy subject, as will be discussed in Chapter 6.

1.2 UNITS OF ANALYSIS

The concentration of dissolved substances in water samples can be presented in different units, depending on the purpose of the presentation and also on tradition. Some common units are listed below:

mg/l   milligrams per liter sample;
ppm   parts per million by weight of the sample;
ppb   parts per billion by weight of the sample;
mmol/l   millimoles per liter sample (= millimolarity);
meq/l   milliequivalents per liter of sample;
epm   equivalents per million, by weight of sample;
M   molality, moles per kg of H_2O;
mM   millimoles per kg of H_2O;
m_i molality of i in this text;
N   normality, equivalents per liter;
EC   Electrical Conductivity, in $\mu S/cm$ (= $\mu mho/cm$), $EC \approx 100 \cdot meq$(anions or cations)/l;
pH $- \log [H^+]$, the log of H^+ activity (see Chapters 3 and 4).

Table 1.4. Common parameters used in the water world (cf. Standard Methods, 1989; Hem, 1985).

Hardness:	sum of the ions which can precipitate as 'hard particles' from water. Sum of Ca^{2+} and Mg^{2+}, and sometimes Fe^{2+}. Expressed in meq/l or mg $CaCO_3$/l or in hardness degrees. 100 mg $CaCO_3$/l $\cong$ 1 mmol Ca^{2+}/l $\cong$ 2 meq Ca^{2+}/l
Hardness degrees:	1 german degree = 17.8 mg $CaCO_3$/l 1 french degree = 10 mg $CaCO_3$/l
Temporary hardness:	part of Ca^{2+} and Mg^{2+} concentrations which are balanced by HCO_3^- (all expressed in meq/l) and can thus precipitate as carbonate
Permanent hardness:	part of Ca^{2+} and Mg^{2+} in excess of HCO_3^- (all expressed in meq/l)
Color:	measured by comparison with a solution of cobalt and platinum
Eh:	redox potential, expressed in Volt. Measured with platinum/reference electrode
pe:	redox potential expressed as $- \log [e^-]$. $[e^-]$ is 'activity' of electrons. pe = $Eh/0.059$ at 25°C
Alkalinity (*Alk*):	acid neutralizing capacity. Determined by titrating with acid down to a pH of about 4.5. Equal to the concentrations of $m_{HCO_3^-} + 2\, m_{CO_3^{2-}}$(mmol/l) in most samples
Acidity:	base neutralizing capacity. Determined by titrating up to a pH of about 8.3. Equal to H_2CO_3 concentration in most samples
TIC:	total inorganic carbon
TOC:	total organic carbon
COD:	chemical oxygen demand. Measured as chemical reduction of permanganate or dichromate solution, and expressed as oxygen equivalents
BOD:	biological oxygen demand

Table 1.5. Recalculation of analysis units.

---

mmol/l = mg/l / (gram formula weight).

mmol/l = ppm · (density of sample)/(gram formula weight).

mmol/l = meq/l / (charge of ion).

$$\text{mmol/l} = \text{molality M} \cdot \text{density} \cdot \frac{(\text{weight solution} - \text{weight solutes})}{(\text{weight solution})} \cdot 1000$$

The official SI-unit is moles/m^3, but not commonly used. It is, however, numerically equal to mmol/l, the unit we often use in this book. Some other units, such as hardness, which are commonly used in the water world are defined in Table 1.4.

Many analyses are expressed in either ppm or mg/l. These are numerically equal when the density of the water sample is 1 g/cm^3 as is the case for dilute freshwaters. The density of salt water is, however, larger than 1; 35‰ seawater, for example, has a density of 1.023 g/cm^3 at 25°C. When the analysis of a salt water has been carried out on a volume base (by pipetting and diluting volumina), recalculation is necessary to convert the results to concentrations by weight.

To recalculate the results of an analysis from mg/l to mmol/l, the figures are divided by the gram formula weight for each species. The gram formula weight is the weight in grams of 1 mol of atoms or molecules (1 mol = 'Avogadro's number', $6 \cdot 10^{23}$ molecules).

Recalculation of the results to meq/l is done by multiplying the values in mmol/l by the charge z of the ion or molecule. Table 1.5 presents equations for the conversion between different units and some examples follow below.

---

1. Gram formula weights are calculated from the periodic system as reproduced in Table 1.6 from the *Handbook of chemistry and physics*.

The mass of 1 mol Ca is 40.08 grams.

1 mol SO_4^{2-} weighs: 32.06 grams from sulfur + 4 × 15.9994 gram from oxygen, in total : 96.06 gram

2. Conversion of mg/l to mmol/l is obtained by dividing by the weight of the element or molecule.

Thus a river water sample contains 1.2 mg Na$^+$/l.

This corresponds to 1.2/22.99 = 0.052 mmol Na$^+$/l.

The sample also contains 0.6 mg SO_4^{2-}/l.

This corresponds to 0.6/96.06 = 0.006 mmol SO_4^{2-}/l.

3. The term mmol/l indicates the number of ions or molecules in the water, when multiplied by Avogadro's number. For Na$^+$ in the river water sample, this yields $0.052 \cdot 10^{-3} \cdot 6 \cdot 10^{23} = 3.1 \cdot 10^{19}$ ions of Na$^+$ in 1 liter of water. (Quite a lot, really!)

4. Ions are electrically charged, and the sums of positive and negative charges in a given water sample must balance. This condition is termed the electroneutrality of the solution. Since mmol/l represents the number of molecules, it should be multiplied by the charge of the ions to yield their total charge in meq/l. Thus:

0.052 mmol Na$^+$ /l × 1 = 0.052 meq/l.

1.8 mmol Ca^{2+}/l × 2 = 3.6 meq/l.

0.41 mmol SO_4^{2-}/l × −2 = −0.82 meq/l.

KEY TO CHART

New notation → / Previous IUPAC form → / CAS version →

(Numbers in parentheses are mass numbers of most stable isotope of that element)

Atomic Number → 50 +2 ← Oxidation States
Symbol → Sn +4
1987 Atomic Weight → 118.71
18 18 4 ← Electron Configuration

At. No.	Symbol	Oxidation States	Atomic Weight	Electron Configuration	Orbit
1	H	+1, −1	1.00794	1	K
2	He	0	4.002602	2	K
3	Li	+1	6.941	2-1	
4	Be	+2	9.012182	2-2	
5	B	+3	10.811	2-3	
6	C	+2, +4, −4	12.011	2-4	
7	N	+1, +2, +3, +4, +5, −1, −2, −3	14.00674	2-5	
8	O	−2	15.9994	2-6	
9	F	−1	18.9984032	2-7	
10	Ne	0	20.1797	2-8	K-L
11	Na	+1	22.989768	2-8-1	
12	Mg	+2	24.3050	2-8-2	
13	Al	+3	26.981539	2-8-3	
14	Si	+2, +4, −4	28.0855	2-8-4	
15	P	+3, +5, −3	30.97362	2-8-5	
16	S	+4, +6, −2	32.066	2-8-6	
17	Cl	+1, +5, +7, −1	35.4527	2-8-7	
18	Ar	0	39.948	2-8-8	K-L-M
19	K	+1	39.0983	-8-8-1	
20	Ca	+2	40.078	-8-8-2	
21	Sc	+3	44.955910	-8-9-2	
22	Ti	+2, +3, +4	47.88	-8-10-2	
23	V	+2, +3, +4, +5	50.9415	-8-11-2	
24	Cr	+2, +3, +6	51.9961	-8-13-1	
25	Mn	+2, +3, +4, +7	54.93085	-8-13-2	
26	Fe	+2, +3	55.847	-8-14-2	
27	Co	+2, +3	58.93320	-8-15-1	
28	Ni	+2, +3	58.69	-8-16-2	
29	Cu	+1, +2	63.546	-8-18-1	
30	Zn	+2	65.39	-8-18-2	
31	Ga	+3	69.723	-8-18-3	
32	Ge	+2, +4	72.61	-8-18-4	
33	As	+3, +5, −3	74.92159	-8-18-5	
34	Se	+4, +6, −2	78.96	-8-18-6	
35	Br	+1, +5, −1	79.904	-8-18-7	
36	Kr	0	83.80	-8-18-8	-L-M-N
37	Rb	+1	85.4678	-18-8-1	
38	Sr	+2	87.62	-18-8-2	
39	Y	+3	88.90585	-18-9-2	
40	Zr	+4	91.224	-18-10-2	
41	Nb	+3, +5	92.90638	-18-12-1	
42	Mo	+6	95.94	-18-13-1	
43	Tc	+7	(98)	-18-13-2	
44	Ru	+3	101.07	-18-15-1	
45	Rh	+3	102.90550	-18-16-1	
46	Pd	+2, +4	106.42	-18-18-0	
47	Ag	+1	107.8682	-18-18-1	
48	Cd	+2	112.411	-18-18-2	
49	In	+3	114.82	-18-18-3	
50	Sn	+2, +4	118.710	-18-18-4	
51	Sb	+3, +5, −3	121.75	-18-18-5	
52	Te	+4, +6, −2	127.60	-18-18-6	
53	I	+1, +5, +7, −1	126.90447	-18-18-7	
54	Xe	0	131.29	-18-18-8	-M-N-O
55	Cs	+1	132.90543	-18-8-1	
56	Ba	+2	137.327	-18-8-2	
57*	La	+3	138.9055	-18-9-2	
72	Hf	+4	178.49	-32-10-2	
73	Ta	+5	180.9479	-32-11-2	
74	W	+6	183.85	-32-12-2	
75	Re	+4, +6, +7	186.207	-32-13-2	
76	Os	+3, +4	190.2	-32-14-2	
77	Ir	+3, +4	192.22	-32-15-2	
78	Pt	+2, +4	195.08	-32-16-2	
79	Au	+1, +3	196.96654	-32-18-1	
80	Hg	+1, +2	200.59	-32-18-2	
81	Tl	+1, +3	204.3833	-32-18-3	
82	Pb	+2, +4	207.2	-32-18-4	
83	Bi	+3, +5	208.98037	-32-18-5	
84	Po	+2, +4	(209)	-32-18-6	
85	At	−1	(210)	-32-18-7	
86	Rn	0	(222)	-32-18-8	-N-O-P
87	Fr	+1	(223)	-18-8-1	
88	Ra	+2	226.025	-18-8-2	
89**	Ac	+3	227.028	-18-9-2	
104	Unq	+4	(261)	-32-10-2	
105	Unp		(262)	-32-11-2	
106	Unh		(263)	-32-12-2	
107	Uns		(262)	-32-13-2	O-P-Q

*Lanthanides

At. No.	Symbol	Oxidation States	Atomic Weight	Electron Configuration	Orbit
58	Ce	+3, +4	140.115	-20-8-2	
59	Pr	+3	140.90765	-21-8-2	
60	Nd	+3	144.24	-22-8-2	
61	Pm	+3	(145)	-23-8-2	
62	Sm	+2, +3	150.36	-24-8-2	
63	Eu	+2, +3	151.965	-25-8-2	
64	Gd	+3	157.25	-25-9-2	
65	Tb	+3	158.92534	-27-8-2	
66	Dy	+3	162.50	-28-8-2	
67	Ho	+3	164.93032	-29-8-2	
68	Er	+3	167.26	-30-8-2	
69	Tm	+3	168.93421	-31-8-2	
70	Yb	+2, +3	173.04	-32-8-2	N-O-P
71	Lu	+3	174.967	-32-9-2	

**Actinides

At. No.	Symbol	Oxidation States	Atomic Weight	Electron Configuration	Orbit
90	Th	+4	232.0381	-18-10-2	
91	Pa	+4, +5	231.03588	-20-9-2	
92	U	+3, +4, +5, +6	238.0289	-21-9-2	
93	Np	+3, +4, +5, +6	237.048	-22-9-2	
94	Pu	+3, +4, +5, +6	(244)	-24-8-2	
95	Am	+3, +4, +5, +6	(243)	-25-8-2	
96	Cm	+3	(247)	-25-9-2	
97	Bk	+3, +4	(247)	-27-8-2	
98	Cf	+3	(251)	-28-8-2	
99	Es		(252)	-29-8-2	
100	Fm		(257)	-30-8-2	
101	Md		(258)	-31-8-2	
102	No		(259)	-32-8-2	O-P-Q
103	Lr	+3	(260)	-32-9-2	

Table 1.6. Periodic table of the elements. Reprinted with permission from *Chem. Eng. News*, february 4, 1985, 63(5), p.26-27. Published in 1985 by the American Chemical Society.

Figure 1.2. Labels of mineral water bottles which suggest a beneficial effect due to their composition.

What is the best unit to use? A study of labels of bottled mineral waters, as shown in Figure 1.2, would suggest that mg/l or g/l is an obvious unit to express the composition of water samples. On the other hand, the unit mmol/l has the advantage that increases and decreases in concentrations can be coupled to 'sources' and 'sinks'. In Example 1.1 the increases of Ca^{2+} and HCO_3^- could be directly related to $CaCO_3$ dissolution because the stoichiometry is reflected by the molar increases. The unit meq/l is useful for checking the accuracy of analysis.

Molality is generally used in equilibrium calculations and has the advantage that it is expressed per kg of H_2O, which makes it independent of density changes with temperature, or with changes in concentrations of other constituents. In normal fresh to slightly brackish water, molality is equal to molarity (mol/l). 35‰ sea water has a density of 1.023 kg/l at 25°C, as compared to 0.997 for air-free pure water, so that differences between molalities and molarities may amount to approximately 2.5%.

1.2.1 *Standards for drinking water*

The labels of the bottled mineral waters shown in Figure 1.2 suggest that the composition of drinking water has an important health aspect. The standards of the World Health Organisation (WHO), and of most national authorities, are consistent with the maximum admissible concentrations which are presented in Table 1.7. The table shows limits for constituents which commonly occur, either naturally or due to pollution or water treatment and water supply systems. The average contribution of drinking water to the daily intake of mineral nutrients is also shown (Safe Drinking Water Comm., 1980). The intake from drinking water is highly significant for F^- and As, and too high concentrations of these elements can in many areas limit the use of groundwater for drinking purposes. Too high fluoride intake leads to painful skeleton deformations termed fluorosis; it is a common disease in African Rift Valley countries such as Kenya and Ethiopia, where volcanic

Table 1.7. Standards for the composition of drinking water and contribution of drinking water to the intake of elements in nutrition.

Constituent	Contribution to mineral nutrition (%)	Highest admissible concentration (mg/l)	Comment
Mg^{2+}	3-10	50	Mg/SO_4 diarrhea
Na^+	1-4	175	
Cl^-	2-15	300	taste; safe < 600 mg/l
SO_4^{2-}		250	diarrhea
NO_3^-		50	blue baby disease
NO_2^-		0.1	
F^-	10-50	1.5	Lower at high water consumption
As	ca. 30	0.05	Black-foot disease
Al	..	0.2	Acidification/Al-flocculation
Cu	6-10	0.1	3 mg/l in new piping systems
Zn	negligible	0.1	5 mg/l in new piping systems
Cd	..	0.005	
Pb	..	0.05	
Cr	20-30	0.05	

sources of F⁻ are important, and in India and West Africa where salts and sedimentary F-bearing minerals are the primary sources (this aspect will be treated more fully in Chapter 3). Elevated arsenic concentrations are commonly associated with sediments partially derived from volcanic rocks of intermediate to acidic composition (Welch et al., 1988). Arsenic is also associated with sulfides, and high As concentrations can sometimes be found in water that also contains appreciable Fe-concentrations. It gives rise to the so-called black foot disease, visible in a blackening of finger and toe-tops, and induces a general lethargy in the patient.

1.3 SAMPLING OF GROUNDWATER

Groundwater sampling and analysis is a costly affair, mainly because of the expense involved in drilling boreholes. It is therefore of importance to evaluate which data are required to solve a specific problem. Conversely, in the interpretation of the chemical analyses, the procedures which were used to collect and analyze the samples should be taken into consideration.

1.3.1 *Depth integrated or depth specific sampling*

Most chemical analyses of groundwater are found in the files of governmental and water works offices. Their prime function is to monitor the groundwater quality in production wells. Such wells are typically installed with a screen of at least several meters length (Figure 1.3). Accordingly, the listed chemical composition of the groundwater reflects an integrated sample over the whole screen interval. Note in Figure 1.3, that the sample obtained is integrated over the length of the screen as well as over the permeability of the formation. Thus the screen interval with fine grained sand will contribute relatively less to the total yield than the coarse grained sand with a much higher permeability.

Chemical groundwater data obtained from depth integrated samples can be useful in identifying regional patterns in groundwater composition and its relation to rock types.

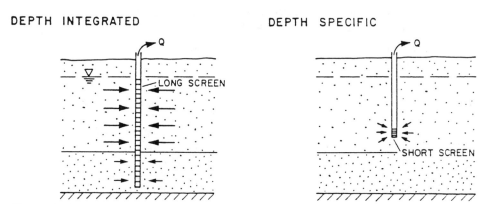

Figure 1.3. Depth integrated versus depth specific groundwater sampling. Coarse dotted areas represent coarse grained sand and densely dotted areas fine grained sand. The size of the arrows reflects the flow rate (modified from Cherry, 1983).

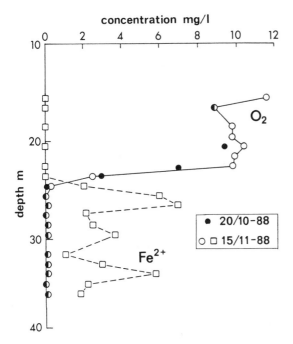

Figure 1.4. The distribution of O_2 and Fe^{2+} in a sandy aquifer.

However, in many cases groundwater compositions show major variations with depth, even on a small scale. Integrated samples over a screen interval of several meters may accordingly represent mixtures of waters with different compositions and the mixing process may also induce chemical reactions during the sampling process. Therefore, depth specific sampling is usually required in order to study chemical processes in detail. An illustration of a groundwater chemistry profile, obtained with a depth specific device, of a sandy aquifer with a well defined oxic zone and an anoxic zone containing Fe^{2+}, is shown in Figure 1.4.

Suppose that a screen is installed so that it draws water from both the oxic and anoxic zones. Then the O_2 containing water will mix and react with Fe^{2+} containing water and alter concentrations both of O_2, Fe^{2+} and pH or alkalinity by the reaction:

$$2Fe^{2+} + \tfrac{1}{2}O_2 + 3H_2O \rightarrow 2FeOOH + 4H^+$$

The composition of the resulting sample will depend on the extent of mixing and reaction. In principle, even a sample which contains both Fe^{2+} and O_2 could be obtained although such a water quality does not occur in the aquifer. In any case the resulting sample will not reflect the in situ conditions in the aquifer very well. It is obviously of great importance to realize whether the data one tries to interpret have been obtained using a depth specific or depth integrated sampling method and also to evaluate how the sampling procedure affected the measured water quality.

1.3.2 *Procedures for sampling of groundwater*

The first concern is contamination and disturbance of natural conditions caused by drilling operations. New materials are brought into the aquifer, comprising drilling fluids, gravel

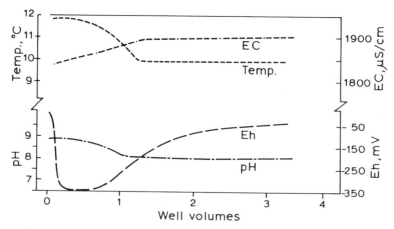

Figure 1.5. The change in chemical composition of discharge during well flushing (modified from Lloyd and Heathcote, 1985).

pack or casing materials. It may take a long time before the influence of such disturbances on groundwater quality is sufficiently diminished to allow representative sampling of the borehole. Water from boreholes where drilling mud has been used displays often cation exchange with the clay in the drilling mud. The time needed to obtain representative samples depends on the groundwater flow rate, the ion exchange capacity etc., and may be in the order of two to three months when the pumping activity on the screen is limited (as is often the case in depth specific sampling).

It is well known that boreholes which have been out of production for some time may yield a water chemistry which is different during production. The main reason is the presence of stagnant water above the screen in the well and it is therefore necessary to empty the well for a number of volumes. On the other hand, excessive pumping or overpumping may draw waters with different composition towards the screen and cause mixing of waters. Thus a balance has to be found between these two aspects. Estimates for the number of well volumes to be emptied vary between 2 and 10 and depend on local hydrological conditions (Barber and Davis, 1987; Robin and Gillham, 1987; Stuyfzand, 1983). In most cases two to four times seems sufficient.

To accomplish effective flushing, the well should be pumped from just below the air-water interface (Robin and Gillham, 1987). The best way to evaluate the degree of flushing needed is to monitor easily measurable field parameters such as the electrical conductivity (*EC*) or pH over time. An example is shown in Figure 1.5 which demonstrates that in this case stationary conditions are obtained after emptying about two well volumes. Problems with sampling from existing wells are often related to faulty well completion and are not always easy to detect. Some examples of common problems are illustrated in Figure 1.6.

In Case A the screen length is too large which causes short circuiting of groundwater from different layers. Case B displays an improperly placed bentonite seal which results in leakage of the upper part of the aquifer into the lower part. In Case C the gravel pack functions as a drain which also may cause short circuiting. Finally, Case D displays the effect of leaky coupling of casing.

Procedures for depth specific groundwater sampling are developing rapidly. Creative minds have invented numerous devices for different purposes and still new ones are added every year. Cheapest are drive-point or hollow stem auger devices which sample water while they are driven down into the aquifer. Their disadvantage is that different depth levels cannot be resampled, except by renewed augering and also the relation of the water samples to the geology remains often uncertain. Permanent installations can be divided (Figure 1.7) into multiple-borehole piezometer nests, multiple-level, single borehole piezometers and single borehole packer samplers (Andersen, 1979; Obermann, 1982). Single hole multiple piezometers can be installed as bundles (Cherry et al., 1983; Appelo et al., 1982; Stuyfzand, 1983; Leuchs, 1988) or enclosed in a casing (Pickens et al., 1978). While the first approach is by far the cheapest, it requires cohesionless aquifer material to ensure sealing between sampling points, but an inherent risk of short circuiting between different sample levels, through the bundles of tubing, remains. This risk increases with the number of sample levels used and differences in hydraulic head in the layers which have been penetrated.

Sample retrieval is done most easily by suction if water table depths are not more than about 9 m. However, the application of vacuum may result in serious degassing of the sample (Suarez, 1987; Stuyfzand, 1983) and thus produce erratic results for dissolved

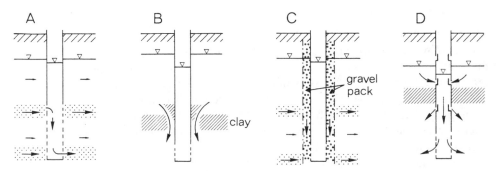

Figure 1.6. Different forms of leakage and short circuiting during groundwater sampling from screened wells (modified from Stuyfzand, 1989). For explanation see text.

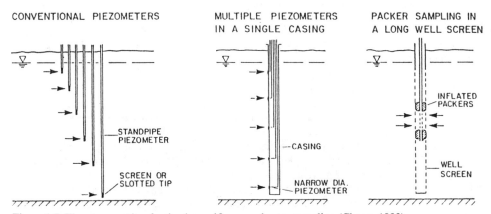

Figure 1.7. Three approaches for depth specific groundwater sampling (Cherry, 1983).

gases, pH and volatile organics. Alternative approaches are pressure operated downhole syringe pumps (Gillham and Johnson, 1981), or double line gas driven sampling devices (Appelo et al., 1982). In the latter approach, a downhole reservoir is filled with water, which subsequently is recovered by pressurizing the sampler with an inert gas like N_2 or Ar. A check valve at the bottom of the reservoir is then closed and the sample transported to the surface.

1.4 CHEMICAL ANALYSIS OF GROUNDWATER

There is probably no such thing as a perfect chemical water analysis carried out by routine procedures. For specific purposes both sampling and analytical procedures have to be adjusted to obtain the best possible results for the critical parameters in the given study.

The analytical chemistry of groundwater in general is beyond the scope of this book. In most cases water analyses are carried out by standard procedures such as described in handbooks like *Standard methods for the examination of water and wastewater* (1989). However, some special aspects concerning collection and conservation of water samples, and the evaluation of the quality of chemical analysis deserve some special attention.

1.4.1 *Field analysis and sample conservation*

When a groundwater sample is brought to the surface, it is exposed to physico-chemical conditions which are different from those in the aquifer. For example, atmospheric oxygen may readily oxidize components like Fe^{2+}, H_2S, etc. which are commonly present in anoxic groundwater. Furthermore, degassing of CO_2 may occur, causing changes in pH, alkalinity and total inorganic carbon, which may also induce carbonate precipitation. Thus, measures are needed to prevent changes in the chemical composition of the sample before analysis. Such measures are of two kinds: conservation and field measurements. An overview of the required treatment is presented in Table 1.8.

Conservation is in most cases done by adding acid to the sample until the pH is < 2. (0.7 ml of 65% HNO_3 is usually enough to neutralize alkalinity and to acidify 100 ml sample).

Table 1.8. Conservation of chemical parameters of water samples.

Parameter	Conservation/field analysis
Ca^{2+}, Mg^{2+}, K^+, Na^+, NH_4^+, Si, PO_4^{3-}	Acidified to pH < 2 in polyethylene (preferably HNO_3 for AAS or ICP-analysis)
Heavy metals	Acidified to pH < 2 in glass
SO_4^{2-}, Cl^-	Not necessary
NO_3^-, NO_2^-	Store cool at 4°C and analyze within 24 hours or add bactericide like thymol. (Note that NO_3^- may form from NH_4^+ in reduced samples. Furthermore NO_2^- may self-decompose even when a bactericide is added)
H_2S	Zn-acetate conservation, or spectrophotometric field measurement
TIC	Dilute the sample to *TIC* < 0.4 mmol/l. (This effectively lowers CO_2 pressure, and prevents the escape of CO_2)
Alkalinity	Field titration with the GRAN method (Stumm and Morgan, 1981)
Fe^{2+}	Spectrophotometric field measurement or in an acidified sample
pH, Temp., *EC*, O_2	Field measurement

Acidification stops most bacterial growth, blocks oxidation reactions, and prevents adsorption or precipitation of cations. Prior to acidification, the water sample has to be filtered to remove suspended material which could dissolve when acid is added.

Pressure filtration, using an inert gas is preferred since vacuum application may seriously degas the sample. The normal procedure is to filtrate through 0.45 μm membrane filters. However, finely dispersed Fe- and Al-oxyhydroxides may pass through 0.45 μm filters and 0.1 μm is a better choice (Kennedy and Zellweger, 1974; Laxen and Chandler, 1982). The effect of improper sample handling on the analytical results is illustrated in Example 1.2.

EXAMPLE 1.2: *Effect of iron oxidation on analytical results*
A groundwater contains 20 mg Fe^{2+}/l. How does the alkalinity change when this iron oxidizes and precipitates as $Fe(OH)_3$ in the sample bottle?
The reaction in the bottle is:

$$2Fe^{2+} + 4HCO_3^- + \frac{1}{2}O_2 + 5H_2O \rightarrow 2Fe(OH)_3 + 4H_2CO_3$$

When 20 mg Fe/l precipitates, this corresponds 20/55.8= 0.36 mmol/l. The reaction consumes twice this amount of HCO_3^-, and the alkalinity is expected to decrease with $2 \times 0.36 = 0.72$ meq/l.

Field analyses are usually carried out for parameters like pH, *EC*, *Eh*, O_2, which are measured by electrode, and sometimes also for alkalinity and Fe^{2+}. Electrode measurements should preferably be carried out in a flow cell in order to prevent air admission. *EC* measurements are particularly useful as a control on analysis and conservation of samples. *Eh* measurements are only qualitative indicators for redox conditions and should be made as sloppy as possible, so that you will not be tempted to relate them to anything quantitative afterwards. The necessity of measuring labile parameters in the field is illustrated in Figure 1.8.

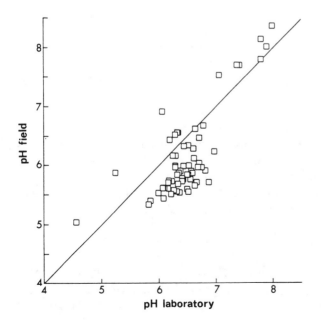

Figure 1.8. Comparison of field measurements of pH, in a carbonate-free sandy aquifer, with those performed in the laboratory (Postma, unpublished results).

The figure shows that substantial differences between field and laboratory measurements are observed which become even more significant when it is remembered that pH is a logarithmic concentration unit (pH = $-$ log[H^+]). pH is a crucial parameter, which has major importance in quantitative calculations of saturation states with respect to minerals and great care should be taken to obtain reliable measurements.

The problems involved in pH measurements of groundwaters are manifold. Some are related to the removal of groundwater from its in situ position during sampling. These include degassing of CO_2, which will increase the pH and also may cause precipitation of $CaCO_3$ (Suarez, 1987). Others concern the oxidation of Fe^{2+} and precipitation as FeOOH, and are analogous for other oxidizable species like H_2S, which may seriously affect pH values. Most of this type of problems can be overcome by careful sampling and field measurement procedures. In particularly the use of flow cells for pH measurement, directly coupled to the sampling system, helps to minimize these problems. Other sources of error, particularly in low ionic strength groundwaters, concern the liquid junction between the calomel electrode and the solution to be measured. The fundamental problem is that the liquid junction potential across the porous ceramic plug of the calomel electrode will vary with the composition of the solution to be measured (Bates, 1973). Liquid junction problems in low ionic strength natural waters have been recognized quite recently (Illingworth, 1981; Brezinski, 1983; Davison and Woof, 1985) and may amount to several tens of a pH unit. A tricky aspect of this type of errors is that they are not revealed by a standard two buffer calibration (Illingworth, 1981; Davison and Woof, 1985), since commercial buffers are all high ionic strength solutions. Electrode performance should therefore be tested in dilute solutions and a good check is to prepare accurately a solution of 10^{-4} HCl or $5 \cdot 10^{-5}$ H_2SO_4 which should yield a pH of 4.00 $\pm$ 0.02. Detailed procedures for testing electrode performance in low ionic strength waters have been described by Davison (1987) and Busenberg and Plummer (1987).

1.4.2 *Accuracy of chemical analysis*

In general, two types of errors are discerned in chemical analyses:

Precision or statistical errors which reflect random fluctuations in the analytical procedure.

Accuracy or systematic errors displaying systematic deviations due to faulty procedures or interferences during analysis.

The precision can be calculated by repeated analysis of the same sample. It is always a good idea to collect a number of duplicate samples in the field as a check on the overall procedure. Systematic errors can be tested only by analyzing reference samples and by interlaboratory comparison of the results. At low concentrations, duplicate analyses may show large variations when the sensitivity of the method is insufficient. The accuracy of the analysis for major ions can be estimated from the Electro Neutrality (E. N.) condition since the sum of positive and negative charges in the water must balance:

$$\text{Electro Neutrality (E. N., \%)} = \frac{(\text{Sum cations} + \text{Sum anions})}{(\text{Sum cations} - \text{Sum anions})} \cdot 100$$

where cations and anions are expressed as meq/l. The sums are taken over the cations Na^+, K^+, Mg^{2+} and Ca^{2+}, and anions Cl^-, HCO_3^-, SO_4^{2-} and NO_3^-. Sometimes other elements contribute significantly, like for example Fe^{2+} or NH_4^+ in reduced groundwater, or H^+ and Al^{3+}

in acid water. The presence of the last two substances in significant amounts requires more accurate calculations of E.N. balances using special computer programs which will be presented in Chapter 3. Differences in E.N. of up to 2% are inevitable in almost all laboratories. Sometimes an even larger error must be accepted, but at deviations of more than 5% the sampling and analytical procedures should be examined.

EXAMPLE 1.3: *Estimating the reliability of water analyses*
Your laboratory returns the following water analysis:

pH = 8.22
$EC = 290\,\mu S/cm$

Na^+	K^+	Mg^{2+}	Ca^{2+}	Cl^-	HCO_3^-	SO_4^{2-}	NO_3^-	
13.7	1.18	3.2	42.5	31.2	79.9	39	1.3	mg/l

Is this a reliable analysis?
  A basic condition for every analysis is that a reasonable charge balance of cations and anions is present: the solution should be electrically neutral. The analysis is recalculated from mg/l to mmol/l and then to meq/l.

	mg/l		Form. wght (grams)		mmol/l	Charge		meq/l
Ca^{2+}	42.5	/	40.08	=	1.06	× 2	=	2.12
Mg^{2+}	3.21	/	24.31	=	0.13	× 2	=	0.26
Na^+	13.7	/	22.99	=	0.60	× 1	=	0.60
K^+	1.18	/	39.1	=	0.03	× 1	=	0.03
						Σ	=	+3.01
Cl^-	31.2	/	35.45	=	0.88	× −1	=	−0.88
SO_4^{2-}	39.	/	96.06	=	0.41	× −2	=	−0.82
HCO_3^-	79.9	/	61.02	=	1.31	× −1	=	−1.31
NO_3^-	1.3	/	62.0	=	0.02	× −1	=	−0.02
						Σ	=	−3.03

The difference between cations and anions is 0.02 meq/l, and this is undoubtedly a 'good' analysis. The electrical balance is:
  E.N. = (3.01 + − 3.03)/(3.01 − − 3.03)·100 = − 2.00/6.04 = − 0.3%, which is very good indeed.

Another useful technique is to compare calculated conductivities with measured electrical conductivity. The electrical conductivity (*EC*) is related to the ions which are present in solution and the relation between *EC* and concentrations of different salts is shown in Figure 1.9. Thus an average $Ca(HCO_3)_2$ freshwater containing some dissolved NaCl will have an equivalent conductance of 100 µS/cm per meq/l.

  Therefore, at 25°C, the *EC* divided by 100 yields a very good estimate of the sum of anions or cations (both in meq/l):

$$\Sigma\,\text{anions} = \Sigma\,\text{cations (meq/l)} = EC/100\,(\mu S/cm)$$

This relation is valid for *EC* values up to around 2000 µS/cm. More complicated

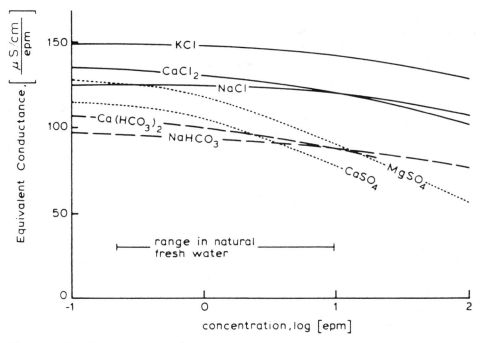

Figure 1.9. The relation between equivalent electrical conductivity and concentration for different salt solutions.

expressions have been derived, by quite a few authors (cf. Stuyfzand, 1983), but their application should only be considered when the accuracy of the *EC* measurement is better than 5%. A few authors (e.g. Hem, 1985) mention relationships between *EC* and *TDS* in mg/l, but these are always less reliable since they contain an additional assumption about average molecular weight of the conducting ions.

EXAMPLE 1.4: *Analytical inaccuracies due to precipitation in the sample bottle*
In samples brought from Portugal to the laboratory in Amsterdam a large charge imbalance was found. A sample consisted of 100 ml water acidified to pH < 2 for the analysis of cations, and 200 ml water for the analysis of HCO_3^-, Cl^-, SO_4^{2-}, NO_3^- as well as pH and *EC*. The samples were transported by car, and were subjected to significant temperature variations. A typical example is given below:

temp. = 18°C
pH_{field} = 7.00 $pH_{lab.}$ = 7.62
EC_{field} = 720 $EC_{lab.}$ = 558 μS/cm

Na^+	K^+	Mg^{2+}	Ca^{2+}	Cl^-	HCO_3^-	SO_4^{2-}	NO_3^-	
0.55	0.01	0.35	3.00	0.64	4.31	0.16	0.12	mmol/l

sum cations = 7.36; sum anions = −5.38 meq/l

The sum of anions and cations in this sample differs by +16%, which could not be improved by duplicate analyses. It was realized, however, that cations and anions were analyzed in different bottles. When the cations were analyzed in the unacidified sample bottle, lower concentrations of

Ca^{2+} were found. The reason for the imbalance was clearly that calcite had precipitated in the bottle without acid, thereby decreasing the HCO_3^- concentration.

The reactions in the bottle can also be discovered when field measurements are compared with laboratory measurements. *EC* of the samples measured in the field was consistently higher by about 100 $\mu S/cm$ than *EC* measured in the laboratory. Note the good agreement between *EC*/100 and the sum of ions.

Calculation of E.N. or *EC* as a check on the chemical analyses is only applicable for major elements. The accuracy of the results for minor elements is much more difficult to estimate. Sometimes the incompatibility of different elements found together in one sample can be a warning that something went wrong. Thus, it is unlikely that O_2 or NO_3^-, which indicate oxidizing conditions, are found together with appreciable concentrations of Fe^{2+} in natural water (see Figure 1.3). Ferrous iron (Fe^{2+}) only occurs in appreciable concentrations (more than 1 $\mu mol/l$) in reduced environments or at low pH, or when organic material is present which acts as a complexing agent. Ferric iron (Fe^{3+}) has a very low solubility in water with pH between 3 and 11, and precipitates very rapidly as $Fe(OH)_3$. Total iron concentrations should therefore be low in water where O_2 or NO_3^- is present. Aluminum has also a low solubility at a pH between 5 and 8, and it is unlikely to find concentrations of more than 1 $\mu mol/l$ in water having a near neutral pH. The best procedure for an estimate of the accuracy of chemical analysis is to include a reference water sample obtained for example from NBS or Community Bureau of Reference-BCR (Rue de la Loi 200, B-1049 Brussels, Belgium).

PROBLEMS

1.1. An analysis of groundwater gave as results:

pH	Na$^+$	K$^+$	Mg^{2+}	Ca^{2+}	Cl$^-$	*Alk*	SO$_4^{2-}$	mg/l
8.5	62	0.1	9	81	168	183	–	

Alkalinity is expressed as HCO_3^-

  a. Is the analysis correct?

  b. Calculate total hardness, in degrees german hardness, and non-carbonate hardness.

1.2. Analyses of three watersamples are listed below. Check the analyses and indicate where they might be wrong. (Concentrations in mmol/l).

	1	2	3
pH (–)	8.01	7.2	7.5
EC ($\mu S/cm$)	750	150	450
Na$^+$	2.3	0.3	0.5
K$^+$	0.1	0.08	<0.001
Mg^{2+}	1.1	0.2	0.2
Ca^{2+}	1.4	1.4	1.5
Cl$^-$	2.0	0.3	0.5
HCO$_3^-$	0.3	0.7	1.5
SO$_4^{2-}$	0.9	0.2	0.5
NO$_3^-$	1.8	0.1	1.5
Al	<0.001	0.2	<0.001
Fe	<0.001	<0.001	0.3

1.3. The analyses listed below reflect a range of water compositions in Denmark.

A = St. Heddinge waterworks, B = Maarum waterworks, C = Haralds mineralwater, D = Rainwater Borris (conc. in mg/l).

	A	B	C	D
pH	7.37	7.8	5.5	4.4
Alk	297.0	1083.0		0.0
SO_4^{2-}	61.0	0		4.48
Cl^-	29.0	201.0	17.0	12.0
PO_4^{3-}	0	0.02		
NO_3^-	35.0	2.0	5.0	4.52
F^-	0.32	1.7		0
NH_4^+	0.012	3.5		1.23
Ca^{2+}	121.0	42.0		1.4
Mg^{2+}	14.0	29.0		0.78
Na^+	18.0	440.0	9.0	7.2
K^+	6.3	14.0	1.0	0.63
EC (µS/cm)	750.0	1900.0	88.0	

   a. Compare the water analyses with the drinking water limits in Table 1.7. Are these waters suitable as drinking water?

   b. Check the quality of each analysis by setting up a charge balance. Compare the total amount of equivalents of ions with the electrical conductivity (*EC*) and identify possible errors in the analysis. Analysis C (Harald's Mineralwater) is incomplete. However, the hardness in german degrees is known to be 1°dH. Use this information to calculate the electrical conductivity. Note that this also sets limits to the missing anions.

   c. Analysis D is rainwater from Borris (C. Jutland). Calculate the contribution of rainwater ions to the groundwater samples A-C, assuming that chloride behaves like a conservative element. When would this assumption become dubious?

   d. Draw a Stiff diagram for each analysis and discuss which processes have influenced the water composition. Consider what kind of geological deposit the analysis has been extracted from.

REFERENCES

Andersen, L.J., 1979, A semi-automatic level-accurate groundwater sampler. *Danmarks. Geol. Unders. Årbog 1978*, 165-171.

Appelo, C.A.J., Krajenbrink, G.J.W., Van Ree, C.C.D.F. and Vasak, L., 1982, *Controls on groundwater quality in the NW Veluwe catchment* (in Dutch). Soil Protection Series 11, Staatsuitgeverij, Den Haag, 140 pp.

Barber, C. and Davis, G.B., 1987, Representative sampling of ground water from short-screened boreholes. *Ground Water 25*, 581-587.

Bates, R.G., 1973, *Determination of pH; theory and practice.* 2nd ed. Wiley, New York, 479 pp.

Biswas, A.K., 1970, *History of hydrology.* Elsevier, Amsterdam, 336 pp.

Brezinski, D.P., 1983, Kinetic, static and stirring errors of liquid junction reference electrodes. *Analyst 108*, 425-442.

Busenberg, E. and Plummer, L.N., 1987, *pH measurement of low conductivity waters.* U.S. Geol. Surv. Water Res. Inv. Rep. 87-4060.

Cherry, J.A., 1983, Piezometers and other permanently-installed devices for groundwater quality monitoring. *Proc. Conf. on Groundwater and Petroleum Hydrocarbons: Protection, Detection and Restoration*, IV-1 IV-39.

Cherry, J.A., Gillham, R.W., Anderson, E.G. and Johnson, P.E., 1983, Migration of contaminants in groundwater at a landfill: a case study. 2. Groundwater monitoring devices. *J. Hydrol.* 63, 31-49.

Davison, W., 1987, Measuring pH of fresh waters. In A.P. Rowland (ed.), *Chemical analysis in environmental research*. ITE symp. no 18, Abbots Ripton, ITE, 32-37.

Davison, W. and Woof, C., 1985, Performance tests for the measurement of pH with glass electrodes in low ionic strength solutions including natural waters. *Anal. Chem.* 57, 2567-2570.

Domenico, P.A. and Schwartz, F.W., 1990, *Physical and chemical hydrogeology*. Wiley & Sons, New York, 824 pp.

Drever, J.I., 1988, *The geochemistry of natural waters*. 2nd ed. Prentice-Hall, 437 pp.

Edmunds, W.M., Darling, W.G. and Kinniburgh, D.G., 1988, Solute profile techniques for recharge estimation in semi-arid and arid terrain. In I. Simmers (ed.), *Estimation of natural groundwater recharge*. NATO ASI Ser. 222, 139-157.

Eriksson, E., 1960, The yearly circulation of chloride and sulfur in nature; meteorological, geochemical and pedological implications, Part II. *Tellus* 12, 63-109.

Gillham, R.W. and Johnson, P.E., 1981, A positive displacement ground-water sampling device. *Groundwater Monitoring Review* 1, 33-35.

Hem, J.D., 1985, *Study and interpretation of the chemical characteristics of natural water*. 3rd ed. U.S. Geol. Survey Water Supply Paper 2254.

Handbook of Chemistry and Physics. Am. Rubber Cy (CRC press).

Illingworth, J.A., 1981, A common source of error in pH measurement. *Biochem. J.* 195, 259-262.

Kennedy, V.C. and Zellweger, G.W., 1974, Filter pore-size effects on the analysis of Al, Fe, Mn, and Ti in water. *Water Resour. Res.* 10, 785-790.

Laxen, D.P.H. and Chandler, I.M., 1982, Comparison of filtration techniques for size distribution in freshwaters. *Anal. Chem.* 54, 1350-1355.

Leeflang, K.W.H., 1938, The chemical composition of precipitation in The Netherlands (in Dutch). *Chem. Wkbld.* 35, 658-664.

Leuchs, W., 1988, Vorkommen, Abfolge und Auswirkungen anoxisch Redoxreaktionen in einem Pleistozänen Porengrundwasserleiter. *Bes. Mitt. Deutsches Gewässerkundl. Jahrb.* 52, 105 pp.

Lloyd, J.W. and Heathcote, J.A., 1985, *Natural inorganic hydrochemistry in relation to groundwater*. Clarendon Press, Oxford, 296 pp.

Obermann P., 1982, Hydrochemische/hydromechanische Untersuchungen zum Stoffgehalt von Grundwasser bei landwirtschaftlicher Nutzung. *Bes. Mitt. Deutsches Gewässerkundl. Jahrb.* 42.

Pickens, J.F., Cherry, J.A., Grisak, G.E., Merrit, W.F. and Risto, B.A., 1978, A multilevel device for ground-water sampling and piezometric monitoring. *Ground Water* 16, 322-323.

Robin, M.J.L. and Gillham, R.W., 1987, Field evaluation of well purging procedures. *Groundwater Monitoring Review* 7, 85-93.

Safe Drinking Water Comm., 1980, Drinking water and health, Vol. 3. Nat. Acad. Press, Washington, 415 pp.

Schoeller, H., 1960, Salinity of groundwater, evapotranspiration and recharge of aquifers (in French). IASH Pub. 52, 488-494.

Standard methods for the examination of water and wastewater, 1985, Joint publication of APHA, AWWA, WPCF. 16th edition, Am. Publ. Health Ass., Washington, 1268 pp.

Stumm, W. and Morgan, J.J., 1981, *Aquatic chemistry*. 2nd ed. Wiley & Sons, New York, 780 pp.

Stuyfzand, P.J., 1983, Important sources of errors in sampling groundwater from multilevel samplers (in Dutch). H_2O 16, 87-95.

Stuyfzand, P.J., 1989, Hydrochemical methods for the analysis of groundwater flow (in Dutch). H_2O 22, 141-146.

Suarez., D.L., 1987, Prediction of pH errors in soil-water extractors due to degassing. *Soil Sci. Soc. Am. J.,* 51, 64-67.

Wedepohl, K.H. (ed.), 1969-1976, *Handbook of geochemistry*. Springer Verlag, Berlin.

Welch A.H., Lico, M.S. and Hughes, J.L., 1988, Arsenic in ground water of the western United States. *Ground Water* 26, 333-347.

CHAPTER 2

From rainwater to groundwater

This chapter follows the evolution of water chemistry through the hydrological cycle, beginning with the composition of rainwater. The chemistry of rainwater is treated in some detail, since it is the source of most groundwater. The processes and reactions which determine the quality of water as it infiltrates into the soil are summarized as an introduction to a more detailed treatment in later chapters. A few examples from the interpretation of trends in water composition are also presented.

2.1 RAINWATER

The composition of rainwater is determined by the source of the water vapor and by the ions which are acquired by water (or lost from it) during transport through the atmosphere. Near the coast, the composition of rainwater strongly resembles diluted seawater. As the distance to the coast increases, the concentrations of ions that are directly derived from seawater decreases. This is illustrated in Figure 2.1 for the Cl^- content in precipitation over the Netherlands; the highest concentrations are found near the coast as the result of a rapid washout of the largest sea salt particles, and the Cl^- concentration decreases exponentially inland to a background concentration of 1-2 mg/l at 200-400 km from the coast.

A major difference is normally observed between the composition of rainwater samples collected by *wet-only* rain gauges which only open during rain, and of *bulk* samplers that are permanently open (Galloway and Likens, 1978). During dry periods bulk samplers collect aerosols as well as anomalies such as bird droppings on the funnel that wash into the sampling bottle with the next rain. The difference between *wet-only* and *bulk* sampling is also fundamental in estimating the atmospheric input to the groundwater reservoir, since dry deposition depends on surface characteristics ('roughness', etc.) which cannot easily be extrapolated from the rain gauge to the natural surface. The importance of dry deposition will be treated more extensively in Section 2.1.2.

Let us return first to the ocean, which is the source of approximately 30% of all water in continental precipitation (Garrels et al., 1975). Experimental studies have shown that the liquid droplets, produced when gas bubbles burst at the sea surface, play a dominant role in generating marine aerosols (Monahan, 1986; Blanchard and Syzdek, 1988). Evaporation of the water from droplets produces aerosols with different size-ranges. The coarsest particles have diameters exceeding ca. 10 μm and retain a composition that closely resembles seawater. From smaller particles, evaporation of chloride occurs, presumably in the form of

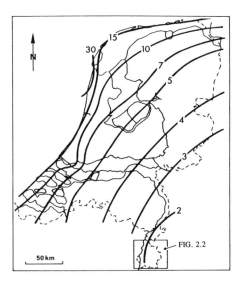

Figure 2.1. Chloride iso-concentration lines in precipitation over the Netherlands (concentrations in mg/l; from Ridder, 1978).

HCl, which results in Cl^- depletions relative to Na^+ of up to 40% (Raemdonck et al., 1986; Möller, 1990). The rainout of Cl^- depleted particles produces a Cl^- depleted rain but since the evaporated HCl also contributes to the total rainwater chemistry, the net effect is that Na/Cl ratios remain similar to seawater ratios in coastal areas of Europe (Möller, 1990), the United States (Wagner and Steele, 1989), and in the Amazonas (Andreae et al., 1990). In industrialized areas additional Cl^- sources are burning of waste plastics (Lightowlers and Cape, 1988), coal fired power plants (Wagner and Steele, 1989), or even the evaporation of salt-water used as coolant in the steel industry (Vermeulen, 1977). In addition to sea salt, Na^+ in aerosols may also originate from soil dust particles in arid climates (Nativ et al., 1983).

The depletion or addition of ions in rainwater is reflected by a change in the concentration ratio with respect to seawater and can be expressed by a *fractionation factor*. For example, for Cl^- relative to Na^+ the fractionation factor F_{Na} is:

$$F_{Na} = \frac{(Cl/Na)_{rain}}{(Cl/Na)_{seaw.}}$$

In this formula, the Cl^- and Na^+ concentrations in rain or seawater can be expressed as mmol/l, mg/l or any other concentration unit. Table 2.1 compiles fractionation factors for the rainwater composition on the oceanic island of Hawaii which for major cations are also valid for other near-coastal stations.

Table 2.1 indicates that major elements in near-coastal areas are not fractionated very

Table 2.1. The range of fractionation factors for seawater components in marine aerosols. Fractionation factors are given relative to Na^+ (F_{Na}). (Data from Duce & Hoffman, 1976).

		Mg^{2+}	Ca^{2+}	K^+	Sr^{2+}	Cl^-	SO_4^{2-}	Br^-	Org. N
F_{Na}	max.	1.07	1.22	1.05	0.89	1.0	?	12	$1 \cdot 10^6$
	min.	0.98	0.97	0.97	0.84	0.93	1	1	$2 \cdot 10^4$

Table 2.2. Rainwater analyses (Concentrations in μmol/l).

	Kiruna Sweden[a]	Hubbard Brook, US[b]	De Kooy Neth.[c]	Beek Neth.[c]	Thumba India[d]	Delhi India[d]
Period	1955-1957	1963-1974	1978-1983	1978-1983	1975	1975
pH	5.6	4.1	4.40	4.75		
Na^+	13	5	302	43	200	30
K^+	5	2	9	7	6	7
Mg^{2+}	5	2	35	9	19	4
Ca^{2+}	16	4	19	47	23	29
NH_4^+	6	12	78	128		
Cl^-	11	7	349	54	229	28
SO_4^{2-}	21	30	66	87	7	4
NO_3^-	5	12	63	63		
HCO_3^-	21	–	–	9		

a) Granat, 1972; b) Likens et al., 1977; c) KNMI/RIV, 1985; d) Sequeira & Kelkar, 1978.

much. For Cl^-, the fractionation factor ranges from 0.93 to 1.0 and perhaps the lower value is the result of some volatilization. For calcium, the fractionation factor ranges from 0.97 to 1.22 indicating some enrichment of Ca^{2+} with respect to Na^+, which suggests the presence of a land based Ca^{2+} source. The elements which are associated with organic matter, Br^- and organic-N, have fractionation factors that are much higher than one and are thus clearly enriched in rainwater. This seems to be due to the concentration of lipid containing organic matter at the interface between water and air, causing an enrichment of these components in the foam that is ejected into the atmosphere.

The absence of extensive fractionation for major components (Table 2.1) leads to the conclusion that rainwater with a purely marine origin is basically strongly diluted seawater. However, during transport over land, the air masses and clouds pick up continental dust and gases, from natural or industrial origin, which modifies the composition of the rainwater. Table 2.2 contains a number of rainwater analyses from different parts of the world that illustrates the observed compositional variations. To facilitate the interpretation of these analyses and to illuminate supplementary sources, the sea water contribution can be subtracted from the rainwater analyses. This calculation requires the selection of a conservative component, for which Cl^- is suitable since it shows little fractionation. The calculation proceeds as follows, for example for Na^+ (Table 2.3). The Na^+ concentration in seawater is 485 mmol/l and the Cl^- concentration is 566 mmol/l. The ratio of Na/Cl in seawater is therefore 0.86. In the Kiruna rain, the Cl^- concentration is 11 μmol/l and the part of the Na^+ that is derived from seawater amounts therefore to $0.86 \cdot 11 = 9$ μmol/l. The observed Na^+ concentration in Kiruna rainwater is 13 μmol/l, and $13 - 9 = 4$ μmol/l Na^+ must have a non-marine origin. This difference (which indicates either a continental source of Na^+, or a depletion of Cl^-) is, however, barely significant compared to the analytical accuracy. Such calculations were performed for all rainwater analyses of Table 2.2, and results are given in Table 2.3.

After subtracting the seawater contribution, the precipitation of the near coastal stations, De Kooy or Thumba, shows low residuals of Na^+, K^+ and Mg^{2+} ions, which indicates that

Table 2.3. Concentrations in rainwater from sources other than seawater. For the Kiruna station both the contribution from seawater and from additional sources is shown, while for the other stations only additional sources are shown. Units are μmol/l, except for seawater.

	seaw. (mmol/l)	Kiruna rain	Kiruna seaw. contr.	Kiruna other sourc.	Hub. Br.	De Kooy	Beek	Thumba	Delhi
Na^+	485	13	9.4	4	−1	3	−3	4	6
K^+	10.6	5	0.2	5	2	3	6	2	6
Mg^{2+}	55.1	5	1	4	1	1	4	−3	1
Ca^{2+}	10.7	16	0.2	16	4	12	46	19	28
NH_4^+	$2 \cdot 10^{-6}$	6	0	6	12	78	128		
Cl^-	566	11	11	–	–	–	–	–	–
SO_4^{2-}	29.3	21	0.6	20	30	48	84	−5	3
NO_3^-	$5 \cdot 10^{-6}$	5	0	5	12	63	63		
HCO_3^-	2.4	21	–	21	0	−1	9		

the content of these ions in rain primarily is derived from seawater. A conspicuous enrichment of the anions NO_3^- and SO_4^{2-} has occurred in the rain of the United States and Europe. These ions originate mainly from industrial and traffic fumes containing gaseous NO_x and SO_2. The gases become oxidized in the atmosphere, producing the acids HNO_3 and H_2SO_4. These acids lower the pH of rainwater and are the cause of *acid rain* which is now a major environmental issue in the industrialized world. On the other hand, the intensification of agriculture produces neutralizing base in the form of NH_4OH, which evaporates from manure that is spread over agricultural land (Van Breemen et al., 1982). Its influence can be seen in the analysis of Beek, located in an area of the Netherlands where agriculture has grown rapidly since 1970. Not much ammonia is present in the rain at Hubbard Brook and the pH is therefore lower at this site.

Natural base is also available in continental dust. Calcite is almost everywhere present in soils, and in the continental rain of India large excesses of Ca^{2+} are likely to be the result of calcite containing dust. In humid climates, the amount of calcite ejected into the atmosphere by natural processes is too small to neutralize the industrially produced acids. In this case, only other anthropogenic activities may effectively neutralize acid rain. The analysis of Beek rainwater in Tables 2.2 and 2.3 shows a large excess of Ca^{2+} when the seawater contribution has been subtracted. This is due to a local cement industry which uses marls from an open pit mine as source material. Such effects of local anthropogenic activities on the pH of rainwater are illustrated in Figure 2.2 which shows the pH of rain in Limburg, the Netherlands, with a high pH around the cement industry in Maastricht, and relatively low pH value around a large chemical plant near Heerlen.

A more complete discussion of sources and processes which lead to acid rain and its effects on vegetation, human health and cultural valuables is beyond the scope of this book and can be found elsewhere (Drabløs and Tollan, 1980; Brimblecombe, 1986). The net effects of acids and bases on the pH of rainwater are very clearly illustrated in a bar diagram by Morgan (1982) showing rainwater analyses from California (Figure 2.3). The anions Cl^- and part of the SO_4^{2-} are of marine origin and are balanced by the cations Na^+ and Mg^{2+}, which also have a marine origin. Additional SO_4^{2-} and NO_3^- comes from industrial sources

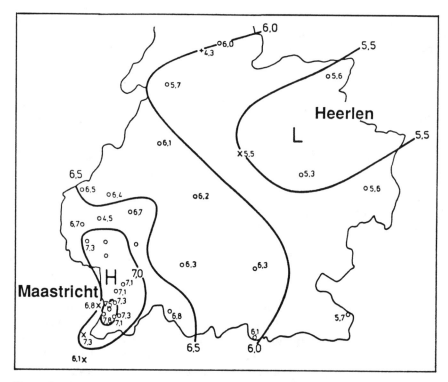

Figure 2.2. The pH of rain in Limburg, the Netherlands, displaying the effects of both the cement industry and chemical industry (from Ridder, 1978). See Figure 2.1 for location.

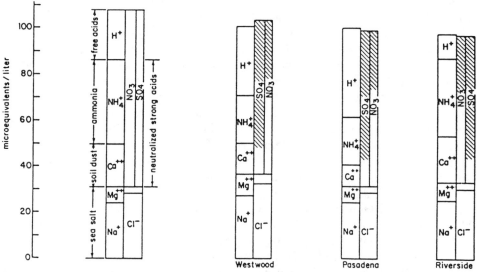

Figure 2.3. The composition of rain in southern California, 1978-1979, interpreted in terms of input components and source type (Morgan, 1982). Hatched areas in the SO_4^{2-} and NO_3^- fields represent immobile sources, unhatched is the contribution from traffic.

and was originally present as sulfuric and nitric acid but has become partly neutralized by $CaCO_3$ from continental dust and NH_4OH from manure evaporation. The unneutralized part of the sulfuric and nitric acid is present as free H^+, and determines the pH of the rainwater.

2.1.1 *Sources and transport of atmospheric pollutants*

Atmospheric pollution operates on an international scale due to rapid dispersion and world-wide transport of local emissions through the atmosphere. The rapidity of dispersion in the atmosphere has been demonstrated in trajectory studies, in which atmospheric particles (or air parcels) are traced back in time based on meteorological observations. Figure 2.4 presents the results of an early study (Newell, 1971) and shows that both in the troposphere and atmosphere, particles or gases travel the distance from the United States to

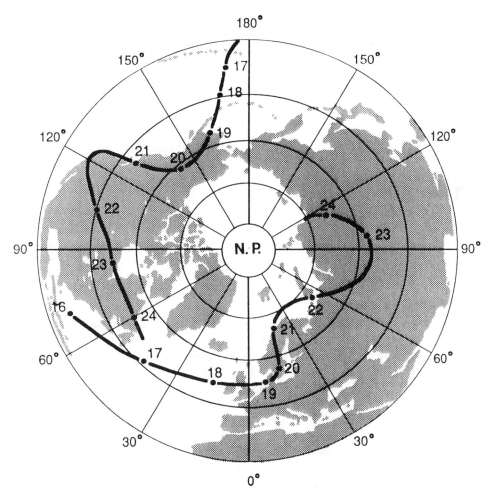

Figure 2.4. Tracks of air parcels around the world. The numbers represent successive days in April, 1964 (Modified from Newell, 1971).

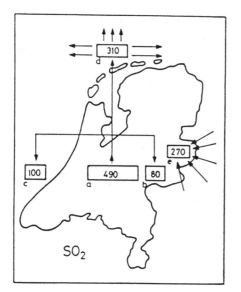
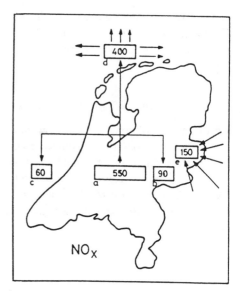

Figure 2.5. Mass balances for SO_2 and NO_x in the Netherlands. The values are given in 1000 tons/year. (a) is total industrial emission from the Netherlands, of which (b) is deposited in the Netherlands, (c) in sea, and (d) in other countries. (e) is deposited in the Netherlands from foreign sources.

Europe in only about 4 days. Industrial pollution of the atmosphere is clearly not limited by national frontiers.

Mass balances for the acidifying gases NO_x and SO_2 have been constructed for individual countries and an example for the Netherlands is shown in Figure 2.5. The figure shows that of a total of 550 000 tons of NO_x (expressed as NO_2) emitted, only 90 000 tons is deposited within the Netherlands. The remainder is dispersed, in part to the sea (60 000 tons), and in part to other countries (400 000 tons). The exported amount exceeds the 150 000 tons deposited as import from foreign sources. Similar numbers apply for SO_2.

2.1.2 *Dry deposition*

Dry deposition, or atmospheric fallout during dry periods, comprises both the deposition of particulate aerosols and the deposition of atmospheric gases by adsorption. It is an important source of elements, as shown already by Garrels and Mackenzie (1972). The dry deposition of trace elements such as Cd, Cu, Pb and Zn is about equal to the wet deposition on the North Sea. The total atmospheric input of Pb into the North Sea is about equal to the total riverine input, while the atmospheric input for Cd and Zn amounts to about 1/3 of the riverine input (Van Aalst et al., 1983). Dry deposition also contributes significant amounts of NO_x and SO_2, and since gaseous adsorption is relatively important for these components, dry deposition is normally greater for conifer than for deciduous canopies (Fowler, 1980). Since Cl^- often is used as a conservative parameter for estimating the groundwater recharge (Schoeller, 1960; Eriksson, 1960; Lerner et al., 1990), it is important to estimate the contribution of dry deposition to atmospheric input for this element.

Unfortunately, dry deposition on vegetation is notoriously difficult to measure. Samples of throughfall under the canopy consist of *wet input*, from which ions may have been lost by adsorption to the canopy, with an admixture of flushed deposits from the previous dry period, and leachates from the vegetation (e.g. Ulrich et al., 1979; Miller and Miller, 1980). Some idea of the importance of dry Cl⁻ deposition can be obtained from chemical balances of watersheds or lysimeters in which both input and output fluxes have been measured over a period of several years. Figure 2.6 shows the ratio of output/input flux for a number of ecosystems with different vegetations. The input flux is the product of the Cl⁻ concentration in precipitation multiplied by the amount of precipitation over the study area (g/year); the output flux is estimated from either chemical analyses of streamflow water and discharge measurements, groundwater analyses and precipitation-surplus, or from lysimeter studies (also in units of g/year). The ratio of output over wet input is plotted against Cl⁻ concentration in precipitation since higher Cl⁻ concentrations in precipitation should be associated with a higher abundancy of Cl⁻ aerosols, leading to increased dry deposition.

Figure 2.6 shows a ratio close to one in areas covered with heather or low shrubs, so that the input with wet rainfall equals output and dry deposition is of minor importance. In forests with very low Cl⁻ concentrations in precipitation, ratios less than one are found, indicating an actual loss of Cl⁻ in the watershed. This could possibly be due to some biological uptake or alternatively to anion adsorption on clay minerals, as suggested by Feth et al. (1964) in a study of the US Sierra Nevada, although this has not been documented in the field studies compiled in Figure 2.6. When the average Cl⁻ concentration in wet deposition of forested areas exceeds 50 μmol/l, a ratio around 2 is found. This suggests a total Cl⁻ deposition that is about twice the amount supplied by rain and underlines the importance of land use for dry deposition. There is no clear difference

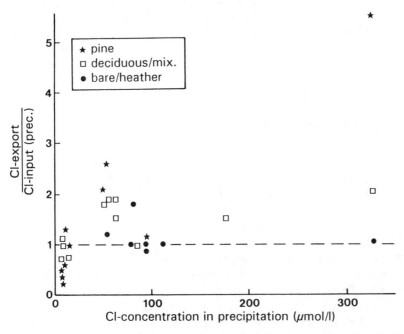

Figure 2.6. The ratio of Cl⁻ ouput/input in watersheds and lysimeters against the Cl⁻ concentration in rain. The Cl-input comprises only wet deposition (rain) (Appelo, 1988).

between deciduous and coniferous forests except for the 3 lysimeters at a coastal site in the Netherlands, with a 40 year long record (Stuyfzand, 1984), that plot at the extreme right of Figure 2.6. Here the pine forest has a ratio exceeding 5, which shows that its perennial canopy effectively collects marine aerosols during winter storms.

The dry deposition calculated from chemical balances can be compared with results from a physical model (Eriksson, 1959). Operational dry deposition might be expressed as:

$$D = 3.15 \cdot 10^5 \, v_d \, C$$

where D is dry deposition ($\mu g/m^2.yr$), v_d is the dry deposition velocity (cm/s), C is concentration in ($\mu g/m^3$), and $3.15 \cdot 10^5$ recalculates cm/s into m/yr.

The value of v_d depends, according to Stokes' law, on the aerosol particle diameter, and could be obtained from particle diameters. However, this appears valid only for aerosols with diameters larger than 10 μm (Mészáros, 1981). The size of aerosol particles almost universally shows a bimodal character, with one fraction between 0.1 and 0.8 μm, and a second fraction between 10 and 100 μm. The large fraction is considered to result from a balance between the production of larger particles (seasalt, soil dust, soot, etc.) and their sedimentation, whereas the fine fraction apparently is the result of coagulation and condensation of particles that become strongly hygroscopic when finer than 0.1 μm (Whitby et al., 1972). The small-size aerosols are most important for transport over distances of more than 10-100 km, and have a considerably higher deposition velocity than calculated from Stokes' law; the velocity has been estimated to range from 0.2 to 2 cm/s (Mészáros, 1981; Möller, 1990). Using v_d = 1 cm/s, and C_{Cl} = 5 $\mu g/m^3$, which is the value observed at 600 m high above the sea (Blanchard et al., 1984), then D = 1.58·10³ mg/m².yr would be the continental dry deposition rate of Cl. This value translates to 2.0 mg/l in 800 mm rain per year and dry deposition therefore approximately doubles the wet input in the areas of the Netherlands and Germany situated at about 100-500 km from the coast. Accordingly, the results from the physical model are in reasonable agreement with estimates from chemical balances for forested areas (Figure 2.6 and Matzner and Hesch, 1981; Appelo, 1988; Mulder, 1988).

2.2 FROM RAINWATER TO GROUNDWATER

The effects of variations in rainwater chemistry, dry deposition, seasonal variations in infiltration etc. on the groundwater chemistry, are seen most clearly in the unsaturated zone and particularly when the rocks contain no highly reactive minerals such as carbonates. This is illustrated by a water chemistry profile through the unsaturated zone of a sandy deposit, near the coast in Denmark in Figure 2.7.

Very large variations in the total dissolved ion concentrations are mostly due to variations in concentrations of Na^+ and Cl^- which both have a marine origin. At this site the average annual water transport rate through the unsaturated zone is about 5 m/yr, so that the whole profile represents the water quality of infiltration through roughly one year, although net infiltration occurs only from October to May. During the fall, infiltration starts and the accumulated salts in the soil from the previous period start to wash down into the unsaturated zone. However, in this same period, westerly storms also set in and may deposit large amounts of sea salt. The effect of these two processes is not easily separated and the

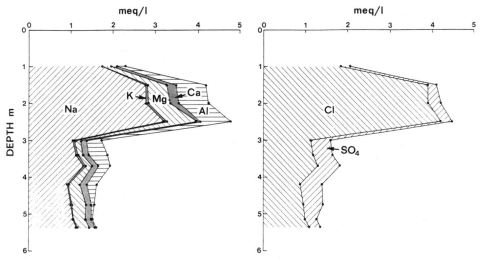

Figure 2.7. The groundwater chemistry in the unsaturated zone of a sandy, carbonate free sediment under a conifer forest in Denmark. Concentrations are plotted cumulatively (Hansen and Postma, 1995).

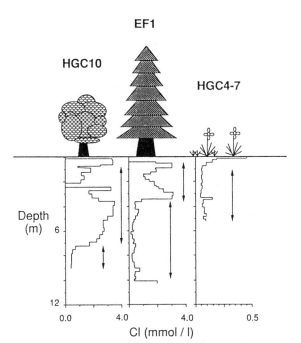

Figure 2.8. Chloride profiles in the unsaturated zone of the Sherwood sandstone below birch woodland, Cyprus Pine and heathland (Moss and Edmunds, 1989).

maximum in concentrations observed in Figure 2.7 may very well be a combination of the two.

The influences of dry deposition, and differences in evapotranspiration, on the chloride content of water in the unsaturated zone of a sandstone are illustrated in Figure 2.8. Below heathland, Cl⁻ concentrations are low and show little variation. Below the forest, chloride concentrations are much higher and vary widely. The high Cl⁻ concentrations below the forest can partly be explained by the effect of dry deposition (Figure 2.6). In addition, evapotranspiration varies distinctly between heath/grassland and forested areas. In particular interception (evaporation from wetted surfaces) is much higher in forested areas and may amount to about 40% of the annual precipitation (Calder, 1979, 1990). Accordingly, the concentration of dissolved ions becomes correspondingly higher. Moss and Edmunds (1989) found that the recharge rate below the heathland was roughly three times higher than below the forest. However, the recharge rates calculated from moisture content agreed poorly with those calculated from a chloride mass balance, which indicates that also dry deposition is of importance here.

Figure 2.8 suggest that total concentrations of ions derived from precipitation must be significantly higher in groundwater downstream of forests than below grass and heathland, but the large variations in concentrations shown in Figures 2.7 and 2.8 are expected to disappear rapidly in the saturated zone due to dispersion. In addition, the total dissolved content in groundwater will of course be affected by dissolution of minerals and other interactions with the rocks, as elaborated in the next section.

2.3 OVERALL CONTROLS ON WATER QUALITY

The concentrations of different species in terrestrial waters of the US is shown in a frequency plot in Figure 2.9. For example, for total dissolved solids, it shows that 10% of the samples contain less than 100 ppm, while 95% of the samples contain less than 1000 ppm. This concentration range is much higher than discussed above for rainwater chemistry and indicates that the water acquires solutes through interactions with soils and rocks.

Closer inspection of Figure 2.9 shows that the distribution curves for different species

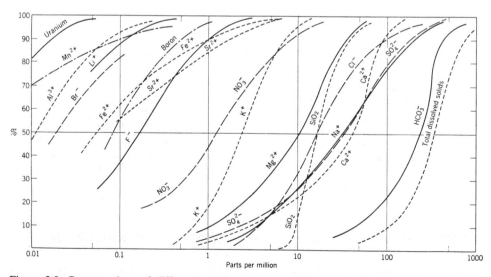

Figure 2.9. Concentrations of different solutions in terrestrial waters in the USA displayed in a frequency plot (Davies and De Wiest, 1966). Reprinted by permission of John Wiley & Sons, Inc.

have variable gradients. For some species, like Ca^{2+}, HCO_3^-, SiO_2, K^+ and F^-, a steep gradient is found which could indicate that the solubility of a mineral places an upper limit on the maximum concentration of the species in natural waters. For most other species, the curve has a lower gradient and the aspect of a normal distribution which suggests that the concentrations of these species depends rather on their availability in rocks, very slowly dissolving minerals, or are controlled by biological processes.

A second approach to obtain a large scale overview of controlling processes on solute compositions of freshwaters has been offered by Gibbs (1970). In Figure 2.10, water compositions are displayed as a function of the dominance of end-member components Na^+ and Cl^- at one end and Ca^{2+} and HCO_3^- at the other. The relative fractions for cations and anions are plotted against the total dissolved solid content. Seawater is dominated by Na^+ and Cl^- and has a high total dissolved solid content; it plots therefore at the terminal part of the upper branch. Rainwater, on the other hand, resembles seawater in relative composition, but has a very low total dissolved solid content; its composition is therefore found at the end of the lower branch. By weathering of rocks, the relative content of Ca^{2+} and HCO_3^- increases and the water composition moves upward to the left central part of the diagram. If subsequent evaporation and precipitation of the least soluble minerals occurs, first Ca^{2+} and HCO_3^- are lost by $CaCO_3$ precipitation and the water composition moves upwards along the upper branch towards the seawater composition dominated by Na^+ and Cl^-.

Figures 2.9 and 2.10 address some overall controlling factors on water chemistry on a worldwide or continental scale. When scaling down to the size of a watershed, the complexity increases and the important processes that may affect the groundwater chemistry are shown schematically in Figure 2.11. Here we give a short introduction to most of them, while subsequent chapters discuss these processes in much more detail.

Evaporation and evapotranspiration lead to increases in concentrations which are proportional to the amount of water that evaporates and have already been discussed in some detail. Cl^- concentrations can be used to calculate a precipitation surplus since Cl^- normally behaves in a conservative manner (and is not affected by other processes); dry deposition must also be accounted for.

Selective uptake of ions by vegetation, and storage in accreting biomass can profoundly influence concentrations of elements. Table 2.4 compares fluxes for different elements in the Hubbard Brook ecosystem on gneiss from the classical study of Likens et al. (1977). Note that the fluxes of elements that enter the soil by precipitation or leave the system with the streamwater are often only a fraction of the fluxes within the biomass. Accordingly most elements are heavily recycled within the soil and biomass. For example the vegetation uptake of K is 64 kg/ha.yr, whereas the streamwater output is only 1.9 kg/ha.yr. Elements such as P are even more heavily cycled in the biomass when compared with the hydrological fluxes, while other elements like Mg and S are less important in the biological cycle. Natural vegetations are, as may be expected from an evolutionary standpoint, masterful in storing essential elements. However, this also implies that a small disturbance in the flux through living matter may greatly disturb the much smaller flux of dissolved constituents leaving the biosphere.

Vegetation can also adsorb gases from the atmosphere, for example SO_2, NH_3 and NO_2. These gases may partly be taken up and incorporated in the plant, while the remainder is flushed away with the rain.

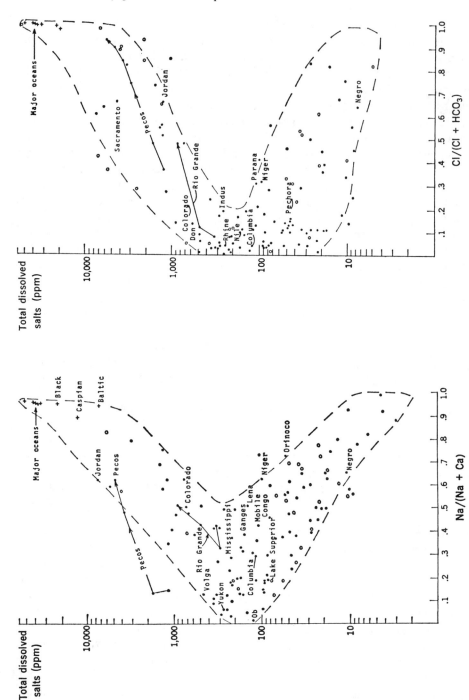

Figure 2.10. The chemistry of world surface waters expressed as a function of rain and seawater chemistry, rock weathering and evaporation (Gibbs, 1970). Copyright 1970 by the AAAS.

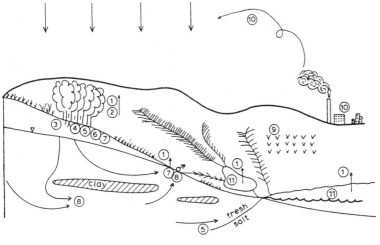

① evaporation
② transpiration
③ selective uptake by vegetation
④ oxidation/reduction
⑤ cation exchange
⑥ dissolution of minerals

⑦ precipitation of secondary minerals
⑧ mixing of water
⑨ leaching of fertilisers, manure
⑩ pollution
⑪ lake/sea biological processes

Figure 2.11. A schematic overview of processes that affect the water quality in the hydrological cycle.

Table 2.4. Yearly fluxes in the biomass compared with fluxes in rain and streamwater output in the Hubbard Brook ecosystem. (Fluxes in kg/ha.yr; Likens et al., 1977).

	Ca	Mg	Na	K	N	S	P	Cl
Bulk precipitation input	2.2	0.6	1.6	0.9	6.5	12.7	0.04	6.2
Streamwater output	13.7	3.1	7.2	1.9	3.9	17.6	0.01	4.6
Vegetation uptake	62	9	35	64	80	25	9	small
Root exudates	4	0.2	34	8	1	2	0.2	1.8
Weathering release	21	4	6	7	0	1	?	small

Decay of organic matter is the reverse process of uptake and storage of elements by growing plants. Plants show seasonal variations in the uptake and release, and young plants incorporate more elements than mature and old vegetation. Decay of organic matter is an oxidation reaction which may occur in the soil but also within aquifers where fossil organic matter can be present as peat, lignite etc. The process uses oxygen, or other electron acceptors, and produces carbonic acid:

$$CH_2O + O_2 \rightarrow H_2O + CO_2$$

In this reaction a carbohydrate, CH_2O, is used as a simplification for organic material. In reality the composition of organic matter is very complex, containing humic and fulvic acids etc, although the overall stoichiometry corresponds roughly to CH_2O. Organic matter also contains minor constituents such as P, K, N, S, etc. which are released upon degradation. Decomposition of organic matter in aquifers may trigger important reactions, such as the reduction of iron oxides, sulfate and nitrate, or the formation of methane. In addition, the production of CO_2 may have a great influence on reactions involving carbonate minerals.

Weathering and dissolution of carbonate, silicate or evaporite minerals releases elements to water. Some minerals, like carbonates and evaporites, dissolve quickly and significantly change the composition of the water already in the soil, while others, like silicates, dissolve slowly and have a less conspicuous effect on the water chemistry. Table 2.5 lists the sources of elements derived from minerals as well as from some additional sources, and the range of concentrations which are to be expected in fresh waters.

The composition of soil water and groundwater will depend on the rock type through which the water flows. For example, ultrabasic rocks are rich in olivines and pyroxenes and Mg^{2+} will be the dominant cation in soil and groundwater. Likewise, Ca^{2+} is the dominant cation in the soil moisture of calcareous soils, and when Ca^{2+} and Mg^{2+} are present in about equal concentrations in soil or groundwater, they are probably derived from dolomite $(CaMg(CO_3)_2)$.

The relations between rock types and groundwater composition are commonly displayed in so-called Piper diagrams. The Piper diagram is an ingenious construction which consists of two triangular diagrams which describe the relative compositions of cations and anions, and a diamond-shaped diagram that combines the compositions of cat- and anions. For a given water analysis, first the compositions of cations and anions are plotted in the triangular diagrams. Subsequently, the data points from the cation and anion triangles are transferred to the diamond diagram by drawing lines parallel to the outer boundary until they unite in the diamond. Figure 2.12 shows a Piper diagram for mineral waters from Europe and their relation to the rock type from which they have been extracted (Zuurdeeg and Van der Weiden, 1985). The diagram shows that limestone and marls produce a Ca, Mg-HCO_3 type water, whereas groundwater in metamorphic rocks (schists, sandstones, etc.) or igneous rocks (granites or magmatic) also contain appreciable amounts of other elements such as Na^+, K^+, and Cl^-. Note that the *EC* ranges of the mineral waters indicate that appreciable amounts of salts can be present (the *EC* divided by 100 yields the total

Table 2.5. Normal ranges of concentrations in unpolluted fresh water and the sources of elements.

Element	Concentrations (mmol/l)	Source
Na^+	0.1 – 2	Feldspar, Rock-salt, Zeolite, Atmosphere
K^+	0.01 – 0.2	Feldspar, Mica
Mg^{2+}	0.05 – 2	Dolomite, Serpentine, Pyroxene, Amfibole, Olivine, Mica
Ca^{2+}	0.05 – 5	Carbonate, Gypsum, Feldspar, Pyroxene, Amfibole
Cl^-	0.05 – 2	Rock-salt, Atmosphere
HCO_3^-	0 – 5	Carbonates, Organic matter
SO_4^{2-}	0.01 – 5	Atmosphere, Gypsum, Sulfides
NO_3^-	0.001 – 0.2	Atmosphere, Organic matter
SiO_2	0.02 – 1	Silicates
Fe^{2+}	0 – 0.5	Silicates, Siderite, Hydroxides, Sulfides
PO_4-total	0 – 0.02	Organic matter, Phosphates

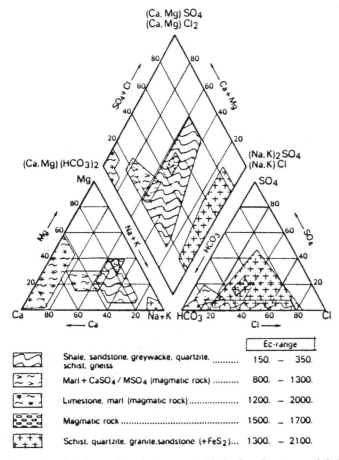

		Ec-range	
	Shale, sandstone, greywacke, quartzite, schist, gneiss	150. –	350.
	Marl + CaSO$_4$ / MSO$_4$ (magmatic rock)	800. –	1300.
	Limestone, marl (magmatic rock)	1200. –	2000.
	Magmatic rock	1500. –	1700.
	Schist, quartzite, granite, sandstone (+FeS$_2$)	1300. –	2100.

Figure 2.12. A Piper plot of European bottled mineral waters and their relation to the rock type from which the water has been extracted (modified from Zuurdeeg and Van der Weiden, 1985).

concentration in meq/l of cations or anions, as suggested in Chapter 1). The amount of salts present in these bottled mineral waters may be higher than is considered advisable for a regular drinking water supply, and in some cases it even exceeds the WHO drinking water limits (Chapter 1). However, the content of specific salts, and their presumed beneficial effect, are at the same time the main attraction for drinking bottled water instead of regular tapwater.

Different mechanisms of rock weathering are illustrated in Figure 2.13, by Walling (1980), based on catchment studies in the USA. Two trends can be identified in the plots of concentration versus annual runoff from watersheds with different bedrock geology. Firstly, the content of total dissolved solids in the water from limestone, volcanics and sand and gravel is almost independent of the amount of runoff, which is in contrast to the other rock types. Secondly, for a specific yearly runoff, the total dissolved solid content varies for different rocks.

These differences are related to solubilities of the minerals present in the parent rock, and the *rate of dissolution* of these minerals. Gravel consists mainly of insoluble quartz pebbles

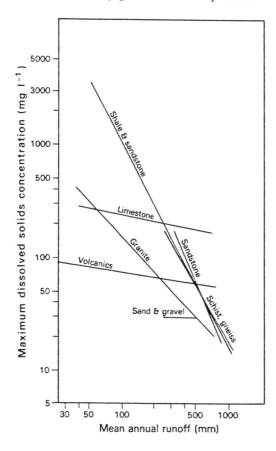

Figure 2.13. Differences in concentration of surface runoff as dependent on rock type and annual runoff (Walling, 1980). Reprinted by permission of John Wiley & Sons, Ltd.

which are highly resistant to dissolution. The concentration levels therefore remain low, and the composition of the groundwater is mainly controlled by rainwater chemistry and evapotranspiration. On the other hand, volcanics and limestones contain more soluble minerals so that the concentrations in runoff become higher. The relatively constant concentrations in runoff from these rock types indicates that dissolution of these minerals is fast compared to the residence time of the water in the drainage area, and the concentration level may well represent the solubility of the minerals. To illustrate this point, it will be shown in Chapter 4 that calcite dissolves so rapidly that water obtains Ca^{2+} concentrations corresponding to the solubility of calcite in most field situations within a few days.

Runoff from granites and sandstones show concentration levels that depend strongly on the annual runoff. This suggests that the dissolution rate of the minerals in granite is in the order of the residence time of water in the drainage basin. The dominant minerals in granite and sandstone are feldspars, micas and quartz and, as will be shown in Chapter 6, these minerals have very slow dissolution kinetics. In other words, Figure 2.13 demonstrates that in some cases, as for limestones, an equilibrium approach is appropriate, while in other cases, as for most silicate rocks, dissolution kinetics must also be considered.

Minerals may precipitate in soils or aquifers. Weathering of silicate minerals leads generally to the formation of secondary clay minerals. The type of clay minerals which

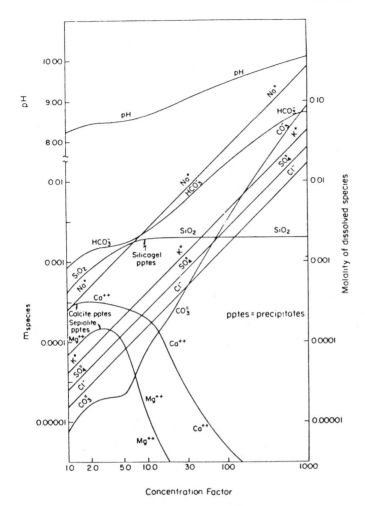

Figure 2.14. Calculated results of evaporation of typical Sierra Nevada spring water at constant temperature in equilibrium with atmospheric CO_2 (Reprinted with permission from Garrels and Mackenzie, Copyright 1967, American Chemical Society).

forms during weathering depends on the composition of the parent rock and the stage of weathering. The clay minerals basically consist of Al-silicates with or without other cations. As the weathering process continues, the clay minerals are stripped first of cations, and then of silicon until the sparingly soluble Al-hydroxide remains. If Fe(II) is present in the parent rock, the formation of Fe-oxides can also be expected.

In arid climates, evaporation may lead to the precipitation of a suite of minerals in the soil. These include calcite, gypsum and chloride-salts. The sequence of minerals that precipitate as a function of the extent of evaporation can be calculated for a given initial water composition. An example of such a calculation is given for Sierra Nevada spring waters in Figure 2.14. It illustrates how the concentrations in the water change when water evaporates and minerals are precipitating from the water. In this case, the initial water is from springs draining a granitic terrain, and the minerals that precipitate are sepiolite (a Mg-silicate), calcite, and silica gel. The final water is a Na-Cl + HCO_3 brine resembling that commonly found in soda lakes.

Ion exchange reactions particularly between dissolved cations and the exchange complex on sediments surfaces may have a significant effect on the water chemistry in non-steady state situations. Such non-steady state situations include intrusion of seawater into an aquifer, the spreading of a pollution plume through an aquifer or the downward movement of acid rain through an aquifer. Ion exchange tends to smooth concentration gradients which are the result of changing conditions in a groundwater reservoir. The exchange complex of a soil consists of clay minerals, weathered minerals and organic material. The exchange complex may contain an amount of cations that is up to 300 times larger than is present in soil moisture (both expressed per water-filled pore volume)!

Mixing of different water qualities often occurs in seepage zones or in springs. An upland area which has locally different characteristics (geology, vegetation) can discharge different water qualities in the same spring. Mixing of different water qualities through dispersion is important when transport of pollutants in aquifers is assessed. Mixing of different water types can lead to subsaturation with respect to calcite, even if the original waters were at saturation before mixing. This may induce renewed dissolution of calcite.

Anthropogenic activities can thoroughly change the water quality. Such changes can be of any kind; from the use of NaCl in households to the leaching of heavy metals or poisonous organic constituents from landfills. Air pollution and acid rain have already been men-

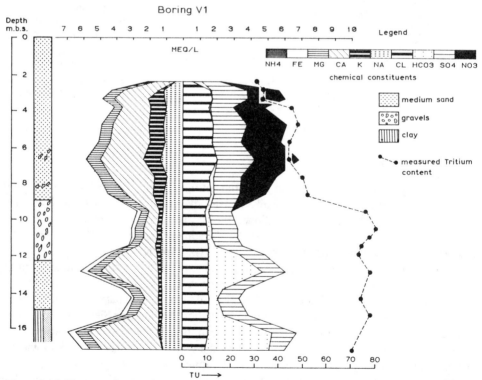

Figure 2.15. The groundwater chemistry below agricultural land. Concentrations are plotted cumulatively (Appelo, 1985).

Table 2.6. Processes which are important as sources of different ions and processes that may limit their concentration ions in fresh water.

Element	Process	Concentration limits
Na^+	Dissolution, Cation exchange in coastal aquifers	Kinetics of silicate weathering
K^+	Dissolution, adsorption, decomposition	Solubility of clay minerals, vegetation uptake
Mg^{2+}	Dissolution	Solubility of clay minerals
Ca^{2+}	Dissolution	Solubility of calcite
Cl^-	Evapotranspiration	None
HCO_3^-	Soil CO_2-pressure, weathering	Organic matter decomposition
SO_4^{2-}	Dissolution, oxidation	Removal by reduction
NO_3^-	Oxidation	Uptake, removal by reduction
Si	Dissolution, adsorption	Chert, chalcedone solubility
Fe	Reduction	Redox-potential, Fe^{3+} solubility, siderite, sulfide
PO_4	Dissolution	Solubility of apatite, Fe, Al phosphates. Biological uptake

tioned. A serious problem in many countries is the large amount of nitrate that is leached from agricultural fields down into aquifers due to excessive fertilizer and manure applications to the soil, though these may be 'optimal' from an agricultural point of view.

The influence of agricultural practices on the water quality is illustrated in Figure 2.15. It shows differences in the groundwater composition with depth in a borehole equipped with multilevel samplers. The high NO_3^- concentrations in the upper part of the borehole are due to a heavy application of manures and fertilizers to fields which are located upstream. The NO_3^- concentrations are far above the WHO drinking water standards (cf. Table 1.7). Source areas of the contaminated water can be traced back by considering groundwater flow paths, and the contamination may thus be used as a tracer for groundwater flow. Questions also arise whether nitrate becomes degraded within the aquifer or whether it is transported as a conservative component.

Thus a variety of processes has to be considered when studying the geochemistry of groundwaters and the rest of this book is dedicated to this subject. We may conclude this chapter with Table 2.6, which gives a first guide to the type of process that is important for the distribution of different elements.

PROBLEMS

2.1. How does the Na^+/Ca^{2+} ratio in rainwater change from the coast to further inland?

2.2. What happens to the Na^+/Ca^{2+} ratio when surface water evaporates?

2.3. Discuss the effect of variations in discharge on the composition of runoff for the following rock types:
   a. gravel;
   b. calcareous rocks;
   c. granite.

2.4. The composition of the rain at 3 locations in the Netherlands is listed in the table (summer averages, 1979, conc. in $\mu mol/l$):

	pH	NH_4^+	Na^+	K^+	Ca^{2+}	Mg^{2+}	Cl^-	NO_3^-	SO_4^{2-}
Vlissingen	4.00	49	265	8	39	34	297	92	85
Deelen	3.95	119	30	3	16	5	37	90	..
Beek	..	143	23	11	50	5	34	87	96
Seawater (meq/l):	–	–	485	10.6	20	110	566	–	59

a. Check the quality of the analysis from Vlissingen. Estimate the missing parameters in the two other analyses.
b. Calculate the influence of seawater on the three samples.
c. Which other influences can be discerned?

2.5. Analyses of 6 groundwater samples are listed below. Consider which natural processes have led to their composition.

	Alps	——— Salt Marsh ———			Brabant	Kenya
pH	7.22	8.31	6.65	7.1	4.50	11.0
Na^+	0.03	194.0	194.0	164.0	0.20	33.2
Mg^{2+}	0.10	22.0	22.0	25.0	0.03	3.6
Ca^{2+}	1.60	2.1	2.3	12.1	0.2	0.1
Cl^-	0.05	230.0	230.0	230.0	0.21	21.0
HCO_3^-	3.1	4.0	28.0	22.0	0	14.8
SO_4^{2-}	0.1	12.0	0	0	0.3	0.1

Hints: The sample from the Alps comes from calcareous rock. Cl^- concentrations in the salt marsh suggest a dominant seawater influence, but which elements have changed with respect to average seawater? The sample from Brabant, The Netherlands was taken from an extremely poor sandy soil. The sample from Kenya was taken in the Rift valley, near evaporative lakes.

REFERENCES

Andreae, M.O., Talbot, R.W., Berresheim, H. and Beecher, K.M., 1990, Precipitation chemistry in Central Amazonia. *J. Geophys. Res.* 95, 16987-16999.

Appelo, C.A.J., 1985, CAC, computer aided chemistry, or the evaluation of groundwater quality with a geochemical computer model (in Dutch). *H₂O* 26, 557-562.

Appelo, C.A.J., 1988, *Water quality in the Hierdensche Beek watershed* (in Dutch), VU, IvA, 100 pp.

Blanchard, D.C. and Syzdek, L.D., 1988, Film drop production as a function of bubble size. *J. Geophys. Res.* 93, 3649-3654.

Blanchard, D.C., Woodcock, A.M. and Cipriano, R.J., 1984, The vertical distribution of the concentration of sea salt in the marine atmosphere near Hawaii. *Tellus* 36B, 118-125.

Brimblecombe, P., 1986, *Air composition and chemistry.* Cambridge Univ. Press, 224pp.

Calder, I.R., 1979, Do trees use more water than grass? *Water services*, Jan. 1979.

Calder, I.R., 1990, *Evaporation in the uplands*, Wiley & Sons, New York, 148 pp.

Davies, S.N. and DeWiest, R.C.M., 1966, *Hydrogeology.* Wiley & Sons, New York, 463 pp.

Drabløs, D. and Tollan, A. (eds), 1980, *Ecological impact of acid precipitation.* SNSF Project, Oslo, Norway.

Duce, R.A. and Hoffman, E.J., 1976, Chemical fractionation at the air/sea interface. *Ann. Rev. Earth Planet. Sci.* 4, 187-228.

Eriksson, E., 1959, The yearly circulation of chloride and sulfur in nature; meteorological, geochemical and pedological implications. I. *Tellus* 11, 375-403.

Eriksson, E., 1960. The yearly circulation of chloride and sulfur in nature; meteorological, geochemical and pedological implcations. II. *Tellus* 12, 63-109.

Feth, J.H., Roberson, C.E. and Polzer, W.L., 1964, Sources of mineral constituents in water from granitic rocks, Sierra Nevada, California and Nevada. US Geol. Surv. Water Supply Pap. 1535 I, 70 pp.

Fowler, D., 1980, Removal of sulphur and nitrogen compounds from the atmosphere in rain and by dry deposition. In D. Drabløs and A. Tollan (eds.), *Ecological impact of acid precipitation.* SNSF Project, Oslo, Norway, 22-32.

Garrels, R.M. and Mackenzie, F.T., 1967, Origin of the chemical compositions of some springs and lakes. In W. Stumm (ed.), *Equilibrium concepts in natural water systems.* Adv. in Chem. Ser. 67, Am. Chem. Soc., 222-242.

Garrels, R.M. and Mackenzie, F.T., 1972, *Evolution of sedimentary rocks.* Norton, 397pp.

Garrels, R.M., Mackenzie, F.T. and Hunt, C., 1975, *Chemical cycles and the global environment.* W. Kaufmann, Inc., Los Altos, Calif., 206pp.

Galloway, J.N. and Likens, G.E., 1978, The collection of precipitation for chemical analysis. *Tellus* 30, 71-82.

Gibbs, R., 1970, Mechanisms controlling world water chemistry. *Science* 170, 1088-1090.

Granat, L., 1972, On the relation between pH and the chemical composition in atmospheric precipitation. *Tellus* 24, 550-560.

Hansen, B.K. and Postma, D., 1995, Acidification, buffering, and salt effects in the unsaturated zone of a sandy aquifer, Klosterhede, Denmark. *Water Res. Res.* 31, 2795-2809.

KNMI/RIVM, 1985, Chemical composition of precipitation over the Netherlands. Ann. Rep. 1983. Royal Met. Inst., De Bilt, the Netherlands.

Lerner, D.N., Issar, A.S. and Simmers, I., 1990, *Groundwater recharge.* Int. Contrib. Hydrogeol. vol. 8, IAH.

Lightowlers, P.J. and Cape, J.N., 1988, Sources and fate of atmospheric HCl in the UK and Western Europe. *Atmos. Environ.* 22, 7-15.

Likens, G.E., Bormann, F.H., Pierce, R.S., Eaton, J.S. and Johnson, N.M., 1977, *Biogeochemistry of a forested ecosystem.* Springer Verlag, Berlin, 146 pp.

Matzner, E. and Hesch, W. 1981, Beitrag zum Elementaustrag mit dem Sickerwasser unter verschieden Ökosystemen im nordwestdeutschen Flachland. *Z. Pflanzenernähr. Bodenkd.* 144, 64-73.

Mészáros, E., 1981, *Atmospheric chemistry, fundamental aspects.* Elsevier, Amsterdam, 201pp.

Miller, H.G. and Miller, J.D., 1980, Collection and retention of atmospheric pollutants by vegetation. In D. Drabløs and A. Tollan (eds), *Ecological impact of acid precipitation.* SNSF Project, Oslo, Norway, 33-40.

Möller, D., 1990, The Na/Cl ratio in rainwater and the seasalt chloride cycle. *Tellus* 42B, 254-262.

Monahan, E.C., 1986, The ocean as a source for atmospheric particles. In P. Buat-Menard, *The role of air-sea exchange in geochemical cycling.* Reidel, Dordrecht, 129-163.

Morgan, J.J., 1982, Factors governing the pH, availability of H^+, and oxidation capacity of rain. In E.D. Goldberg (ed.), *Atmospheric chemistry.* Springer Verlag, Berlin, 17-40.

Moss, P.D. and Edmunds, W.M., 1989, Interstitial water-rock interaction in the unsaturated zone of a Permo-Triassic sandstone aquifer. In D.L. Miles (ed.), *Water-rock interaction. Proc. 6th water-rock interaction symp., Malvern, UK.* Balkema, Rotterdam, 495-499.

Mulder, J., 1988, *Impact of acid atmospheric deposition on soils.* Ph. D. Thesis, Wageningen, 163 pp.

Nativ, R. and Issar, A., 1983, Chemical composition of rainwater and floodwaters in the Negev desert. Israel. *J. Hydrol.* 62, 201-223.

Newell, R.E., 1971, The global circulation of atmospheric pollutants. *Scientific Amercican*, 224, 32-42.

Raemdonck, H., Maenhout, W. and Andreae, M.O., 1986, Chemistry of marine aerosol over the tropical and equatorial Pacific. *J. Geophys. Res.* 91, 8623-8636.

Ridder, T.B., 1978, *On the chemistry of precipitation* (in Dutch). Royal Met. Inst. Rep. 78-4, 45 pp.

Schoeller, H., 1960, *Salinity of groundwater, evapotranspiration and recharge of aquifers* (in French). IASH Publ. 52, 488-494.

Sequeira, R. and Kelkar, D., 1978, Geochemical implications of summer monsoonal rainwater composition over India. *J. Appl. Meteor.* 17, 1390-1396.

Stuyfzand, P.J., 1984, Effects of vegetation and air pollution on groundwater quality in calcareous dunes near Castricum: measurements from lysimeters (in Dutch). H_2O 17, 152-159.

Ulrich, B., Mayer, R. and Khanna, P.K., 1979, *Die Deposition von Luftverunreinigungen und ihre Auswirkungen in Waldökosystemen im Sölling.* Forstl. Fak. Univ. Göttingen u. Nieders. Forstl. Vers. Anst., Band 58, 291 pp.

Van Aalst, R.M., Van Aardenne, R.A.M., De Kreuk, J.F. and Lems, Th., 1983, *Pollution of the North Sea from the atmosphere.* TNO-Report CL 82/152, TNO, Delft, 124pp.

Van Breemen, N., Burrough, P.A., Velthorst, E.J., Van Dobben, H.F., de Wit, T., Ridder, T.B. and Reijnders, H.F.R., 1982, Soil acidification from atmospheric ammonium sulphate in forest canopy throughfall. *Nature* 299, 548-550.

Vermeulen, A.J., 1977, *Immision measurements with rain gauges* (in Dutch). Prov. North Holland, Haarlem.

Wagner, G.H. and Steele, K.F., 1989, Na^+/Cl^- ratios in rain across the USA, 1982-1986. *Tellus* 41B, 444-451.

Walling, D.E., 1980, Water in the catchment ecosystem. In A.M. Gower (ed.), *Water quality in catchment ecosystems.* Wiley & Sons, New York. 1-48.

Whitley, K.T., Husar, R.B. and Liu, B.Y.H., 1972, Aerosol size distribution of Los Angeles smog. *J. Colloid Interface Sci.* 39, 177-204.

Zuurdeeg, B.W. and Van der Weiden, M.J.J., 1985, Geochemical aspects of European bottled waters. *Geothermics, thermal-mineral waters and hydrogeology.* Theophrastus Publ. Athens, 235-264.

CHAPTER 3

Solutions, minerals and equilibria

The title of this chapter is also the title of a famous book by Garrels and Christ (1965) that describes how the toolbox of physical chemistry can be used to quantify relations between minerals and dissolved species in natural systems. The book has influenced a generation of geochemists and hydrogeologists who were searching for ways to develop traditional descriptive groundwater geochemistry in a more quantitative direction. The equilibrium approach outlined by Garrels and Christ (1965) has proved to be very useful to set limits to concentrations of dissolved substances in natural systems. The purpose of this chapter is to recapitulate some basics of equilibrium chemistry and to demonstrate its application to groundwater geochemistry. Kinetics are discussed at the end of this chapter; the importance of this subject is widely recognized, although at present it is much more difficult to apply to groundwater systems.

Fundamental to any description of equilibria in water is the *law of mass action*. It states that for a reaction of the generalized type:

$$aA + bB \quad \leftrightarrow \quad cC + dD$$

the distribution at equilibrium between the species at the left and right is described by:

$$K = \frac{[C]^c[D]^d}{[A]^a[B]^b} \tag{3.0}$$

Here K is the equilibrium constant and the bracketed quantities denote activities or 'effective concentrations'. For the present we will equate activities in aqueous solutions with molal concentrations, but return to the difference between these two units in Section 3.2. The law of mass action is generally applicable, both for dissolution/precipitation reactions of minerals, relations between dissolved species, dissolution of gases in water etc.

3.1 SOLUBILITY OF MINERALS

Equations for dissolution of minerals are usually written as dissociation reactions. Thus for a mineral like fluorite (CaF_2) we may write:

$$CaF_2 \leftrightarrow Ca^{2+} + 2F^-$$

For which the solubility constant is:

$$K_{\text{fluorite}} = [\text{Ca}^{2+}][\text{F}^-]^2 = 10^{-10.57} \qquad \text{at } 25°\text{C} \tag{3.1}$$

This is in fact a direct application of the law of mass action, except that $[\text{CaF}_2]$ is omitted in Equation (3.1) since the activity of a pure solid is equal to one by definition. The exact value of K_{fluorite} is somewhat uncertain (Nordstrom and Jenne, 1977). Here we use the value of Handa (1975) in order to be consistent with the following example. Relation (3.1) can be rewritten in logarithmic form as:

$$\log K_{\text{fluorite}} = \log [\text{Ca}^{2+}] + 2 \log [\text{F}^-] = -10.57 \tag{3.2}$$

Thus, on a logarithmic plot, the equilibrium condition between fluorite and the solution (3.2) is given by a straight line, as illustrated in Figure 3.1. All combinations of $[\text{Ca}^{2+}]$ and $[\text{F}^-]$ which plot below the line are subsaturated, while those above the straight line are supersaturated with respect to fluorite. Together with the fluorite solubility line, a number of groundwater analyses from India are plotted. It is apparent from Figure 3.1 that equilibrium with fluorite places an upper limit to the F^- (and Ca^{2+}) concentrations in the ground water. Although F^- at low concentration in drinking water has been considered as beneficial in candy bar consuming countries, it constitutes a health hazard at concentrations above 3 ppm, causing dental fluorosis (tooth mottling) and more seriously skeletal fluorosis (bone deformation and painful brittle joints in older people). High F^- concentrations in groundwater are found in many places of the world, but particularly in Asia and Africa. The origin of the F^- is basically leaching from minerals in the rocks. In particular, volcanic rocks have high natural fluor contents but also the dissolution of fossil shark teeth, containing fluorapatite, has been reported (Zack, 1980). In addition, in arid areas evaporation may increase F^- concentrations in groundwaters.

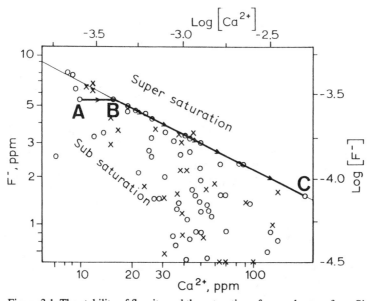

Figure 3.1. The stability of fluorite and the saturation of groundwaters from Sirohi, W. Rajasthan, India (modified from Handa, 1975). The evolution in water chemistry upon addition of gypsum is described by the pathway A, B to C as discussed in the text.

Figure 3.1 shows that equilibrium with fluorite places an upper limit to F^- concentrations in groundwater. Accordingly, it is expected that groundwater with a high natural Ca^{2+} concentration generally will contain less F^- than groundwater with a low Ca^{2+} concentration. This seems substantiated by the field observations of Handa (1975) (Figure 3.1). The fluorite equilibrium control on F^- concentrations can be exploited to remove F^- during water treatment. At saturation for fluorite, an increase of $[Ca^{2+}]$ will, according to Equation (3.1), decrease the $[F^-]$. Therefore it has been proposed (Schuiling, pers. comm.) to add gypsum to high fluoride waters which constitute a health hazard. For gypsum ($CaSO_4 \cdot 2H_2O$) dissolution we may write, neglecting crystal water for simplicity:

$$CaSO_4 \leftrightarrow Ca^{2+} + SO_4^{2-}$$

For which the solubility product is:

$$K_{gypsum} = [Ca^{2+}][SO_4^{2-}] = 10^{-4.60} \qquad \text{at } 25°C \tag{3.3}$$

Note that gypsum is much more soluble than fluorite. The amount of gypsum which has to be added to a given water sample, with for example composition A in Figure 3.1, in order to reach equilibrium with fluorite, can be calculated following Example 3.1.

EXAMPLE 3.1: *Gypsum addition to high fluor groundwater*
A groundwater sample contains 10 ppm Ca^{2+} and 5.5 ppm F^-. Is this water saturated with respect to fluorite, and if not how much gypsum should be added in order to reach saturation for fluorite?
First we recalculate ppm to molal concentrations, which yields $[Ca^{2+}] = 10^{-3.60}$ and $[F^-] = 10^{-3.54}$. A quick answer is obtained by plotting these values directly into Figure 3.1 which results in point A. Alternatively we may calculate the product:

$$[Ca^{2+}] [F^-]^2 = [10^{-3.60}] [10^{-3.54}]^2 = 10^{-10.68}$$

The value obtained is slightly lower than the solubility product for fluorite (Equation 3.2) and the sample is subsaturated for fluorite.
When gypsum is added, the water composition will change from point A (Figure 3.1) parallel to the Ca-axis until it meets the fluorite saturation line at point B. From the fluorite solubility product (Equation 3.2) we find that:

$$[Ca^{2+}]_B = K_{fluorite} / [F^-]^2 = [10^{-10.57}] / [10^{-3.54}]^2 = 10^{-3.49}$$

Thus, to reach equilibrium with fluorite, the amount of gypsum to be added is equal to $[Ca^{2+}]_B - [Ca^{2+}]_A = 10^{-3.49} - 10^{-3.60} = 0.072 \text{ mmol/l}$.

Once equilibrium with fluorite is reached, addition of more gypsum will modify the composition of the water down along the fluorite solubility line in Figure 3.1. The change in water chemistry is described quantitatively by Equation (3.2) and results in decreasing F^- concentrations due to fluorite precipitation. However, gypsum dissolution cannot continue indefinitely, since at some point saturation for gypsum is also reached. Simultaneous saturation for fluorite and gypsum is indicated in Figure 3.1 as point C and its location can be calculated as follows.

When the composition of the solution changes from B to C, dissolution of gypsum and precipitation of fluorite occurs simultaneously. We may define:

$x = $ mol/l gypsum dissolved
$y = $ mol/l fluorite precipitated.

Using these two variables we may write the following mass balance equations for the concentrations of Ca^{2+} and F^- at point C:

$$[Ca^{2+}]_C = [Ca^{2+}]_B + x - y \tag{3.4}$$

$$[F^-]_C = [F^-]_B - 2y \tag{3.5}$$

$$[SO_4^{2-}]_C = [SO_4^{2-}]_B + x \tag{3.6}$$

Together with the solubility product for fluorite (Equation 3.1) and gypsum (Equation 3.3) this yields 5 equations with 5 unknowns, which in principle can be solved. The result is a rather awkward quadratic equation and a much simpler method is to inspect the mass balance equations for a parameter which can be neglected. From Figure 3.1 we can expect that during change in water composition from B to C, the amount of Ca^{2+} released by gypsum dissolution is much larger than the amount of Ca^{2+} precipitated as fluorite. We introduce this as a simplifying assumption, and check its validity afterwards:

$$y \ll x$$

Which reduces Equation (3.4) to:

$$[Ca^{2+}]_C = [Ca^{2+}]_B + x \tag{3.7}$$

Substitution of this relation, together with Equation (3.6), in the solubility product for gypsum (Equation 3.3) results in:

$$K_{gypsum} = ([Ca^{2+}]_B + x)([SO_4^{2-}]_B + x)$$

Rearranging yields:

$$x^2 + ([Ca^{2+}]_B + [SO_4^{2-}]_B)x + ([Ca^{2+}]_B \cdot [SO_4^{2-}]_B - K_{gypsum}) = 0$$

And substitution of known values (from Example 3.1):

$$x^2 + 10^{-3.40}x - 10^{-4.60} = 0$$

This is a neat standard equation that solves to:

$$x = 4.84 \cdot 10^{-3} = 10^{-2.32}$$

Thus, 4.8 mmol/l gypsum must dissolve to reach saturation for both gypsum and fluorite. Substituting x in Equation (3.7) produces:

$$[Ca^{2+}]_C = 10^{-2.29}$$

Which again is substituted in the solubility product of fluorite (Equation 3.2):

$$[F^-]_C = \left(\frac{K_{fluorite}}{[Ca^{2+}]_C} \right)^{1/2} = 10^{-4.14}$$

This corresponds to 1.4 ppm fluoride, so that gypsum addition has lowered the fluoride content of the water by 75%. Experimental work in progress, where gypsum is included in the gravel pack of the well, seems to confirm the use of this principle as a feasible water treatment technique (Schuiling pers. comm.). Some caution is required, however, since the resulting sulfate concentration in our example amounts to 468 ppm, which exceeds the drinking water limit (Table 1.7).

It remains to check the validity of the initial assumption by substitution in Equation (3.5):

$$y = \tfrac{1}{2}([F^-]_B - [F^-]_C) = 1.08 \cdot 10^{-4}$$

Thus y is about fifty times smaller than x and our initial assumption, $y \ll x$, is quite reasonable. This type of simplification is often very useful in chemical calculations. Any reasonable assumption can be made, but take care that assumptions are always tested afterwards. Note that exclusion of terms is only allowed in mass balance equations and never in mass action equations.

At point C there exists simultaneous equilibrium between fluorite and gypsum, for which we can write the equilibrium:

$$SO_4^{2-} + CaF_2 \leftrightarrow CaSO_4 + 2F^- \tag{3.8}$$

The corresponding equilibrium constant is easily derived from the solubility products of fluorite and gypsum:

$$K = \frac{[F^-]^2}{[SO_4^{2-}]} = \frac{[Ca^{2+}][F^-]^2}{[Ca^{2+}][SO_4^{2-}]} = \frac{K_{fluorite}}{K_{gypsum}} = \frac{10^{-10.57}}{10^{-4.60}} = 10^{-5.97} \tag{3.9}$$

Equation (3.9) shows the interesting property that when simultaneous equilibrium between gypsum and fluorite is established, the ratio of the squared F^- over SO_4^{2-} concentration is invariant. If by some other process, for example oxidation of pyrite (FeS_2), the SO_4^{2-} concentration increases while equilibrium is maintained for both fluorite and gypsum, then the F^- concentration also increases due to conversion of fluorite to gypsum (Equation 3.8).

In summary, we have seen in this section how to operate with solubility products of minerals and how by simple calculations we can predict the evolution of water chemistry as a function of dissolution and precipitation of minerals.

3.2 CORRECTIONS FOR SOLUBILITY CALCULATIONS

So far we have equated activities and molal concentrations in calculations of mineral solubility. In reality, total concentrations of ions in aqueous solution have to be corrected for the effect of electrostatic shielding and for the presence of aqueous complexes. These corrections are discussed in the next two sections.

3.2.1 *Concentration and activity*

Activities which appear in solubility products reflect the chance that two ions react and form a precipitate. In pure water an ion like Ca^{2+} is surrounded by a shield of water molecule dipoles. In the presence of charged solutes additional electrostatic shielding occurs, and the reactivity of the ion is reduced. The activity of the ion approaches its molal concentration as the concentration of all solutes approaches zero. A correction term, the activity coefficient, is needed that relates activities to molal concentrations:

$$[i] = \gamma \cdot m_i / m_i^0 = \gamma_i \cdot m_i \tag{3.10}$$

where: $[i]$ is the activity of ion i (dimensionless), γ_i is the activity coefficient (dimensionless), m_i is the molality, (mol/kg H_2O), m_i^0 is the standard state, i.e. 1 mol/kg H_2O.

Activities are defined thermodynamically with respect to a standard state, which for a solute is an ideal unimolal solution (where $[i] = m_i$). The factor $1/m_i^0$ is unity for all species and cancels for practical purposes in (3.10). Activity coefficients vary between zero and one, and $\gamma_i \rightarrow 1$ as $m_i \rightarrow 0$. A more elaborate explanation of the activity concept is given in Appendix A. In this book the activity of an ion is denoted by square brackets, but in other texts also the notation a_i is used.

Activity coefficients are calculated using the electrostatical Debije-Hückel theory (Lewis and Randall, 1961). In this theory, first the ionic strength, I, is defined which describes the number of electrical charges that are present in solution:

$$I = \tfrac{1}{2}\Sigma m_i \cdot z_i^2 \tag{3.11}$$

where: z_i = charge of ion i, m_i = mol/l.

The ionic strength of freshwater is normally less than 0.02 while seawater has an ionic strength of about 0.7. Different equations have been proposed to derive activity coefficients from the ionic strength (Stumm and Morgan, 1981). One of the most generally applicable relations is the Davies equation which is valid to $I \approx 0.5$.

$$\log \gamma_i = -Az_i^2 \left(\frac{\sqrt{I}}{1 + \sqrt{I}} - 0.3I \right) \tag{3.12}$$

Here, A is a temperature dependent coefficient; values of A are listed in Table 3.1.

As an example we may calculate the solubility of fluorite in an electrolyte solution containing 10 mmol/l NaCl. First calculate the ionic strength:

$$I = \tfrac{1}{2}[m_{Na^+} \cdot (1)^2 + m_{Cl^-} \cdot (-1)^2] = 0.01$$

Then calculate the activity coefficient for Ca^{2+}:

$$\log \gamma_{Ca^{2+}} = -0.5085\,(2)^2 \left(\frac{\sqrt{0.01}}{1 + \sqrt{0.01}} - 0.3\,(0.01) \right) = -0.179 \tag{3.13}$$

So that:

$$\gamma_{Ca^{2+}} = 0.66$$

Table 3.1. Values of A in the Davies equation as a function of temperature (Berner, 1971)

Temp °C	A
0	0.4883
5	0.4921
10	0.4960
15	0.5000
20	0.5042
25	0.5085
30	0.5130
35	0.5175
40	0.5221
50	0.5319
60	0.5425

Likewise for F^-:

$$\log \gamma_{F^-} = -0.5085 \, (-1)^2 \left(\frac{\sqrt{0.01}}{1 + \sqrt{0.01}} - 0.3 \, (0.01) \right) = -0.0447 \qquad (3.14)$$

which gives:

$$\gamma_{F^-} = 0.90$$

Note that the activity coefficient of the monovalent ion, F^-, is much higher than of the divalent ion, Ca^{2+}. The solubility product of fluorite is given by Equation (3.1):

$$K_{\text{fluorite}} = [Ca^{2+}][F^-]^2 = 10^{-10.57} \qquad \text{at } 25^\circ C \qquad (3.15)$$

Substituting (3.10) in (3.15) produces:

$$\gamma_{Ca^{2+}} \, m_{Ca^{2+}} \gamma_{F^-}^2 \, m_{F^-}^2 = K_{\text{fluorite}} \qquad (3.16)$$

Rearranging yields:

$$m_{Ca^{2+}} \, m_{F^-}^2 = \frac{1}{\gamma_{Ca^{2+}} \gamma_{F^-}^2} K_{\text{fluorite}} \qquad (3.17)$$

And substituting activity coefficients

$$m_{Ca^{2+}} \, m_{F^-}^2 = \frac{1}{(0.66)(0.90)^2} K_{\text{fluorite}} = 1.87 \, K_{\text{fluorite}} \qquad (3.18)$$

In distilled water the concentrations of calcium and fluoride in equilibrium with fluorite are so low that the ionic strength approaches zero and the activity coefficients unity. However, in a 10 mmol/l NaCl solution our calculation shows that the solubility of fluorite has increased by almost a factor of two. In these calculations, the contributions of Ca^{2+} and F^- concentrations to I were omitted for simplicity; their inclusion would give an additional slight increase in the solubility of fluorite.

3.2.2 Complexes

Apart from electrostatic shielding, the reactivity of ions in solution is further reduced by the presence of ion pairs or aqueous complexes. Examples are $CaSO_4^0$, CaF^+ etc. The formation of aqueous complexes can be described by equilibria of the type:

$$Ca^{2+} + SO_4^{2-} \leftrightarrow CaSO_4^0$$

Its distribution is simply obtained by applying the law of mass action (3.0):

$$K = \frac{[CaSO_4^0]}{[Ca^{2+}][SO_4^{2-}]} = 10^{2.5} \qquad (3.19)$$

The reaction is written as an ion-association reaction and the corresponding mass action constant is therefore termed a *stability constant*. Sometimes equilibria for aqueous complexes are also written as dissociation reactions and the *dissociation constant* for the $CaSO_4^0$ complex is obviously the reverse of Equation (3.19). Clearly, these two formulations should not be confused. Values of stability constants (as well as of solubility products) are normally obtained from tableworks like the WATEQ/PHREEQE database (Nordstrom et al., 1990) in Appendix B. Alternatively, the constants can be calculated from thermodynamic tables as demonstrated in Section 3.3.1.

3.2.3 *Combined complexes and activity corrections*

Considering the various possible aqueous complexes of Ca^{2+}, the total amount of Ca^{2+} in solution is described by a mass balance equation of the type:

$$\Sigma Ca^{2+} = m_{Ca^{2+}} + m_{CaF^+} + m_{CaSO_4^0} + m_{CaOH^+} + \dots \tag{3.20}$$

Similar mass balance equations can be written for all other substances in solution. Thus, in order to calculate the activities of say $[Ca^{2+}]$ and $[SO_4^{2-}]$ we have to solve simultaneously a set of mass action equations like (3.19), where concentrations are expressed as activities, and a set of mass balance equations where concentrations are in molal units. Such calculations are usually carried out in an iterative procedure as displayed graphically in Figure 3.2. From the analytical data, first activity coefficients and activities are calculated, which are used to calculate activities of complexes. Then molal concentrations of complexes are obtained, which enable us to calculate new mass balances and thereby a better estimate of the ionic strength. This procedure continues until no further significant improvement is obtained.

The importance of complexing and activity corrections, as percentage of total concentrations, is illustrated for seawater in Figure 3.3. Note that the effect of complexing is small for Na^+ and Cl^- and around 10% for Ca^{2+} and Mg^{2+}. However, aqueous complexes constitute a significant part of the total concentrations of dissolved SO_4^{2-} and HCO_3^-, and completely dominate dissolved aluminum. As expected from (3.12), activity corrections have a much stronger effect for divalent ions than for monovalent ions.

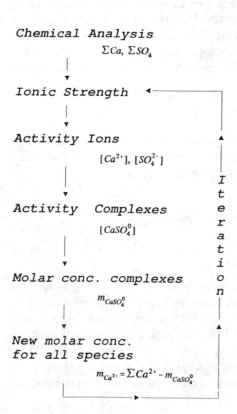

Figure 3.2. A flow chart for iterative calculation of $[Ca^{2+}]$ and $[SO_4^{2-}]$ from analytical data for total concentrations in solution.

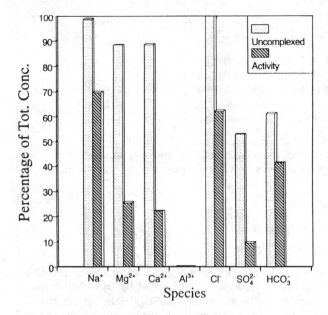

Figure 3.3. The importance of complexing and activity corrections as percentage of total concentrations for 35‰ seawater with a pH of 8.22.

EXAMPLE 3.2: *Solubility of gypsum*

The solubility of gypsum in water, considering both aqueous complexes and activity corrections, can be calculated as follows:

For equilibrium between gypsum and water we write:

$$CaSO_4 \cdot 2H_2O \leftrightarrow Ca^{2+} + SO_4^{2-} + 2H_2O$$

And the mass action expression is:

$$K = \frac{[Ca^{2+}][SO_4^{2-}][H_2O]^2}{[CaSO_4 \cdot 2H_2O]} = 10^{-4.6}$$

Since the activity of a pure solid like gypsum is unity, and for dilute electrolyte solutions $[H_2O] = 1$, this simplifies to the solubility product:

$$K_{gypsum} = [Ca^{2+}][SO_4^{2-}] = 10^{-4.60} \quad \text{at } 25°C$$

Substituting molal concentrations for activities yields:

$$K_{gypsum} = (\gamma_{Ca^{2+}} \cdot m_{Ca^{2+}})(\gamma_{SO_4^{2-}} \cdot m_{SO_4^{2-}}) = 10^{-4.60}$$

Since equal amounts of Ca^{2+} and SO_4^{2-} are released during gypsum dissolution, $m_{Ca^{2+}} = m_{SO_4^{2-}}$ and according to Equation (3.12) also $\gamma_{Ca^{2+}} = \gamma_{SO_4^{2-}}$, the solubility of gypsum is simplified to

$$m_{Ca^{2+}}^2 = (10^{-4.60})/\gamma_{Ca^{2+}}^2$$

or

$$m_{Ca^{2+}} = (5.01 \cdot 10^{-3})/\gamma_{Ca^{2+}}$$

$\gamma_{Ca^{2+}}$ can be estimated from the Davies equation (3.12) and the ionic strength, I:

$$I = \tfrac{1}{2} \Sigma m_i \cdot z_i^2 = \tfrac{1}{2}(m_{Ca^{2+}} \cdot 4 + m_{SO_4^{2-}} \cdot 4) = 4m_{Ca^{2+}}$$

The problem now is that $m_{Ca^{2+}}$ depends on $\gamma_{Ca^{2+}}$ which depends on I and again on $m_{Ca^{2+}}$. It is solved by an iterative procedure similar to that in Figure 3.2. In first approximation $\gamma_{Ca^{2+}}$ is set equal to 1, which enables a first estimate of I and so forth. In tabulated form the calculation proceeds as follows:

Iteration	$m_{Ca^{2+}}$	I	$\gamma_{Ca^{2+}}$
0			1
1	$\rightarrow 5.01 \cdot 10^{-3}$	$\rightarrow 0.0201$	$\rightarrow 0.58$
2	$\rightarrow 8.71 \cdot 10^{-3}$	$\rightarrow 0.0348$	$\rightarrow 0.50$
3	$\rightarrow 9.97 \cdot 10^{-3}$	$\rightarrow 0.0399$	$\rightarrow 0.49$
4	$\rightarrow 10.3 \cdot 10^{-3}$	$\rightarrow 0.0413$	$\rightarrow 0.48$
5	$\rightarrow 10.4 \cdot 10^{-3}$	$\rightarrow 0.0417$	$\rightarrow 0.48$
6	$\rightarrow 10.4 \cdot 10^{-3}$		

Note that the application of activity corrections has doubled the calculated solubility of gypsum. However, we also know that the aqueous complex $CaSO_4^0$ is of importance. The stability constant of this complex (Equation 3.19) can be rewritten as:

$$[CaSO_4^0] = 10^{2.5} \cdot [Ca^{2+}][SO_4^{2-}]$$

Substitution of the solubility product of gypsum gives:

$$[CaSO_4^0] = 10^{2.5} \cdot 10^{-4.60} = 10^{-2.10} = 7.94 \cdot 10^{-3}$$

Since for uncharged species the activity coefficient is close to unity, this is also the molal concentration of the complex. Thus the total solubility of gypsum is $10.4 \cdot 10^{-3} + 7.94 \cdot 10^{-3} = 18.3 \cdot 10^{-3}$ mol/l. Note that 40% of the solubility is accounted for by the complex $CaSO_4^0$, while 30% is due to activity corrections. These results are easily transcribed in grams of gypsum dissolved per liter by multiplication with the molecular weight of gypsum (172.1). Thus the solubility product alone predicts that $(5.01 \cdot 10^{-3})(172.1) = 0.86$ gram gypsum can dissolve per liter water, while including complexes and activity corrections leads to a solubility of $(18.3 \cdot 10^{-3})(172.1) = 3.15$ g/l.

The methods described here work well for electrolyte solutions of up to about seawater strength ($I = 0.7$). For more concentrated solutions, which are found especially in association with evaporites, other methods are available. So-called Pitzer equations (Pitzer, 1981; Harvie et al., 1982; Monnin and Schott, 1984) have been very successful in describing mineral equilibria in highly concentrated solutions.

3.2.4 *Calculation of saturation states*

Now that we are able to calculate the activities of ions in solution, the state of saturation of a groundwater sample for any mineral can be obtained. One way to do this has been shown in Figure 3.1 where the analytical data, or more correctly the activities are plotted in a stability diagram for fluorite. Another approach is to compare the solubility product K with the analogue product of activities derived from the groundwater analyses. The latter is often termed the ion activity product (*IAP*). Thus for gypsum:

$$K_{gypsum} = [Ca^{2+}][SO_4^{2-}] \qquad \text{(activities at equilibrium)}$$

and

$$IAP_{gypsum} = [Ca^{2+}][SO_4^{2-}] \qquad \text{(activities in the water sample)}$$

An example is shown in Figure 3.4. Here oxidation of pyrite (FeS_2) in the upper part of the profile produces large amounts of sulfate and equilibrium with gypsum appears to place an upper limit to dissolved calcium and sulfate concentrations.

Saturation conditions may also be expressed as the ratio between *IAP* and *K*, for example

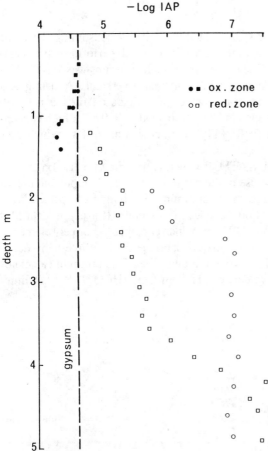

Figure 3.4. The IAP_{gypsum} compared with the solubility product of gypsum at the site of pyrite oxidation in porewaters of swamp sediments Reproduced by permission of Postma (1983), copyright Blackwell Sci. Publ.

as the *saturation state* Ω:

$$\Omega = IAP/K$$

Thus for $\Omega = 1$ there is equilibrium, $\Omega > 1$ supersaturation and $\Omega < 1$ subsaturation. For larger deviations from equilibrium, a logarithmic scale can be useful like the *saturation index SI*

$$SI = \log(IAP/K) \tag{3.21}$$

For $SI = 0$, there is equilibrium between the mineral and the solution; $SI < 0$ reflects subsaturation, and $SI > 0$ supersaturation. Different types of information can be obtained from saturation data. In few cases an actual control of solutes by equilibrium with a mineral can be demonstrated as clearly as in Figure 3.4. Normally, however, equilibrium is not found and then the saturation state merely indicates in which direction the processes may go; for subsaturation dissolution is expected, and supersaturation suggests precipitation.

3.2.5   *Computer programs for equilibrium calculations*

Calculation of ion activities and saturation states are tedious and time consuming except for simple systems. Fortunately computer programs are available which perform such calculations quickly, even on small computers. One of the first of such programs was WATEQ by Truesdell and Jones (1973) which is now available in many versions. A listing and comparison of a number of different programs which are all based on the activity-correction complexing approach is contained in Nordstrom et al. (1979). For high concentrations of dissolved ions, the program PHRQPITZ (Plummer et al., 1988), which uses Pitzer equations, can also be useful.

A Pascal version of WATEQ, named WATEQP, which is designed for use on PCs, is available with this book. Instructions for use of the program and sample input and output are presented below. WATEQP calculates the equilibrium distribution of species in an aqueous solution and the state of saturation for relevant minerals. It uses the data files WATEQP.ELE, WATEQP.SPE, and WATEQP.MIN, which contain respectively elements, species (and complexes), and mineral thermodynamic data; the files are internally documented. Elements and species can easily be added to these files. The program calculates TIC/Alkalinity from measured pH, or calculates pH from CO_2 /HCO_3^-/TIC/Alkalinity combinations depending on options:

Options		Input values	
pH/nopH	Alk/TIC/CO_3^{2-}/CO_2	pH	HCO_3^-
pH	Alk	pH	Alk
pH	TIC	pH	TIC
nopH	Alk	! not possible !	
nopH	TIC	TIC	Alk
nopH	CO_3^{2-}	CO_3^{2-}	HCO_3^- (titration values)
nopH	CO_2	CO_2	Alk

Specification of calculation:
- The program uses continuous fraction iteration and stops when the total mass balance error (for the sum of all elements) is less than 0.05%.
- Activity coefficients are calculated with the Davies equation
- The temperature dependency of equilibrium constants is calculated with the Van 't Hoff equation, or optionally with an analytical expression of the form:

$$\log K = a_1 + a_2 \cdot T + a_3/T + a_4 \cdot \log T + a_5/T^2$$

- The density of the solution is assumed to be 1 kg/l.
- For calculation of pH values from TIC/Alk, CO_3^{2-}/HCO_3^- or CO_2/Alk-combinations, the complexes of HCO_3^- are included in HCO_3^-, the complexes of CO_3^{2-} in CO_3^{2-}.
- The program does not calculate redox equilibria. Different oxidation levels of the same element must be entered in the form of separate species.

INPUT-file

WATEQP requires an input-file into which the analyzed concentrations are typed. The input file can be constructed with an editor (Sidekick, WordPerfect, etc.), or with WIP, the WATEQP Input Processor that is included on the program disk (cf. p. 520). An example input file is present in the file WATEQP.TES, see below.

```
!!! NOTE each WATEQP input file starts with 12 lines of text input !!!
Each sample has the following input (see also WATEQP.MAN):
1) a line with  sample identifier. 80 characters.
2) a line with  concentration-units (mmol, meq , mg  , use 4 chars)
   +      2 options to indicate pH/nopH (4 chars) and Alk/TIC/CO2/CO3 (3 chars)
   +      pH (real), or TIC/CO3/CO2 (real)      Temp (real).
   three lines with  concentrations of elements in fixed order of WATEQP.ELE:
3) Na   K     Mg    Ca    NH4    Cl   Alk(or TIC/HCO3)   SO4    NO3    (9 elements)
4) Fe   Mn    Al    F     SiO2   PO4  H2S  (and more, if in ....ELE)   (7 elements)
5) 999 (to signify end of sample).
   Use in all cases free format, separated by spaces; zero if not analysed.
   The following examples are in the file WATEQP.TES
Seawater test case #1.- - - - - - - - - - - - - - - - - - - - - -
mg  pH  alk  8.22  25.0
10768.0   399.1   1291.8    412.3   0.03   19353.0   156.53   2712.0   0.29
0.002   2e-4   2e-3    1.39   4.28   0.06      1e-3
999 - - - - - - - - - - - - - - - - - - - - - - - - - - - - - -
```

Running the program

After you have made your input file, the program is started by typing WATEQP in answer to the DOS prompt >. The program asks for the name of the input file and also two output files into which results are written. One output file contains saturation indexes in a format readable by spreadsheet programs. The program displays on the screen the sample identifiers, and the electroneutrality balance of the samples.

OUTPUT-file

The output file can be read with an editor, or printed with a DOS command and is listed below for the example input file. The output contains first the analyzed concentrations in mmol/l, meq/l and mg/l. Then follows a block with calculated activities, the logarithm of activity, and the activity coefficients for the elements and the complexes.

A block of text follows with the saturation indices for relevant minerals. 'Log IAP' = logarithm of ion activity product, 'log KT' = logarithm of the solubility product of the mineral, and 'L. IAP/KT' = $\log IAP - \log K$, or the saturation index. The output for the first test-case is shown below:

```
Seawater test case #1.- - - - - - - - - - - - - - - - - - - - - -
Temp.= 25.0    pH =  8.220    Alkalinity used for C-species

Conc. in  mmol/l     meq/l       mg/l

Na+       468.378   468.378  10768.000
K+         10.207    10.207    399.100
Mg++       53.160   106.321   1291.800
Ca++       10.287    20.574    412.300
NH4+        0.002     0.002      0.030
Cl-       545.924   545.924  19353.000
HCO3-       1.571     1.571     95.876
SO4--      28.232    56.465   2712.000
NO3-        0.005     0.005      0.290
Fe++        0.000     0.000      0.002
Mn++        0.000     0.000      0.000
Al          0.000     0.000      0.002
F-          0.073     0.073      1.390
SiO2        0.071     0.000      4.280
PO4-3       0.001     0.002      0.060
H2S         0.000     0.000      0.001

TIC         2.332              28.008
Alk                   2.565    156.530
```

```
Sum +ion = 578.66    Sum -ion = -578.22   Electrical Balance=  0.0%
```

Element	Activity	Log Act.	Act.coef	Element	Activity	Log Act.	Act.coef
Na+	3.4E-0001	-0.466	0.743	K+	7.4E-0003	-2.130	0.743
Mg++	1.4E-0002	-1.857	0.305	Ca++	2.6E-0003	-2.579	0.305
NH4+	1.1E-0006	-5.950	0.743	Cl-	4.1E-0001	-0.392	0.743
HCO3-	1.2E-0003	-2.933	0.743	SO4--	3.4E-0003	-2.471	0.305
NO3-	3.5E-0006	-5.459	0.743	Fe++	9.3E-0009	-8.033	0.305
Mn++	6.0E-0010	-9.221	0.305	Al	7.0E-0018	-17.155	0.069
F-	2.8E-0005	-4.553	0.743	SiO2	8.0E-0005	-4.096	1.160
PO4-3	2.3E-0012	-11.645	0.069	H2S	1.2E-0009	-8.907	1.160
CO3--	9.1E-0006	-5.043	0.305	H2CO3	1.6E-0005	-4.800	1.160
NH3	1.1E-0007	-6.970	1.160	CaOH+	1.1E-0007	-6.959	0.743
CaCO3	4.0E-0005	-4.398	1.160	CaHCO	3.9E-0005	-4.407	0.743
CaSO4	1.8E-0003	-2.741	1.160	CaPO4	1.7E-0008	-7.765	0.743
CaHPO	4.4E-0008	-7.354	1.160	CaH2P	2.0E-0010	-9.704	0.743
CaF+	6.4E-0007	-6.192	0.743	MgOH+	3.7E-0006	-5.427	0.743
MgCO3	1.2E-0004	-3.919	1.160	MgHCO	1.9E-0004	-3.721	0.743
MgSO4	8.4E-0003	-2.078	1.160	MgPO4	1.2E-0007	-6.913	0.743
MgHPO	3.2E-0007	-6.502	1.160	MgH2P	1.3E-0009	-8.872	0.743
MgF+	2.6E-0005	-4.590	0.743	NaCO3	5.7E-0005	-4.241	0.743
NaHCO	2.2E-0004	-3.649	1.160	NaSO4	6.1E-0003	-2.217	0.743
NaHPO	2.0E-0008	-7.691	0.743	KSO4-	1.8E-0004	-3.754	0.743
KHPO4	4.4E-0010	-9.355	0.743	FeOH+	4.9E-0010	-9.313	0.743
FeOH2	6.9E-0013	-12.163	1.160	FeOH3	4.2E-0015	-14.373	0.743
FeSO4	5.6E-0009	-8.254	1.160	FeHPO	1.1E-0012	-11.948	1.160
FeH2P	1.4E-0014	-13.868	0.743	MnOH+	2.6E-0012	-11.591	0.743
MnOH3	4.4E-0020	-19.361	0.743	MnCl+	9.9E-0010	-9.005	0.743
MnHCO	1.3E-0011	-10.883	0.743	MnSO4	3.7E-0010	-9.432	1.160
MnF+	1.2E-0013	-12.923	0.743	AlOH+	1.2E-0014	-13.925	0.305
AlOH2	1.5E-0011	-10.815	0.743	AlOH3	3.2E-0009	-8.495	1.160
AlOH4	5.3E-0008	-7.275	0.743	AlSO4	2.5E-0017	-16.607	0.743
AlSO4	6.6E-0018	-17.178	0.743	AlF+	2.0E-0015	-14.698	0.305
AlF2+	3.1E-0014	-13.511	0.743	AlF3	1.6E-0014	-13.794	1.160
AlF4-	2.3E-0016	-15.646	0.743	H3SiO	1.6E-0006	-5.808	0.743
NH4SO	4.8E-0008	-7.321	0.743	HPO4-	3.1E-0008	-7.515	0.305
H2PO4	2.9E-0009	-8.535	0.743	HF	2.5E-0010	-9.604	1.160
HF2-	2.7E-0014	-13.576	0.743	HSO4-	2.0E-0009	-8.706	0.743
H2SiO	5.2E-0010	-9.284	0.305	OH	1.7E-0006	-5.778	0.743
HS-	2.1E-0008	-7.677	0.743	S--	4.2E-0013	-12.377	0.305
FeHS2	3.6E-0015	-14.448	1.160	FeHS3	8.0E-0021	-20.095	0.743
FeHS3	8.0E-0021	-20.095	0.743				

Printout of SI also in Spreadsheet file wateqp.spr

Mineral	Log IAP	Log KT	L.IAP/KT	Mineral	Log IAP	Log KT	L.IAP/KT
Calcite	2.71	1.85	0.86	Dolomite	6.14	3.64	2.50
Siderite	-2.75	-0.22	-2.53	Rhodochros	-3.93	-0.09	-3.84
Gypsum	-5.05	-4.60	-0.45	OH-Apatite	-39.61	-40.47	0.86
F-Apatite	-52.38	-70.49	18.11	Fluorite	-11.68	-10.96	-0.72
Chalcedony	-4.10	-3.52	-0.57	Quartz	-4.10	-4.01	-0.09
Gibbsite	7.50	8.77	-1.27	Kaolinite	6.82	9.08	-2.26
Sepiolite	16.88	15.92	0.96	Vivianite	-47.39	-36.00	-11.39
P-CO2	-11.15	-7.82	-3.33	Aragonite	2.71	1.99	0.71
FeS	-0.50	3.07	-3.57				

It is important to emphasize that the results obtained by such programs are never better than the quality of the analytical input and the constants used. Analytical sources of error have been discussed in Chapter 1 and pH errors in particular may affect saturation calculations significantly. For example we may write for equilibrium between gibbsite ($Al(OH)_3$) and H_2O:

$$3H^+ + Al(OH)_3 \leftrightarrow Al^{3+} + 3H_2O$$

$$\log (IAP) = \log [Al^{3+}] + 3pH$$

Thus an error in pH of about 0.33 unit, which is not uncommon, would affect log (IAP) by a whole unit! Also in carbonate equilibria (see Chapter 4), pH errors may affect saturation calculations seriously and the uncertainty introduced by such errors should be evaluated carefully in each individual case.

The mass action constants which are used in the programs should be internally consistent. For example, the solubility product of gypsum listed in the data base of the program is probably derived from experiments during which gypsum has been equilibrated with water. In the calculations to derive K_{gypsum}, a stability constant for the aqueous complex $CaSO_4^0$ has been used. Obviously, the same stability constant should be used in all subsequent calculations involving K_{gypsum}. For major components in natural systems, consistent databases (like the WATEQ/PHREEQE database listed in Appendix B) are available, but care should be taken when new constants are added. For trace components serious deviations may arise when the same input is calculated with computer programs using different databases (Nordstrom et al., 1979). A final word of caution concerns the description of the redox state of water samples, a subject that will be taken up again in Chapter 7.

3.3 MASS ACTION CONSTANTS AND THERMODYNAMICS

Thermodynamics is the science concerned with energy distributions among substances in a system. It offers an impressive framework of formulas which can be derived from a few basic laws. A rigorous treatment of thermodynamics is outside the scope of this book and can be found elsewhere (Lewis and Randall, 1961; Denbigh, 1971; Nordstrom and Munoz, 1986). Here we confine ourselves to more practical aspects and in particular to the calculation of mass action constants and their dependency on temperature from thermodynamic tabulations.

3.3.1 *The calculation of mass action constants*

For the general reaction:

$$aA + bB \leftrightarrow cC + dD$$

we may write

$$\Delta G_r = \Delta G_r^0 + RT \ln \frac{[C]^c[D]^d}{[A]^a[B]^b} \qquad (3.22)$$

where ΔG_r is the change in Gibbs free energy (kJ/mol) of the reaction, ΔG_r^0 the standard Gibbs free energy of the reaction and equal to ΔG_r when each product or reactant is present at unit activity (so that the last term becomes zero) at a specified standard state (25°C and 1 atm), $[i]$ denotes the activity of i, R is the gas constant ($8.314 \cdot 10^{-3}$ kJ/mol.deg), T the absolute temperature, Kelvin (Kelvin = °C + 273.15).

In older texts, energy is often expressed in kcal/mol which can be converted to kJ/mol by multiplying with 4.184 ($R = 1.987 \cdot 10^{-3}$ kcal/mol.deg). The prefix Δ is used because energy can be measured only as relative amounts. The direction in which the reaction will proceed is indicated by ΔG_r:

$\Delta G_r > 0$ the reaction proceeds to the left;
$\Delta G_r = 0$ the reaction is at equilibrium;
$\Delta G_r < 0$ the reaction proceeds to the right.

Therefore, in the case of equilibrium, Equation (3.22) reduces to:

$$\Delta G_r^0 = -RT \ln \frac{[C]^c[D]^d}{[A]^a[B]^b} \tag{3.23}$$

Note that the activity product in the last term is equal to the mass action constant, K:

$$\Delta G_r^0 = -RT \ln K \tag{3.24}$$

Back substitution of Equation (3.24) in (3.22) results in

$$\Delta G_r = -RT \ln K + RT \ln \frac{[C]^c[D]^d}{[A]^a[B]^b} \tag{3.25}$$

In (3.25) the distance from equilibrium is expressed in terms of the mass action constant and the composition of the solution, which is in fact an analogue to the saturation index SI (Equation 3.21).

Equation (3.24) has the practical application that it allows us to calculate the mass action constant for any reaction from tabulated data of ΔG_f^0 for dissolved substances, minerals, and gases. ΔG_f^0 is the free energy of formation i.e. the energy needed to produce one mole of a substance from pure elements in their most stable form. These latter and the H^+ ion have by definition zero values. Tabulations are normally given for 25°C and 1 atm pressure. ΔG_r^0 is calculated as follows:

$$\Delta G_r^0 = \sum \Delta G_{f \, products}^0 - \sum \Delta G_{f \, reactants}^0$$

EXAMPLE 3.3: *Calculation of solubility products from Gibbs free energy data*
Calculate the solubility product of calcite at 25°C.
   The Gibbs free energies of formation at 25°C are (Wagman et al., 1982):

$$\Delta G_{f \, CaCO_3}^0 = -1128.8 \text{ kJ/mol}$$
$$\Delta G_{f \, Ca^{2+}}^0 = -553.6 \text{ kJ/mol}$$
$$\Delta G_{f \, CO_3^{2-}}^0 = -527.8 \text{ kJ/mol}$$

For the reaction

$$CaCO_3 \leftrightarrow Ca^{2+} + CO_3^{2-}$$

we obtain

$$\Delta G_r^0 = \Delta G_{f \, Ca^{2+}}^0 + \Delta G_{f \, CO_3^{2-}}^0 - \Delta G_{f \, calcite}^0;$$
$$\Delta G_r^0 = -553.6 - 527.8 + 1128.8 = 47.4 \text{ kJ/mol};$$
$$\Delta G_r^0 = -RT \ln K;$$
$$47.4 = -(8.314 \cdot 10^{-3})(298.15)(2.3) \log K = -5.701 \log K;$$
$$\log K = 47.4/ -5.701 = -8.31$$

A number of compilations of thermodynamic data concerning mineral water systems are available (Helgeson et al. 1978; Robie et al. 1978; Wagman et al. 1982; Cox et al. 1989; Woods and Garrels, 1987 etc.). Preferably, consistent sets of data (such as are present in the first four references) should be used as otherwise erratic values can be obtained. For example, the listed $\Delta G_{f \, calcite}^0$ could have been calculated from solubility measurements using values of $\Delta G^0_{fCO_3^{2-}}$ and $\Delta G^0_{fCa^{2+}}$ by the reverse procedure of Example 3.3. When $K_{calcite}$ subsequently is calculated from the thermodynamic tables with a different $\Delta G^0_{f\,CO_3^{2-}}$ value, then the result will clearly be erroneous. Even when these precautions are taken you will discover that different compilations may give slightly different constants. For example, the value of $K_{calcite}$, calculated in Example 3.3, is slightly higher than the currently accepted value of $10^{-8.48}$ (Plummer and Busenberg, 1982).

3.3.2 *Calculation of mass action constants at different temperature*

In situ groundwater is generally not found under standard conditions of 25°C and 1 atm pressure. While variations in pressure have little effect on the values of the mass action constants, temperature variations are important. Variations of mass action constants with temperature are usually calculated with the Van't Hoff equation:

$$\frac{d \ln K}{dT} = \frac{\Delta H^0_r}{RT^2} \tag{3.26}$$

Here ΔH^0_r is the reaction enthalpy which is negative for exothermic reactions and positive for endothermic reactions. The calculation of ΔH^0_r and the definition of the standard state is analogous to that of ΔG^0. At 25°C, the value of the reaction enthalpy, ΔH^0_r, is calculated from the formation enthalpies, ΔH^0_f, values, which are listed in thermodynamic tables. ΔH^0_r varies only slightly with temperature and within a range of a few tens of degrees, it can be considered as a constant. In that case, Equation (3.26) can be integrated to give for two temperatures:

$$\log K_{T_1} - \log K_{T_2} = \frac{-\Delta H^0_r}{2.303\, R} \left(\frac{1}{T_1} - \frac{1}{T_2} \right) \tag{3.27}$$

This equation is generally applicable in groundwater environments.

EXAMPLE 3.4: *Temperature dependency of the solubility product*
Calculate the solubility product of calcite at 10°C from the following enthalphies of formation (Wagman et al. 1982):

$\Delta H^0_{fCaCO_3} = -1206.9 \text{ kJ/mol}$

$\Delta H^0_{fCa^{2+}} = -542.8 \text{ kJ/mol}$

$\Delta H^0_{fCO_3^{2-}} = -677.1 \text{ kJ/mol}$

Thus for the reaction

$CaCO_3 \leftrightarrow Ca^{2+} + CO_3^{2-}$

we may write,

$\Delta H^0_r = (-542.8) + (-677.1) - (-1206.9) = -13.0 \text{ kJ/mol}$

which means that the reaction is exothermal: heat is lost when calcite dissolves. The difference in log K between 25°C and 10°C, according to Equation (3.27) is:

$$\log K_{25} - \log K_{10} = \frac{-\Delta H^0_r}{2.303\, R} \left(\frac{1}{298.15} - \frac{1}{283.15} \right)$$

$$= \frac{13.0}{2.303 \cdot 8.314 \cdot 10^{-3}} \left(\frac{1}{298.15} - \frac{1}{283.15} \right) = -0.12 \qquad (3.28)$$

In Example 3.3 it was calculated that $\log K_{calcite}$ at 25°C is -8.31, so that at 10°C $\log K_{calcite} = -8.31 + 0.12 = -8.19$. Thus for an exothermal reaction the solubility increases with decreasing temperature and vice versa for an endothermal reaction.

An alternative way of expressing the variation of K with temperature is to fit the experimental data for different temperatures with a polynomial function. For example a description for the variation of $K_{calcite}$ with temperature is given by the following equation (Plummer and Busenberg, 1982):

$$\log K_{calcite} = -171.9065 - 0.077993T + 2839.319/T + 71.595 \log T \qquad (3.29)$$

3.4 SOLID SOLUTION

In the preceding part of this chapter, minerals found in aquifer materials have been considered as pure phases. However, the analysis of naturally occurring minerals by microanalytical devices, like the microprobe, shows that pure minerals are the exception rather than the rule. Minerals of variable composition are considered as mixtures between pure end-member minerals and are termed *solid solutions*. These are found among detrital minerals, originally formed at high temperatures, such as the plagioclases which are solid solutions between Ca-feldspar and Na-feldspar, amphiboles, pyroxenes, etc. Low temperature phases like carbonates also form extensively solid solutions, the best studied example is that of Mg-calcites which we treat in more detail in Chapter 4. Numerous other cases are known such as Fe and Mn replacement in calcite, Ca in rhodochrosite, Mn in siderite, Sr in aragonite etc. Additional examples are substitution of fluorapatite in hydroxyapatite (as happens in tooth enamel) which affects fluoride concentrations in groundwater (Zack, 1980), and aluminum substitution in goethite which influences groundwater Al concentrations. An example of the latter is illustrated in Figure 3.5.

Solid solution is often neglected in groundwater chemistry, because it is difficult to obtain good data on the composition of minerals with variable composition that occur in aquifers. In additions, the physico-chemical relations between the solid solution and the aqueous solution are not easily derived. There is, however, little doubt that solid solution is important in regulating the chemical composition of groundwater.

3.4.1 *Basic theory*

The calcite/rhodochrosite system can be used as an example of a binary solid solution. Equilibrium between the aqueous solution and the solid solution is defined by two mass action equations which have to be fulfilled simultaneously:

For calcite

$$CaCO_3 \leftrightarrow Ca^{2+} + CO_3^{2-}$$

$$K_{cc} = \frac{[Ca^{2+}][CO_3^{2-}]}{[CaCO_3]_{ss}} \qquad (3.30)$$

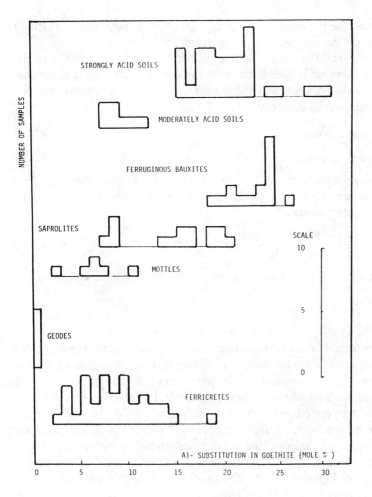

Figure 3.5. The aluminum content of goethites in various soil environments (Fitzpatrick and Schwertmann, 1982).

and for rhodochrosite

$$MnCO_3 \leftrightarrow Mn^{2+} + CO_3^{2-}$$

$$K_{rhod} = \frac{[Mn^{2+}][CO_3^{2-}]}{[MnCO_3]_{ss}} \tag{3.31}$$

Here $[CaCO_3]_{ss}$ and $[MnCO_3]_{ss}$ are the activities of the two components in the solid phase. Until now the activity of the solid phase has not appeared in the mass action expression because the activity of a pure solid phase by definition is equal to one. However, for a solid solution the activity of each component, that is of $[CaCO_3]_{ss}$ and $[MnCO_3]_{ss}$, will depend on the composition of the solid solution. Rearranging Equations (3.30) and (3.31) by eliminating $[CO_3^{2-}]$ gives:

$$\frac{[Mn^{2+}]}{[Ca^{2+}]} = \frac{[MnCO_3]_{ss}}{[CaCO_3]_{ss}} \cdot \frac{K_{rhod}}{K_{cc}} \tag{3.32}$$

Equation (3.32) illustrates the important property that for a given $[Mn^{2+}]/[Ca^{2+}]$ ratio in solution, the composition of the solid solution is also fixed.

The standard state of a solid is defined as the pure phase and the activities of the components in the solid phase are related to the mole fraction, X, with respect to the pure mineral. As for solutes, an activity coefficient λ is used to correct for non-ideal behavior. Thus for our rhodochrosite/calcite example:

$$[MnCO_3]_{ss} = \lambda_{MnCO_3} \cdot X_{MnCO_3} \tag{3.33}$$

$$[CaCO_3]_{ss} = \lambda_{CaCO_3} \cdot X_{CaCO_3} \tag{3.34}$$

Note that the sum of the two mole fractions must be equal to one. Substitution of Equations (3.33) and (3.34) in Equation (3.32) and replacing activities of dissolved species by molal concentration yields after rearranging:

$$\frac{K_{rhod} \cdot \gamma_{Ca^{2+}} \cdot \lambda_{MnCO_3}}{K_{cc} \cdot \gamma_{Mn^{2+}} \cdot \lambda_{CaCO_3}} = \frac{m_{Mn^{2+}}}{m_{Ca^{2+}}} \cdot \frac{X_{CaCO_3}}{X_{MnCO_3}} \tag{3.35}$$

The left hand term is often collected in a single parameter, the distribution coefficient D:

$$D = \frac{m_{Mn^{2+}}}{m_{Ca^{2+}}} \cdot \frac{X_{CaCO_3}}{X_{MnCO_3}} \tag{3.36}$$

Thus D relates molal concentrations in solution directly to mole fractions in the solid phase. However, D is only a constant when the activity coefficients of aqueous and solid phase components in Equation (3.35) are constant and the latter is not generally true.

Distribution coefficients are mostly used semi-empirically to describe experimentally observed partitioning of trace components between solutions and solids. In such cases the controlling mechanisms are not always well defined. For example, experimentally obtained D values may depend on the precipitation rate and are then controlled by kinetic mechanisms (e.g. Lorens, 1981; Mucci, 1988; Plummer and Busenberg 1987) instead of true equilibrium as required by Equation (3.35). Distribution coefficients are also used to describe adsorption processes as discussed in Chapter 5.

The simplest solid solution model is that of ideal solid solution. In this model, the activities of components in the solid phase are equal to the mole fractions and λ is equal to one. Unfortunately ideal solid solution is rare, although the dominant component in a solid solution ($X > 0.9$) often behaves close to ideal. Other solid solution models are based on the extent of crystal lattice deformation and the degree of ordering. In such models, solid phase activity coefficients can be fitted to the following equations (Glynn, 1990) (illustrated for the rhodochrosite/calcite example):

$$\ln \lambda_{MnCO_3} = X_{CaCO_3}^2 [\alpha_o - \alpha_1 (3X_{MnCO_3} - X_{CaCO_3}) +$$
$$\alpha_2 (X_{MnCO_3} - X_{CaCO_3}) (5X_{MnCO_3} - X_{CaCO_3}) + ...]$$

$$\ln \lambda_{CaCO_3} = X_{MnCO_3}^2 [\alpha_o + \alpha_1 (3X_{CaCO_3} - X_{MnCO_3}) +$$
$$\alpha_2 (X_{CaCO_3} - X_{MnCO_3}) (5X_{CaCO_3} - X_{MnCO_3}) + ...] \tag{3.37}$$

In many cases the first term with the parameter α_o is sufficient to describe the variation of the activity coefficient with the composition and this is known as the regular solid solution model. More complicated models, such as the subregular model or the Margules activity coefficient series, require the use of additional terms in Equation (3.37) (Glynn, 1990; Glynn and Reardon, 1990).

EXAMPLE 3.5: *Cadmium partitioning in calcite*

Davis et al. (1987) and Fuller and Davis (1987) studied partitioning of cadmium between aqueous solution and calcareous aquifer material. They found that Cd was dominantly sorbed by calcite, which after a fast initial adsorption stage could be described by a regular solid solution model between calcite and otavite ($CdCO_3$) with an interaction parameter of $\alpha_o = -0.8$. What would be the Cd^{2+} concentration in contact with calcite containing 1 mol % $CdCO_3$ and a Ca^{2+} concentration of 3 mmol/l?

First rewrite Equation (3.35) for the calcite/otavite system:

$$\frac{K_{ota} \cdot \gamma_{Ca^{2+}} \cdot \lambda_{CdCO_3}}{K_{cc} \cdot \gamma_{Cd^{2+}} \cdot \lambda_{CaCO_3}} = \frac{m_{Cd^{2+}}}{m_{Ca^{2+}}} \cdot \frac{X_{CaCO_3}}{X_{CdCO_3}}$$

The activities of solid phase components are calculated from Equation (3.37), which for a regular solid solution simplifies to:

$$\ln \lambda_{CdCO_3} = \alpha_o X^2_{CaCO_3}$$
$$\ln \lambda_{CaCO_3} = \alpha_o X^2_{CdCO_3}$$

Substitution of $\alpha_o = -0.8$ and 1 mol % $CdCO_3$ yields:

$$\ln \lambda_{CdCO_3} = -0.8\,(0.99)^2 = -0.78 \qquad \lambda_{CdCO_3} = 0.46$$
$$\ln \lambda_{CaCO_3} = -0.8\,(0.01)^2 = -8.0 \cdot 10^{-5} \qquad \lambda_{CaCO_3} = 1.00$$

According to the Davies Equation (3.12), $\gamma_{Ca^{2+}} = \gamma_{Cd^{2+}}$. Furthermore, $K_{ota} = 10^{-13.74}$ and $K_{cc} = 10^{-8.48}$ which allows us to calculate $m_{Cd^{2+}}$:

$$m_{Cd^{2+}} = \frac{K_{ota} \cdot \lambda_{CdCO_3} \cdot X_{CdCO_3}}{K_{cc} \cdot \lambda_{CaCO_3} \cdot X_{CaCO_3}} \cdot m_{Ca^{2+}}$$

$$= \frac{10^{-13.74} \cdot 0.46 \cdot 0.01}{10^{-8.48} \cdot 1.00 \cdot 0.99} \cdot 3 \cdot 10^{-3} = 7.66 \cdot 10^{-11} \text{ mol/l}$$

This result suggest that solid solution formation between otavite and calcite would be a very effective process for cadmium removal from groundwater.

Some caution is warranted in the application of these simple solid solution models since exchange equilibrium according to Equation (3.32) is not often attained in experiments at low temperature. In response to these difficulties, less restrictive equilibria criteria such as stoichiometric saturation (Thorstenson and Plummer, 1977) have been formulated which will be elaborated in Chapter 4.

Figure 3.6 illustrates solid solution behavior in a natural system for Ca-rhodochrosites in marine sediments of the Baltic Sea and displays the activity ratio of ions in solution versus the mole fraction of calcite in rhodochrosite. Non-equilibrium behavior is indicated by the large compositional variation of the Ca-rhodochrosites, since according to Equation (3.32) only one composition of the carbonate is stable at a given water composition. Figure 3.6 also shows the relation between the $[Ca^{2+}]/[Mn^{2+}]$ in solution and the mole fraction of calcite in rhodochrosite in the cases of ideal solid solution and for a regular solid solution ($\alpha_o = -3.2$). Clearly more $CaCO_3$ is incorporated in rhodochrosite than predicted by ideal solid solution. The regular solid solution model represents here merely a data fit to field data, since it is well known from high temperature data (Capobianca and Navrotsky, 1987)

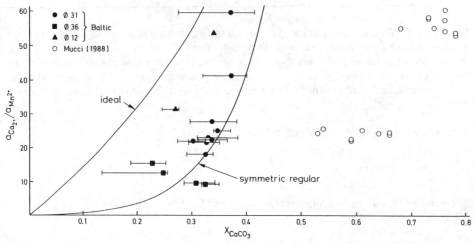

Figure 3.6. Ideal and regular solid solution ($\alpha_o = -3.2$) in the system rhodochrosite/calcite. Data shown are from natural sediments of the Baltic, where bars indicate compositional variation in each sample, and experimental precipitates by Mucci (1988). (Jakobsen and Postma, 1989. Reprinted with permission from Pergamon Press PLC).

that the calcite rhodochrosite system requires a two parameter fit of Equation (3.37). Finally, as an example of the deviations that may result, the kinetically controlled compositions of Ca-rhodochrosites from experiments by Mucci (1988) are also shown in Figure 3.6.

3.5 KINETICS OF GEOCHEMICAL PROCESSES

3.5.1 *Kinetics and equilibrium*

If the concentrations of dissolved constituents in groundwater are governed by chemical interactions with the solids in the aquifer, then two different types of controls can be considered. Either the concentration is controlled by equilibrium or it is determined by reaction kinetics in combination with flow velocities or residence time. Recognition of the type of control which is operational is extremely important for the approach to the problem. This can be illustrated in the following example.

If a simple mineral like halite (NaCl) dissolves in distilled water, and the concentrations are followed as a function of time, you may find curves as shown in Figure 3.7. Thus [Na$^+$] increases with time until equilibrium between water and mineral is attained at time t_2. From t_2 onwards, [Na$^+$] becomes independent of time and is determined by an equilibrium constraint:

$$K_{halite} = [Na^+][Cl^-]$$

K_{halite} can be measured directly in the laboratory or calculated from thermodynamic tables. As we have seen, the dependence of K_{halite} on temperature can also be calculated. In many cases, the equilibrium concept gives a satisfactory explanation for the observed ground-

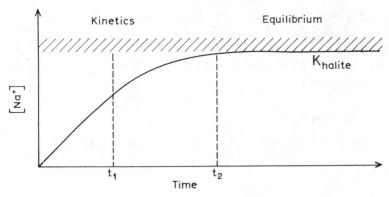

Figure 3.7. The realms of kinetics and equilibrium, illustrated in a hypothetical dissolution experiment with halite.

water chemistry. Examples are ion exchange or carbonate chemistry in groundwaters. Due to its firm foundation and ease in calculations, it should always be the first approach to any problem.

However, a number of phenomena in aquifers is insufficiently explained by equilibrium chemistry. Many silicate minerals, like feldspars, amphiboles, pyroxenes etc. are thermodynamically unstable at the earth's surface (groundwaters/surface waters are strongly subsaturated for these minerals). Still they persist for thousands or tens of thousands of years in aquifers. Other examples are the observed subsaturation for calcite in some aquifers as well as the survival of aragonite in aquifers for thousands of years. There are two fundamental causes for the lack of equilibrium at the earth's surface. Firstly, tectonic activity transports minerals from the environments where they were formed and stable, often deep down in the earth at high pressures and temperatures, to environments at the earth's surface where they are unstable. Second, photosynthesis converts solar energy into thermodynamically unstable organic matter; its subsequent decomposition affects the equilibria of redox sensitive compounds such as iron. Biological activity also disturbs carbonate equilibria through the production or consumption of CO_2 and through biomineralization. To understand, and to describe quantitatively, such deviations from equilibrium requires the use of reaction kinetics.

In the example of halite dissolution (Figure 3.7), the first part of the reaction, before equilibrium with halite is attained, should clearly be studied from a kinetic point of view. Qualitatively, one can expect that the rate of dissolution of halite depends on factors such as grain size (small crystals have a larger surface area than large ones per unit weight), the amount of stirring, temperature and the distance from equilibrium. The link to equilibrium chemistry is of course, to tell us in which direction the reaction will go. At time t_1, equilibrium chemistry predicts that dissolution will take place rather than precipitation. It does not tell us, however, whether it will take 10 seconds or 10 million years before equilibrium is attained.

Our understanding of reaction kinetics is at a much lower level than that of equilibrium chemistry even though rapid progress is being made (e.g. Sparks, 1989; Hochella and White, 1990). There is no unifying theory, and different notations are used to describe even the same problem. The approach to a kinetic problem is usually divided into two steps:

1. To describe quantitatively the rate data measured in the laboratory or in the field.
2. To interpret the quantitative description of the rate data in terms of mechanisms.

It is often found that different mechanisms fit equally well to the same rate data. Therefore, kinetic data derived from solution chemistry may at best be demonstrated as consistent with a given mechanism.

3.5.2 *Rate laws and chemical reactions*

Consider a simple reaction where compound A is converted to compound B by the reaction:

$$A \rightarrow B$$

This reaction can be followed by recording the concentration of A as a function of time as shown in Figure 3.8. The reaction rate is the change of A with time. Thus at time x_1, the rate can be determined from the slope of the tangent, C_{x1}/t_{x1}, at this point. For the whole curve, this is equal to:

$$\text{rate} = -\,dC_A/dt$$

The units of the rate depends on the definition of the reaction and could for example be mol/l.s. For a decrease in concentration of a reactant, the rate is given a negative sign (the slope of the tangent is negative), while for the increase of a product, the rate is positive (the slope of the tangent is positive). Thus:

$$\text{rate} = -\,dC_A/dt = dC_B/dt$$

For the more complicated reaction

$$A + 2B \rightarrow 3C$$

$$\text{rate} = dC_C/dt = -3dC_A/dt = -\tfrac{3}{2}\,dC_B/\,dt$$

This shows the importance of defining the reaction rate explicitly. The manner in which the reaction rate varies with the concentrations of reacting substances is referred to as the *reaction order*. For example, it is found experimentally that the reaction rate is proportional to the αth power of the concentration of reactant A and to the βth power of reactant B etc.:

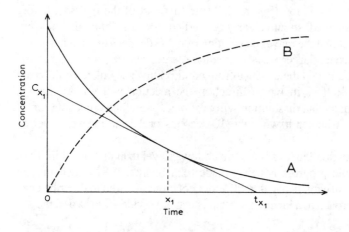

Figure 3.8. Derivation of rates from concentration/time data.

$$\text{rate} = k \cdot C_A^\alpha \cdot C_B^\beta \dots \tag{3.38}$$

Such a reaction is said to be αth order with respect to [A] and βth order for [B]. The overall order of this reaction is simply:

$$n = \alpha + \beta + \dots$$

The coefficient k in Equation (3.38) is the rate constant, also known as the specific rate, and is numerically equal to the reaction rate when all reactants are present at unit concentrations. In that case all variables in Equation (3.38) reduce to one so that the rate $= k$. Since the rate has fixed units (for example mol/l.s), the units of the rate constant depend on the overall order of the reaction. In general for an n-order reaction, the units of k are $mol^{1-n} \cdot l^{n-1} \cdot s^{-1}$.

The effect of the reaction orders on relations between concentration and time, and rate and concentration is illustrated in Figure 3.9 for the reaction

$$A \rightarrow B$$

Thus for a zero order reaction, the rate is independent of the concentration of the reactant and for a first order reaction the rate is proportional to the concentration of the reactant etc. Remember that the rate at a given time or concentration is given by the tangent on the concentration-time plot. An example of a first order reaction is radioactive decay and the first order rate law forms the basis of tritium or ^{14}C dating of groundwater.

The Fe(II) oxidation example illustrates the effect of reaction order on the rate, which

---

EXAMPLE 3.6: *Oxidation of Fe(II)*

For oxidation of Fe^{2+} in aqueous solutions at 20°C and 1 atm., the following rate law has been reported by Stumm and Morgan (1981):

$$\text{rate} = -\frac{dm_{Fe^{2+}}}{dt} = k \cdot m_{Fe^{2+}} \cdot [OH]^2 \cdot P_{O_2}$$

where $k = 8.0 \, (\pm 2.5) \cdot 10^{13} \, min^{-1} \cdot atm^{-1}$.

Thus the rate equation is formulated as a decrease in $m_{Fe^{2+}}$ and is first order for $m_{Fe^{2+}}$ and P_{O_2} and second order for [OH$^-$]. What is the rate of Fe^{2+} oxidation if $m_{Fe^{2+}} = 1$ mmol/l and $P_{O_2} = 0.2$ atm, for pH values 5 and 7?

ANSWER:

pH = 5;

$$K_w = [H^+][OH^-] = 10^{-14} \rightarrow [OH^-] = 10^{-9}$$

$$-\frac{dm_{Fe^{2+}}}{dt} = 8.0 \cdot 10^{13} \cdot 10^{-3} \cdot (10^{-9})^2 \cdot 0.2 = 1.6 \cdot 10^{-8} \, mol/l.min$$

$$= .001 \, mmol/l.hour$$

pH = 7;

$$-\frac{dm_{Fe^{2+}}}{dt} = 8.0 \cdot 10^{13} \cdot 10^{-3} \cdot (10^{-7})^2 \cdot 0.2 = 2.0 \cdot 10^{-4} \, mol/l.min$$

$$= 9.4 \, mmol/l.hour$$

Note the very large difference in rates at pH 5 and 7, due to the second order rate dependence on [OH$^-$].

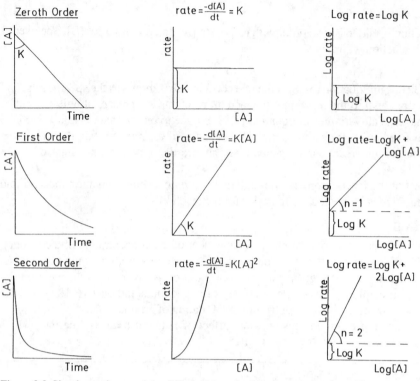

Figure 3.9. Simple rate laws and the differential method for the reaction A → B.

qualitatively also can be read from Figure 3.9. Note that by doubling the concentrations of $m_{Fe^{2+}}$ or P_{O_2}, with first order dependencies, the rate increases with a factor of two, while for the second order dependence on [OH⁻] the rate becomes four times higher when the concentration is doubled. It is thus clear that increasing pH is a much more effective way to accelerate Fe(II) oxidation than increasing P_{O_2}.

For zero order kinetics, the rate is independent of the reactant concentration and this implies that other factors or reactants control the rate. An example is shown in Figure 3.10. for bacterial reduction of sulfate by organic matter, which shows zero order behavior as long as the $[SO_4^{2-}]$ is higher than 2 mmol/l. The obvious conclusion from the zero order dependence on $[SO_4^{2-}]$ is that the amount of reactive organic matter is rate controlling rather than $[SO_4^{2-}]$.

In these examples, reaction orders have been whole numbers. There is, however, no a priori reason why they could not be fractional numbers and the latter is in fact often found. Rate laws for a given reaction can be derived from experimental or field data in many different ways (see Laidler 1965; Lasaga, 1981). One of the easiest, and most general procedures is the differential method. This is illustrated in Figure 3.9 and consists of measuring rates from concentration versus time plots, followed by plotting log rate versus log concentration. This plot yields the reaction order as the slope of the line, and the rate constant as the intercept on the y-axis. An obvious alternative method is to integrate zero, first and second order rate laws and then simply to fit concentration/time data by trial and error.

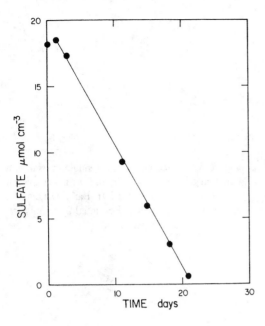

Figure 3.10. Dissolved sulfate versus time for marine sediment incubated in the laboratory (reprinted from Berner, 1981). The straight line indicates a zero order dependence on the sulfate concentration.

3.5.3 Crystal formation

Crystal formation is a two step process: the first step is the formation of a crystal embryo by nucleation, and the second step is crystal growth. The distinction is based on the fact that very small crystals are dominated by edges and corners, and since atoms located at such positions are not charge compensated, such very small crystals are less stable than larger crystals. In terms of energy this can be described as:

$$\Delta G_n = \Delta G_{bulk} + \Delta G_{interf}$$

Where ΔG_n is the free energy of formation of a crystal as a function of the number of atoms n precipitated in the crystal, ΔG_{bulk} the bulk free energy identical to the one used in Section 3.3.1, and ΔG_{interf} the interfacial free energy which reflects the excess free energy of very small crystals. A direct result of this equation is that very small crystals are more soluble than larger crystals. However, at crystal sizes larger than 2 μm, the effect of the interfacial free energy becomes negligible.

The change in ΔG_n with increasing crystal size is shown in Figure 3.11. During the growth of the crystal embryo an energy barrier has to be passed in order to allow further growth. The crystal size corresponding to this energy barrier is known as the critical nucleus and its size is dependent on the saturation state Ω. For increasing Ω both the energy barrier and the size of the critical nucleus decreases. Once the size of the critical nucleus has been passed, the crystal enters the domain of crystal growth.

Qualitatively these relations may explain why a high degree of supersaturation yields poor crystals. In that case the nucleation threshold is insignificant which allows many crystal nuclei to form but few to grow larger. Typical natural examples are the formation of Fe-oxyhydroxides and Al-hydroxides. On the other hand, at low supersaturation, crystal growth will dominate over nucleation and result in few but larger crystals. The energy barrier also explains why minerals in aquifers rarely form by spontaneous nucleus

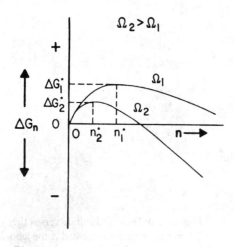

Figure 3.11. Free energy of a single crystal as a function of a number of atoms (n) in the crystal (Nielsen, 1964; Berner, 1980. Early Diagenesis. Copyright 1980 by PUP. Reprinted by permission of Princeton University Press).

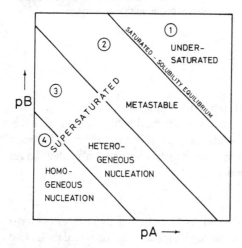

Figure 3.12. Regimes of crystal growth for ionic substances A and B that may form mineral AB. pA = – log [A] and pB = – log [B]. (Reprinted from Nielsen and Toft, 1984).

formation from solution (homogeneous nucleation), but preferentially on pre-existing surfaces (heterogeneous nucleation) which decrease the energy barrier considerably.

The different regimes of crystal formation are illustrated schematically in Figure 3.12 for the activities of ionic substances A and B that may form a mineral AB. In subsaturated solutions, region (1), only dissolution may take place. The line separating region (1) and (2) corresponds to the solubility product. In region (2), existing crystals may grow but no nucleation can take place. With increasing supersaturation, region (3), crystal growth can be accompanied by nucleation on existing surfaces of other minerals etc. And finally in region (4) the degree of supersaturation has become high enough to allow homogeneous nucleation.

3.5.4 *Dissolution and crystal growth*

Presently, the kinetics of minerals dissolution or growth have in only few cases been applied to groundwater studies. However, the potential influence of kinetics of mineral dissolution on the composition of natural waters on a regional scale is suggested by the relationship between runoff and total dissolved solid concentrations as a function of rock type (Figure 2.13).

Dissolution and crystal growth processes show many similarities and are therefore treated together. In its most general form dissolution or growth of crystals from aqueous solution can be described by the overall rate expression:

$$J = dn_{cr}/dt \qquad \text{for crystal growth} \tag{3.39a}$$

and

$$J = -dn_{cr}/dt \qquad \text{for crystal dissolution} \tag{3.39b}$$

Here, J is the overall rate of crystal growth or dissolution expressed in mol/s and n_{cr} the number of mole formula units of the solid. Confining ourselves in the following to crystal growth, we may define the flux density j as the flux of substance into the surface area A per unit time:

$$j = J/A \qquad (\text{mol/m}^2.\text{s})$$

The flux density j is related to the growth rate of crystals by the relation (Nielsen, 1984):

$$dr/dt = V_m \cdot j = V_m \cdot J/A \qquad (\text{m/s}) \tag{3.40}$$

Here A is the surface area of the crystals at time t, r is the radius of the crystal and V_m is the molar volume of the crystals. In order to relate the change in concentration of an ion in solution to the overall crystal growth rate we must express the latter in terms of the volume of solution in which the crystals grow, and introduce a stoichiometric factor v which indicates the number of ions per formula unit of the crystal.

$$-dC_i/dt = v_i J/V \tag{3.41}$$

Where V is the volume of the solution. For example for precipitation of fluorite:

$$-d[Ca^{2+}]/dt = J/V, \qquad \text{and}$$

$$-d[F^-]/dt = 2J/V$$

Substitution of Equation (3.41) in (3.40) yields the following result:

$$-\frac{dC_i}{dt} = \frac{A}{V} \frac{v_i}{V_m} \frac{dr}{dt} \tag{3.42}$$

This relation shows the important property that the rate of concentration change is proportional to the ratio between the surface area and the volume of the solution. In terms of aquifers, the volume is related to the porosity while A is related to the surface area for the mineral concerned. Surface areas are in experimental studies usually measured with the BET method which is based on gas adsorption (Adamson, 1982). This is impossible in sediments and rocks which consist of mixtures of different minerals. Only geometric surfaces can be estimated using petrographic methods. A comparison of BET surface areas with geometric surface area measurements is given in Table 3.2 for some minerals, and shows that BET surface areas, due to the inclusion of sub-microscopic surface roughness and crevices, are much higher than geometric surface areas.

During dissolution or crystal growth, the rate is expected to be influenced by the amount of crystals present, their change of shape during the process, and the composition of the

Table 3.2. A comparison of geometric and BET surface areas for some common minerals. (Sverdrup and Warfvinge, 1988).

Mineral	Geometric surface area (m^2/g)	BET surface area (m^2/g)
Orthoclase	0.032	0.12
Albite	0.072	0.30
Hornblende	0.057	0.24
Zoisite	0.025	0.30
Epidote	0.068	0.11
Biotite	0.032	0.53
Almandine	0.027	0.22

solution. The introduction of these parameters in the rate equation can be formulated as follows (Christoffersen and Christoffersen, 1979).

$$J = \left| dn_{cr}/dt \right| = k \cdot m_o \cdot F(m/m_o) \cdot g(C) \tag{3.43}$$

Here J is the overall rate, k is the rate constant, m_o is the initial mass of crystals at the start of the experiment and m is the mass of crystals at time t. $F(m/m_o)$ is a term representing the influence of crystal morphology, size and so forth on the rate, while $g(C)$ represents the influence of solution composition on the rate.

For ideal dissolving spheres which do not change geometry, it can be derived from the relation between volume and surface area that $F(m/m_o) = (m/m_o)^{2/3}$ (Liu and Nancollas, 1971). However, $F(m/m_o)$ may become a complex function of both changing crystal geometry and changing reaction mechanism on the crystal surface. In many studies the experimental conditions are therefore designed so that only an insignificant amount of the initial mineral mass dissolves. In that case, the term $m_o \cdot F(m/m_o)$ is directly proportional to the surface area A and the surface effect is included in the rate constant k. When equilibrium is approached, either from sub- or supersaturation, the term $g(C)$ becomes a function of the distance from equilibrium. This type of reaction kinetics is found for carbonates, phosphates, gypsum, fluorite etc.

Under experimental conditions very far from equilibrium, as usually is the case in dissolution studies of detrital silicate minerals, but also in many redox reactions, only the forward reaction is of importance and the rate becomes independent of the saturation state although the rate will still depend on the solution composition, for example on the pH of the solution.

3.5.5 *Mechanisms of dissolution and crystallization*

Mineral dissolution or growth proceeds through a chain of processes. These include the transport of solutes from bulk solution to the mineral surface, the adsorption at the surface, the chemical reactions at the surface, the desorption of product solutes from the surface and finally the transport of product solutes back into the bulk solution. These processes must take place simultaneously but some may proceed at a faster rate than others. Since the overall rate will depend on the rate of the slowest process, it is of importance to determine the *rate limiting* process or mechanism.

An essential distinction is whether transport of solutes from the bulk solution to and from

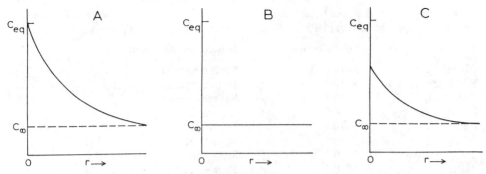

Figure 3.13. Schematic concentration profiles from the mineral surface to the bulk of solution for different rate controlling processes during dissolution. C_∞ is the concentration in the bulk solution and C_{eq} the concentration in equilibrium with the mineral. A. Transport, B. Surface reaction, C. Mixed control (after Berner, 1978. Reprinted by permission of American Journal of Science).

the mineral surface, or a process at the mineral surface is rate controlling. The difference is illustrated in Figure 3.13 for a dissolving mineral. If transport of ions to and from the mineral surface is rate limiting, then equilibrium will be approached at the mineral surface and a concentration gradient develops from the mineral surface to the bulk of the solution. On the other hand when a surface reaction is rate limiting, any ion released from the crystal lattice will quickly be carried away and no concentration gradient developes. Also intermediate cases of surface and transport controlled reactions can be found.

How to decide whether a mineral dissolution reaction is transport or surface reaction controlled? Experimentally, the easiest way is to change the stirring rate, since this affects convection at the mineral surface and thereby the concentration gradient. If the rate changes, the reaction is transport controlled. However, the reverse is not always true, since very small crystals (< about 5 μm) are carried along with the water, so that the convection at the mineral surface remains the same (Nielsen, 1984). A more elegant way is to calculate whether the reaction is slower or faster than molecular diffusion. Since molecular diffusion is the slowest form of transport possible, slower rates are necessarily surface reaction controlled.

EXAMPLE 3.7: *Dissolution of hydroxyapatite; transport or surface reaction controlled?*
Christoffersen and Christoffersen (1979) studied the dissolution kinetics of hydroxyapatite (HAP) using suspensions of fine crystals. The overall dissolution rate is expressed as $J = -dn/dt$, where n is the amount of undissolved HAP. For steady state diffusion through a sphere circumscribing the crystal we may use the equation:

$$J = 4\pi D_{app} Nr(C_s - C)$$

Where D_{app} is the apparent diffusion coefficient, N the number of crystals, r the radius of the crystals, C_s the concentration at equilibrium and C the concentration in the bulk solution. Substituting measured values of J, N, r, C_s and C, yield a value of $D_{app} \approx 10^{-9}$ cm^2/s. Directly measured values for diffusion of ion in aqueous solution are in the order 10^{-5} cm^2/s. Since the apparent diffusion coefficient D_{app}, is four orders of magnitude lower, one must conclude that dissolution of HAP is much slower than predicted by molecular diffusion. Therefore, the reaction must be surface reaction controlled.

Table 3.3. Rate controlling dissolution mechanism and solubility for various substances (Berner, Early Diagenesis, Copyright 1980 by PUP reproduced by permission of Princeton University Press).

Substance	Solubility (mol/l)	Dissolution rate control
$Ca_5(PO_4)_3OH$	$2 \cdot 10^{-8}$	Surface-reaction
$KAlSi_3O_8$	$3 \cdot 10^{-7}$	Surface-reaction
$NaAlSi_3O_8$	$6 \cdot 10^{-7}$	Surface-reaction
$BaSO_4$	$1 \cdot 10^{-5}$	Surface-reaction
$AgCl$	$1 \cdot 10^{-5}$	Transport
$SrCO_3$	$3 \cdot 10^{-5}$	Surface-reaction
$CaCO_3$	$6 \cdot 10^{-5}$	Surface-reaction
Ag_2CrO_4	$1 \cdot 10^{-4}$	Surface-reaction
$PbSO_4$	$1 \cdot 10^{-4}$	Mixed
$Ba(IO_3)_2$	$8 \cdot 10^{-4}$	Transport
$SrSO_4$	$9 \cdot 10^{-4}$	Surface-reaction
Opaline SiO_2	$2 \cdot 10^{-3}$	Surface-reaction
$CaSO_4 \cdot 2H_2O$	$5 \cdot 10^{-3}$	Transport
$Na_2SO_4 \cdot 10H_2O$	$2 \cdot 10^{-1}$	Transport
$MgSO_4 \cdot 7H_2O$	$3 \cdot 10^0$	Transport
$Na_2CO_3 \cdot 10H_2O$	$3 \cdot 10^0$	Transport
KCl	$4 \cdot 10^0$	Transport
$NaCl$	$5 \cdot 10^0$	Transport
$MgCl_2 \cdot 6H_2O$	$5 \cdot 10^0$	Transport

The solubility of minerals appears to be related to the dissolution mechanism (Berner, 1980). The dissolution of sparingly soluble minerals is generally controlled by surface processes whereas the dissolution of soluble minerals is predominantly controlled by transport processes (Table 3.3). Silicate minerals, like feldspars, pyroxenes and amphiboles are also insoluble and their dissolution controlled by surface reactions Thus we may predict, to some extent, the dissolution mechanism of a mineral from its solubility. Caution is, however, warranted since both the presence of inhibitors, the composition of the solution and other factors may affect the rate mechanism.

In the case of surface controlled reactions, it is essential to realize that dissolution or growth only occurs at energetically favorable sites on the mineral surface. A schematic picture of a mineral surface (Figure 3.14) identifies the favorable sites as irregularities on the crystal face. Clearly, ions or atoms which are not surrounded by other crystal units on three sides have a lower binding energy and are the first to be released upon dissolution. Also crystal growth will take place preferentially at those sites where bonds in several directions can be established. Selective dissolution will yield beautiful etch patterns on mineral surfaces that can be observed by SEM on a microscopic level. In fact such etch patterns can be used as an indicator for surface reaction dissolution control (Berner, 1978).

Various models have been derived to describe crystal growth or dissolution mechanisms at reactive surface sites. These include spiral growth or dissolution, polynuclear growth/ dissolution etc. which are discussed in more detail in texts like Nielsen (1964, 1984) Zhang and Nancollas (1990). Such reaction mechanisms are usually inferred from empirical rate laws of the type:

$$J = k(\Omega - 1)^n$$

(3.44)

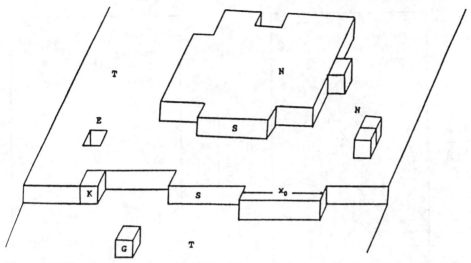

Figure 3.14. The schematic surface of a simple cubic crystal. Energetically favorable sites include kinks (K), steps (S), clusters (N) and etch pits (E) (Zhang and Nancollas, 1990. Reprinted by permission, copyright by the Mineralogical Society of America).

Where k is the rate constant, the term $(\Omega - 1)^n$ represents the $g(C)$ in Equation (3.43) and m_o and $F(m/m_o)$ are assumed constant. Ω is the saturation state, equal to $(IAP/K)^{1/\nu} \approx C/C_{eq}$. The relationship between the reaction order n and the probable mechanism of dissolution or precipitation is illustrated in Table 3.4.

As an example of how to derive a mechanism from analytical data we discuss the growth of fluorite crystals (Christoffersen et al. 1988) in some detail. Figure 3.15A displays the rate of fluorite growth as a function of the distance from equilibrium $(\Omega - 1)$. The straight line describing the data points has a slope of 3 which corresponds to exponent $n = 3$ in Equation (3.44). This suggests according to Table 3.4 a polynuclear mechanism, but this hypothesis needs some further testing. For polynuclear growth it can be derived (Nielsen, 1984; Christoffersen et al. 1988) that the concentration term $g(C)$ (Equation 3.43) is described by

Table 3.4. The relation between the reaction order n in Equation (3.44) and the probable mechanism of precipitation or dissolution (Zhang and Nancollas, 1990 reprinted by permission, copyright Mineralogical Society of America).

n	Probable mechanisms
1	Volume diffusion; Adsorption; Volume diffusion + adsorption
1–2	Combined mechanisms such as: Adsorption + surface diffusion; adsorption + integration; volume diffusion + surface diffusion; volume diffusion + integration; volume diffusion + polynucleation
2	Surface diffusion; integration; surface diffusion + integration
>2	Polynuclear growth; polynuclear + spiral growth. Polynuclear dissolution; spiral dissolution controlled by surface diffusion and/or detachment; polynuclear + spiral dissolution

Note: Except the case of $n > 2$, all the mechanisms are valid for both growth and dissolution.

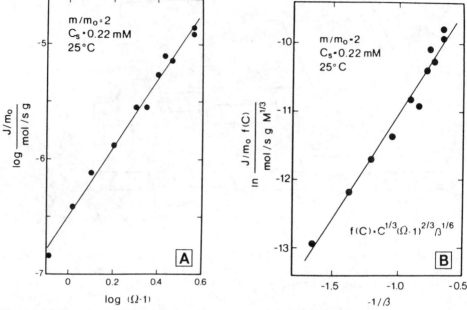

Figure 3.15. The rate of fluorite crystal growth against distance from equilibrium (A) and the same data replotted according to a polynuclear growth mechanism (B). The rates are measured in separate experiments when the original fluorite crystals have doubled their mass. (Reprinted from 'Kinetics of dissolution and growth of calcium fluoride and effects of phosphate' by Christoffersen et al. from Acta Odontol. Scand. 1988, vol. 46, by permission of Scandinavian University Press).

the equation:

$$g(C) = C^{1/3} (\Omega - 1)^{2/3} \beta^{1/6} \exp(-\alpha/\beta)$$

$$g(C) = f(C) \exp(-\alpha/\beta) \tag{3.45}$$

Where $\beta = \ln(IAP/K_{\text{fluorite}})^{1/\nu}$ and α depends on the ion size, surface energy etc. Equation (3.45) in combination with Equation (3.43) indicates that for constant values of m/m_0 a plot of $\ln(J/m_0 f(C))$ against $-1/\beta$ should yield a straight line if the rate of growth is controlled by a polynuclear mechanism. Such a plot is shown in Figure 3.15B which confirms that a polynuclear mechanism is consistent with the data.

The recognition that dissolution and growth only takes place at specific sites on the crystal surface explains why substances in very low concentrations may have a strongly inhibiting effect, since such substances may block active sites on the surface. An example is shown for fluorite dissolution in Figure 3.16 where the addition of only 0.5 μmol/l phosphate markedly inhibits the process. The ratio J/J_0 indicates the ratio of the rate after and before phosphate addition and demonstrates, for example, at pH 5.15 that the rate of fluorite dissolution is reduced by a factor of 10 due to phosphate inhibition.

Another well known example is the strongly inhibiting effect of phosphate on the dissolution of calcite. Figure 3.17 shows that for a given degree of subsaturation the presence of only a few micromolar phosphate decreases the dissolution rate of calcite very strongly. Cations may also function as inhibitor. For example Al does strongly inhibit the

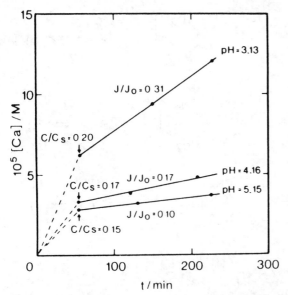

Figure 3.16. The inhibitory effect of phosphate on fluorite dissolution. 0.5 µmol/l phosphate was added at the time indicated by the arrows. J/J_o gives the rate after phosphate addition divided by the rate before addition. C/C_s indicates the measured Ca concentration over the concentration at equilibrium. (Reprinted from 'Kinetics of dissolution and growth of calcium fluoride and effects of phosphate' by Christoffersen et al. from Acta Odontol. Scand. 1988, vol. 46, by permission of Scandinavian University Press).

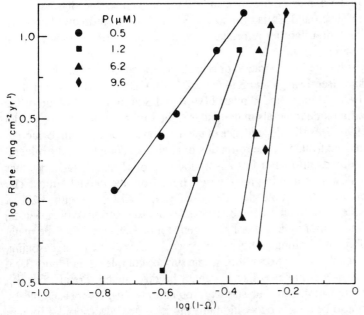

Figure 3.17. Inhibition of calcite dissolution kinetics in seawater by phosphate. $\Omega = IAP/K$ (Reprinted with permission from Morse and Berner, copyright 1979, American Chemical Society).

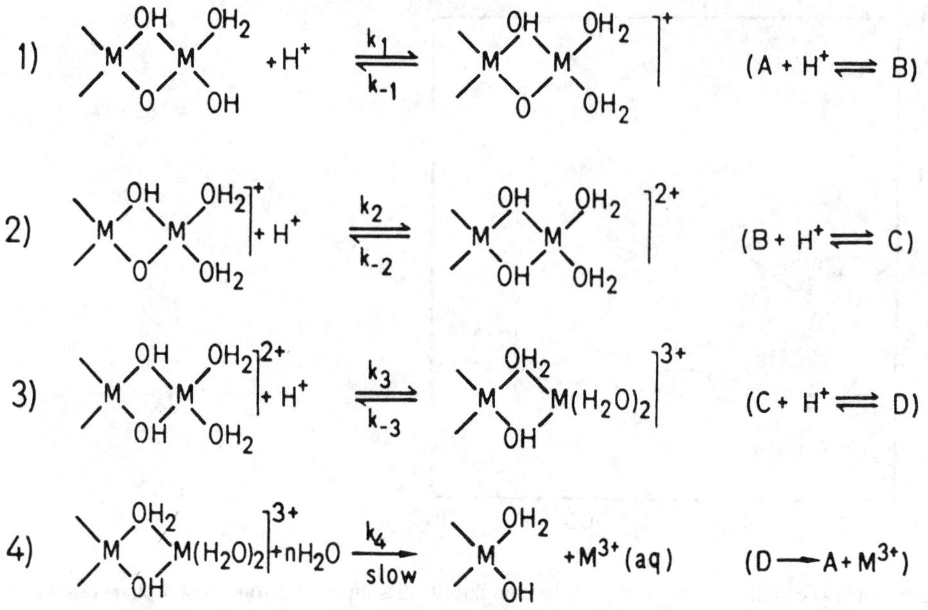

Figure 3.18. Schematic representation of the proton-promoted dissolution of a metal-oxide using the surface-complex model. The reaction mechanism is given for an oxide of the type M_2O_3 (for example Fe_2O_3 or Al_2O_3). The protonation steps 1), 2) and 3) are fast, while the detachment reaction 4) is slow and rate determining (Furrer and Stumm, 1986. Reprinted with permission from Pergamon Press PLC).

reductive dissolution of Fe-oxides (Banwart et al. 1989) as well as the dissolution of hydroxyapatite (Christoffersen and Christoffersen, 1985) and there are many other well documented cases of inhibition. Further information on mechanisms of inhibition can be found in Zhang and Nancollas (1990).

Different models exist for the incorporation or release processes for ions at the reactive sites of the crystal surface (see Lasaga and Kirkpatrick (1981), Davis and Hayes (1986), Hochella and White (1990), Stumm and Wollast (1990) and Stumm (1990)). One of the more succesful models for oxide minerals in particular is the surface complexation model (Stumm and Wollast, 1990). Here the dependence of the reaction rate on the composition of the solution is considered controlled by the formation of a reactive complex at the mineral surface. The process proceeds through a fast adsorption step of H_2O, H^+, OH^- or other ligands, followed by a slow, rate limiting release of cations from the crystal lattice. The process is illustrated for dissolution of metal oxides, like Fe_2O_3 or Al_2O_3 in Figure 3.18.

The surface of the oxide is protonated stepwise in a series of fast reactions which weaken the bonds of the metal atom (M) to neighboring oxygen atoms. Detachment of the metal atom is slow and therefore rate limiting in the overall process. According to this reasoning, the rate must be proportional to the concentration of the surface complex D in Figure 3.18. The concentration of surface complexes like D cannot be determined directly, but can indirectly be obtained from the degree of surface protonation which is measurable by titration. Inhibition can also be understood within this model, since adsorption of foreign ions, like phosphate, may block the protonation process.

PROBLEMS

3.1. The following table lists the composition of some fluoride rich waters from A. Maarum, Denmark, B. Rajasthan, India, C. Lake Abiata Ethiopia. Concentrations in mmol/l.

Sample	A	B	C
pH	7.8	7.3	9.62
NO_3^-	0.032	7.80	–
HCO_3^-	17.8	14.8	138
SO_4^{2-}	0.00	5.20	0.15
Cl^-	5.67	17.4	53.9
F^-	0.089	0.356	6.28
Ca^{2+}	1.05	0.675	0.042
Mg^{2+}	1.19	0.785	0.023
Na^+	19.1	47.9	194
K^+	0.358	0.150	4.91

    a. Calculate the saturation state of these waters for fluorite without making any corrections for ionic strength and complexes.
    b. Do the same but correct for the ionic strength effect.

3.2. The mineral villiaumite (NaF) has also been proposed as a possible solubility control on dissolved fluoride. While the solubility of fluorite can be found in any geochemistry book, this is not the case for villiaumite. Fortunately, the NBS (1982) tables list the following thermodynamic data for villiaumite and associated species.

	ΔG_f^0	ΔH_f^0
villiaumite	–543.5 kJ/mol	–573.6 kJ/mol
Na^+	–261.9 kJ/mol	–240.1 kJ/mol
F^-	–278.8 kJ/mol	–332.6 kJ/mol

    a. Calculate $K_{villiaumite}$ at 25°C.
    b. Calculate $K_{villiaumite}$ at 10°C.
    c. Check the saturation state of the water samples in problem 3.1 for villiaumite and consider whether this mineral could limit dissolved fluoride concentrations.

3.3. One could consider whether it is possible to decrease the fluoride content of the Rajasthan water in problem 3.1 by addition of gypsum.
    a. Calculate the saturation state of the Rajasthan water for gypsum. Would it increase or decrease when complex $CaSO_4^0$ is included. (The complete calculation considering both ionic strength and complexes is too cumbersome to carry out by hand calculations).
    b. To what level could the fluoride concentration be reduced by gypsum addition if you neglect the effects of ionic strength and complexes? Would these additional corrections increase or decrease the molar fluoride concentration?

3.4. It has been argued in the literature that ion exchange of Ca^{2+} replacing adsorbed Na^+ may increase fluoride concentrations in groundwater. Inspect the water analyses of problem 3.1 to find evidence that supports such a mechanism. Which additional condition is required?

3.5. Calculate the solubility (mg/l) of atmospheric O_2 in water at: a. 5°C, b. 15°C and c. 25°C.

	$O_{2(g)}$	$O_{2(aq)}$	
ΔG_f^0	0	16.5	kJ/mol
ΔH_f^0	0	−10	kJ/mol

3.6. Redraw Figure 3.13 for mineral precipitation.

3.7. Pyrite forms in sediments mostly by a two step process: First metastable FeS is formed, which is then transformed to FeS_2 by the overall reaction:

$$FeS + S° \rightarrow FeS_2$$

Two reaction mechanisms have been proposed: 1) A solid state reaction between FeS and S°. 2) Dissolution of both FeS and S° and subsequent reaction between Fe^{2+} and polysulfides. (Polysufides are $S_n - S^{2-}$ compounds which form by the reaction between H_2S and S°). Experimental data of Rickard (1975) are listed below. Rates are given as $d(FeS_2)/dt$.

1) P_{H_2S} variable (pH = 7, FeS-surf. area = $1.6 \cdot 10^5$ cm², S°-surf. area = $1.4 \cdot 10^3$ cm²)

P_{H_2S} (atm)	Rate (mol/l.s)
1.00	$8.3 \cdot 10^{-7}$
0.50	$4.8 \cdot 10^{-7}$
0.25	$2.3 \cdot 10^{-7}$
0.10	$9.5 \cdot 10^{-8}$
0.05	$4.9 \cdot 10^{-8}$

2) S° surf. area variable (pH = 7, P_{H_2S} = 1 atm, FeS-surf. area = $1.6 \cdot 10^5$ cm²).

S°-surf. area (cm²)	Rate (mol/l.s)
$1.4 \cdot 10^3$	$8.3 \cdot 10^{-7}$
$7.0 \cdot 10^2$	$4.8 \cdot 10^{-7}$
$3.5 \cdot 10^2$	$1.9 \cdot 10^{-7}$
$1.4 \cdot 10^2$	$5.4 \cdot 10^{-8}$

3) FeS-surf. area variable (pH = 7, P_{H_2S} = 1 atm S°-surf. area = $1.4 \cdot 10^3$ cm²).

FeS-surf. area (cm²)	Rate (mol/l.s)
$2.5 \cdot 10^5$	$2.0 \cdot 10^{-6}$
$1.6 \cdot 10^5$	$8.3 \cdot 10^{-7}$
$8.2 \cdot 10^4$	$1.5 \cdot 10^{-7}$
$4.1 \cdot 10^4$	$4.8 \cdot 10^{-8}$

4) pH variable (P_{H_2S} = 1 atm, S°-surf. area = $1.4 \cdot 10^3$ cm², FeS-surf. area = $1.6 \cdot 10^5$ cm²).

pH	Rate (mol/l.s)
7.0	$8.3 \cdot 10^{-7}$
7.5	$7.0 \cdot 10^{-7}$
8.0	$6.4 \cdot 10^{-7}$

  a. Determine the reaction order for P_{H_2S}, pH, FeS-surface area and S°-surface area.
  b. Construct a rate equation for the reaction.
  c. Which of the two mechanisms is supported by the experimental data?

REFERENCES

Adamson, A.W., 1982, *Physical Chemistry of Surfaces*. 4th ed. Wiley & Sons, New York, 664 pp.

Banwart, S., Davies, S. and Stumm, W., 1989, The role of oxalate in accelerating the reductive dissolution of hematite ($\alpha - Fe_2O_3$) by ascorbate. *Colloids Surf.* 39, 303-309.

Berner, R.A., 1971, *Principles of Chemical Sedimentology*. McGraw-Hill, New York, 240 pp.

Berner, R.A., 1978, Rate control of mineral dissolution under earth surface conditions. *Am. J. Sci.* 278, 1235-1252.

Berner, R.A., 1980, *Early Diagenesis – A theoretical Approach*. Princeton University Press, Princeton, N.J., 241 pp.

Berner, R.A., 1981, Authigenic mineral formation resulting from organic matter decomposition in modern sediments. *Fortschr. Miner.* 59, 1, 117-135.

Capobianco, C. and Navrotsky, A., 1987, Solid-solution thermodynamics in $CaCO_3 - MnCO_3$. *Am. Mineral.* 72, 312-318.

Christoffersen, J. and Christoffersen, M.R., 1979, Kinetics of dissolution of calcium hydroxy apatite. *J. Crystal Growth* 47, 671-679.

Christoffersen, M.R. and Christoffersen, J., 1985, The effect of aluminium on the rate of dissolution of calciumhydroxyapatite – A contribution to the understanding of aluminium-induced bone diseases. *Calcif. Tissue Int.* 37, 673-676.

Christoffersen, J., Christoffersen, M.R., Kibalczyc, W. and Perdok, W.G., 1988, Kinetics of dissolution and growth of calcium fluoride and effects of phosphate. *Acta Odontol. Scand.* 46, 325-336.

Cox, J.D., Wagman, D.D. and Medvedev, V.A., 1989, *CODATA Key values for thermodynamics*. Hemisphere Publ. Corp., Washington, D.C.

Davis, J.A. and Hayes, K.F. (eds.), 1986, *Geochemical Processes at Mineral Surfaces*. ACS Symp. Ser. 323, Am. Chem. Soc., Washington, 683 pp.

Davis, J.A., Fuller, C.C. and Cook, A.D., 1987, A model for trace metal sorption processes at the calcite surface: Adsorption of Cd^{++} and subsequent solid solution formation. *Geochim. Cosmochim. Acta* 51, 1477-1490.

Denbigh, K., 1971, *The Principles of Chemical Equilibrium*. 3rd ed. Cambridge Univ. Press, London, 496 pp.

Fitzpatrick, R.W. and Schwertmann, U., 1982, Al-substituted goethite – an indicator of pedogenic and other weathering environments in South Africa. *Geoderma* 27, 335-347.

Fuller, C.C. and Davis, J.A., 1987, Processes and kinetics of Cd^{++} sorption by a calcareous aquifer sand. *Geochim. Cosmochim. Acta* 51, 1491-1502.

Furrer, G. and Stumm, W., 1986, The coordination chemistry of weathering, I. Dissolution kinetics of δ-Al_2O_3 and BeO. *Geochim. Cosmochim. Acta* 50, 1847-1860.

Garrels, R.M. and Christ, C.L., 1965, *Solutions, Minerals, and Equilibria*. Harper and Row, New York, 450 pp.

Glynn, P.D., 1990, Modeling solid-solution reactions in low-temperature aqueous systems. In D.C. Melchior and R.L. Basset (eds.), *Chemical modeling of aqueous systems II*. ACS Symp. Ser. 416, 74-86.

Glynn, P.D. and Reardon, E.J., 1990, Solid-solution aqueous-solution equilibria: thermodynamic theory and representation. *Am. J. Sci.* 290, 164-201.

Handa, B.K., 1975, Geochemistry and genesis of fluoride-containing ground waters in India. *Ground Water* 13, 275-281.

Harvie, C.E., Eugster, H.P. and Weare, J.H., 1982, Mineral equilibria in the six-component seawater system, Na-K-Mg-Ca-SO_4-Cl-H_2O at 25°C. II: Compositions of the saturated solutions. *Geochim. Cosmochim. Acta* 46, 1603-1618.

Helgeson, H.C., Delany, J.M., Nesbitt, W.H. and Bird, D.K., 1978, Summary and critique of the thermodynamic properties of rock forming minerals. *Am. J. Sci.* 278-A, 229 pp.

Hochella, M.F. and White, A.F. (eds.), 1990, *Mineral-water interface geochemistry*. Reviews in Mineralogy 23, Mineral. Soc. Am., 603 pp.

Jakobsen, R. and Postma, D., 1989, Formation and solid solution behavior of Ca-rhodochrosites in marine muds of the Baltic deeps. *Geochim. Cosmochim. Acta* 53, 2639-2648.

Laidler, K.J., 1965, *Chemical kinetics*. 2nd ed. McGraw-Hill, New York, 566 pp.

Lasaga, A.C., 1981, Rate laws of chemical reactions. In A.C. Lasaga and R.J. Kirkpatrick (eds.), *Kinetics of Geochemical Processes*. Reviews in Mineralogy 8, Mineral. Soc. Am., pp. 1-68.

Lewis, G.N. and Randall, M., 1961, *Thermodynamics*. 2nd ed. McGraw- Hill, New York, 723 pp.

Liu, S-T. and Nancollas, G.H., 1971, The kinetics of dissolution of calcium sulfate dihydrate. *J. Inorg. Nucl. Chem.* 33, 2311-2316.

Lorens, R.B., 1981, Sr, Cd, Mn and Co distribution coefficients in calcite as a function of calcite precipitation rate. *Geochim. Cosmochim. Acta* 45, 553-561.

Monnin, C. and Schott, J., 1984, Determination of the solubility products of sodium carbonate minerals and an application to trona deposition in Lake Magadi (Kenya). *Geochim. Cosmochim. Acta* 48, 571-581.

Morse, J.W. and Berner, R.A., 1979, Chemistry of calcium carbonate in the deep oceans. In E.A. Jenne (ed.), *Chemical Modelling in Aqueous Systems*. ACS Symp. Ser. 43, 499-535.

Mucci A., 1988, Manganese uptake during calcite precipitation from seawater: Conditions leading to the formation of a pseudokutnahorite. *Geochim. Cosmochim. Acta* 52, 1859-1868.

Nielsen, A.E., 1964, *Kinetics of precipitation*. Pergamon Press, Oxford, 151 pp.

Nielsen, A.E., 1984, Electrolyte crystal growth mechanisms. *J. Crystal Growth*, 67, 289-310.

Nielsen, A.E. and Toft, J.M., 1984, Electrolyte crystal growth kinetics. *J. Crystal Growth* 67, 278-288.

Nordstrom, D.K. and Jenne, E.A., 1977, Fluorite solubility equilibria in selected geothermal waters. *Geochim. Cosmochim. Acta* 41, 175-188.

Nordstrom, D.K. and Munoz, J.L., 1986, *Geochemical thermodynamics*. Blackwell, Oxford, 477 pp.

Nordstrom, D.K., Plummer, L.N., Wigley, T.M.L., Wolery, T.J., Ball, J.W., Jenne, E.A., Bassett, R.L., Crerar, D.A., Florence, T.M., Fritz, B., Hoffman, M., Holdren Jr., G.R., Lafon, G.M., Mattigod, S.V., McDuff, R.E., Morel, F., Reddy, M.M., Sposito, G. and Thrailkill, J., 1979, A comparison of computerized chemical models for equilibrium calculations in aqueous systems. In E.A. Jenne (ed.), ACS Symp. Ser. 93, 857-892.

Nordstrom, D.K., Plummer, L.N., Langmuir, D., Busenberg, E., May, H.M., Jones, B.F. and Parkhurst, D.L., 1990, Revised chemical equilibrium data for major water-mineral reactions and their limitation. In D.C. Melchior and R.L. Basset (eds.), *Chemical modeling of aqueous systems II*. ACS Symp. Ser. 416, 398-413.

Pitzer, K.S., 1981, Characteristics of very concentrated aqueous solutions. In F.E. Wickman and D.T. Rickard (eds.), *Chemistry and geochemistry of solutions at high temperatures and pressures*. Physics and Chemistry of the Earth, 3 and 4, 249-272.

Plummer, L.N. and Busenberg, E., 1982, The solubilities of calcite, aragonite and vaterite in $CO_2 - H_2O$ solutions between 0 and 90°C, and an evaluation of the aqueous model for the system $CaCO_3 - CO_2 - H_2O$. *Geochim. Cosmochim. Acta* 46, 1011-1040.

Plummer, L.N. and Busenberg, E., 1987, Thermodynamics of aragonite- strontianite solid solutions: Results from stoichiometric solubility at 25 and 76°C. *Geochim. Cosmochim. Acta* 51, 1393-1411.

Plummer, L.N., Parkhurst, D.L., Fleming, G.W. and Dunkle, S.A., 1988, A computer program incorporating Pitzer's equations for calculation of geochemical reactions in brines. U.S. Geol. Surv. Water Resour. Inv. Rep. 88-4153.

Postma, D., 1983, Pyrite and siderite oxidation in swamp sediment. *J. Soil Sci.* 34, 163-182.

Rickard, D.T., Kinetics and mechanism of pyrite formation at low temperatures. *Am. J. Sci.* 275, 636-652.

Robie, R.A., Hemingway, B.S. and Fisher, J.R., 1978, *Thermodynamic Properties of Minerals and Related Substances at 298.15 K and 1 Bar (105 Pascals) Pressure and at Higher Temperatures*. Geological Survey Bulletin 1452, US Gov. Printing Off., Washington, 456 pp.

Sparks, D.L., 1989, *Kinetics of Soil Chemical Processes*. Academic Press, Inc., San Diego, California, 210 pp.

Stumm, W. (ed.), 1990, *Aquatic Chemical Kinetics*. Wiley & Sons, New York, 545 pp.

Stumm, W. and Morgan, J.J., 1981, *Aquatic Chemistry*. 2nd ed. Wiley & Sons, New York, 780 pp.

Stumm, W. and Wollast, R., 1990, Coordination chemistry of weathering: Kinetics of the surface controlled dissolution of oxide minerals. *Rev. Geophysics* 28, 53-69.

Sverdrup, H. and Warfvinge, P., 1988, Weathering of primary silicate minerals in the natural soil environment in relation to a chemical weathering model. *Water Air Soil Pollution* 38, 387-408.

Thorstenson, D.C. and Plummer, L.N., 1977, Equilibrium criteria for two-component solids reacting with fixed composition in an aqueous phase-example: the magnesian calcites. *Am. J. Sci.* 277, 1203-1223.

Truesdell, A.H. and Jones, B.F., 1973, *Wateq, a computer program for calculating chemical equilibria of natural waters.* US Geol. Surv., National Technical Information Service, PB-220 464, 73 pp.

Wagman, D.D., Evans, W.H., Parker, V.B., Schumm, R.H., Halow, I., Bailey, S.M., Churney, K.L. and Nuttall, R.L., 1982, The NBS tables of chemical thermodynamic properties: Selected values for inorganic and C_1 and C_2 organic substances in SI units. *J. Phys. Chem. Ref. Data Suppl.* 392 pp.

Woods, T.L. and Garrels, R.M., 1987, *Thermodynamic values at low temperature for natural inorganic materials.* Oxford Univ. Press, New York, 242 pp.

Zack, A.L., 1980, Geochemistry of fluoride in the Black Creek Aquifer system of Horry and Georgetown counties, South Carolina and its physiological implications. U.S. Geol. Surv. Water-Supply paper 2067, 40 pp.

Zhang, J.W. and Nancollas, G.H., 1990, Mechanisms of growth and dissolution of sparingly soluble salts. In M.F. Hochella and A.F. White (eds.), *Mineral-water interface geochemistry*. Reviews in Mineralogy 23, Mineral. Soc. Amer. 365-396.

CHAPTER 4

Carbonates and carbon dioxide

Carbonate reactions are very important in controlling the composition of groundwater. Rocks made of carbonates, such as limestones and dolomites, are often aquifers that have a high productivity and offer favorable conditions for groundwater extraction. The main minerals in these rocks are Ca- and Mg-carbonates which react easily with groundwater and give water its 'hard' character. On a world-wide basis, the effect of carbonate dissolution on water compositions is quite conspicuous. Figure 4.1 compares, for average river water compositions from different continents, the relation between total dissolved solids (*TDS*) and the concentrations of individual ions.

It is clear that high *TDS* values are mainly related to increases in the content of Ca^{2+} and HCO_3^-, which again for the major part are due to carbonate dissolution. Note also that the abundance of carbonate rocks in Europe is clearly reflected in the river water composition. Prolonged dissolution of carbonates in aquifers over periods of thousands of years may result in dramatic effects such as the development of caves and karst landscapes. In that case carbonate dissolution significantly alters the hydraulics of the system due to the development of subsurface channels. In general, carbonate rocks also show a special type of hydrogeology. In most cases, they consist of extensively recrystallized biological material with a high porosity, but an often low permeability. Flow through fracture zones and karst channels is therefore an important aspect.

Apart from carbonate rocks which consist exclusively of carbonate minerals, sandy or sandstone aquifers may contain abundant carbonate minerals as accessory minerals or cement around the more inert grains. Due to their high reactivity, carbonate minerals still have a dominant effect on the water chemistry over much slower processes such as silicate dissolution. The chemistry of carbonates is coupled to the biological cycle, first of all to the consumption or production of CO_2 by biological processes, but also because most carbonate rocks consist of originally accumulated skeletal remains of organisms.

This chapter presents aspects of the mineralogy and aqueous chemistry which must be understood in order to appreciate the carbonates as important regulators of natural water quality. Recent results of laboratory investigations and theoretical studies in the field of kinetics and solid solutions have been included as well, and these exemplify the great efforts devoted to understanding more fully the chemistry of carbonate reactions in natural environments.

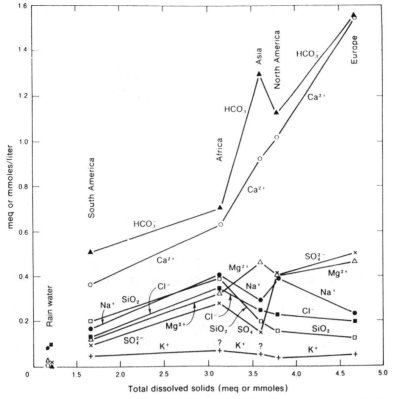

Figure 4.1. The relation between total dissolved solids and concentrations of individual ions in average river water compositions from different continents (adapted from Garrels and Mackenzie, 1971).

4.1 CARBONATE MINERALS

Groundwaters in carbonate aquifers obtain their chemical characteristics from interactions with the carbonate minerals in the aquifer. The presence of different carbonate minerals will yield groundwaters with characteristic compositions and some basic information concerning carbonate mineralogy is therefore appropriate. Also, some simple considerations about the stability of different carbonate minerals allow predictions of the type of reaction to be expected within aquifers. Some carbonate minerals and their solubility products are listed in Table 4.1.

Rock-forming carbonate minerals are calcite, dolomite and aragonite. However, the presence of other carbonates in small amounts, like siderite or rhodochrosite, may also exercise important controls on dissolved concentrations of, for example, Fe^{2+} or Mn^{2+} in aquifers (see Chapter 7). Carbonate minerals crystallize with a trigonal or an orthorhombic structure. The chemical compound $CaCO_3$ can be found in two different minerals, calcite or aragonite, with a different crystal structure, and a different solubility. The most stable form can be found by comparing the solubility products. The reaction:

$$CaCO_3 \leftrightarrow Ca^{2+} + CO_3^{2-}$$

Table 4.1. Mineralogy and solubility of some carbonates (thermodynamic data from Nordstrom et al., 1990).

Trigonal	Formula	$-\log K$	Orthorhombic	Formula	$-\log K$
Calcite	$CaCO_3$	8.48	Aragonite	$CaCO_3$	8.34
Magnesite	$MgCO_3$	8.24	Strontianite	$SrCO_3$	9.27
Siderite	$FeCO_3$	10.89	Cerussite	$PbCO_3$	13.1
Rhodochrosite	$MnCO_3$	11.13			
Dolomite	$CaMg(CO_3)_2$	17.09			

has

$$K_{calcite} = [Ca^{2+}]\,[CO_3^{2-}] = 3.3 \cdot 10^{-9} \quad \text{at } 25°C$$

or

$$K_{aragonite} = [Ca^{2+}]\,[CO_3^{2-}] = 4.6 \cdot 10^{-9} \quad \text{at } 25°C$$

The solubility of aragonite is larger than of calcite. In other words, water in equilibrium with aragonite is supersaturated for calcite so that aragonite should dissolve and calcite precipitate: calcite is the more stable mineral.

Sometimes solid solutions form of two or more end-member minerals. Solid solution can take place among minerals that belong to the same crystal group, within limits determined by the degree of deformation allowed in the crystal lattice. For example, Mg atoms may

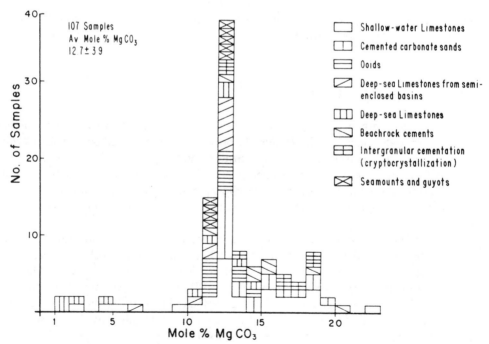

Figure 4.2. The composition of Mg-calcite cements precipitated from seawater (Mucci 1987. Reprinted with permission from Pergamon Press PLC).

substitute for Ca in calcite, and Sr for Ca in aragonite, but Sr will not replace Ca in calcite because the crystal radii are too different. Solid solutions between calcite and magnesite are called Mg-calcites. They are commonly subdivided into low Mg-calcites with < 5 mole% and high Mg-calcites with 5-30 mole% Mg. The composition of Mg-calcites in recent marine sediments is displayed in Figure 4.2 and shows that most naturally occurring Mg-calcites, precipitated from seawater, fall within the range of 10-15 mole% $MgCO_3$. The stability of Mg-calcites will be discussed in more detail in Section 4.5.

Another important rock forming carbonate mineral is dolomite, with its structure displayed in Figure 4.3. Dolomite shows a high degree of ordering with Ca and Mg atoms separated in different layers within the crystal structure. This is different from high Mg-calcites in which Ca and Mg atoms are distributed more or less at random in the Ca-layer (Figure 4.3). A 50 mole% Mg-calcite is therefore not a dolomite. Although the difference may appear rather academic, it has important implications since chemically a dolomite should be treated as a pure phase of more or less constant composition while Mg-calcites are solid solutions. The highly ordered structure of dolomite complicates its precipitation from aqueous solutions at low temperatures and the formation of dolomite is still a matter of lively debate (Hardie, 1987).

Originally, most carbonate rocks were formed in the marine environment. Recent marine carbonates consist mainly of aragonite and high Mg-calcite with only subordinate amounts of low Mg-calcite. This mineralogy is still preserved in young fresh water aquifers which have been raised only recently (10,000-20,000 yrs) above sea level. Typical examples are the Bermuda aquifer and the Yucatan aquifer in Mexico. Circulation of fresh groundwater provokes a spectacular recrystallization of the whole rock until only stable minerals like low Mg-calcite and dolomite are left. Pre-Pleistocene (>10^6 yr) sediments are indeed

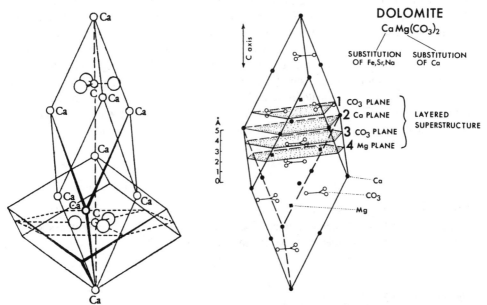

Figure 4.3. The structures of calcite and dolomite. Note the separation of Ca and Mg atoms in different layers in the dolomite (from Morrow, 1982; Deer, Howie and Zussman, 1966).

virtually free of aragonite, which shows that the thermodynamically more stable mineral dominates ultimately, given enough time. The recrystallization process not only completely changes the mineralogy but also affects permeability and porosity of the rock (Hanshaw and Back, 1979).

4.2 CARBONATE EQUILIBRIA

If we place small pieces of calcite in a beaker with distilled water, and return, for kinetic reasons, not the following day but the following week to sample and analyze the solution, we would find from the dissolution reaction:

$$CaCO_3 \rightarrow Ca^{2+} + CO_3^{2-}$$

that:

$$[Ca^{2+}] = [CO_3^{2-}],$$

and therefore

$$K_{calcite} = [Ca^{2+}]^2 = 10^{-8.3}$$

or

$$[Ca^{2+}] = 10^{-4.15} = 0.06 \text{ mmol/l}.$$

Note that $K_{calcite}$ used here is slightly higher than listed in Table 4.1, as this compensates to some extent for aqueous complexes.

Field data from carbonate aquifers show, however, that Ca^{2+} concentrations easily can be as high as 1-5 mmol/l, which is almost a hundred times higher than predicted by the above dissolution reaction. The higher Ca^{2+} concentrations observed in the field are a result of reaction with carbonic acid (H_2CO_3) derived from respiration of organic matter. The acid provides protons (H^+) which associate with the carbonate-ion (CO_3^{2-}) from calcite to form bicarbonate (HCO_3^-). This is similar to complexation of Ca^{2+} and SO_4^{2-} ions leading to an increase in the solubility of gypsum (Example 3.2 in Chapter 3). We must therefore study the forms of dissolved carbonate in water, before returning to the solubility of calcite under natural conditions. The concentrations or relative proportions of the dissolved carbonate species are obviously influenced by the pH of water, and we will consider in the following sections how pH is used as a *master variable*.

4.2.1 *CO₂ species in water*

Let us begin with CO_2 gas which dissolves in water, for example atmospheric CO_2 from the air in rainwater or soil water. Gaseous $CO_{2(g)}$ forms aqueous $CO_{2(aq)}$, which associates to some extent with water molecules to form carbonic acid, H_2CO_3:

$$CO_{2(g)} \rightarrow CO_{2(aq)}$$

and subsequently:

$$CO_{2(aq)} + H_2O \rightarrow H_2CO_3$$

Table 4.2. Approximate equilibrium constants for dissolved CO_2 species at 25°C. (Precise values and temperature dependance are provided in Appendix B).

$H_2O \leftrightarrow H^+ + OH^-$	$K_w = [H^+][OH^-]$	$K_w = 10^{-14.0}$	(4.1)
$CO_{2(g)} + H_2O \leftrightarrow H_2CO_3^*$	$K_H = [H_2CO_3^*]/P_{CO_2}$	$K_H = 10^{-1.5}$	(4.2)
$H_2CO_3^* \leftrightarrow H^+ + HCO_3^-$	$K_1 = [H^+][HCO_3^-]/[H_2CO_3^*]$	$K_1 = 10^{-6.3}$	(4.3)
$HCO_3^- \leftrightarrow H^+ + CO_3^{2-}$	$K_2 = [H^+][CO_3^{2-}]/[HCO_3^-]$	$K_2 = 10^{-10.3}$	(4.4)

Although $CO_{2(aq)}$ at 25°C is about 250 times more abundant than H_2CO_3 it is for more convenient calculations included in H_2CO_3. The overall reaction becomes then:

$$CO_{2(g)} + H_2O \rightarrow H_2CO_3^*$$

where $H_2CO_3^* = CO_{2(aq)} + H_2CO_3$.

Carbonic acid can dissociate in two steps, releasing one proton in each step. The relevant reactions and mass action constants that are needed to calculate the composition of the solution are given in Table 4.2. (Note that approximate values are tabulated, adequate for the manual calculations in this chapter). All that is necessary now is some manipulation of the reactions and some arithmetic. Recall that the addition of two reactions requires multiplication of the two mass action constants, and subtraction requires similarly that the two be divided. For example, to obtain the reaction whereby carbonate ion is produced directly from undissociated carbonic acid, we must add two reactions in Table 4.2

$$H_2CO_3^* \rightarrow H^+ + HCO_3^- \qquad K_1 = 10^{-6.3}$$
$$HCO_3^- \rightarrow H^+ + CO_3^{2-} \qquad K_2 = 10^{-10.3}$$

+ _____ Π _____

$$H_2CO_3^* \rightarrow 2H^+ + CO_3^{2-} \qquad K = 10^{-16.6} \qquad (4.5)$$

When the gas pressure of CO_2 is known, the activity of carbonic acid can be calculated (which is independent of pH), and subsequently the activity of the other species (which are a function of pH). A log-transformation linearizes the mass action formulae and makes calculations easier. For example, the activity of HCO_3^- as a function of carbonic acid and pH is:

$$\log [HCO_3^-] = -6.3 + \log [H_2CO_3^*] + pH \qquad (4.6)$$

which shows that, at a constant activity of carbonic acid, the logarithm of the activity of HCO_3^- increases *once* with a pH-unit increase. Similarly for the activity of CO_3^{2-}, using Equation (4.5):

$$\log [CO_3^{2-}] = -16.6 + \log [H_2CO_3^*] + 2pH \qquad (4.7)$$

which indicates that the log-activity of the carbonate ion increases *twice* for each unit of pH increase. Figure 4.4 shows a pH-log[activity] diagram in which the activities of the CO_2 species are given for a constant CO_2 gas pressure of 0.01 atm.

The total amount of CO_2 (*TIC*, or $\Sigma CO_2 = m_{H_2CO_3^*} + m_{HCO_3^-} + m_{CO_3^{2-}}$) which can dissolve in water can be calculated if we assume that activity coefficients are equal to one; it is then simply the numerical addition of the activities of the different species, as shown in Example 4.1.

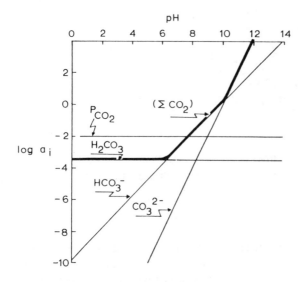

Figure 4.4. The activity of CO_2 species in water as function of pH, at a constant CO_2 pressure of 0.01 atm.

EXAMPLE 4.1. Calculate *TIC* at pH = 7 and pH = 10, when the CO_2 pressure is 0.01 atm.

ANSWER.
We calculate the concentrations of individual species, and add them up:

	pH = 7	pH = 10
$[H_2CO_3^*] = 10^{-1.5} P_{CO_2}$	$10^{-3.5}$	$10^{-3.5}$
$[HCO_3^-] = 10^{-6.3} [H_2CO_3^*]/[H^+]$	$10^{-2.8}$	$10^{0.2}$
$[CO_3^{2-}] = 10^{-10.3} [HCO_3^-]/[H^+]$	$10^{-6.1}$	$10^{-0.1}$
$\Sigma CO_2 =$	$10^{-2.72}$	$10^{0.38}$ mol/l

Activities can be easily calculated with mass action formulae, and used to predict concentrations under certain conditions of equilibrium. However, the recalculation of activity into concentration requires rather lengthy arithmetic of activity coefficients and complexes in an iterative procedure, and nowadays a number of computer programs are available which can do this job. (Note that the recalculation of activity from analyzed concentration as is done by WATEQP (Chapter 3), is the reverse of the problem stated here; we are now *predicting* concentrations for known conditions, rather than *calculating* the conditions for a given water composition). We will discuss the use and applicability of one of these more elaborate programs in Chapter 10, but insight and understanding can already be gained by assuming that activity coefficients and complexes are correction terms, which do not invalidate the general trends of our calculations when they are neglected. These correction terms lead to an activity which may be 40% of analyzed quantities in fresh water with variable composition. Accordingly, we may be wrong by a factor of about two, if we assume that activity is equal to molal concentration.

If, for example, *TIC* remains constant (instead of the activity of carbonic acid), we can also calculate the distribution of the CO_2 species as a function of pH. The line which

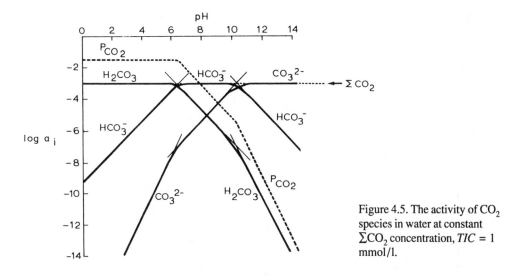

Figure 4.5. The activity of CO_2 species in water at constant ΣCO_2 concentration, $TIC = 1$ mmol/l.

indicates TIC (or $\Sigma\,CO_2$) in Figure 4.4 should now be parallel to the pH-axis (Figure 4.5). The CO_2-species are still in identical proportions with respect to each other, since the mass action relationship among the species remains valid. At different pH's, different species are dominant, as dictated by the dissociation constants. Carbonic acid is dominant at pH < 6.3, CO_3^{2-} becomes dominant at pH > 10.3, and at intermediate pH-values HCO_3^- is the major CO_2 species in water. Above pH = 6.3, the activity of carbonic acid decreases one log-unit with each unit of pH increase; above pH = 10.3, it decreases two log-units for each unit of pH increase (log $TIC \approx$ log $[CO_3^{2-}]$ in Equation (4.7). At pH's in the near vicinity of the equivalence pH (where two species have equal activity) a light rounding of the curves indicates that two species have a significant contribution to TIC.

The pH-log[activity] diagrams offer an easy way to depict effects of pH on relative concentrations of different ions. A more quantitative approach to finding the activity of a single species which forms a part of given total concentration, makes use of a *fractionation factor* α. For example for HCO_3^-:

$$m_{HCO_3^-} = \alpha \cdot \Sigma CO_2$$

The fractionation factor thus calculates the fraction of total CO_2 made up by HCO_3^-. It can be expressed as a function of $[H^+]$ only.

The fractionation-factor $\alpha = HCO_3^- / \Sigma CO_2$, or:

$$\alpha^{-1} = \frac{\Sigma\,CO_2}{m_{HCO_3^-}} = \frac{m_{H_2CO_3^*}}{m_{HCO_3^-}} + \frac{m_{HCO_3^-}}{m_{HCO_3^-}} + \frac{m_{CO_3^{2-}}}{m_{HCO_3^-}}$$

$$= \frac{[H_2CO_3^*]\gamma_1}{[HCO_3^-]\gamma_0} + \frac{[HCO_3^-]}{[HCO_3^-]} + \frac{[CO_3^{2-}]\gamma_1}{[HCO_3^-]\gamma_2}$$

$$= \frac{[H^+]\gamma_1}{K_1\gamma_0} + 1 + \frac{K_2\gamma_1}{[H^+]\gamma_2}$$

where γ_1 is the the activity coefficient for the single-charged ion, γ_2 for the double-charged ion, and γ_0 for the neutral molecule.

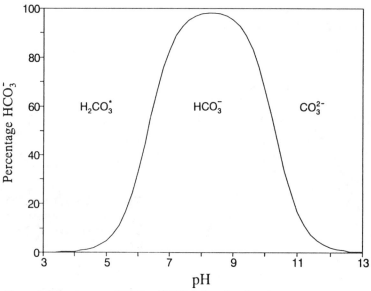

Figure 4.6. Percentage of HCO_3^- of ΣCO_2, plotted as function of pH with a fractionation factor.

With this equation we can easily depict the percentage HCO_3^- of total CO_2 as function of pH, as shown in Figure 4.6. Fractionation factors are often used for trace elements, to show how complexes of the trace elements behave as a function of pH (Stumm and Morgan, 1981). The strict application, in which activity corrections are included, requires that the changing proportions of the complexes do not influence the activity coefficients. The variable must therefore be at trace concentration in solution.

Alkalinity analysis
The alkalinity of a water sample is equal to the number of equivalents of dissociated weak acids. It is often determined by titration with a HCl or H_2SO_4 solution of known normality towards an endpoint pH of ca. 4.5, as indicated by a color reaction. Although phosphoric acid and other minor weak acids may contribute to some extent, in practice only the dissolved carbonic acid is of quantitative importance for the measured alkalinity. It is thus equal to:

$$Alk = m_{HCO_3^-} + 2m_{CO_3^{2-}}$$

At pH values less than 8.3, less than 1% of the carbonic acid is present as CO_3^{2-}, so that only the first term is of importance. In that case $[CO_3^{2-}]$, which we need for carbonate saturation calculations, must be calculated from measured pH, K_2 and the proper activity coefficients (as is done by WATEQP). A reliable estimate of $[CO_3^{2-}]$ requires a good measurement of pH and alkalinity.

The most accurate way to determine alkalinity is the socalled *Gran titration* (Stumm and Morgan, 1981). The principle of the method is that past the point where all HCO_3^- has been converted to H_2CO_3 (at a pH of ca. 4.3), the concentration of H^+ increases linearly with the amount of H^+ added. In practice, the volume of HCl added is plotted against the Gran function

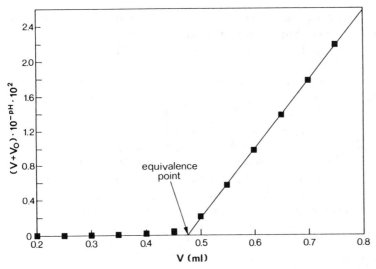

Figure 4.7. Illustration of the Gran plot method for determining alkalinity by titration with acid.

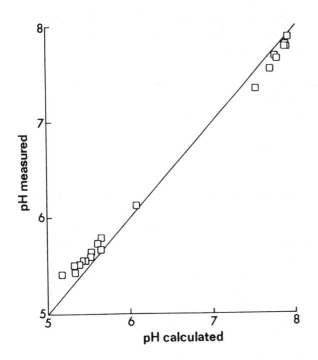

Figure 4.8. A comparison of pH calculated with WATEQP from *TIC*/Alkalinity measurements, with pH-values measured in the field.

$$F = (V + V_0) \cdot 10^{-pH}$$

where V is volume of acid added, and V_0 is start volume of sample.

The equivalence point is then obtained by extrapolating backwards from the linear part of the curve as shown in Figure 4.7. The advantages of this method are several; firstly, the equivalent point need not be determined accurately during the titration as such; secondly, the equivalent point is determined by linear regression on a number of pH readings; and thirdly, calibration errors of the pH-meter are not important since only changes in [H⁺] are used.

Quite often, the pH measurement of field samples is not reliable due to temperature differences of electrode, buffers and sample, degassing of CO_2, sluggish or faulty electrodes, etc. Errors in measured pH directly influence the calculation of CO_3^{2-} activity from alkalinity. An alternative may be to measure *TIC* and alkalinity, and to calculate both the pH and the distribution of CO_2 species, including the CO_3^{2-} activity (cf. Problem 4.2). A comparison of measured field pH values and pH values calculated with WATEQP according to this method is shown in Figure 4.8. *TIC* is either obtained as the sum of titrated values of alkalinity and of CO_2, or as total CO_2 which evolves upon addition of acid. The titrations can be carried out in the field, which is generally preferable. The total CO_2 determination must be performed in the laboratory by gas chromatography or infrared analysis, and requires that outgassing of CO_2 be prevented, e.g. by diluting the sample with distilled water until CO_2 pressure is near the atmospheric value.

4.2.2 CO_2 in soil water

Respiration, or oxidation of organic matter is the major source of CO_2 in soils. Consider soil water as pure water in which CO_2 dissolves, and let us calculate the composition of this 'soil moisture' in equilibrium with a CO_2 pressure (P_{CO_2}) of $10^{-1.5}$ atm. According to the equations given in Table 4.2, the CO_2 species in water are: H_2CO_3, HCO_3^- and CO_3^{2-}. When CO_2 dissolves in water these species have to be considered in addition to H⁺ and OH⁻. The equilibrium constants for the reactions give activity ratios among the species, and electroneutrality gives another equation that can be used to calculate the water composition.

The electrical neutrality condition is in this case :

$$m_{H^+} = m_{HCO_3^-} + 2m_{CO_3^{2-}} + m_{OH^-} \tag{4.8}$$

since these are the ions which are present in our 'pure' system. At different pH-ranges, different CO_2 species have been found dominant; if an initial estimate of pH could be obtained, we might neglect a CO_2 species to simplify the calculations. If for example pH is far below 10.3, then $2m_{CO_3^{2-}}$ in Equation (4.8) can be neglected; if pH is far below 7, then m_{OH^-} can be neglected compared with m_{H^+}.

The addition of CO_2 to water produces carbonic acid, and we may guess (and check afterwards) that the pH is so low that CO_3^{2-} and OH⁻ can be neglected in comparison to HCO_3^-.

The electroneutrality is then simply:

$$m_{H^+} = m_{HCO_3^-} \tag{4.9}$$

We can express [HCO_3^-] as a function of [H⁺] and P_{CO_2} with Equations (4.2) and (4.3), to find:

$$[HCO_3^-] = 10^{-7.8} P_{CO_2}/[H^+] \tag{4.10}$$

and hence, by combining with Equation (4.9), and neglecting differences between activity and molality, we obtain the concentration of H⁺:

Table 4.3. Calculated CO_2 species in soilwater and rainwater. Values in mol/l and atm.

P_{CO_2}	$10^{-1.5}$	$10^{-3.5}$
H^+	$10^{-4.6}$	$10^{-5.6}$
pH	4.6	5.6
HCO_3^-	$10^{-4.6}$	$10^{-5.6}$
CO_3^{2-}	$10^{-10.3}$	$10^{-10.3}$
OH^-	$10^{-9.4}$	$10^{-8.4}$

$$m_{H^+} = \sqrt{10^{-7.8}\,P_{CO_2}} \qquad\qquad (4.11)$$

Back substitution in the equations of Table 4.2 gives the concentrations entered in Table 4.3. The concentrations for a CO_2 pressure of $10^{-3.5}$ atm are entered for comparison; this is the CO_2 pressure of the present-day atmosphere, which would give unpolluted rainwater the slightly acid pH of 5.6 (compare with values in Table 2.2 in Chapter 2). Table 4.3, but also Figure 4.4, show that our assumption to neglect CO_3^{2-} and OH^- in the mass balance has been a reasonable one.

CO_2 pressures in the soil
Protons are essential for weathering and dissolution of minerals, but since their activity is influenced by many processes it is difficult to assign a value of H^+ or pH to different environments. The protons are mostly derived from $H_2CO_3^*$, and CO_2 pressure may be used as the master variable for calculating the dissolution of carbonates. The CO_2 is generated in soils by decay of organic material and by root respiration. Land use and biological productivity are therefore important in determining CO_2 pressure in soils. Figure 4.9 shows how CO_2 in the soil atmosphere varies seasonally, and with depth in the soil profile for different vegetations in North Germany. Note that agricultural land has the highest P_{CO_2}, followed by grass-vegetation. CO_2 pressures are highest when respiration is at maximum

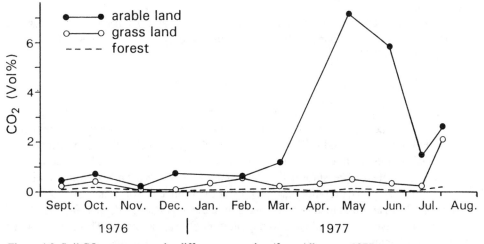

Figure 4.9. Soil CO_2 pressures under different vegetation (from Albertsen, 1977).

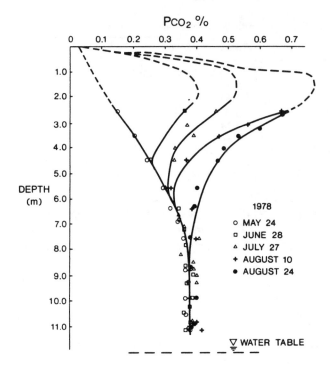

Figure 4.10. Seasonal fluctuations of soil CO_2 pressures in the unsaturated zone (Reardon et al., 1979).

during summer, and decrease in autumn and winter. The seasonal fluctuations are also clearly reflected in measurements by Reardon et al. (1979) shown in Figure 4.10; the data shown in this figure indicate that CO_2 diffusion in the summer is downward to the water table, but that a backward diffusion from groundwater during the non-production season is also possible. Atmospheric CO_2 contents also fluctuate seasonally, mainly as result of photosynthesis of land plants so that atmospheric concentrations are lowest during the late summer (September in the northern hemisphere, and March in the southern hemisphere; Bolin et al., 1979).

Temperature increases CO_2 production. In calcareous springs in the United States the following relation was found between CO_2 pressure and average yearly temperature (Harmon et al., 1975):

$$\log P_{CO_2} = -3.3 + 0.08T \, (°C)$$

Similar relationships have been found by Drake and Wigley (1975) and Drake (1983).

However, temperature cannot be the only controlling variable for biological productivity, but availability of water and soil conditions must also play a role. It has been suggested to use, therefore, the actual evapotranspiration as a measure of biological production in an area, and hence of CO_2 pressures in the soil. Figure 4.11 shows the relationship obtained by Brook et al. (1983). The relation has been used to produce a world map of CO_2 pressure (Figure 4.12). In addition to aerobic respiration in the upper soil, a deeper origin of CO_2 in water bearing layers is possible due to degradation of organic matter. This subject will be taken up in Chapter 7.

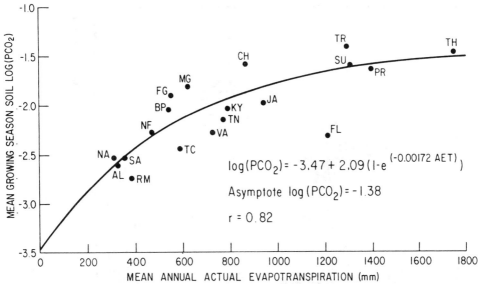

The relationship between mean growing-season soil $\log(PCO_2)$ and mean annual actual evapotranspiration predicted by equation (7). NA = Nahanni, Canada; SA = Saskatchewan, Canada; RM = Rocky Mountains, Canada; NF = Newfoundland, Canada; BP = Bruce Peninsula, Canada; TC = Trout Creek, Ontario, Canada; AL = Alaska, USA; VA = Reston, Virginia, USA; TN = Sinking Cove, Tennessee, USA; KY = Mammoth Cave, Kentucky, USA; FL = south Florida, USA; FG = Frankfurt-Main, W. Germany; MG = Müllenbach, W. Germany; JA = Jamaica; TR = Trinidad; PR = Puerto Rico; CH = Yunan, China; SU = Sulawesi; TH = Phangnga, Thailand

Figure 4.11. Relationship between soil CO_2 pressure and evapotranspiration (Brook et al., 1983. Reprinted by permission of John Wiley & Sons, Ltd).

4.2.3 *Open system dissolution of calcite*

When calcite dissolves in the soil moisture, the equilibrium concentration can be found in the same way as in the case of CO_2 dissolving in water without calcite. The dissolution of calcite adds Ca^{2+} to the solution, so that the electroneutrality equation must be extended to:

$$2m_{Ca^{2+}} + m_{H^+} = m_{HCO_3^-} + m_{OH^-} + 2m_{CO_3^{2-}} \qquad (4.12)$$

Again, we start by simplifying, and assume that the pH of soil water will be much lower than 10, so that $m_{CO_3^{2-}}$ and m_{OH^-} can be neglected in Equation (4.12). Furthermore we assume that m_{H^+} is negligible in comparison to $m_{Ca^{2+}}$ so that electroneutrality becomes:

$$2m_{Ca^{2+}} = m_{HCO_3^-} \qquad (4.13)$$

If the dissolution reaction is written in terms of HCO_3^-, the system is easily solved:

$$CaCO_3 \leftrightarrow Ca^{2+} + CO_3^{2-} \qquad K_{cc} = 10^{-8.3}$$

$$CO_3^{2-} + H^+ \leftrightarrow HCO_3^- \qquad K_2^{-1} = 10^{10.3}$$

$$H_2CO_3^* \leftrightarrow H^+ + HCO_3^- \qquad K_1 = 10^{-6.3}$$

$$+ \quad \underline{CO_{2(g)} + H_2O \leftrightarrow H_2CO_3^* \qquad \qquad \prod K_H = 10^{-1.5}}$$

$$CO_{2(g)} + H_2O + CaCO_3 \leftrightarrow Ca^{2+} + 2HCO_3^- \qquad K_{4.14} = 10^{-5.8}$$

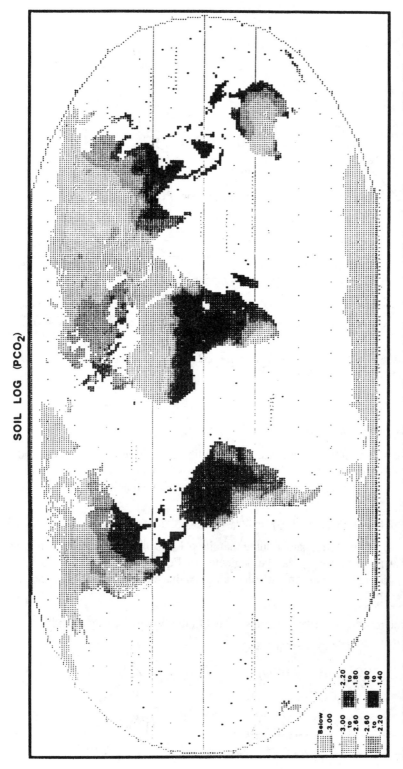

Figure 4.12. Map of soil CO_2 pressure produced with the relationship shown in Figure 4.11 (Brook et al., 1983. Reprinted by permission of John Wiley & Sons, Ltd).

Table 4.4. Calculated CO_2 species and Ca^{2+} in soil water in contact with CO_2 and calcite, in an open system. Values in mol/l and atm.

P_{CO_2}	$10^{-1.5}$	$10^{-3.5}$
Ca^{2+}	$10^{-2.6}$	$10^{-3.3}$
CO_3^{2-}	$10^{-5.6}$	$10^{-5.0}$
HCO_3^-	$10^{-2.3}$	$10^{-3.0}$
H^+	$10^{-7.0}$	$10^{-8.3}$
OH^-	$10^{-7.0}$	$10^{-5.7}$
pH	7.0	8.3

The corresponding mass action equation is:

$$\frac{[Ca^{2+}] [HCO_3^-]^2}{P_{CO_2}} = 10^{-5.8} \tag{4.14}$$

Combination with Equation (4.13), and assuming that activity is molality, gives:

$$m_{Ca^{2+}} = \sqrt[3]{10^{-5.8} P_{CO_2}/4} \tag{4.15}$$

Thus we have obtained a very simple relationship between the Ca^{2+} concentration and the CO_2 pressure for calcite dissolution. As soon as $m_{Ca^{2+}}$ ($\simeq [Ca^{2+}]$) is known, $[CO_3^{2-}]$ can be calculated from calcite equilibrium; the amount of HCO_3^- can be calculated with Equation (4.13), and all the other species follow with the equations provided in Table 4.2. The results of such calculations have been entered in Table 4.4 for two CO_2 pressures, of $10^{-1.5}$ atm (which is about the maximal value that one may expect in a normal soil, see Figure 4.11), and of $10^{-3.5}$ atm (which is the value of the present-day atmosphere). Note that concentrations of H^+, CO_3^{2-} and OH^- are small compared to concentrations of HCO_3^- and Ca^{2+}, and were rightly neglected in the mass balance equations.

Decalcification

We have noted before that calcite dissolution has a major effect on the Ca^{2+} and HCO_3^- concentrations of water. The rapid dissolution of calcite as compared to other minerals which make up the solid earth, leads easily to depletion of calcite from soils, sediments and rocks. This process is known as *decalcification* when other, less soluble minerals (quartz, gibbsite) stay behind, and *denudation* if the carbonate rock completely dissolves. Knowledge of Ca^{2+} concentrations in water can be used to calculate geological timescales of decalcification (Hissink, 1937; Spears and Reeves, 1975; White, 1984). The calculation requires assumptions to be made on the CO_2 pressure in the soil, which gives the maximal Ca^{2+} concentration (Equation 4.15), and the flux of water over the years, which gives the amount of Ca^{2+} that can be carried away in solution.

EXAMPLE 4.2: *Denudation rates in limestone area (White, 1984)*
Calculate the denudation rate in mm/yr when P_{CO_2} is 10^{-2} atm, and the precipitation surplus is 300 mm/yr. Density of the limestone is 2.0 g/cm^3.

ANSWER
With Equation (4.15), we obtain $Ca^{2+} = 1.6 \, mol/m^3$, or $0.47 \, mol/m^2$ .yr leaves the area. This is 47 g $CaCO_3$, or $24 \, cm^3/m^2$ .yr. Denudation rate is therefore 24 mm/1000 yr.

The calculation of denudation and decalcification timescales must rely on geological assumptions, which are not easily verifiable. However, another perhaps more fundamental point is, that as decalcification proceeds in the soil to deeper levels, the source of CO_2 (in the root zone) and the calcite (deeper down) become separated in space. This means that we cannot use an equation with a fixed CO_2 pressure to calculate Ca^{2+} concentrations in water, but must instead solve the equations for a *closed* system.

4.2.4 Closed system dissolution of calcite

Let us consider the other case, where the CO_2 pressure is not a constant. An extreme form is the system *closed* with respect to CO_2 gas, which means that flow of CO_2 into or out of the system is not possible. A sealed laboratory vessel which contains water and a certain amount of CO_2, and to which calcite is added, represents such a system. In nature it is found in a calcite-free (or decalcified) rootzone where soil moisture obtains its CO_2. Upon further downward percolation, the water may encounter calcite in deeper horizons. The zone where calcite dissolves, and this is important, should be different from the zone of CO_2 production for a closed system. It is clear from the open system dissolution, that CO_2 is used up when calcite dissolves. When CO_2 is not replenished, the CO_2 pressure drops as calcite dissolves, and eventually less calcite will have dissolved than may be estimated for constant CO_2 pressure of the upper soil. An extreme case of closed system dissolution is encountered in groundwater in Iceland. The barren climate supports no vegetation, and dissolution of basaltic rock consumes CO_2 that comes from the atmosphere only. The rivers in which this low-CO_2 water discharges, slowly gain CO_2 by diffusion from the atmosphere (Gíslason, 1989).

The water composition for closed system dissolution of calcite can be calculated using a *mass balance* on total dissolved carbon dioxide species in water when calcite has dissolved. The ΣCO_2 is the sum of the amount dissolved in the upper soil, *and* the amount which derives from calcite dissolution:

$$\Sigma CO_2 = m_{H_2CO_3^*} + m_{HCO_3^-} + m_{CO_3^{2-}} = (\Sigma CO_2)_{root} + m_{Ca^{2+}} \qquad (4.16)$$

where we have equated the amount of the CO_2 species which originates from calcite dissolution with the Ca^{2+} concentration. The amount of root zone $(\Sigma CO_2)_{root}$ was calculated before in Section 4.2.2, e.g. for a CO_2 pressure of $10^{-1.5}$ atm:

$$(\Sigma CO_2)_{root} = m_{H_2CO_3^*} + m_{HCO_3^-} + m_{CO_3^{2-}} = 10^{-3.0} + 10^{-4.6} + 10^{-10.3} = 10^{-3.0}$$

And the mass balance of Equation (4.16) for this initial CO_2 pressure is:

$$\Sigma CO_2 = 10^{-3.0} + m_{Ca^{2+}}$$

If we assume that the pH increases to more than 8 when calcite has dissolved, $H_2CO_3^*$ can be neglected as species in ΣCO_2 (see for example, Figure 4.4). In the electroneutrality equation H^+ and OH^- can be neglected as usual when other ions are present, and we have the following equations:

Table 4.5. Calculated CO_2 species and Ca^{2+} in soilwater in contact with CO_2 and calcite, in a closed system. Values in mol/l and atm.

$P_{CO_2 \text{ (initial)}}$	$10^{-1.5}$	$10^{-3.5}$
HCO_3^-	$10^{-2.7}$	$10^{-4.6}$
Ca^{2+}	$10^{-3.0}$	$10^{-4.1}$
CO_3^{2-}	$10^{-5.3}$	$10^{-4.2}$
H^+	$10^{-7.7}$	$10^{-10.7}$
H_2CO_3	$10^{-4.1}$	$10^{-9.0}$
OH^-	$10^{-6.3}$	$10^{-3.3}$
pH	7.7	10.7
$P_{CO_2 \text{ (final)}}$	$10^{-2.6}$	$10^{-7.5}$

$$m_{HCO_3^-} + 2m_{CO_3^{2-}} = 2m_{Ca^{2+}} \qquad \text{(electroneutrality)} \qquad (4.17)$$

$$m_{HCO_3^-} + m_{CO_3^{2-}} = m_{Ca^{2+}} + (\Sigma CO_2)_{root} \qquad \text{(mass balance)} \qquad (4.18)$$

Subtracting Equation (4.18) once from Equation (4.17) gives

$$m_{Ca^{2+}} = (\Sigma CO_2)_{root} + m_{CO_3^{2-}}$$

and if we subtract Equation (4.18) twice from Equation (4.17) we obtain

$$m_{HCO_3^-} = 2(\Sigma CO_2)_{root}$$

In other words, the final concentration of HCO_3^- in closed system dissolution of calcite, is twice the initial charge of CO_2 in the water. In the equation for Ca^{2+}, it can be noted that CO_3^{2-} can be neglected when pH < 9, which thus allows for a direct calculation of Ca^{2+}. With pH > 9 it can be combined with the equilibrium equation for calcite:

$$[Ca^{2+}][CO_3^{2-}] = 10^{-8.3} \qquad \text{or} \qquad [CO_3^{2-}] = 10^{-8.3}/[Ca^{2+}]$$

which gives a quadratic equation that can be simply solved. The other parameters follow thereafter with equations from Table 4.2. The results of the full calculation for, again, the two CO_2 pressures which are about the limiting values for normal fresh water are given in Table 4.5.

We note, on back-checking our assumptions, that the concentration of H_2CO_3 in the case with the higher CO_2 pressure is about 4% of *TIC*, and may be on the verge of influencing the calculation. When the calculation is done with H_2CO_3 taken into account, the outcome is: $Ca^{2+} = 10^{-3.03}$ mol/l, pH = 7.76, etc. which is not really much different from the values entered in Table 4.5. More serious in the case with low CO_2 pressure is that OH^- has been neglected in the electroneutrality equation (4.17), whereas it is now found to be higher than HCO_3^- and CO_3^{2-} concentrations. The calculated result gives a Ca^{2+} concentration only slightly larger than without initial CO_2 ($Ca^{2+} = 10^{-4.15}$ mol/l). The correct calculation without the neglect of OH^- yields: $Ca^{2+} = 10^{-4.08}$ mol/l, pH = 10.0, etc.

4.2.5 *Dolomite dissolution*

Up to now, the calculations have been performed for calcite, but these can be easily extended for dolomite dissolution (or any other carbonate). An additional constraint must be obtained for Mg^{2+} since that variable is now added to the system. The mass balance yields $m_{Ca^{2+}} = m_{Mg^{2+}}$, which allows the equations to be solved.

The dissolution of dolomite is written:

$$CaMg(CO_3)_2 \leftrightarrow Ca^{2+} + Mg^{2+} + 2\,CO_3^{2-} \tag{4.19}$$

The electroneutrality Equation (4.13) becomes:

$$2m_{Ca^{2+}} + 2m_{Mg^{2+}} = m_{HCO_3^-} \tag{4.13a}$$

or with $m_{Ca^{2+}} = m_{Mg^{2+}}$:

$$4m_{Ca^{2+}} = m_{HCO_3^-} \tag{4.13b}$$

Rewriting Equation (4.19) in terms of $CO_{2(g)}$ and HCO_3^- gives:

$$2\,CO_{2(g)} + 2H_2O + CaMg(CO_3)_2 \leftrightarrow Ca^{2+} + Mg^{2+} + 4\,HCO_3^-$$

The open system dissolution of dolomite can thus be calculated with the formula:

$$[Ca^{2+}]\,[Mg^{2+}]\,[HCO_3^-]^4 = K_{dol}(P_{CO_2} \cdot K_H \cdot K_1/K_2)^2$$

which combines with Equation (4.13b) to

$$m_{Ca^{2+}} = \sqrt[6]{\frac{1}{256} K_{dol}\left(P_{CO_2} K_H \frac{K_1}{K_2}\right)^2} \tag{4.20}$$

where $K_{dol} = 10^{-16.9}$ at $10°C$

The closed system dissolution of dolomite uses identical arguments, and yields:

$$m_{HCO_3^-} = 2\,(\Sigma CO_2)_{root}$$

$$m_{Ca^{2+}} = m_{Mg^{2+}} = \tfrac{1}{4}\,m_{HCO_3^-}, \text{ etc.}$$

EXAMPLE 4.3: *Dissolution of dolomite in the Italian Dolomites*
The Dolomites in North Italy form isolated, barren plateaus at levels up to 3000 m. Scree slopes which are vegetated at lower altitudes, extend from the plateaus. The timberline is at around 2000 m. This situation, depicted in Figure 4.13, gives low concentrations in water infiltrated on the plateaus

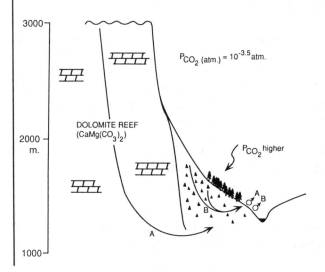

Figure 4.13. Cross section through a dolomite plateau, showing flowlines and water qualities.

(flowline A), and higher concentrations in water which infiltrates in the forests on the slopes (flowline B). The difference suggests a rapid dissolution of dolomite to saturation levels dictated by the CO_2 pressure of the infiltration area.

When the discharge of springs increases in response to rain, it is often observed that concentrations in discharge increase as well, which appears in conflict with common knowledge. However, it becomes understandable when primarily the contribution of runoff from the scree slopes increases spring discharge. The corresponding flowline goes through highly permeable scree slopes which can give very rapid response to rainfall, while at the same time the soil slopes have a higher CO_2 pressure than the plateaus at 3000 m.

When CO_2 pressure at the plateau is $10^{-3.5}$ atm, and in the forested scree slope 10^{-2} atm, the calculated concentrations in flowline A and B are as given in the table, in close correspondance to observed concentrations.

flowline	A	B	
P_{CO_2}	$10^{-3.5}$	$10^{-2.0}$	atm
$m_{Ca^{2+}} \equiv m_{Mg^{2+}}$	0.33	1.04	mmol/l
$m_{HCO_3^-}$	1.31	4.16	mmol/l
pH	8.42	7.42	
EC	131	416	µS/cm

4.3 FIELD CONDITIONS FOR CALCITE/DOLOMITE DISSOLUTION

Dissolution and precipitation of calcite under field conditions, can be summarized in a single equation:

$$CaCO_3 + CO_2 + H_2O \leftrightarrow Ca^{2+} + 2HCO_3^- \qquad (4.21)$$

If a source of CO_2 is available dissolution of calcite will take place, while degassing of CO_2 will cause precipitation of calcite. The most dramatic results of this reaction, both in the forward and the backward sense, are visible in caves. The downward transport of CO_2 charged water enlarges fissures, etc., and results in the formation of caves. Degassing of CO_2 from water that enters the cave at a later stage and already has dissolved calcite leads to precipitation of calcite and the formation of stalactites and stalagmites.

As discussed above, one may distinguish between the case of *open* system dissolution (P_{CO_2} = constant) and *closed* system dissolution, where an initial content of CO_2 is consumed by Reaction (4.21) and not replenished. The ideal calculation schemes for open and closed system dissolution can be translated in terms of field conditions as follows. Open system calcite dissolution takes place in the soil which contains calcite, and where a continuous source of CO_2 is available due to degradation of organic matter. The open system extends through the whole unsaturated zone due to rapid gas exchange through the gas phase (Laursen, 1991). However, if calcite is not available in the soil or unsaturated zone, the water may become isolated from the source of CO_2 production. When further down the flow path calcite is encountered and dissolved, the P_{CO_2} will rapidly decrease according to (4.21). The paths of chemical evolution for water that reacts with calcite under open and closed conditions until equilibrium with calcite is reached, are illustrated in Figure 4.15.

For an open system, the P_{CO_2} remains constant during dissolution with calcite. The HCO_3^- content increases log-linearly with pH so that the final bicarbonate concentration at

Figure 4.14. Cave formation in a limestone on Puerto Rico (courtesy N.Stentoft).

equilibrium is completely controlled by the initial P_{CO_2} and the solubility of calcite. In the case of a closed system the initially available amount of CO_2 becomes depleted through the reaction with calcite and P_{CO_2} decreases during the reaction. For a given initial P_{CO_2} the amount of bicarbonate in solution will be much higher in the case of open system dissolution, than in the case of closed system dissolution. Note that according to the stoichiometry of Reaction (4.21), the concentration of Ca^{2+} is half that of HCO_3^- and the evolution of Ca^{2+} concentrations in open and closed systems is therefore directly analogous to that of HCO_3^-.

These relationships can be quantified more precisely using the computer program PHREEQE (Parkhurst et al., 1980), including all activity coefficient corrections and complexes. This program will be discussed extensively in Chapter 10, but we use it here and in following chapters to present the ideal equilibrium compositions of water in which all chemical corrections are rigorously included. The results of PHREEQE calculations for open and closed systems are presented in Table 4.6.

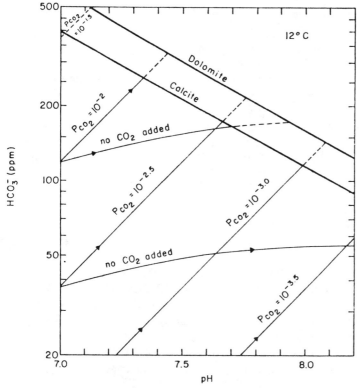

Figure 4.15. The evolution of groundwater composition towards equilibrium with calcite or dolomite in the case of a system open CO_2 (P_{CO_2} = constant) or closed (no CO_2 added) with respect to CO_2. Solid lines indicate equilibrium with calcite or dolomite (Langmuir, 1971. Reprinted with permission from Pergamon Press PLC).

The influence of CO_2 pressure on final concentrations is again clearly illustrated. CO_2 pressures of $10^{-1.5}$ to $10^{-2.5}$ atm are commonly found in soils, and may be expected in water which has equilibrated with calcite in the soil proper (open system dissolution). In the case of closed system dissolution, the CO_2 content derived from the soil decreases substantially, and the final CO_2 pressures are lower than $10^{-2.5}$ atm, and sometimes even lower than the P_{CO_2} of the atmosphere.

The clear difference in Ca^{2+} concentrations (or bicarbonate) between open and closed systems has been used as a diagnostic criterion for the type of environment where Ca^{2+} has entered the water, and hence as a method to deduce the infiltration area (Langmuir, 1971; Pitman, 1978; Hoogendoorn, 1983). For example, groundwater in the Veluwe area (ice-pushed hills in The Netherlands, of which a cross section is shown in Figure 4.16) has low Ca^{2+} concentrations of less than 1 mmol/l, and a pH higher than 8. This water infiltrated in sandy hills where the sediments are either calcite-free at the time of deposition, or have been leached free from calcite since deposition. The CO_2 pressure in these soils may well be above 0.01 atm, but since the soil is calcite-free, the water only obtains a high CO_2 pressure in the soil together with small concentrations of elements which are leached from silicates in the solid matrix of the soil. In deeper layers calcite dissolution occurs, but since

Table 4.6. Summary of concentrations in water in which calcite dissolves at 15°C (PHREEQE calculations, which include complexes and activity corrections).

	Open system with constant P_{CO_2} (Section 4.2.2)		Open system with constant P_{CO_2} and calcite (Section 4.2.3)		Closed system with known initial P_{CO_2} and calcite (Section 4.2.4)		
P_{CO_2} initial	$10^{-1.5}$	$10^{-3.5}$	$10^{-1.5}$	$10^{-3.5}$	$10^{-1.5}$	$10^{-3.5}$	atm
P_{CO_2} final	$10^{-1.5}$	$10^{-3.5}$	$10^{-1.5}$	$10^{-3.5}$	$10^{-2.5}$	$10^{-6.4}$	atm
pH	4.63	5.63	6.98	8.29	7.62	10.06	
Ca^{2+}	–	–	2.98	0.58	1.32	0.12	mmol/l
Alk	0.02	0.002	5.96	1.16	2.65	0.24	mmol/l
EC	2	1	600	120	265	25	µS/cm

there is no CO_2 production, the initial CO_2 content is rapidly used up. This closed environment hence leads to low Ca^{2+} concentrations, low CO_2 pressure in water, and a high pH.

The soils in the discharge area still contain calcite and the water that infiltrates here displays higher Ca^{2+} concentrations (> 1 mmol/l), higher CO_2 pressures (> 0.01 atm), and a pH lower than 8. Since the water that seeps up into the brooks and rivers surrounding the ice-pushed hills often shows the high Ca^{2+} concentrations typical for open system dissolution of calcite, it has been suggested that most of this water must have infiltrated locally near the discharge point, where the sediment still contains calcite. Figure 4.16 shows the hydrogeological situation. It is assumed that there are no deeper sources of CO_2 available

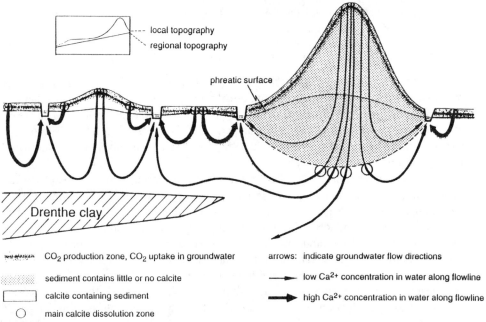

Figure 4.16. A cross section through Dutch landscape with indications of water quality and water contributions (Hoogendoorn, 1983).

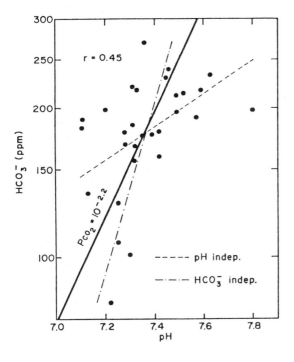

Figure 4.17. A plot of pH versus HCO_3^- in spring waters of Paleozoic rocks in Pennsylvania. Regression lines are drawn assuming HCO_3^- or pH as the independent variable. The solid line indicates a $P_{CO_2} = 10^{-2.2}$ atm at 12°C (Langmuir, 1971. Reprinted with permission from Pergamon Press PLC).

which might increase the CO_2 pressure of water below the upper soil. If this assumption is true (we will discover in later chapters that deeper CO_2 production is quite well possible), then the suggestion about a local infiltration source of upward seeping water is justified.

A second example is the classical study by Langmuir (1971) of the groundwater geochemistry in Paleozoic carbonate rocks in Pennsylvania, USA, that contain both calcite and dolomite. It was found that deeper groundwaters generally are close to equilibrium with calcite or dolomite. On the other hand, spring waters were often found to be subsaturated with respect to calcite. Whether the composition of these spring waters evolves under open or closed system dissolution of calcite or dolomite, can be evaluated by plotting pH against the bicarbonate content, similarly to Figure 4.15. The results are shown in Figure 4.17 and show a great deal of scatter, which is easily understood, considering the variability of CO_2 production in the soil (Figures 4.9 and 4.10). Nevertheless, the data seem to group roughly around the line for a constant P_{CO_2} of $10^{-2.2}$ which is the average of all spring samples, and the data are thus in rough agreement with ongoing dissolution of calcite in an open system (the waters are not in equilibrium with calcite).

The composition of the deeper well waters can also be displayed in such a diagram. These well waters are with few exceptions close to equilibrium with calcite or dolomite (Figure 4.18). The observed range in pH and bicarbonate concentrations is quite considerable. Part of the variation can be explained by the pathway through which the water has attained equilibrium with calcite (Figure 4.17). However, processes within the aquifer may also change the water composition up and down the equilibrium line. For example, exsolution of CO_2, in fissures or caves, will cause precipitation of carbonate and the water composition changes along the equilibrium line towards higher pH and lower bicarbonate

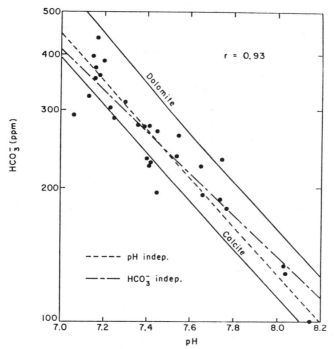

Figure 4.18. A plot of pH versus HCO_3^- in well waters of Paleozoic rocks in Pennsylvania. Regression lines are drawn assuming HCO_3^- or pH as the independent variable. The solid lines indicate the solubilities of calcite and dolomite at 12°C (Langmuir, 1971. Reprinted with permission from Pergamon Press PLC).

concentrations. On the other hand CO_2 production within the aquifer, for example by oxidation of organic matter in the sediment, will cause additional carbonate dissolution and the water composition changes in the opposite direction.

Flow mechanisms in carbonate aquifers

A distinction has to be made between conduit flow along fissures and larger openings in carbonate rock, and diffuse flow through the pores of the carbonate rock. Conduit flow is connected with rapid, sometimes even turbulent flow, whereas diffuse flow is slow, giving long residence times of water in the rock. Seasonal variations in water qualities of carbonate rock springs have been related to the type of flow mechanism. Shuster and White (1972) observed that conduit flow springs have large variations in concentrations, temperature and discharge over the year, whereas diffuse flow springs have in all respects a more stable regime (Figure 4.19).

There is, of course, always a connection between conduits and micropores in the rock, and a combination of flow mechanisms is likely to be operative. Atkinson (1975) has shown from hydrograph-analysis and other parameters in the Mendip Hills, England, that conduit flow probably accounts for between 60% and 80% of the transmission of water in the aquifer, although conduits contain only 1/29 of all the phreatic groundwater when the aquifer is fully recharged.

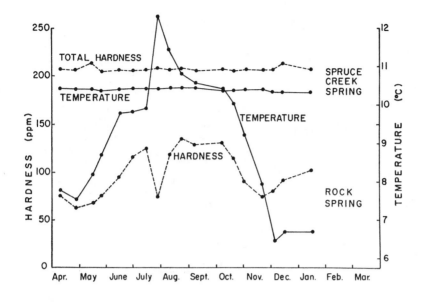

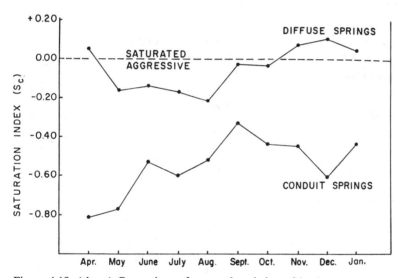

Figure 4.19. (above) Comparison of seasonal variation of hardness and temperature for a typical diffuse-flow spring (Spruce Creek spring) and a typical conduit-flow spring (Rock spring). (below) Average seasonal variation in saturation index for the two types of springs. Reprinted from Shuster and White, 1972. Copyright by the Am. Geophys. Union.

Figure 4.20 shows a number of flowpaths in carbonate karst in New Zealand, together with mean Ca^{2+} and Mg^{2+} concentrations, and standard deviations (Gunn, 1981). Water in the subcutaneous zone showed the highest concentrations, and this water seems to feed the larger karst-pipes (termed shafts in Figure 4.20). It is normal that concentrations are lower in overland flow and throughflow, but the presence of lower concentrations in smaller fissures seems illogical. It is suggested in Figure 4.20 that these waters are also fed from the

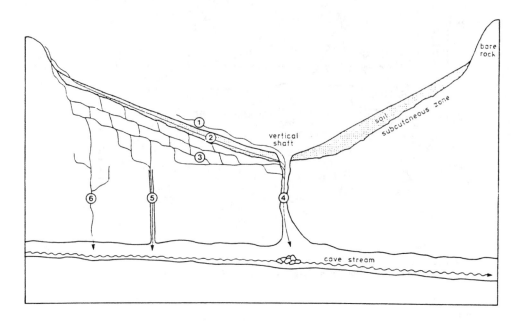

Flow component (key below)	Number of observations	Mean	Standard deviation	Coefficient of variation (%)	Range
(a) Calcium ion concentrations (mg/l)					
1	6	7.5	2.49	33.2	4.2-10.4
2	162	18.7	5.04	26.9	8.0-31.7
3	154	46.5	6.75	14.5	34.8-61.4
4	61	46.8	6.66	14.2	30.0-56.4
5	186	30.6	17.86	58.3	10.4-61.4
6	1714	36.1	6.72	18.6	20.6-59.0
(b) Magnesium ion concentrations (mg/l)					
1	6	0.6	0.308	47.4	0.2-1.0
2	154	1.1	0.270	23.9	0.5-1.8
3	139	1.3	0.212	16.2	0.8-1.5
4	58	1.3	0.224	16.9	0.7-1.8
5	183	1.0	0.294	30.4	0.5-1.5
6	1754	1.5	0.427	28.0	0.7-3.0
(c) Temperature (°C)					
2	78	10.3	2.43		4.9-14.7
3	95	10.7	0.58		9.4-12.2
4	34	10.0	1.33		8.3-12.8
5	105	11.1	1.53		8.1-13.4

Key to flow components: 1. Overland flow; 2. Throughflow; 3. Subcutaneous flow; 4. Shaft flow; 5. Vadose flow. 6. Vadose seepage.

Figure 4.20. Types of flow in New Zealand karst, and summary statistics of flow components (Gunn, 1981).

subcutaneous zone, and one expects then similar concentrations. Perhaps there is a rapid flow component involved, which brings water rapidly downwards in particular fissures.

Dedolomitization

Dedolomitization may occur in aquifers containing limestones, dolostones (dolomite-type carbonates), in combination with gypsiferous layers (Back and Hanshaw, 1970; Wigley, 1973; Atkinson, 1983; Back et al., 1984; Plummer et al, 1990). The overall process is, that as result of gypsum or anhydrite dissolution, dolomite dissolves and calcite precipitates. The reactions have been deduced in most cases from trends in water chemistry, but they were also confirmed in a petrographic study by Deike (1991).

In an aquifer containing dolomite and calcite, we might expect that the groundwater is close to equilibrium for both minerals:

$$Ca^{2+} + CaMg(CO_3)_2 \leftrightarrow 2CaCO_3 + Mg^{2+} \tag{4.22}$$

and accordingly the ratio $[Mg^{2+}]/[Ca^{2+}]$ remains fixed as

$$K_{4.22} = \frac{[Mg^{2+}]}{[Ca^{2+}]} = \frac{K_{dol}}{K_{cc}^2} = \frac{10^{-17.09}}{(10^{-8.48})^2} = 0.8 \qquad \text{at } 25°C \tag{4.23}$$

Dissolution of anhydrite or gypsum takes place according to

$$CaSO_4 \rightarrow Ca^{2+} + SO_4^{2-} \tag{4.24}$$

The increasing Ca^{2+} concentration due to gypsum dissolution causes calcite to precipitate, which is similar to the effect of gypsum dissolution on fluorite precipitation discussed in Chapter 3. The CO_3^{2-} concentration decreases as calcite precipitates, and this provokes the dissolution of dolomite and an increase of the Mg^{2+} concentration. When Mg^{2+} increases, Ca^{2+} increases as well because of relation (4.23). The net result is therefore that dissolution of gypsum induces the transformation of dolomite to calcite in the rock and produces waters with increased Mg^{2+}, Ca^{2+} and SO_4^{2-} concentrations, and decreased alkalinity.

This situation is illustrated (Plummer et al., 1990) for the Madison aquifer, USA, in Figures 4.21 and 4.22. The sulfate concentration was used to monitor the extent of gypsum dissolution. Figure 4.21 shows that the groundwater is at saturation or slightly supersaturated for calcite throughout the aquifer, regardless of the sulfate concentrations. The data scatter for saturation with respect to dolomite is somewhat larger, but still within a *SI* variation of ± 0.5 from equilibrium. For gypsum, the saturation state ranges from strongly subsaturated to equilibrium. When equilibrium with calcite and dolomite is maintained, the ratio of $[Mg^{2+}]/[Ca^{2+}]$ must remain about 0.8. Since the two ions behave similarly as regards complexation and activity coefficient corrections, the ratio of total concentrations of the two ions should also remain about 0.8. The total mass transfer is thus given by the reaction:

$$1.8\,CaSO_4 + 0.8\,CaMg(CO_3)_2 \rightarrow 1.6\,CaCO_3 + Ca^{2+} + 0.8\,Mg^{2+} + 1.8\,SO_4^{2-} \tag{4.25}$$

Accordingly, 2 moles of calcite precipitate for each mole of dolomite that dissolves. The reaction scheme predicts, that the SO_4^{2-} concentration must increase with both the concentration of Mg^{2+} and Ca^{2+}, as is illustrated to occur in the Madison aquifer in Figure 4.22. The ideal stoichiometry conforming to Reaction (4.25) is indicated by lines, labelled with

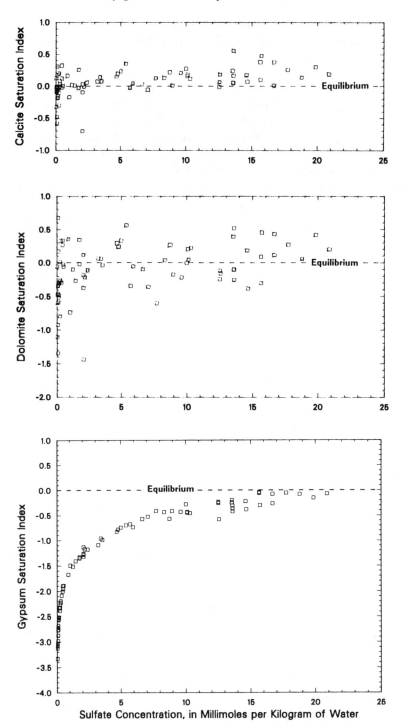

Figure 4.21. Comparison of calcite, dolomite and gypsum saturation indices as a function of total dissolved sulfate content for wells and springs in the Madison aquifer (Plummer et al., 1990. Copyright by the Am. Geophys. Union).

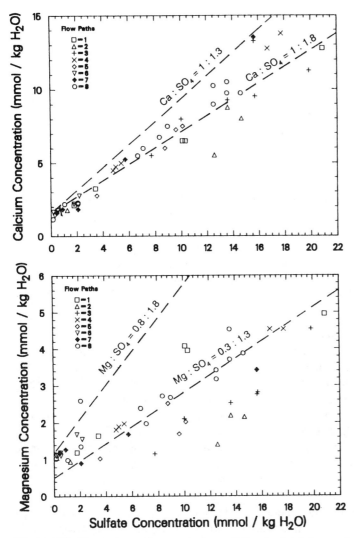

Figure 4.22. Comparison of concentrations of dissolved calcium and magnesium as a function of dissolved sulfate content for waters from the Madison aquifer. Broken lines indicate expected trends when groundwater are in exact equilibrium with calcite and dolomite (line with 1.8 SO_4^{2-}), or when $SI_{cc} = 0.2$ (line with 1.3 SO_4^{2-}) (modified from Plummer et al., 1990. Copyright by the Am. Geophys. Union).

1.8 as the coefficient for SO_4^{2-}. The suggested trend is closely followed by Ca^{2+}, but less so by Mg^{2+}. However, the apparent state of supersaturation of the groundwater with respect to calcite also influences the ratio of $[Mg^{2+}]$ over $[Ca^{2+}]$. If we take into account that the waters have, on the average, a $SI_{cc} = 0.2$, and a $SI_{dol} = 0$, this ratio becomes:

$$\frac{[Mg^{2+}]}{[Ca^{2+}]} = \frac{10^{-17.09}}{(10^{-8.48 + 0.2})^2} = 0.3 \qquad (4.23a)$$

The overall mass transfer now changes to:

$$1.3CaSO_4 + 0.3CaMg(CO_3)_2 \rightarrow 0.6CaCO_3 + Ca^{2+} + 0.3Mg^{2+} + 1.3SO_4^{2-}$$

$$(4.25a)$$

Lines according to this stoichiometry have also been drawn in Figure 4.22, and are labelled with 1.3 as the coefficient for SO_4^{2-}. This relation presents an upper boundary for the increase of Ca^{2+} with SO_4^{2-}, but the increase of Mg^{2+} with SO_4^{2-} is fitted quite well. Note that the intercept in the plot of Mg^{2+} vs. SO_4^{2-} has been adapted to 0.45 mmol/kg H_2O, to obtain the required $[Mg^{2+}]/[Ca^{2+}]$ ratio in the lower left part of the graph.

It thus appears that the mass transfer coefficients are determined by failure to attain complete equilibrium with calcite, when the groundwaters become supersaturated. This may well be caused by precipitation-inhibitors which are present in the water or on the aquifer rock, as will be discussed in Section 4.4.3. Plummer et al. (1990) found that the average apparent rates remain very low in this aquifer: 0.59 µmol/1.yr for calcite precipitation, 0.25 µmol/1.yr for dolomite dissolution and 0.95 µmol/1.yr for gypsum dissolution. Other reactions, such as ion exchange, were calculated to be of minor importance except for flow paths 2 and 3 (flow paths are indicated with symbols in Figure 4.22).

Sources of constituents in mixed carbonate terrain

The Ahrntal (Zillertal Alps) in Northern Italy shows the typical central alpine sequence of micaschists, limestones, dolomites, gypsiferous layers, and serpentinite rock. The rocks dip vertically and are incised by alpine trough valleys (Figure 4.23). Water quality of springs and the evolution in river water quality reflects the contribution of different lithologies (Appelo et al., 1983). Figure 4.24 shows individual ion concentrations downstream along GrossKlausenbach which is one of the tributary streams to the Ahrn. The concentrations increase when runoff from serpentinite and carbonate rocks contributes more to streamflow. Cl^- and Na^+ concentrations remain more or less constant compared to Mg^{2+}, Ca^{2+}, HCO_3^- and SO_4^{2-} and the following discussion is therefore limited to these *reactive* ions. Minerals which actively contribute to water composition will be gypsum, sulfides, calcite, dolomite and Mg-bearing minerals (notably talc, serpentine, chrysotile). Feldspar and mica are relatively unimportant, as reflected by the low and uniform concentrations of Na^+ and K^+. The reaction scheme is similar to the process of dedolomitization in calcite-dolomite-gypsum carbonate aquifers, with the additional reaction of the Mg-silicates.

Sulfate in this water is likely to be derived from gypsum. Dissolution of gypsum releases equal amounts of Ca^{2+} and SO_4^{2-}. Subtracting $m_{SO_4^{2-}}$ from m_{Ca_t} should give therefore the Ca^{2+} which comes from other sources:

$$Ca_c = m_{Ca_t} - m_{SO_4^{2-}}$$

These other sources have characteristic Ca/Mg ratios, and $(Mg - Ca_c)$ numbers:

Mineral	Ca/Mg ratio	$(Mg - Ca_c)$
Dolomite	1:1	0
Calcite	1:0	$0 - x$
Serpentinite	0:1	$x - 0$

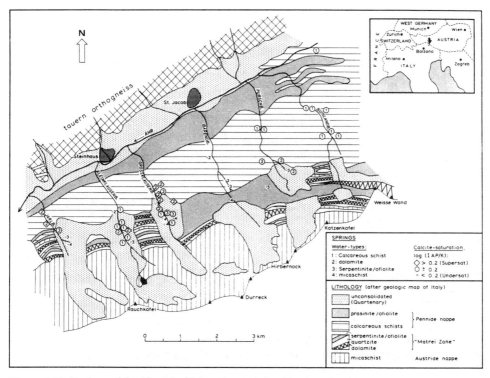

Figure 4.23. Geology of the Ahrntal with location and classification of springs.

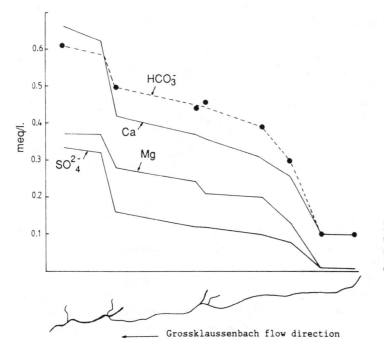

Figure 4.24. Concentrations of ions in water along the course of GrossKlausenbach.

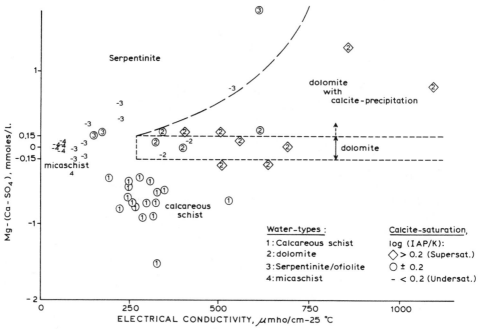

Figure 4.25. Plot of the difference ($Mg - Ca_c$) in springwaters versus electrical conductivity. Calcium concentrations are corrected for dissolution of gypsum. Different water types are symbolized by numbers.

Figure 4.25 shows a plot of the difference ($Mg - Ca_c$) in springwaters versus electrical conductivity. The plot shows the water associated with different rocks and mineral reactions:

1. *$CaCO_3$ water*: is derived from calcareous schists. The water has a low SO_4^{2-} concentration, and therefore also a low *EC*, and is only slightly supersaturated with respect to calcite (log (IAP/K) = 0 to 0.3).

2. *Dolomite water*: has in principle ($Mg - Ca_c$) numbers of 0. The SO_4^{2-} content is high in samples with a high *EC*. Gypsum is the likely source of SO_4^{2-} in these waters. Part of the Ca^{2+} released from gypsum dissolution precipitates as $CaCO_3$ and ($Mg - Ca_c$) is then relatively high. Calcite precipitates at the spring hollow, and forms thick travertine deposits. This water has high *EC*, and is strongly supersaturated with respect to calcite: log (IAP/K) = 0.3 – 0.7.

3. *Serpentinite water*: has a lower *EC* than group 2, and an identical or occasionally even higher ($Mg - Ca_c$) number. This water is always subsaturated with respect to calcite.

4. *Micaschist water* with an *EC* below 100 μS/cm. This water is very strongly subsaturated with respect to calcite and dolomite, and it is unlikely that it has been in contact with these minerals.

Table 4.7 gives a few characteristic analyses for the four groups. The group codes have also been entered in Figure 4.25 at the site where the spring discharges. It can be seen that the grouping does not in all cases coincide with the hard rock geology. Possibly, the geological map is not accurate enough (this is very difficult geological terrain!), or

Table 4.7. Typical composition of spring-waters. 'mcs' = micaschist; 'serp' = serpentine; 'pras' = prasinite/ofiolite; 'dol' = dolomite; 'cc' = calcareous schist. (mmol/l).

Group code	4	4	3	3	2	2	1
Geol. unit	'mcs'	'mcs'	'serp'	'pras'	'dol'	'dol'	'cc'
pH	7.63	7.70	8.47	7.41	8.00	7.93	8.12
EC (µS/cm)	52	100	152	610	325	505	250
Temp (°C)	3.7	4.0	4.0	4.8	5.5	7.5	2.0
Na^+	0.04	0.03	0.02	0.06	0.04	0.04	0.04
K^+	0.016	0.026	0.011	0.015	0.011	0.008	0.011
Mg^{2+}	0.11	0.12	0.44	2.30	0.67	0.93	0.31
Ca^{2+}	0.17	0.35	0.33	1.20	0.77	1.95	0.94
Cl^-	0.05	0.05	0.06	0.05	0.04	0.06	0.05
HCO_3^-	0.37	0.86	1.35	5.54	2.10	3.34	2.23
SO_4^{2-}	0.08	0.04	0.05	0.72	0.14	1.19	0.15
Si	0.05	0.05	0.06	0.13	0.06	0.09	0.05
$\log P_{CO_2}$	−3.37	−3.08	−3.67	−2.01	−3.00	−2.74	−3.11
$\log (IAP_{cc}/K_{cc})$	−1.8	−1.1	−0.19	−0.18	−0.11	+0.37	+0.08
$\log (IAP_{dol}/K_{dol})$	−4.1	−2.9	−0.48	−0.29	−0.47	+0.27	+0.82

alternatively surface deposits (moraines, slumps, slope deposits) control the water chemistry and obscure the effects of underlying hard rock geology on water chemistry. It may also occur that a spring originates from a fault or fissure zone which transects the lithological units.

4.4 KINETICS OF CARBONATE REACTIONS

Karst phenomena are a clear macroscopic indicator of the rapid solution of carbonate rocks. Fissures widen by dissolution and become conduits for water flow, sinkholes (dolines) and caves develop, and solution features are visible on the surface of all barren limestones. Figure 4.26 shows the typical solution flutes (denoted with the German term 'Rillenkarst') that are often observed. Surface solution phenomena as well as cave formations are likewise, though more seldom, found in dolomite rock (Zötl, 1974; White, 1988; Ford and Williams, 1989; Figure 4.27).

We have seen in the previous sections that pH will increase during solution of carbonates; a similar increase would occur when silicate minerals dissolve and release cations to the solution. However, dissolution of the carbonate minerals is fast in comparison with other rock (or soil) forming minerals, so that a pH increase is more clearly manifest in soils with carbonates. The pH increase is already observed at small carbonate contents, as shown in Figure 4.28. The figure gives pH's of the B-horizon of decalcifying dune soils as a function of calcite percentage (Rozema, et al., 1985). The pH in the soils which still contain calcite may be related to a CO_2 pressure of ca. $10^{-3.0}$ atm (see Problem 4.7). The fact that pH is near-uniform, and independent of the calcite concentration in the soil above a certain limit, suggests that dissolution kinetics are sufficiently fast to reach equilibrium already

Figure 4.26. (left) Rillenkarst on Limestone in the N. Italian Alps (Appelo, priv. coll.).
Figure 4.27. (right) Karstic development of shafts on dolomite (from the N. Italian Dolomites (Appelo, priv. coll.).

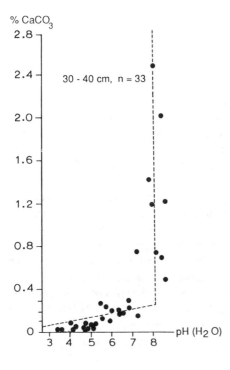

Figure 4.28. Soil-pH versus calcite content of B-horizon of dune soils (Rozema et al., 1985).

when ca. 0.3% calcite is present in the soil. It is of interest to relate such field observations to laboratory experiments and theoretical studies which have been performed in the past decade.

4.4.1 *Dissolution*

The general pattern of calcite dissolution kinetics as a function of pH and CO_2 pressure is shown in Figure 4.29 (Plummer et al., 1978). Three regions are discerned as a function of pH. The first region is confined to pH values below 3.5, where the rate is proportional to $[H^+]$. At these low pH values a strong dependence on the stirring rate is found which indicates that transport of H^+ to the calcite surface is rate controlling (see Section 3.5.5). At higher pH, the rate becomes less dependent of pH but instead increasingly dependent on the P_{CO_2}. Here, the rate appears to be controlled by both transport and surface reaction (Rickard and Sjöberg, 1983). The third region shows a sharp drop in dissolution rate when the pH comes close to saturation.

Different approaches have been used to model these general results. Sjöberg (1978), Rickard and Sjöberg (1983) and Morse (1978) described calcite dissolution kinetics in seawater by empirical rate expressions of the type

$$R = k \frac{A}{V} (1 - \Omega)^n$$

where A is the surface area, V is the volume of the solution, Ω is the saturation state IAP/K, and k and n are coefficients which depend on the composition of the solution, and are obtained by fitting with observed rates.

Plummer, Wigley and Parkhurst (1978) developed a mechanistic rate model for calcite dissolution. The process is interpreted in terms of three distinct dissolution reactions:

$$CaCO_3 + H^+ \quad \rightarrow Ca^{2+} + HCO_3^-$$

$$CaCO_3 + H_2CO_3^* \rightarrow Ca^{2+} + 2HCO_3^-$$

$$CaCO_3 + H_2O \quad \rightarrow Ca^{2+} + HCO_3^- + OH^-$$

The first reaction reflects the dominating process of proton attack at pH < 3.5, and the second reaction incorporates the effect of $H_2CO_3^*$ at higher pH. It was observed that the rate is independent of both pH and P_{CO_2} in the still higher pH range above ca. pH = 7, and there the third reaction becomes important, which reflects simple hydrolysis of calcite. Finally the backward precipitation reaction was added:

$$Ca^{2+} + HCO_3^- \rightarrow CaCO_3 + H^+$$

These reactions form the basis for the rate equation which covers both dissolution and precipitation:

$$R = \underbrace{k_1[H^+] + k_2[H_2CO_3^*] + k_3[H_2O]}_{R_{fw}} - \underbrace{k_4[Ca^{2+}][HCO_3^-]}_{R_b} \tag{4.26}$$

where R is the rate of calcite dissolution ($mmol/cm^2.s$), separated in a forward rate R_{fw} and a backward rate R_b, and $k_1, .., k_4$ are the rate constants. The forward rate constants have been

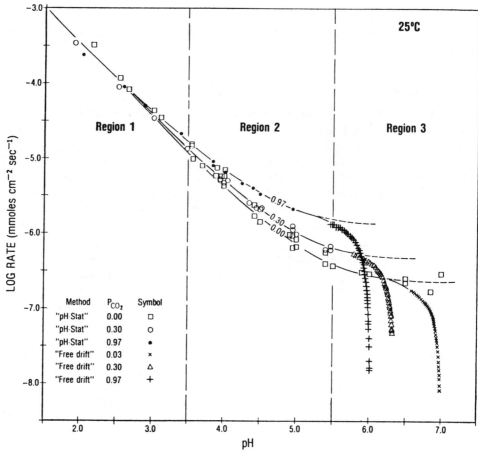

Figure 4.29. Dissolution rates of calcite as a function of pH and P_{CO_2} (Plummer et al., 1978).

fitted to the experimental data as function of temperature (T, in K):

$$\log k_1 = 0.198 - 444/T$$
$$\log k_2 = 2.84 - 2177/T$$

when $T < 298$: $\quad \log k_3 = -5.86 - 317/T$

when $T > 298$: $\quad \log k_3 = -1.1 - 1737/T$

The backward rate has been derived from the principle of microscopic reversibility of the individual forward reactions as a function of $[H^+]_s$ and $[H_2CO_3^*]_s$, which are the activities on the crystal surface. It is assumed that saturation exists at the surface during the whole of the dissolution process, so that $[H^+]_s$ and $[H_2CO_3^*]_s$ are given by their saturation values in the bulk solution. This rather complicated reasoning was necessary, since the backward reaction was *observed* to depend on $[Ca^{2+}]$ and $[HCO_3^-]$ only, i.e. the backward reaction of the first dissolution reaction. Strict microscopic reversibility would require a similar dependence of the backward reaction on $[OH^-]$ from the third term, and a quadratic dependence on $[HCO_3^-]$ from the second term. Anyhow, Plummer et al. (1978) could argue that as long as $[H^+]_s$ and $[H_2CO_3^*]_s$ remain constant during the dissolution process, the value of k_4 remains constant as well. This allows a straightforward calculation of k_4 since at saturation the net rate $R = R_{fw} - R_b = 0$. Constant $[H_2CO_3^*]_s$ is true when CO_2 pressure is constant (open system dissolution). For example, at constant $P_{CO_2} = 10^{-1.5}$ atm, the water composition at equilibrium with calcite was calculated to be (Section 4.2.3): $[H^+]_s = 10^{-7.0}$; $[H_2CO_3^*] = 10^{-3.0}$; $[Ca^{2+}]_s = 10^{-2.6}$, and $[HCO_3^-]_s = 2[Ca^{2+}]_s$. Hence we find, at 10°C (= 283 K):

$$
\begin{aligned}
k_1 \cdot [H^+]_s &= 10^{-1.4} \cdot 10^{-7.0} & &= 10^{-8.4} \\
k_2 \cdot [H_2CO_3^*] &= 10^{-4.8} \cdot 10^{-3.0} & &= 10^{-7.8} \\
k_3 \cdot [H_2O] &= 10^{-7.0} \cdot 1 & &= 10^{-7.0} \\
\hline
& & R_{fw} &= 10^{-6.9}
\end{aligned}
$$

and with $R_b = k_4 \cdot 2 \cdot [Ca^{2+}]_s^2 = R_{fw}$, we obtain $k_4 = 10^{-6.9}/(2 \cdot (10^{-2.6})^2) = 10^{-2.0}$. We furthermore define $k_4' = 2 \cdot k_4 = 10^{-1.7}$.

We observe that the contribution from the first term, $k_1[H^+]$, rapidly diminishes as pH increases. The term can be neglected when pH is more than 6: the forward rate then becomes a constant which depends on P_{CO_2} only. The overall rate which describes calcite dissolution in a system with a constant CO_2 pressure of $10^{-1.5}$ atm is therefore (assuming $[Ca^{2+}] = m_{Ca^{2+}}$):

$$
R = \frac{dm_{Ca^{2+}}}{dt} = R_{fw} - k_4' m_{Ca^{2+}}^2 = 10^{-6.9} - 10^{-1.7} m_{Ca^{2+}}^2 \tag{4.27}
$$

Equation (4.27) can be integrated to provide the Ca^{2+} concentration as a function of time:

$$
\int_0^{m_{Ca_f}} \frac{dm_{Ca^{2+}}}{(R_{fw} - k_4' m_{Ca^{2+}}^2)} = \int_0^t dt \tag{4.28}
$$

where m_{Ca_f} is the Ca^{2+} concentration reached after a time t has elapsed. The solution for Equation (4.28) is:

$$
\frac{1}{2\sqrt{R_{fw}k_4'}} \ln \left[\frac{(\sqrt{R_{fw}} + \sqrt{k_4'} \cdot m_{Ca_f})}{(\sqrt{R_{fw}} - \sqrt{k_4'} \cdot m_{Ca_f})} \right] = t \tag{4.29}
$$

If we insert values for R_{fw} and k_4', the timespace can be obtained that is needed to reach a concentration in solution of, say, 95% of the value at saturation. (The time to reach 100% saturation is infinite). In the case of $P_{CO_2} = 10^{-1.5}$ atm, for which we calculated R_{fw} and k_4, it is:

$$
t_{.95} = 38\,000 \text{ s} = 10.7 \text{ hr}
$$

We may try to apply this equation to a real world situation, for example a shaft in limestone along which water flows in small films (Weyl, 1958; Dreybrodt, 1981), or a thin film of water which covers calcite fragments in a soil. The reaction rate is given in units mmol/cm².s, and must be multiplied with the available surface area A (cm²) of calcite, and divided by the solution volume V (cm³) to obtain the actual rate in mol/l.s. Take a waterfilm of 0.05 cm thickness, which gives $A/V = 20$/cm, so that $t_{.95}$ reduces to 32 minutes. In that timespan a distance is covered which can be calculated from the hydrological situation. A net precipitation of 0.3 m/yr in a soil with 10% water filled porosity, yields a percolation velocity of 3 m/yr. A flowpath of 0.2 mm in soil is therefore sufficient to reach concentrations corresponding to 95% of calcite saturation values. The average flow velocity in a shaft can be calculated from:

$$v = \frac{\delta^2 \rho g}{3\eta} \frac{dh}{dz} \qquad (4.30)$$

where δ is waterfilm thickness (m), ρ is density of water ($1000 \, kg/m^3$), g is the acceleration in the earth's gravitation field ($10 \, m/s^2$), η is viscosity of water ($1.4 \cdot 10^{-3} \, Pa.s$), and dh/dz is the hydraulic gradient (m H_2O/m). The average flow velocity in a waterfilm of 0.5 mm, with a hydraulic gradient of 1, is $v = 0.6 \, m/s$ (or ca. 2 km/hr). The distance, covered in 32 minutes is accordingly ca. 1100 m.

Clearly, the dissolution kinetics of calcite are fast enough to reach equilibrium in a soil which contains calcite fragments, whereas waterflow in and over barren rock may be so fast that water enters the inner realms of the rock while still undersaturated. Such undersaturation is necessary for karst development. The *saturation length*, defined as the length of flow required to reach a certain percentage of complete saturation, decreases with the square of film-thickness, as we have calculated above for flow in a joint. Although that saturation length seems realistic, for example compared with field results obtained by Gunn (1981, cf. Figure 4.20), karst hydrologists maintain that smaller joints and smaller hydraulic gradients are more common (Davis, 1968; Milanović, 1981). The saturation length may easily become as small as a few cm, and cave formation is then difficult to explain.

However, it has been calculated that the dissolution rate is limited by diffusion of reactants to and from the calcite surface when flow is laminar (Weyl, 1958). Buhmann and Dreybrodt (1985) have incorporated the rate equation of Plummer et al. (1978) as well as the diffusion process in a numerical model, to calculate the actual dissolution rate in stagnant waterfilms. They found that the dissolution rate is an almost linear function of $(m_{Ca_s} - m_{Ca^{2+}})$, where m_{Ca_s} is the Ca^{2+} concentration at equilibrium:

$$R = \frac{dm_{Ca^{2+}}}{dt} = k'(m_{Ca_s} - m_{Ca^{2+}}) \qquad (4.31)$$

where k' is a function of temperature, P_{CO_2}, open or closed conditions, and also of film thickness of the stagnant layer of water. Interestingly, the numerical model that includes activity coefficients and complexes, gives rates for turbulent flow which are closely approximated by the quadratic dependence on Ca^{2+} as predicted by Equation (4.27). For laminar flow in water films with a thickness of $0.01 - 0.1$ cm, under open conditions, k' is ca. $1.5 \cdot 10^{-5}$ cm/s (and little dependent on P_{CO_2}). Integrating Equation (4.31) gives the time to reach 95% saturation, with $A/V = 1/cm$, as:

$$t_{.95} = 1/k' \cdot \ln 20$$

or $t_{.95} = 1/(1.5 \cdot 10^{-5}) \cdot 3 = 200\,000 \, s = 56 \, hr$, which is ca. 5 times longer than in the case of turbulent flow. This timelength to reach 95% saturation may be long enough to explain cave formation. The dissolution process takes a few hours, and flow lengths may be considerable in fractured rock during that time. We just note here that another, probably even more powerful mechanism to explain carbonate dissolution is based on mixing of waters with different CO_2 pressures, as will be discussed in Chapter 8.

So far we have considered calcite only, but what about dolomite? The field diagnosis is to discern, with a few drops of 1% HCl, between limestone and dolomite rock, because the latter dissolves much slower and does not as fiercely release CO_2. Other contrasting field experience shows that karst phenomena are identical in both rock-types, and that springs and streams have similar concentrations. Karst rivers fed by dolomite may in fact show higher and more steady concentrations during runoff peaks, which has been attributed to a faster dissolution process for dolomite (Zötl, 1974). Laboratory experiments of dolomite dissolution have been interpreted in a rate expression similar to calcite dissolution by Busenberg and Plummer (1982). The reaction rates were based on weight loss of dolomite crystals, and the experiments were performed as pH-stats far from equilibrium ($SI < -3$). The suggested rate equation predicts negative rates while the solution is still undersaturated, and allows no extrapolation from the highly subsaturated pH-stat experiments to the 'free-drift' conditions which apply in nature. Free-drift experiments at low CO_2 pressure indicate that dissolution rate can be appreciable, and may be not much smaller than of calcite under similar conditions (James, 1981; Schulz, 1988). Dissolution rates found by Appelo et al. (1984) in free drift experiments at $P_{CO_2} = 10^{-3.5}$ atm, and 22°C, are compared with rates calculated with the equations given by Busenberg and Plummer (1982) in fig. 4.30.

Appelo et al. (1984) related the dissolution rate to the SI of the solution:

$$R = \frac{dm_{Ca^{2+}}}{dt} = -k_d \cdot SI = -k_d \cdot \ln\left(\frac{IAP}{K_{dol}}\right) \tag{4.32}$$

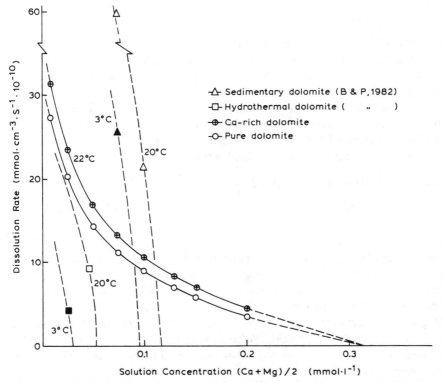

Figure 4.30. Dolomite dissolution rates in free drift experiments, compared with rates given by Busenberg and Plummer (1982). Free drift observations continue until ca. 70% of saturation. Data are normalized to $A/V = 1/cm$.

At a constant CO_2 pressure, one may express IAP and K_{dol} in terms of Ca^{2+} concentration if the difference between activity and concentration is neglected (cf. Equation 4.20):

$$IAP = 256 \cdot [Ca^{2+}]^6 /(P_{CO_2} \cdot K_H K_1/K_2)^2$$

and

$$K_{dol} = 256 \cdot [Ca_s]^6 /(P_{CO_2} \cdot K_H K_1/K_2)^2$$

Substituting in (4.32) and equating activities with molal concentrations gives:

$$dm_{Ca^{2+}}/dt = -k_d \cdot 6 \ln (m_{Ca^{2+}}/ m_{Ca_s})$$

which can be integrated:

$$\int_0^t dt = -\frac{m_{Ca_s}}{6k_d} \int_0^{m_{Ca_f}/m_{Ca_s}} \frac{d\,(m_{Ca^{2+}}/m_{Ca_s})}{\ln\,(m_{Ca^{2+}}/m_{Ca_s})}$$

to give:

$$t = -\frac{m_{Ca_s}}{6k_d} \, \text{Ei}[\ln(m_{Ca_f}/m_{Ca_s})]$$

Here $\text{Ei}(\ln x)$ is the exponential integral, defined as:

$$\text{Ei}(\ln x) = \int_0^x \frac{ds}{\ln s}$$

which is tabulated in Abramowitz and Stegun (1968). Note that for negative arguments ($\ln x < 0$, such as is the case here, since $m_{Ca_f}/ m_{Ca_s} < 1$), the value of the exponential integral is obtained from $-\text{Ei}(-x) = \text{E1}(x)$, which is likewise tabulated, or can be acquired with a numerical approximation.

The equation provided an excellent linear fit to the (curved) concentration increases in solution. The value of k_d was found to vary slightly for dolomites with different Ca/Mg-ratios as shown in Figure 4.30, but was about $1.2 \cdot 10^{-10}$ mmol/cm^2.s at 22°C. We calculate the time to reach 95% saturation with $-\ln (0.95) = 0.0513$ and for $A/V = 1$/cm:

$$t_{.95} = \text{E1}\,(0.0513) \cdot m_{Ca_s}/ (6k_d) = 2.44\, m_{Ca_s}/(6 \cdot 1.2 \cdot 10^{-10})$$

When $P_{CO_2} = 10^{-1.5}$ atm, $m_{Ca_s} = 1.3 \cdot 10^{-3}$ mmol/cm^3 (from Equation 4.20), hence

$$t_{.95} = 4.4 \cdot 10^6 \, s$$

which is more than 100 times longer than required for calcite under similar conditions. How then, are the observations of karst hydrologists reconciled with such a sluggish reaction rate? There is principally one important reason. Dolomite rock weathers (or rather, falls apart) into a grainy, friable material which consists almost of individual crystals. This material gives the magnificent scree slopes which are renowned in the Dolomites of Northern Italy and alps all over the world. Percolation rates in such deposits are fast, but the contact time of water and dolomite is longer than in a barren limestone area. The scree, by its grainy nature, offers furthermore a high ratio of surface area over water volume, that also increases the dissolution rate.

4.4.2 *Precipitation*

The dissolution rate law of Plummer et al. (1978) contains a backward term, which has been used with considerable success to describe the precipitation rate (Plummer et al., 1979; Reddy et al., 1981; Inskeep and Bloom, 1985). The backward rate (Equation 4.26) incorporates activities of ions at the surface where crystallization takes place, that can be expressed as function of the local CO_2 pressure. This local P_{CO_2} may differ from the experimentally accessible value in bulk solution, which introduces another variable in the rate equation. Inskeep and Bloom (1985) used the third term of the rate equation only, since it is dominant when pH > 7.5, and P_{CO_2} < 0.1 atm, as in their experiments. The overall precipitation reaction is then

$$Ca^{2+} + HCO_3^- + OH^- \rightarrow CaCO_3 + H_2O \tag{4.33}$$

with a reaction rate

$$R_p = k_5[Ca^{2+}]\,[HCO_3^-][OH^-] - k_3[H_2O] \tag{4.34}$$

Expression (4.34) is directly derived from Reaction (4.33), and does not require surface activities to be defined. With the mass action constants of Table 4.2, we can rewrite $[HCO_3^-][OH^-] = [CO_3^{2-}] \cdot K_w/K_2$. At equilibrium, $[Ca^{2+}][CO_3^{2-}] = K_{cc}$ (the solubility product of calcite), the forward and backward reaction balance ($R_p = 0$), and we can equate $k_5 = k_3 \cdot [H_2O] \cdot K_2/(K_{cc} \cdot K_w)$. If we use concentrations and activity coefficients, and set the activity of water to one, Equation (4.34) is finally rewritten as:

$$R_p = k_3 \cdot \gamma_2^2/K_{cc} \cdot \{(m_{Ca^{2+}} \cdot m_{CO_3^{2-}}) - K_{cc}/\gamma_2^2\} \tag{4.35}$$

where γ_2 is the activity coefficient for a double-charged ion. Inskeep and Bloom (1985) have noted that this equation is equal to the empirical rate equation of Nancollas and Reddy (1971):

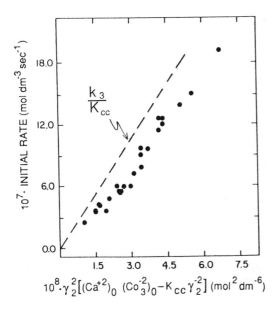

Figure 4.31. Initial rates of calcite precipitation on seed crystals at 25°C, compared with the rate calculated from the Plummer et al. (1978) formula. Experimental data from Inskeep and Bloom (1985).

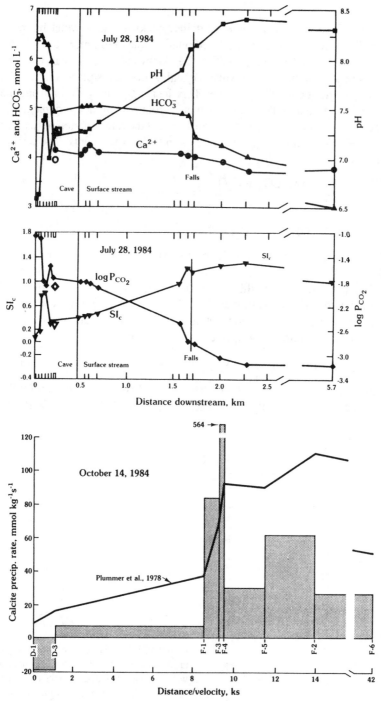

Figure 4.32. Calcite precipitation along Falling Spring Creek. Above: Change of concentrations along the reach. Below: Observed mass which is deposited, compared with predicted mass according to the Plummer et al. (1978) rate equation (from Herman and Lorah, 1988. Reprinted with permission from Pergamon Press PLC).

$$R_p = k_3 \cdot (\Omega - 1) \tag{4.35a}$$

Figure 4.31 shows the rates observed by Inskeep and Bloom (1985) in a plot with the parameters from Equation (4.35) as axes. Included in the plot is the theoretical slope $k_3/K_{cc} = 35.5$, where $k_3 = 10^{-6.93}$ is obtained from the Plummer et al. (1978) equations, and $K_{cc} = 10^{-8.48}$ at 25°C. A reasonable agreement with the slope of 30.35 ± 2.94 found in the experiments by Inskeep and Bloom can be observed.

One of the problems with kinetic data is to estimate the reactive surface area of the mineral. In carbonate dissolution experiments it is either estimated from geometrical calculation of the surface of the grains, or by gas adsorption with the BET technique. It is well known that the surface characteristics may change during the dissolution process, whereby etch pits and etched lines form (e.g. Baumann et al., 1985). However, it is rarely checked how the BET surfaces compare before and after the dissolution experiment, although etching might actually increase the surface during dissolution and influence observed rates. Estimation of the surface area for field situations has hardly passed the stage of, well, reasonable guessing. An interesting study by Herman and Lorah (1988) compared the mass of calcite deposited by a karst stream with the amounts predicted by the Plummer et al. (1978) rate equation. The actual deposited mass was calculated from decline of concentrations at sampling points along the stream, as shown in Figure 4.32. The theoretical mass was obtained by multiplying the rate, as calculated for the streamwater composition, with the wetted perimeter of the streambed. The comparison is shown in Figure 4.32, and we may note some discrepancies; the predicted amounts are lower where a waterfall increases outgassing of CO_2 and precipitation, and are higher in the other stretches of the stream. However, agreement is generally within a factor of 2 to 3, also for sampling times in other seasons, and is surprisingly reasonable.

4.4.3 *Inhibitors*

The Plummer-Wigley-Parkhurst rate equation is firmly based on a large number of measurements in different solution compositions, and incorporation of all the species in solution which contribute to the dissolution process. As long as water compositions in the field are akin to the experimental conditions, one can expect that the rate equation offers a good description of the field rate. Fresh waters in karst areas are mainly $Ca(HCO_3)_2$ solutions and thus should behave well. Calcium and HCO_3^- are relatively minor elements in seawater, and large discrepancies are manifest there. Surface seawater shows supersaturation with respect to calcite, aragonite, and dolomite, but the first two are slow to precipitate, and dolomite does not precipitate at all in the surface layers of the ocean. This is one of the remarkable non-equilibrium features of nature.

The discrepancies are explained as due to inhibitors. Seawater contains about five times as much Mg^{2+} as Ca^{2+} on a molar base. Berner (1975) showed in a series of precipitation experiments with calcite and aragonite, that the rate of calcite precipitation in the presence of Mg^{2+} becomes negligible. Apparently, the presence of large amounts of Mg^{2+} disturbs the crystal growth of calcite. In contrast, aragonite is not affected by the presence of Mg^{2+} (Figure 4.33). Note in Figure 4.33 that seawater is largely supersaturated for calcite, but much less so for aragonite.

Inhibition by Mg^{2+} is due to adsorption of the ion on the crystal surface, where it blocks the subsequent crystal growth of pure calcite. The Mg^{2+} ion can become incorporated in the

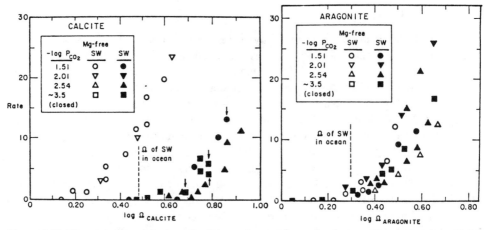

Figure 4.33. The rate of precipitation of calcite and aragonite vs the degree of saturation in artificial seawater with or without Mg^{2+}. $\log \Omega = SI = \log IAP/K$, or the saturation index (Berner, 1975. Reprinted with permission from Pergamon Press PLC).

structure when supersaturation is very high and the resulting calcites may contain up to 30% $MgCO_3$. Organic acids (Berner et al., 1978) and phosphates (Walter and Hanor, 1979; Mucci, 1986) are also well known inhibitors, and likewise form crystalline solids with Ca^{2+} in nature. They can inhibit the carbonate reactions in fresh water environments and soils, but have been reported to do so mainly with the precipitation reaction (Inskeep and Bloom, 1986). The inhibitory action already starts at low concentrations of phosphates (< 1.0 μmol/l, cf. Figure 3.17) and organic acids (< 10 μmol/l, Reddy, 1977; Tomson, 1983) but the diversity and combination of several inhibitors that may be found in nature have only recently been subjected to quantitative treatment (Tomson, 1983).

As a rule of thumb, dissolution of carbonates is generally fast enough to reach thermodynamic equilibrium, while precipitation of calcite in natural settings is sluggish when the saturation ratio Ω is smaller than 5, corresponding to SI values smaller than 0.3. Supersaturation with respect to calcite, ($SI = 0 - 0.3$), is no exception in fresh groundwater and may truly reflect the process which has led to supersaturation: dissolution of gypsum, production of additional alkalinity through SO_4^{2-} reduction, or dissolution of high Mg-calcites.

4.5 SOLID SOLUTION: Mg-CALCITES

Recent marine carbonate sediments are extensively found in tropical areas like the Caribbean and the South Pacific. They consist of aragonite and high Mg-calcite of biogenic origin, such as shells, skeletal debris, carbonate mud, etc. A minor part is played by abiotic precipitation at high supersaturation which also gives aragonite and high Mg-calcite (Given and Wilkinson, 1985). These minerals are unstable relative to low Mg-calcite, and recrystallize in time, thereby affecting the hydrological properties of the rock (Hanshaw and Back, 1979). The chemistry of the recrystallization process can be simulated by dissolving high Mg-calcite in distilled water, and the results of such an experiment are shown in Figure 4.34. Initially both the Mg^{2+} and Ca^{2+} concentrations increase, cor-

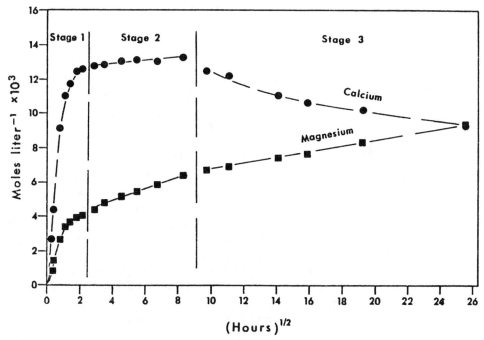

Figure 4.34. The dissolution of a high Mg-calcite in distilled water (Plummer and Mackenzie, 1974).

responding to congruent dissolution. However, after some time, the Ca^{2+} concentration starts to fall. Apparently a new Mg-calcite with a lower $MgCO_3$ content precipitates, and the process can be formalized as:

$$Ca_{1-x}Mg_xCO_3 \rightarrow aCa_{1-y}Mg_yCO_3 + (x - ay)Mg^{2+} + (1 - x - a + ay)Ca^{2+} \\ + (1 - a)CO_3^{2-}$$

where $y < x$.

In this fashion, a row of Mg-calcites with slowly decreasing $MgCO_3$ contents will precipitate, and the water composition becomes enriched in Mg. The process of dissolution whereby another, less soluble solid precipitates, is termed *incongruent dissolution*.

In order to predict the stable compositions of Mg-calcites at different solution compositions, we need to know the solid solution properties of the Mg-calcites. Figure 4.34 shows that it is quite impossible to measure the solubility of a Mg-calcite with a specific composition from such dissolution experiments since the solid phase composition changes during the experiment. This inability to reach equilibrium in experiments with Mg-calcite, and which probably also is valid in many natural settings, has led Thorstenson and Plummer (1977) to reevaluate the equilibria criteria for two-component solid solutions. They propose the concept of stoichiometric saturation, which defines equilibrium between an aqueous solution and a homogeneous solid solution of fixed composition. For Mg-calcite the stoichiometric solubility expression becomes:

$$Ca_{1-x}Mg_xCO_3 \rightarrow (1 - x)Ca^{2+} + xMg^{2+} + CO_3^{2-}$$

for which the stoichiometric solubility product is:

$$K_{(x)} = [Mg^{2+}]^x [Ca^{2+}]^{1-x} [CO_3^{2-}] \tag{4.36}$$

Note the similarity between this type of solubility constant and that of dolomite, which has a true fixed composition. For stoichiometric saturation as expressed by Equation (4.36) the $[Mg^{2+}]/[Ca^{2+}]$ ratio in solution is decoupled from the $MgCO_3$ content of the solid, which is in contrast to full exchange equilibrium (3.30), (3.31) and (3.32), where the $[Mg^{2+}]/[Ca^{2+}]$ ratio of the solution determines the composition of the solid. Stoichiometric saturation can be considered as a less restrictive equilibrium condition than exchange equilibrium and the latter can be shown to be a special case of the first. The stoichiometric saturation concept has been applied with some success to the dissolution reaction of Mg-calcite (Busenberg and Plummer, 1989) and the strontianite-aragonite system (Plummer and Busenberg, 1987). Criteria for attaining true stoichiometric saturation are that dissolution should be congruent as visible in stage 1 of Figure 4.34, that a constant *IAP* (corresponding to Equation 4.36) is maintained over a considerable range of time, and that no apparent changes in solution composition occurs when the crystals are placed in a solution with the same *IAP* = $K_{(x)}$. Busenberg and Plummer (1989) prevented reprecipitation of lower Mg-calcite, such as happens during stage 3 in Figure 4.34, by adding phosphate to their experiments which strongly inhibits calcite precipitation. The stability of Mg-calcite, obtained with the stoichiometric saturation concept, is shown in Figure 4.35.

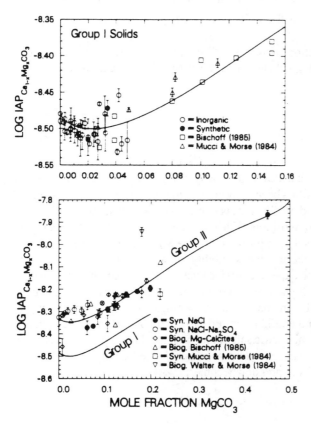

Figure 4.35. The stoichiometric solubility of well crystallized Mg-calcite (Group I) and poorly crystallized Mg-calcite (Group II) (from Busenberg and Plummer, 1989. Reprinted with permission from Pergamon Press PLC). Note the difference in scale.

The Mg-calcites were subdivided in well crystallized material, Group I, consisting of high temperature and slowly precipitated crystals, and poorly crystalline material, Group II, embracing biogenic material and rapidly precipitated synthetic Mg-calcite. Figure 4.35 shows indeed that high Mg-calcite, such as present in recent marine carbonates, are unstable relative to low Mg-calcite and in addition that calcite containing a few mole percent $MgCO_3$ is more stable than pure calcite.

It is possible to derive the true exchange equilibrium properties of a solid solution, according to Equations (3.30) and (3.31), from stoichiometric saturation data. Using the Gibbs-Duhem equation, Thorstenson and Plummer (1977) showed that the following expressions between the activity of the solid components at exchange equilibrium and the stoichiometric saturation constants $K_{(x)}$ are valid:

$$\log [CaCO_3]_{ss} = -x \cdot \frac{d (\log K_{(x)})}{dx} + \log K_{(x)} - \log K_{cc} \tag{4.37}$$

and

$$\log [MgCO_3]_{ss} = (1 - x) \cdot \frac{d (\log K_{(x)})}{dx} + \log K_{(x)} - \log K_{magn} \tag{4.38}$$

Here K_{cc} and K_{magn} are the solubilities of pure calcite and magnesite respectively. In principle, the values of $[CaCO_3]_{ss}$ and $[MgCO_3]_{ss}$ can be obtained from the slope of the curve in Figure 4.35 and the values of $K_{(x)}$. Busenberg and Plummer (1989) found that the Mg-calcites are best described when using dolomite instead of magnesite as the Mg-rich end-member. They modeled their experimental data using a two parameter subregular solid solution model (see Section 3.4.1). The results are shown in Figure 4.36.

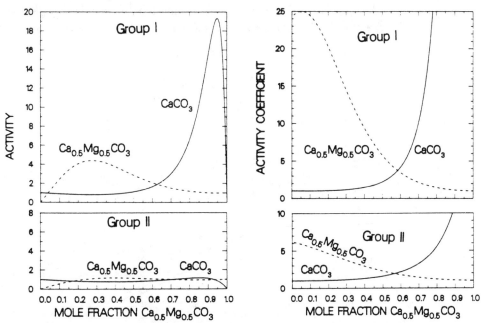

Figure 4.36. The activity and activity coefficients of components in the solid solution using calcite and dolomite as end members (Busenberg and Plummer, 1989. Reprinted with permission from Pergamon Press PLC).

Remember that for ideal solid solution, solid phase activities should be between 0 and 1 and obviously large deviations from this ideal solid solution are found. This is also reflected in the activity coefficients. For ideal solid solution, the solid phase activity coefficient λ is equal to one and it is clear that near the end-members ideal solid solution is most closely approximated for the dominant component, while the minor component shows very large deviations. Ideal solid solution requires that no ordering in the crystal lattice is found, and it is interesting to note that group II Mg-calcites, which are the most disordered ones, approximate ideal solid solution more closely than the better crystalline group I Mg-calcites. Through the concept of stoichiometric saturation it appears possible to derive thermodynamic properties for solid solutions from non-equilibrium data.

Although this description of Mg-calcites provides some insight in the system, the basic problem remains, however, that Mg-calcites under field conditions show non-equilibrium behavior. Neither is stoichiometric saturation a workable concept for field studies, since the criteria for attaining stoichiometric saturation cannot be controlled. The important conclusion to be learned from the laboratory studies is that Mg-calcite containing a few mole percent $MgCO_3$ is the most stable phase, which at least indicates in which direction the reactions may go.

4.6 PLEISTOCENE CARBONATE AQUIFERS

Coastal marine carbonate sediments rich in aragonite and high Mg-calcite may become elevated above the seawater level and enter the freshwater zone. Sea level variations, for example related to glaciations and deglaciations, may amount to several hundreds of meters and are an obvious mechanism to lift marine sediments into the freshwater zone. Tectonic activity may have the same effect. Carbonate aquifers of this type are found many places in the tropics, for example in the Caribbean and the Pacific. Once introduced into the

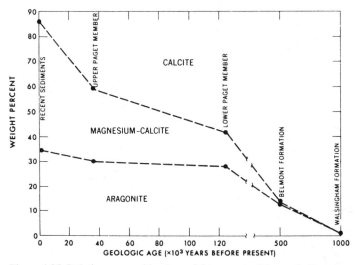

Figure 4.37. Relative composition by weight percent of Bermuda limestones as a function of geological age (Plummer et al., 1976).

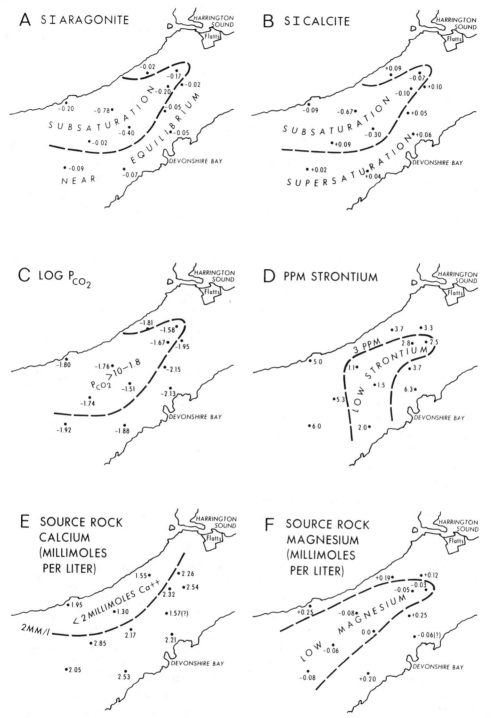

Figure 4.38. The groundwater chemistry of the Bermuda aquifer. *SI* is the saturation index (*SI* = log (*IAP/K*)). Source rock concentrations of Ca and Mg reflect the contribution derived from limestone dissolution (Plummer et al., 1976).

freshwater environment high Mg^{2+} concentrations no longer inhibit low Mg-calcite precipitation and a tremendous recrystallization process starts. First aragonite will recrystallize to low Mg-calcite according to the reaction:

$$CaCO_{3\ aragonite} \rightarrow Ca^{2+} + CO_3^{2-} \rightarrow CaCO_{3\ calcite}$$

The reaction is written as a two step process in order to emphasize that in reality a dissolution/precipitation process is taking place.

One of the best studied aquifers in Pleistocene carbonates is the Bermuda aquifer. The sediments of Bermuda consist of Pleistocene calcarenites (beach deposits, dunes, etc.). The sediments are oldest in the central part of the island and become younger towards the edges. The change in mineralogy as a function of time is shown in Figure 4.37. Note that the recent sediments contain about 35% aragonite and 50% high Mg-calcite with 14 mole% magnesite on average. At the other end of the scale, 10^6 years old limestones consist exclusively of low Mg-calcite. In between, a huge recrystallization process takes place which is reflected in the groundwater chemistry. Bermuda groundwaters consist of several freshwater lenses in which a few percent of seawater is mixed. The groundwater chemistry in the freshwater lenses of Bermuda is shown in Figure 4.38. First one may note (A and B) that the groundwater in the central parts of Bermuda is subsaturated for both calcite and aragonite. This is due to the presence of marshes on central Bermuda, which produce CO_2-rich waters that enter the aquifer. Part C displays the regional distribution of P_{CO_2} and shows that the highest values are found in the central part of the island and coincide with the area where subsaturation for calcite and aragonite is found.

In the marginal parts, the groundwaters are close to equilibrium with aragonite, but supersaturated for calcite. This can be interpreted as dissolution of aragonite and precipitation of calcite. Apparently, dissolution of aragonite is faster than precipitation of calcite, so that groundwaters are kept close to equilibrium with aragonite. Thus calcite precipitation is the overall rate limiting step in the recrystallization of aragonite to calcite. An additional indication for aragonite dissolution, are the high strontium concentrations found in the marginal groundwaters (part D). Small amounts of strontium can easily be incorporated in aragonite (Plummer and Busenberg, 1987), but not in calcite (see Chapter 3). Thus when aragonite recrystallizes to calcite, Sr^{2+} remains in the groundwater. In our case, the Sr/Ca ratio in groundwaters is three times greater than in the aragonitic skeletal grains which are being dissolved.

Finally, the amounts of Ca^{2+} and Mg^{2+} in the groundwaters which are derived from limestone dissolution, can be calculated by substraction of the part which comes from seawater using the cation/Cl$^-$ ratio. The contribution from precipitation is negligible. The results, parts E and F, show that much of the Ca^{2+} in solution is derived from rock dissolution but very little Mg^{2+}. Source rock Mg^{2+} would indicate the ongoing dissolution of high Mg-calcite and the petrographic data (Figure 4.37), strongly indicates that high Mg-calcite dissolution should take place. However, since no Mg^{2+} enrichment is found in the groundwater, high Mg-calcite dissolution is probably a slow process compared to the residence time of the groundwater which is about 100 years in the Bermuda aquifer.

In conclusion then, it is recrystallization of aragonite to low Mg-calcite that particularly influences the groundwater chemistry of the Bermuda aquifer. It produces groundwater supersaturated for low Mg-calcite with a high strontium content. The presence of marshes in the central part of the island appears to be the cause of extensive limestone dissolution in the central parts of the aquifer.

PROBLEMS

4.1. Calculate *TIC* in Example 4.1, at pH = 7, when the solution contains 10 mmol NaCl/l, Na^+ balances alkalinity. Correct for activity coefficients. Why is the correction problematic when pH = 10?

4.2. Calculate pH and $[CO_3^{2-}]$ in a sample with a *TIC* content of 1.7, and $m_{HCO_3^-}$ of 1.3 mmol/l. (Hint: consider as a first estimate that $TIC = m_{H_2CO_3^*} + m_{HCO_3^-}$, i.e. pH < 8.5).

4.3. Calculate the activity and concentration of CO_3^{2-} at pH = 10.5 and pH = 6.3, when *TIC* = 2.5 mmol/l. Ionic strength is fixed to I = 0.1 mol/l.

4.4. Calculate pH and CO_2 species in pure water at a P_{CO_2} of 10^{-2} atm.

4.5. Calculate Ca^{2+} concentration when CO_2 pressure is 0.01 atm and calcite dissolves until saturation, also calculate the pH of this solution.

4.6. Which CO_2 pressure is associated with a Ca^{2+} concentration of 2 mmol/l, assuming equilibrium with calcite, and absence of other reactions?

4.7. Soil-water in equilibrium with calcite has a pH = 7.6. Calculate the CO_2 pressure? (Hint: find $m_{HCO_3^-}$ as a function of P_{CO_2} with (4.13) and (4.15); also for $m_{CO_3^{2-}}$ as function of P_{CO_2} with (4.15) and calcite equilibrium).

4.8. Calculate the time required to reach 95% saturation with calcite under open conditions, initial $P_{CO_2} = 10^{-3.5}$ atm. Calculate the distance covered in this timespan in a limestone fracture of width δ = 0.05 cm.

4.9. Groundwater samples from the Veluwe (Netherlands) are listed below (values in mmol/l):

	1	2	3	4
pH	4.55	7.54	7.86	6.24
Na^+	0.16	0.41	0.59	0.81
K^+	0.03	0.02	0.05	0.10
Mg^{2+}	0.03	0.20	0.39	0.26
Ca^{2+}	0.05	1.55	2.62	1.43
Cl^-	0.17	0.32	0.51	1.11
HCO_3^-	0.07	3.57	2.75	0.76
SO_4^{2-}	0.22	0.01	1.86	0.55
NO_3^-	0.17	<.002	0.008	1.47
Fe	<.001	0.02	0.02	<.001
Al	0.14	<.001	0.008	<.001

a. Calculate P_{CO_2}.
b. Which water was in contact with calcite? Did dissolution occur in an open or in a closed system?
c. Do you see effects of contamination?

4.10. Pouhon is the local name for a spring with water rich in Fe and CO_2 in Spa, Belgium. An analysis is given below (values in mmol/l):

pH	?	Cl^-	.54
Na^+	1.0	Alk	6.38
K^+	.15	SO_4^{2-}	.43
Mg^{2+}	.54	Fe	.34
Ca^{2+}	.77	CO_2	66.0

a. Calculate pH from HCO_3^-/CO_2 equilibrium.

b. What would be pH of Pouhon-water, if P_{CO_2} decreases to atmospheric level?

c. Check the analysis of the Pouhon-water. Alkalinity and CO_2 were titrated at the spring, Fe was analyzed in the laboratory, in a non-acidified sample. Explain the error in the charge balance.

d. What would be the change in alkalinity, if all Fe precipitates?

e. CO_2 in Pouhon-water might originate from Eifel-vulcanism, and is called 'juvenile'. What is 'juvenile'?

4.11. Write reactions which take place when the following substances are added to a water sample. Does alkalinity increase, remain constant, or decrease?

1. CO_2	4. $MnCl_2$	7. Na_2CO_3
2. H_2SO_4	5. $MnCl_2 + O_2$	8. $NaHCO_3$
3. $FeCl_3$	6. NaOH	9. Na_2SO_4

4.12. An analysis of water from a playa (evaporating lake) gave as results: pH = 10.0, SiO_2 = 60 mg/l. What is the contribution of Si to titrated alkalinity?

Data: Alkalinity $= m_{HCO_3^-} + 2m_{CO_3^{2-}} + m_{OH^-} - m_{H^+} + \text{base} - \text{acid}$.

$$H_4SiO_4^0 \leftrightarrow H_3SiO_4^- + H^+ \qquad \log K = -9.1.$$

REFERENCES

Abramowitz, M. and Stegun, I.A., 1964, *Handbook of mathematical functions.* Dover Pub., New York, 7th print, 1046 pp.

Albertsen, M., 1977, *Labor- und Felduntersuchungen zum Gasaustausch zwischen Grundwasser und Atmosphäre über natürlichen und verunreinigten Grundwässern.* Thesis, Univ. Kiel, 145 pp.

Appelo, C.A.J., Groen, M., Heidweiler, G. and Smit, P., 1983, Hydrochemistry of springs in an alpine carbonate/serpentinite terrain. *4th. Int. Symp. Water-Rock Interaction, Misasa, Japan*, pp. 26-31.

Appelo, C.A.J., Beekman, H.E. and Oosterbaan, A.W.A., 1984, Hydrochemistry of springs from dolomite reefs in the southern Alps of Northern Italy. *IAHS Pub.* 150, 125-138.

Atkinson, T.C., 1975, Diffuse flow and conduit flow in limestone terrain in the Mendip Hills, Somerset, England. *IAH Congress, Huntsville, Alabama 1975.*

Atkinson, T.C., 1983, Growth mechanism of speleothems in Castleguard cave, Columbia icefields, Alberta, Canada. *Arct. Alpine Res.* 15, 523-536.

Back, W. and Hanshaw, B.B., 1970, Comparison of chemical hydrogeology of the carbonate peninsula of Florida and Yucatan. *J. Hydrol.* 10, 330-368.

Back, W., Hanshaw, B.B., Plummer, L.N., Rahn, P.H., Rightmire, C.T. and Rubin, M., 1984, Process and rate of dedolomitisation: mass transfer and ^{14}C dating in a regional carbonate aquifer. *Geol. Soc. Am. Bull.* 94, 1415-1429.

Baumann, J., Buhmann, D., Dreybrodt, W. and Schulz, H.D., 1985, Calcite dissolution kinetics in porous media. *Chem. Geol.* 53, 219-228.

Berner, R.A., 1975, The role of magnesium in the crystal growth of calcite and aragonite from sea water. *Geochim, Cosmochim. Acta* 39, 489-504.

Berner, R.A., Westrich, J.T., Graber, R., Smith, J. and Martens, C.S., 1978, Inhibition of aragonite precipitation from supersaturated seawater: a laboratory and field study. *Am. J. Sci.* 278, 816-837.

Bögli, A., 1978, *Karsthydrographie und physische Speläologie.* Springer Verlag, Berlin, 292 pp.

Bolin, B., Degens, E.T., Duvigneaud, P. and Kempe, S., 1979, The global biogeochemical carbon cycle. In B. Bolin et al. (eds), *The global carbon cycle*, SCOPE 13. Wiley & Sons, New York, 1-56.

Brook, G.A., Folkoff, M.E. and Box, E.O., 1983, A world model of soil carbon dioxide. *Earth Surf. Proc.* 8, 79-88.

Buhmann, D. and Dreybrodt, W., 1985, The kinetics of calcite dissolution and precipitation in geologically relevant situations of karst areas. 1, Open systems. *Chem. Geol.* 48, 189-211.

Busenberg, E. and Plummer, L. N., 1982, The kinetics of dissolution of dolomite in $CO_2 - H_2O$ systems at 1.5 to 65°C and 0 to 1 atm P_{CO_2}. *Am. J. Sci.* 282, 45-78.

Busenberg, E. and Plummer, L. N., 1989, Thermodynamics of magnesium calcite solid-solutions at 25° and 1 atm total pressure. *Geochim. Cosmochim. Acta* 53, 1189-1208.

Davis, S. N., 1968, Initiation of groundwater flow in jointed limestone. *Bull. Nat. Speleol. Soc.* 28, 111-118.

Deer, W. A., Howie, R. A. and Zussman, J., 1966, *An introduction to the rock forming minerals*. Longman, London, 528 pp.

Deike, R. G., 1991, Comparative petrology of cores from two test wells in the Eastern part of the Edwards aquifer, South-Central Texas. US Geol. Surv. Water Resour. Inv. Rep. 87-4266, 142 pp.

Drake, J. J., 1983, The effects of geomorphology and seasonality on the chemistry of carbonate groundwater. *J. Hydrol.* 61, 223-236.

Drake, J. J. and Wigley, T. M. L., 1975, The effect of climate on the chemistry of carbonate groundwater. *Water Resour. Res.* 11, 958-962.

Dreybrodt, W., 1981, Kinetics of the dissolution of calcite and its applications to karstification. *Chem. Geol.* 31, 245-269.

Ford, D. C. and Williams, P. W., 1989, *Karst geomorphology and hydrology*. Unwin Hyman Inc., Winchester MA, 601 pp.

Garrels, R. M. and Mackenzie, F. T., 1971, *Evolution of sedimentary rocks*. Norton, New York, 397 pp.

Gíslason, S. R., 1989, Kinetics of water-air interaction in rivers: a field study in Iceland. In D. L. Miles (ed.), *Water-Rock Interaction: Proc. 6th Water-Rock Interaction Symp., Malvern*. Balkema, Rotterdam, 263-266.

Given, R. and Wilkinson, B. H., 1985, Kinetic control of morphology, composition and mineralogy of abiotic sedimentary carbonates. *J. Sed. Petrol.* 55, 109-119.

Gunn, J., 1981, Hydrological processes in karst depressions. *Z. Geomorphol. N.F.* 25, 313-331.

Hanshaw, B. B. and Back, W., 1979, Major geochemical processes in the evolution of carbonate-aquifer systems. *J. Hydrol.* 43, 287-312.

Hardie, L. A., 1987, Dolomitization: a critical view of some current views. *J. Sed. Petrol.* 57, 166-183.

Harmon, R. S., White, W. B., Drake, J. J. and Hess, J. W., 1975, Regional hydrochemistry of North American carbonate terrains. *Water Resour. Res.* 11, 963-967.

Herman, J. S. and Lorah, M. M., 1988, Calcite precipitation rates in the field: Measurement and prediction for a travertine-depositing stream. *Geochim. Cosmochim. Acta* 52, 2347-2355.

Hissink, D. J., 1937, The reclamation of the Dutch saline soils (Solonchak) and their further weathering under the humid climatic conditions of Holland. *Soil Sci.* 45, 83-94.

Hoogendoorn, J. H., 1983, *Hydrochemistry of Eastern Netherlands* (in Dutch). DGV-TNO, Delft.

Inskeep, W. P. and Bloom, P. R., 1985, An evaluation of rate equations for calcite precipitation kinetics at pCO_2 less than 0.01 atm and pH greater than 8. *Geochim. Cosmochim. Acta* 49, 2165-2180.

Inskeep, W. P. and Bloom, P. R., 1986, Kinetics of calcite precipitation in the presence of water-soluble organic ligands. *Soil Sci. Soc. Am. J.* 50, 1167-1172.

James, A. N., 1981, Solution parameters of carbonate rocks. *Bull. Int. Ass. Eng. Geol.* 24, 19-25.

Langmuir, 1971, The geochemistry of some carbonate ground waters in central Pennsylvania, *Geochim. Cosmochim. Acta* 35, 1023-1045.

Laursen, S., 1991, On gaseous diffusion of CO_2 in the unsaturated zone. *J. Hydrol.* 122, 61-69.

Milanović, P. T., 1981, *Karst hydrogeology*. Water Resour. Pub., Littleton CO, 434 pp.

Morrow, D. W., 1982, The chemistry of dolomitization and dolomite precipitation. *Geoscience Canada* 9, 5-13.

Morse, J. W., 1983, The kinetics of calcium carbonate dissolution and precipitation. In R. J. Reeder (ed), Reviews in Mineralogy, Miner. Soc. Am. 11, 227-264.

Mucci, A., 1986, Growth kinetics and composition of magnesian calcite overgrowths precipitated from seawater: quantitative influence of orthophosphate ions. *Geochim. Cosmochim. Acta* 50, 217-233.

Mucci, A., 1987, Influence of temperature on the composition of magnesian calcite overgrowths precipitated from seawater. *Geochim. Cosmochim. Acta* 51, 1977-1984.

Nancollas, G.H. and Reddy, M.M., 1971, The crystallization of calcium carbonate II. Calcite growth mechanism. *J. Colloid Interface Sci.* 37, 824-829.

Nordstrom, D.K., Plummer, L.N., Langmuir, D., Busenberg, E., May, H.M., Jones, B.F and Parkhurst, D.L., 1990, Revised chemical equilibrium data for major water-mineral reactions and their limitations. In D.C Melchior and R.L Bassett (eds), *Chemical modeling of aqueous systems* II, ACS Symp. Ser. 416, 398-413.

Parkhurst, D.L., Thorstenson, D.C. and Plummer, L.N., 1980, PHREEQE – A computer program for geochemical calculations. US Geol. Surv. Water Resour. Inv. 80-96, 210 pp.

Pitman, J.I., 1978, Carbonate chemistry of groundwater from chalk, Givendale, East Yorkshire. *Geochim. Cosmochim. Acta* 42, 1885-1897.

Plummer, L. N. and Busenberg, E., 1987, Thermodynamics of aragonite strontianite solid-solutions: Results from stoichiometric solubility at 25° and 76°C. *Geochim. Cosmochim. Acta* 51, 1393-1411.

Plummer, L.N. and MacKenzie, F.T., 1974, Predicting mineral solubility from rate data: application to the dissolution of magnesian calcites. *Am. J. Sci.* 274, 61-83.

Plummer, L.N., Vacher, H.L., Mackenzie, F.T., Bricker, O.P. and Land, L.S., 1976, Hydrogeochemistry of Bermuda: a case history of groundwater diagenesis of biocalcarenites. *Geol. Soc. Am. Bull.* 87, 1301-1316.

Plummer, L.N., Wigley, T.M.L. and Parkhurst, D.L., 1978, The kinetics of calcite dissolution in CO_2 water systems at 5° to 60°C and 0.0 to 1.0 atm CO_2. *Am. J. Sci.* 278, 179-216.

Plummer, L.N., Parkhurst, D.L. and Wigley, T.M.L. 1979, Critical review of the kinetics of calcite dissolution and precipitation. In E.A. Jenne (ed.), *Chemical modeling in aqueous systems*, ACS Symp. Ser. 93, 537-573.

Plummer, L.N., Busby, J.F., Lee, R.W. and Hanshaw, B.B., 1990, Geochemical modeling in the Madison aquifer in parts of Montana, Wyoming and South Dakota. *Water Resour. Res.* 26, 1981-2014.

Reardon, E.J., Allison, G.B. and Fritz, P., 1979, Seasonal chemical and isotopic variations of soil CO_2 at Trout Creek, Ontario. *J. Hydrol.* 43, 355-371.

Reddy, M.M., 1977, Crystallization of calcium carbonate in the presence of trace concentrations of phosphorous containing anions. *J. Cryst. Growth* 41, 287-295.

Reddy, M.M., Plummer, L.N. and Busenberg, E., 1981, Crystal growth of calcite from calcium bicarbonate solutions at constant P_{CO_2} and 25°C: a test of a calcite dissolution model. *Geochim. Cosmochim. Acta* 45, 1281-1289.

Rickard, D. and Sjöberg, E.L., 1983, Mixed kinetic control of calcite dissolution rates. *Am. J. Sci.* 283, 815-830.

Rozema, J., Laan, P., Broekman, R., Ernst, W.H.O. and Appelo, C.A.J., 1985, On the lime transition and decalcification in the coastal dunes of the province of North Holland and the island of Schiermonnikoog. *Acta Bot. Neerl.* 34, 393-411.

Shuster, E.T. and White, W.B., 1972, Source areas and climatic effects in carbonate groundwaters determined by saturation indices and carbon dioxide pressures. *Water Resour. Res.* 8, 1067-1073.

Shulz, H.D., 1988, Labormessung der Sättigungslänge als Mass für die Lösungskinetik von Karbonaten im Grundwasser. *Geochim. Cosmochim. Acta*, 52, 2651-2657,

Sjöberg, E.L., 1978, Kinetics and mechanism of calcite dissolution in aqueous solutions at low temperatures. *Stockholm Contrib. Geol.* v. XXXII, 1, 1-92.

Spears, D.A. and Reeves, M.J., 1975, The infiltration rate into an aquifer determined from the dissolution of calcite. *Geol. Mag.* 112, 585-591.

Stumm, W. and Morgan, J.J., 1981, *Aquatic chemistry.* 2nd ed. Wiley & Sons, New York, 780 pp.

Thorstenson, D.C. and Plummer, L.N., 1977, Equilibrium criteria for two-component solids reacting with fixed composition in an aqueous phase – example: the magnesian calcites. *Am. J. Sci.* 277, 1203-1223.

Tomson, M.B., 1983, Effect of precipitation inhibitors on calcium carbonate scale formation. *J. Cryst. Growth* 62, 106-112.

Walter, L.M. and Hanor, J.S., 1979, Effect of orthophosphate on the dissolution kinetics of biogenic magnesian calcites. *Geochim. Cormochim. Acta* 43, 1377-1385.

Warfvinge, P.H. and Sverdrup, H. 1989, Modeling limestone dissolution in soils. *Soil Sci. Soc. Am. J.* 53, 44-51.

Weyl, P.K., 1958, Solution kinetics of calcite. *J. Geol.* 66, 163-176.

White, W.B., 1984, Rate processes: chemical kinetics and karst landform development. In R.G. LaFleur (ed), *Groundwater as a geomorphic agent*. Allen & Unwin, London.

White, W.B., 1988, *Geomorphology and hydrology of karst terrains*. Oxford Univ. Press, New York, 464 pp.

Wigley, T.M.L., 1973, The incongruent dissolution of dolomite. *Geochim. Cosmochim. Acta* 37, 1397-1402.

Zötl, J.G., 1974, *Karsthydrogeologie*. Springer Verlag, Berlin, 291 pp.

CHAPTER 5

Ion exchange and sorption

Soils and aquifers contain abundant materials which are able to sorb chemicals from water. Sorption is, in abstract terms, the change in concentration of a chemical in the solid matter as a result of mass transfer among solution and solid. Sorption processes may be subdivided as illustrated in Figure 5.1. *Adsorption* indicates that a chemical adheres to the surface of the solid, *absorption* suggests that the chemical is taken up *into* the solid, and *exchange* involves a replacement of one chemical for another one at the solid surface.

Sorption processes are modeled with equations that are derived from the law of mass action. The major distinction between *sorption* (adsorption and absorption) and *exchange* is that model equations for sorption use the concentration of one chemical only, and neglect effects of other solutes. In other words, a simple linear or non-linear relation between sorbed and solute concentrations gives the distribution of the chemical over solid and solution. Distribution coefficients are often applied to describe absorption of hydrophobic organic chemicals (as will be discussed in Chapter 9), and to model adsorption of trace elements. Ion exchange equations, on the other hand, do explicitly account for all ions which compete for the exchange sites. This complicates the calculations of ion exchange for natural waters which always contain a variety of ions, but we will show in this chapter that the algebra is straightforward.

It is sometimes difficult to discern between sorption (or exchange) and other reactions with solids such as precipitation and dissolution (Sposito, 1984). A major difference is that sorption and exchange reactions are limited by the *sorption* or *exchange capacity* of the solid. Clay minerals, organic matter and oxides/hydroxides all have a certain exchange capacity for cations (and also for anions). Sorption and exchange has become an important topic for hydrologists, since it largely regulates transport of pollutant chemicals in aquifers and soils. Cation exchange acts as a *temporary buffer* in non-steady state situations which are the result of pollution, acidification, or moving salt/fresh water interfaces. It tends to smoothen changes in water composition, and on the other hand can completely alter cation concentrations through a process known as *ion-chromatography*. A hydrologist who tries to decipher hydrogeological conditions may use the hydrochemical patterns which are a result of cation exchange. As aquifers become more exploited and polluted, flowpatterns and water quality along flowlines vary but cation exchange allows the interpretation of former, displaced water quality at sampling points. This chapter demonstrates the principles of ion exchange in aquifers and soils. We start with examples of cation exchange as indicator of salinization. Then we treat the exchanger materials, exchange equations, and give practical examples, amongst others on salt water intrusions, and chromatographic

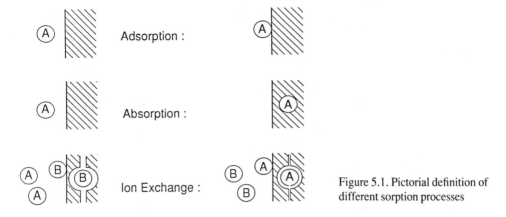

Figure 5.1. Pictorial definition of different sorption processes

patterns in aquifers. The last part of this chapter gives a more detailed discussion of electrostatic theories for the interaction of charged particles and solutions, and of the surface complexation model that is applied more and more to describe adsorption of trace elements.

5.1 CATION EXCHANGE IN SALT/FRESH WATER INTRUSIONS

Fresh water is often dominated by Ca^{2+} and HCO_3^- ions, as a result of dissolution of calcite. Cation exchangers in the aquifer therefore have mostly Ca^{2+} adsorbed on the surfaces. In sea water Na^+ and Cl^- are the dominant ions, and sediment in contact with sea water will have adsorbed Na^+ for a large part. When seawater intrudes in a coastal fresh water aquifer an exchange of cations takes place:

$$Na^+ + \tfrac{1}{2} Ca\text{-}X_2 \rightarrow Na\text{-}X + \tfrac{1}{2} Ca^{2+}$$ (5.1)

where X indicates the soil exchanger.

Sodium is taken up by the exchanger, and Ca^{2+} is released. The water quality thus changes from NaCl to $CaCl_2$ type water. The reverse process takes place with refreshening, i.e. when fresh water flushes a salt water aquifer:

$$\tfrac{1}{2} Ca^{2+} + Na\text{-}X \rightarrow \tfrac{1}{2} Ca\text{-}X_2 + Na^+$$ (5.2)

where Ca^{2+} is taken up from water, in return for Na^+, with a $NaHCO_3$ type water as result. Water quality can thus indicate upconing of sea water, or conversely that salt water is flushed by fresh water.

An example is given in Figure 5.2, showing a section through the Dutch coastal dunes (Stuyfzand, 1985). Groundwater had been pumped abundantly from the area for drinking water until a few decades ago an alarming upconing of salt water threatened to salinize all the pumping wells. Fresh water from the river Rhine was subsequently infiltrated via numerous canals and recovered by shallow wells. More water was deliberately infiltrated than was produced, to flush the salt water backwards. This strategy has been succesful, as indicated by the 'freshening' $NaHCO_3$ water type now observed (Figure 5.2). The quality pattern of Figure 5.2 has been constructed on the basis of a detailed network of sampling wells, and the pattern allows for an interpretation of flowlines in the aquifer. When Na^+ has

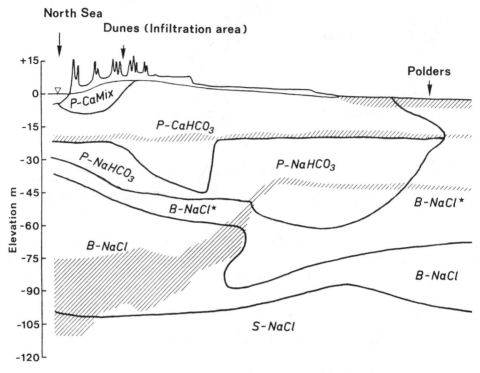

Figure 5.2. Section through the Dutch coastal dunes near Castricum, showing refreshening of a brackish water aquifer. The prefixes P, B and S before the water types indicate fresh, brackish and salt water. Low permeability layers (clays) are shown hatched. (Modified from Stuyfzand, 1985).

been flushed from the sediment, the $Ca(HCO_3)_2$ type fresh water dominates again, and it will appear sooner where flushing is more rapid, i.e. along the more rapid flowlines.

Another example is from the Nile delta in Egypt, shown in Figure 5.3. Stiff diagrams of the groundwater quality indicate salt water intrusion in boreholes 121 and 129, as evidenced from the relative increase of Ca^{2+} when compared with sea water (also shown in Figure 5.3). The boreholes near the Ismailya canal, which runs from Caïro to the Red Sea, show on the other hand a relative decrease of Ca^{2+} when compared with Cl^- in Nile water. Chloride is a conservative element which increases in this groundwater primarily as a result of evapotranspiration of Nile water used for irrigation. The relative decrease of Ca^{2+} is partly a result of calcite precipitation, but the concomitant relative increase of Na^+, suggests that cation exchange of Ca^{2+} for Na^+ from the soil is also important. An explanation can be found in Na-containing loess that is blown in from the desert, to settle in the wetter area of the delta, as has been observed in the Negev desert (Nativ et al., 1983).

The effects of cation exchange are particularly clear when analyses are plotted in a Piper diagram, as shown in Figure 5.4. Remember that in a Piper diagram, percentages of cations and anions from a water analysis are plotted in two separate triangles (lower triangles in Figure 5.4). The two points of a watersample are combined into one point in the central diamond shaped field by lines from the points parallel to the outer sides until these intersect in the central field. 'Average' fresh water and seawater have been plotted in Figure 5.4.

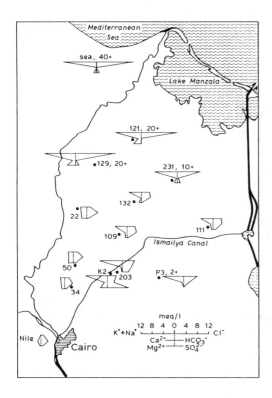

Figure 5.3. Stiff diagrams of groundwater in the Nile delta, Egypt, indicating salt water upconing in borehole 121 and 129, and Ca^{2+}/Na^+ exchange near the Ismailya canal.

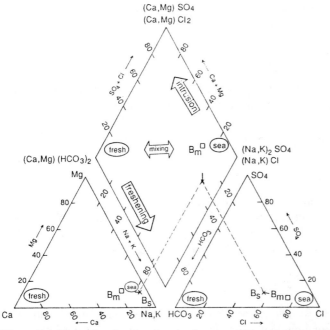

Figure 5.4. Piper plot showing 'average' compositions of fresh water and seawater, and NaHCO$_3$-sample (B$_s$) from Table 5.2. The tail at B$_s$ points to the calculated composition of a conservative mixture B$_m$.

Table 5.1. Groundwaters from the Netherlands showing 'mixed water' composition, and composition influenced by cation exchange ('CaCl$_2$' and 'NaHCO$_3$' types). Concentrations in mmol/l.

	Mixed water		CaCl$_2$ water		NaHCO$_3$ water	
pH	7.5	7.2	6.91	6.6	8.7	8.3
Na$^+$	24.53	54.33	341.	124.	40.	2.0
K$^+$	0.82	1.41	2.8	2.4		
Mg^{2+}	2.9	7.1	27.9	30.7	2.8	1.1
Ca^{2+}	3.0	7.9	39.6	47.2	0.7	0.65
Cl$^-$	27.2	70.8	440.	271.	25.6	1.4
HCO$_3^-$	9.2	15.3	7.0	3.8	14.4	4.0
SO$_4^{2-}$	0.07	0	18.8	4.7	2.7	0.2
Dutch borehole code			48E, 45-1	48H, 35-1	48E, 69-1	48A, 3-3

Groundwater samples in a coastal area can show a surplus of Ca^{2+} which indicates seawater intrusion, or a surplus of Na$^+$ which indicates fresh water intrusion. They may also plot in between fresh and seawater compositions in the Piper diagram, which indicates that conservative mixing has occurred. A few typical groundwaters that show these trends are given in Table 5.1.

The chemical reactions during fresh/salt water displacements can be deduced more specifically by calculating a composition based on conservative mixing of salt water and fresh water, and comparing the conservative concentrations with those actually found in the water analysis. The concentration of an ion i, by conservative mixing of seawater and fresh water is:

$$m_{i,\mathrm{mix}} = f_{\mathrm{sea}} \cdot m_{i,\mathrm{sea}} + (1 - f_{\mathrm{sea}}) \cdot m_{i,\mathrm{fresh}} \qquad (5.3)$$

where m_i is concentration of i (mmol/l), f_{sea} is fraction of seawater in the mixed water, and subscripts $_{\mathrm{mix}}$, $_{\mathrm{sea}}$, and $_{\mathrm{fresh}}$ indicate the conservative mixture, and end-members seawater and fresh water.

Any change in concentration $m_{i,\mathrm{react}}$ as a result of reactions (not of mixing) then becomes simply:

$$m_{i,\mathrm{react}} = m_{i,\mathrm{sample}} - m_{i,\mathrm{mix}} \qquad (5.4)$$

where $m_{i,\mathrm{sample}}$ is the actually observed concentration in the sample.

The fraction of seawater is normally based on Cl$^-$ concentration of the sample. Chloride is assumed to be a conservative parameter just as it was used in Chapter 2 to find the seawater contribution to rainwater. The Cl$^-$ based fraction of seawater is:

$$f_{\mathrm{sea}} = \frac{m_{\mathrm{Cl}^-,\mathrm{sample}} - m_{\mathrm{Cl}^-,\mathrm{fresh}}}{m_{\mathrm{Cl}^-,\mathrm{sea}} - m_{\mathrm{Cl}^-,\mathrm{fresh}}} \qquad (5.5)$$

Ions in fresh water near the coast are often derived from seaspray, and only Ca^{2+} and HCO$_3^-$ are added due to calcite dissolution. In this case $m_{i,\mathrm{fresh}} = 0$ for all components except Ca^{2+} and HCO$_3^-$. If there is no source of Cl$^-$ but seawater, we can write $m_{\mathrm{Cl}^-,\mathrm{fresh}} = 0$ as well, which simplifies Equation (5.5) to:

Table 5.2. Recalculated water analyses of Table 5.1, showing the extent of reactions (cation exchange). Values in mmol/l.

	Seawater	CaCl$_2$ water			NaHCO$_3$ water			Fresh
		Sample	Mix	React	Sample	Mix	React	water
Na$^+$	485.	341.	377.	−36.	40.	22.	18.	0
K$^+$	10.6	2.8	8.2	− 5.4				0
Mg^{2+}	55.1	27.9	42.8	−14.9	2.8	2.5	0.3	0
Ca^{2+}	10.7	39.6	9.0	30.6	0.7	3.3	−2.6	3.0
Cl$^-$	566.	440.	440.	−	25.6	25.6	−	0
HCO$_3^-$	2.4	7.0	3.2	3.8	14.4	5.8	8.6	6.0
SO$_4^{2-}$	29.3	18.8	22.8	− 4.0	2.7	1.3	1.4	0
% seawater	100	78			5			0
Symbol in Figure 5.4:					B$_s$	B$_m$		

Sample: given water composition from Table 5.1. Mix: calculated water composition based on mixing. React: Sample − Mix, showing effects of cation exchange and other reactions.

$$f_{sea} = m_{Cl^-,sample}/566 \tag{5.5a}$$

where Cl$^-$ concentration is expressed in mmol/l, and the Cl$^-$ concentration of 35‰ seawater is set at 566 mmol/l. (Notice that the ratio between major ions in ocean water is near constant everywhere on earth).

Table 5.2 illustrates the results of this calculation for analyses of Table 5.1. The relative increase of Na$^+$ in NaHCO$_3$ water, and of Ca^{2+} in CaCl$_2$ water is quite obvious here. Other conspicuous changes are an increase of HCO$_3^-$ in NaHCO$_3$ type water. This relative increase is commonly found, and may easily amount to 10 mmol/l or more; it can be attributed to calcite dissolution that occurs when Ca^{2+} is removed by cation exchange with Na$^+$ (Back, 1966; Chapelle, 1983). The calcite dissolution is driven by subsaturation due to loss of Ca^{2+} from solution. The pH of the resulting NaHCO$_3$ water often becomes high, above 8, also as the result of dissolution of calcite. The CaCl$_2$ type waters on the other hand, often have a low pH, below pH = 7. This may be the result of calcite precipitation, driven by the increase of Ca^{2+} due to cation exchange. However, other reactions also take place, as can be seen in Table 5.2.

The other reactions can be visualized in a Piper diagram, by plotting a tail from the plotted point of the water analysis towards the composition resulting from conservative mixing of seawater and fresh water. The NaHCO$_3$ water of Table 5.2 has been entered in Figure 5.4, as point B$_s$ (the actual sample) and point B$_m$ (the calculated mixture). When only cation exchange takes place, the tail is parallel to the outer side of the Piper diamond. When other reactions occur, the tail deviates from the side. Dissolution of CaCO$_3$ brings the mixture towards the Ca^{2+} + HCO$_3^-$ corner; reduction of SO$_4^{2-}$ gives a shift parallel to the anion-axis; deposition of CaCO$_3$ leads towards the Na$^+$ and (SO$_4^{2-}$ + Cl$^-$) corner. These shifts have been indicated in Figure 5.5. Groundwater analyses from Zeeland and Western Brabant in the Netherlands have been plotted, areas where inundations have been common in the past centuries. The tail given with each plotted point is directed towards the location on the mixing line based on chloride content of the water samples.

It is clear from Figure 5.5, that reactions with calcite occur upon refreshening of the

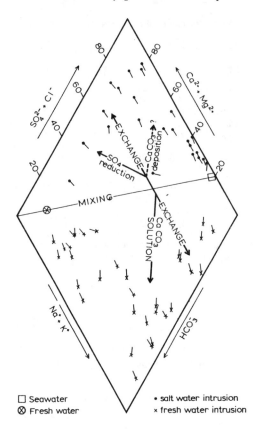

Figure 5.5. Piper plot showing compositions of groundwater from Zeeland and Western Brabant in The Netherlands, in which cation exchange is visible. The tail at plotted points is directed towards position on the mixing line based on Cl⁻ concentration of the groundwater sample.

aquifer (relative increase of HCO_3^-), but are less important during salinization. In fact, it seems that water composition evolves in a direction opposite to what would be expected for an exchange of Na^+ for Ca^{2+} accompanied by $CaCO_3$ deposition. It is possible that $CaCO_3$ deposition (though occurring) is masked in the Piper diagram by SO_4^{2-} reduction, since it lowers the percentage of (SO_4^{2-} + Cl^-), and also produces HCO_3^- by the reaction:

$$SO_4^{2-} + 2\,CH_2O \rightarrow H_2S + 2\,HCO_3^- \tag{5.6}$$

Sulfate reduction is a common process and may be related to the genesis of these waters: seawater often intrudes via peat which started to grow at the initial rise of sea level. It is also possible that the higher salt concentrations of sea water induce an exchange of protons from the sediment. This effect is observed when the pH of a soil is measured in distilled water and in 1 M KCl solution: the latter invariably gives a lower pH by as much as 2 pH-units. With seawater intrusion, the effect would lower the pH of water so that it might actually become subsaturated with respect to calcite, despite the high Ca^{2+} concentrations which are due to cation exchange.

5.2 ADSORBENTS IN SOILS AND AQUIFERS

All solid surfaces in soils and aquifers can act as adsorbers. However, solids with a large specific surface may adsorb more, so that the grain size influences the adsorption capacity

Table 5.3. Cation exchange capacities of common soil and sediment materials.

	CEC, meq/100g
Kaolinite	3 – 15
Halloysite	5 – 10
Montmorillonite	80 – 120
Vermiculite	100 – 200
Glauconite	5 – 40
Illite	20 – 50
Chlorite	10 – 40
Allophane	up to 100
Goethite & Hematite:	up to 100
Organic matter (C)	150 – 400 (at pH = 8)
or, accounting for pH-dependence:	51·pH – 59 (= *CEC* per 100 g organic carbon), (Scheffer and Schachtschabel, 1970)

considerably. Solids with a large specific surface reside in the clay fraction (< 2 μm), and the coarser grains in a sediment are often coated with organic matter and with oxyhydroxides of iron. Adsorption capacity is therefore linked to clay content (fraction < 2 μm), clay minerals, organic matter (% C), and oxide or hydroxide content. Usually the *CEC*, or Cation Exchange Capacity of a soil, is given in meq/100 g. An empirical formula which relates *CEC* to percentage clay (< 2 μm) and organic carbon is (e.g. Breeuwsma et al. 1986):

$$CEC \text{ (meq/100 g)} = 0.7 \cdot (\% \text{ clay}) + 3.5 \cdot (\% \text{ C}) \tag{5.7}$$

Table 5.3 gives cation exchange capacities of common soil constituents, and Example 5.1 shows how these values can be compared with amounts of cations in solution.

EXAMPLE 5.1: *Recalculate CEC (meq/100 g soil) to concentration (meq/l pore water)*
The cation exchange capacity (*CEC*) of a sediment is expressed in units meq per 100 g of dry soil. Express a *CEC* of 1 meq/100 g of two sediments with porosities ε = 0.2 and 0.3, in meq/l pore water. The sediment consists mainly of quartz-grains with specific weight of 2.65 g/cm³.

ANSWER
Since quartz has a specific weight of 2.65 g/cm³, the 100 g of dry sediment contains 100/2.65 = 37.7 ml 'grains'. The amount of pore water in 100 g sediment can be calculated and *CEC* expressed in units meq/l pore water:

For ε = 0.2: Total sediment volume is 37.7/(1 – ε) = 47.2 ml; pore water = 47.2 × ε = 9.4 ml;
         CEC = 1 meq/100g = 1 / 9.4 meq/ml = 106 meq/l pore water;
         bulk density ρ_b = 100 / 47.2 = 2.12 g/cm³.

For ε = 0.3: Pore water volume = 37.7 (ε / (1 – ε)) = 16.2 ml;
         CEC = 1 meq/100 g = 1 / 16.2 meq/ml = 62 meq/l pore water;
         bulk density ρ_b = 100 /(37.7 + 16.2) = 1.86 g/cm³.

Note that *CEC* can be expressed in meq/l pore water by multiplying the concentration in meq/100 g with $10 \rho_b/\varepsilon$.

Example 5.1 also serves to illustrate that adsorbed amounts are often quite large in comparison to amounts in solution, even though the concentration in the solid phase appears to be low. Fresh water normally contains less than 10 meq/l cations, which is 6 to 10 times less than is present in exchangeable (adsorbed) form in this example of a quartz-rich sediment.

5.2.1 Clay minerals

Clay minerals can have a deficit of (positive) charge which arises from substitutions of cations in the structure of the crystal. A substitution of Al^{3+} for Si^{4+} reduces the positive charge by one in an otherwise not much affected structure. The possible substitutions can be understood most easily with Pauling's rules for building crystal structures (Pauling, 1960). The crystal structures of the alumino-silicates are built from oxygens which form coordination units around a central cation. The first rule entails that in filling the sites of cations in the framework of oxygens, the length of the cation-oxygen bond is most important. It is even more important than the charge on the cation. Charge imbalance can, if needed, be compensated in other units some distance away from the point of imbalance (although Pauling's second rule states that the most stable structure is obtained when charges are balanced closely). The basic coordination units for clay minerals are tetrahedra and octahedra of which oxygens form the corners, and in which the cation resides in the centre (Figure 5.6; note that the oxygens are larger and 'touch' over the side). Tetrahedral coordination means that a central cation is surrounded by four oxygens, octahedral coordination means that a cation is surrounded by six oxygens.

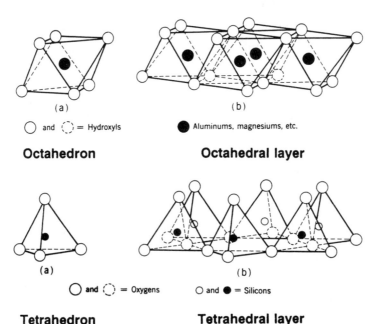

Figure 5.6. A perspective drawing of octahedra and tetrahedra as coordinating units in minerals (from Grim, 1968).

The oxygens in the coordination units are envisaged as spheres with a radius of ca. 1.4 Å which should just touch. Tetrahedra therefore have smaller metal ions such as Si or Al, and octahedra have larger metal ions such as Al, Mg, Fe, Mn. (Still larger metal ions such as Ca, Na, K, etc. have even larger coordination polyhedra). In the clay minerals, tetrahedra and octahedra form layers which share oxygens, and stacking of the layers determines the type of clay mineral. In *kaolinite* the layers of tetrahedra (T) and octahedra (O) show the repetition: ⊢O ⊢O ⊢O ⊢O. The chemical formula of kaolinite is:

$$[Al_2]^{vi} \, [Si_2]^{iv} \, O_5(OH)_4$$

where superscripts ^{vi} and ^{iv} indicate six- and four-coordination with oxygen. Stacking of a tetrahedral layer of ca. 3 Å with an octahedral layer of ca. 4 Å gives a repetition of the ⊢O structure every 7 Å. This is the characteristic *c*-axis spacing (or 001 reflection) observed in X-ray diffraction analysis. The outer oxygens of the octahedra of one layer share protons with the tetrahedra from the next layer.

In *mica* or soil clay minerals such as *smectite* or *montmorillonite* the octahedra are sandwiched between two layers of tetrahedra, and the structure repeats as: ⊢O⊣ ⊢O⊣ ⊢O⊣. The ⊢O⊣ structure is ca. 10 Å thick, and cohesion of the ⊢O⊣ layers is obtained with *interlayer cations*. Another group, of which *chlorite* is an example, has the the interlayer filled with an octahedral layer, and these structures can be indicated as repetitions of ⊢O⊣ O ⊢O⊣O (of 14 Å each).

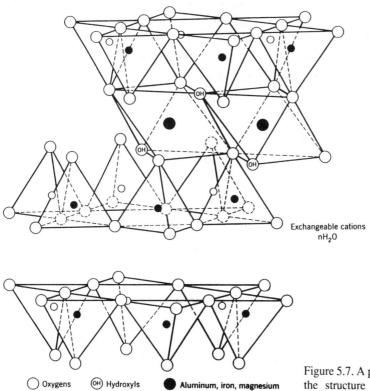

Exchangeable cations
nH₂O

○ Oxygens   ⊙ Hydroxyls   ● Aluminum, iron, magnesium

○ and ● Silicon, occasionally aluminum

Figure 5.7. A perspective drawing of the structure of the clay mineral smectite (from Grim, 1968).

All these minerals are found in the finest fraction of soils and aquifers. A charge imbalance is mostly limited to the mica-type clay minerals. The charge imbalance can be easily calculated from the structural formula, starting from the mica *muscovite*:

$$K^{xii} [Si_3Al]^{iv} [Al_2]^{vi}O_{10}(OH)_2 \tag{5.8}$$

The 10 oxygens and 2 hydroxyls give 22 negative charges; four tetrahedral metal sites, and two octahedral metal sites must be filled in this unit formula. When the charge of the cations adds up to 21, there is just a single charge left for the interlayer cation, which is K^+ in muscovite. This K^+ is actually in twelve-coordination with oxygens from the two adjacent tetrahedral layers. When Si takes a larger share in the tetrahedral cation sites, the interlayer charge is diminished. For example,

$$K_{0.7}^{xii} [Si_{3.3}Al_{0.7}]^{iv} [Al_2]^{vi}O_{10} (OH)_2 \tag{5.9}$$

which is the typical formula for the clay mineral *illite*. A further decrease of the interlayer charge is possible, but this also reduces the cohesion of the tetrahedral layers around the interlayer cation. The interlayer cation may then become hydrated, and the ⊢ o ⊣ layers separate as more and more layers of water are incorporated in the structure.

Clay minerals with a mica structure, which expand to 18 Å with Mg-ions as interlayer cations, are called *smectites*. Figure 5.7 shows such a mineral with hydrated cations in the interlayer space. The cations in the interlayer space can freely move into solution to be exchanged for other ions. Smectites (or *montmorillonites*) typically have an interlayer charge that ranges from 0.6 to 0.3. The effect of interlayer charge on swelling can be calculated with an electrostatic model (Appelo, 1979). Figure 5.8 shows how the energy of expansion of separate layers is related to the interlayer charge.

Substitutions also occur in the octahedral layer, for example in *vermiculite*:

$$K_{0.9}^{xii} [Si_{3.8}Al_{0.2}]^{iv} [Al_{1.3}Mg_{0.7}]^{vi} O_{10}(OH)_2 \tag{5.10}$$

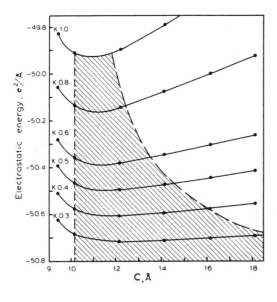

Figure 5.8. The electrostatic energy of mica as function of interlayer charge and of expansion of the interlayer space. The x-axis gives c or the 001 spacing of the unit cell. Hatched area shows where expansion is more favorable than the contracted structure. 1 $e^2/Å = 332.1$ kcal/mol.

Here, the low tetrahedral charge of 0.2 is surpassed by an octahedral charge of 0.7, and the full interlayer charge is 0.9. Vermiculites are able to swell to 14 Å with strongly hydrated cations such as Mg^{2+}, despite the large interlayer charge, since the charge imbalance is predominantly in the octahedral layer, i.e. at a relatively large distance from the oxygens in the basal tetrahedral plane to which the interlayer cations are bonded. However, most soil clay minerals with a high charge will tend to collapse when ions of suitable size enter the interlayer space, such as K^+, NH_4^+ or ions of similar size and low hydration number. A collapsed structure can only exchange through solid state diffusion, which is 3 to 4 orders of magnitude slower than diffusion in solution.

EXAMPLE 5.2. *Structural charge of smectite*
A smectite has the chemical formula $Na_{0.3}$ $[Si_{3.8}Al_{0.2}]^{iv}$ $[Al_{1.9}Mg_{0.1}]^{vi}$ $O_{10}(OH)_2$. Calculate the amount of exchangeable cations per gram of mineral, and calculate the charge density on the basal plane in $\mu eq/m^2$. The size of the unit cell is $a \times b \times c = 5.2 \times 9 \times 10 Å^3$, and $a \times b = 5.2 \times 9 = 46.8 Å^2$ is the size of the basal plane where the interlayer cations reside.

ANSWER
The gram formula weight of the smectite is 366.75 g/mol. Hundred grams of the smectite contain 0.27 moles of the unit cell formula, hence the exchange capacity is $0.3 \cdot 0.27 \cdot 1000 = 81.8$ meq/100 g. The charge of 0.3 is divided over the two basal planes of a completely separated ⊢ O ⊣ layer, or 0.15 q_e per 46.8 $Å^2$. This is $3.2 \cdot 10^{17}$ q_e/m^2 (q_e is the elementary charge). Multiply with the elementary charge $1.6 \cdot 10^{-19}$ C, and obtain 0.05 C/m^2; or divide by Avogadro's number $N_a = 6 \cdot 10^{23}$/mole, and obtain the charge density as 0.53 $\mu eq/m^2$.

5.2.2 *Charge from surface reactions*

Clay minerals also have a variable charge due to protonation of surface oxygens, and deprotonation of hydroxyls. The charge on the clay mineral kaolinite is normally attributed to surface reactions, and so is the charge of oxides and hydroxides such as goethite (FeOOH), pyrolusite (MnO_2), as well as silicates. The protonation reactions can be written as:

$$\equiv SO^- + 2H^+ \leftrightarrow \equiv SOH + H^+ \leftrightarrow \equiv SOH_2^+ \tag{5.11}$$

where $\equiv SO$ indicates a surface oxygen. This surface oxygen may either be part of the structure of the solid, or it may be an adsorbed water molecule. It is also possible that specific adsorption of ions imparts a surface charge. For example, adsorption of Pb^{2+} on the surface of goethite gives a positive charge:

$$Pb^{2+} + \equiv SOH \leftrightarrow \equiv SOPb^+ + H^+ \tag{5.12}$$

The surface charge depends on the pH of the solution. It can be zero at the pH where the surface oxygens are protonated just enough to compensate broken bonds and a small internal charge. The pH where the charge of the mineral is zero is called *Point of Zero Charge*, *PZC* or pH_{PZC}; recent literature uses more accurate expressions such as *Point of Zero Net Proton Charge* (indicating that the mineral may still have a structural charge, but that at pH_{PZNPC} the dissociation of surface hydroxyls adds no charge), or *Pristine Point of Zero Charge* (indicating that other effects are absent, such as charge due to specific adsorption) (cf. Bolt, 1982; Sposito, 1984; Davis and Kent, 1990). Table 5.4 lists *PZC*'s, for

Table 5.4. Point of zero charge pH_{PZC} of clay minerals and common soil oxides and hydroxides (Parks, 1967; Stumm and Morgan, 1981; Davis and Kent, 1990).

	pH_{PZC}
Kaolinite	4.6 (Parks)
Montmorillonite	< 2.5 (Parks)
Corundum, α-Al_2O_3	9.1 (Stumm and Morgan)
γ-Al_2O_3	8.5 (Stumm and Morgan)
α-$Al(OH)_3$	5.0 (Stumm and Morgan)
Hematite, α-Fe_2O_3	8.5 (Davis and Kent)
Goethite, α-$FeOOH$	7.3 (Davis and Kent)
$Fe(OH)_3$	8.5 (Stumm and Morgan)
Birnessite, δ-MnO_2	2.2 (Davis and Kent)
Rutile, TiO_2	5.8 (Davis and Kent)
Quartz, SiO_2	2.9 (Davis and Kent)
Calcite, $CaCO_3$	9.5 (Parks)
Hydroxyapatite, $Ca_5OH(PO_4)_3$	7.6 (Davis and Kent)

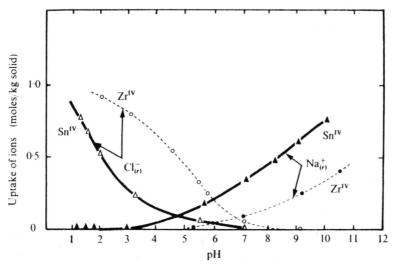

Figure 5.9. pH-dependent sorption on SnO_2 and ZrO_2 oxides: Cl^- is adsorbed at low pH, Na^+ at high pH (from Paterson, 1970).

a number of minerals. It is measured as the pH where a suspension of particles has lowest mobility in an electric field. Alternatively it can be obtained from acid-base titrations as the pH where buffering capacity is smallest, or where titration curves in background solutions of different normality intersect. Minerals have a capacity for anion exchange when the pH of water is below the *PZC*, and a cation exchange capacity when pH is above *PZC*. The magnitude of the exchange capacity depends on the difference between the *PZC* and the pH of the solution. A clear example is given in Figure 5.9, showing adsorption of Na^+ and Cl^- on the oxides SnO_2 and ZrO_2. The *PZC* of SnO_2 is found at pH = 4.8, and is slightly lower than the *PZC* of ZrO_2 at 6.8. When pH is above *PZC*, the oxide surface adsorbs Na^+ from solution, and when pH is below *PZC*, Cl^- is adsorbed. Both ions are adsorbed in small, equal quantities at the *PZC* of the oxide.

The pH_{PZC} of the clay minerals kaolinite and montmorillonite (Table 5.4) is lower than the pH range of 6.5 to 8.5 that is found normally in groundwater, and these minerals therefore behave as cation exchangers only. Anion exchange capacity might be present on the Fe- and Al-oxides, or on carbonates of aquifer sediments (Table 5.4). If anion exchange is important, the sequence of high to low sorption is from $PO_4^{3-} > F^- > SO_4^{2-} > HCO_3^-$ $> Cl^-$, but this sequence may be affected by the presence of cations and of complexes. Column experiments with aquifer sands and non-aggregated soils generally show that transport of radioactive Cl^- is not retarded with respect to tritium, which indicates that adsorption of Cl^- is absent at pH > 6 (Nkedi-Kizza et al., 1982; Griffioen, 1992; in fact, Cl^- often shows faster flow rates due to anion exclusion, as we will discuss later on). In more acid soils, or in acid aquifer sediments with pH < 5, there can be more substantial adsorption of SO_4^{2-} (Van Stempvoort et al., 1990).

5.3 EXCHANGE EQUATIONS

Ion exchange can be described as a reaction with an equilibrium constant. The exchange of Na^+ for K^+ is for example:

$$Na^+ + K\text{-}X \leftrightarrow Na\text{-}X + K^+ \tag{5.13}$$

with the equilibrium constant:

$$K_{Na\backslash K} = \frac{[Na\text{-}X]\,[K^+]}{[K\text{-}X]\,[Na^+]} \tag{5.14}$$

Note that we write the elements in subscript form below the equilibrium constant in the order in which they appear as solute ions in the reaction. If we write Reaction (5.13) the other way round, the equilibrium constant gets the inverse value:

$$K_{Na\backslash K} = \frac{1}{K_{K\backslash Na}} \tag{5.15}$$

which also means that:

$$K_{I\backslash I} = 1 \tag{5.16}$$

for any ion *I*.

Square brackets in the equations denote activities. The Debije-Hückel theory offers a straightforward model to relate concentrations and activities in water (Chapter 3). For the adsorbed cations the matter is more complicated, and different conventions are in common usé. The use of a given convention is largely a personal choice, prompted primarily by the best fit that can be obtained with experimental data. The lack of agreement among researchers is caused by the absence of a unifying theory to calculate activity coefficients, which the Debije-Hückel theory has so brilliantly provided for solutes in water. Since the convention has a bearing on the results of exchange calculations we must discuss this matter.

Activities of exchangeable ions are sometimes calculated as molar fractions (as for water), but more often as equivalent fractions. These fractions are taken either as fractions of the number of exchange sites, or as fractions of the number of exchangeable cations. The standard state, i.e. where the activity of the exchangeable ion is equal to 1, is in all cases an exchanger which is fully covered with one cation only.

An equivalent fraction β_I for ion I^{i+} is calculated as:

$$\beta_I = \frac{\text{meq } I\text{-}X_i \text{ per 100 g sediment}}{CEC} = \frac{meq_{I\text{-}X_i}}{\sum\limits_{I,J,K,...} meq_{I\text{-}X_i}} \tag{5.17}$$

where $I,J,K,...$ are the exchangeable cations, with charges i,j,k, and $meq_{I\text{-}X_i}$ indicate meq $I\text{-}X_i$ per 100 g.

A molar fraction β_I^M is likewise obtained from:

$$\beta_I^M = \frac{\text{mmol } I\text{-}X_i \text{ per 100 g sediment}}{TEC} = \frac{(meq_{I\text{-}X_i})/i}{\sum\limits_{I,J,K,...} (meq_{I\text{-}X_i})/i} \tag{5.18}$$

where TEC are total exchangeable cations, in mmol/100 g sediment. Use of fractions gives $\sum \beta = 1$. We denote an activity that is calculated with respect to the number of exchanger sites as $[I_{1/i}\text{-}X]$, and when it is calculated with respect to the number of exchangeable cations as $[I\text{-}X_i]$.

For homovalent exchange it makes no difference what convention is used, but for heterovalent exchange the effect is quite notable. A heterovalent exchange is, for example, the exchange of Ca^{2+} for Na^+ such as occurs with seawater intrusion:

$$Na^+ + \tfrac{1}{2} Ca\text{-}X_2 \leftrightarrow Na\text{-}X + \tfrac{1}{2} Ca^{2+} \tag{5.19}$$

with

$$K_{Na\backslash Ca} = \frac{[Na\text{-}X][Ca^{2+}]^{0.5}}{[Ca\text{-}X_2]^{0.5}[Na^+]} = \frac{\beta_{Na}[Ca^{2+}]^{0.5}}{\beta_{Ca}^{0.5}[Na^+]} \tag{5.20}$$

The way of writing of Reaction (5.19) conforms to the *Gaines-Thomas* convention, after Gaines and Thomas (1953) who were among the first to give a rigorous definition of a thermodynamic standard state of exchangeable cations. It uses the equivalent fraction of the exchangeable cations for the activity of the adsorbed ions. Use of molar fractions in Equation (5.20), would follow the *Vanselow* convention (Vanselow, 1932).

If, on the other hand, the activities of the adsorbed ions are assumed proportional to the number of exchange sites (X^-) which are occupied by the ion, Reaction (5.19) is written as:

$$Na^+ + Ca_{0.5}\text{-}X \leftrightarrow Na\text{-}X + \tfrac{1}{2} Ca^{2+} \tag{5.19a}$$

with

$$K_{Na\backslash Ca}^G = \frac{[Na\text{-}X][Ca^{2+}]^{0.5}}{[Ca_{0.5}\text{-}X][Na^+]} = \frac{\beta_{Na}[Ca^{2+}]^{0.5}}{\beta_{Ca}[Na^+]} \tag{5.20a}$$

Reaction (5.19a) agrees with the *Gapon* convention (Gapon, 1933). The molar and equivalent fraction are identical in the Gapon convention: both are based on a single exchanger-site with charge -1.

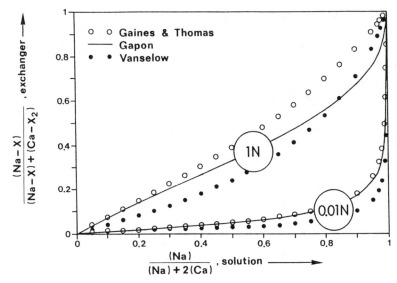

Figure 5.10. Three options to calculate the exchanger composition from solution concentrations give different results. $K_{Na\backslash Ca} = 0.5$; two solution normalities, 1N and 0.01N.

The effects of the three conventions for Na\Ca exchange are shown for two solution normalities of 1N and 0.01N in Figure 5.10. Both adsorbed Na^+ and solute Na^+ are expressed as fractions in this figure. The value for the equilibrium constant in this example has been assumed $K_{Na\backslash Ca} = 0.5$ for all three equations, which means that $[Ca-X_2]$ or $[Ca_{0.5}-X]$ is higher than $[Na-X]$ when both ions have an activity of 1 in solution. The Ca^{2+} ion is thus preferred or selected with respect to Na^+. The higher selectivity for Ca^{2+} is also visible in Figure 5.10. A particular aspect of exchange with heterovalent ions is the effect of total solute concentrations. The higher charged ion is preferred more strongly if total solute concentrations decrease, as is illustrated by the curves for two different normalities in Figure 5.10. When a suspension containing Na^+, Ca^{2+} and clay which has these ions adsorbed, is diluted, the effect is that adsorbed Ca^{2+} increases, and solute Ca^{2+} decreases. This effect is a consequence of the exponent that is used in the mass action equation, and the fact that the clay forms a separate phase in the thermodynamical sense, even though clay and water together may exist in a homogeneous suspension (Vanselow, 1932).

We noted already that the choice for a particular convention is largely a matter of personal choice that is determined by the goodness of fit for experimental data. However, it can be seen in Figure 5.10 that the difference among the equations is small at low total solute concentrations, and one might consider practical aspects of calculation as a factor when making the choice. The *CEC* of a sediment can often be considered constant, whereas *TEC* in a heterovalent system varies, since it depends on the amounts of cations with different charges that neutralize the (constant) *CEC*. This means that in most situations the activities of exchangeable cations are calculated more conveniently (the argument of Gaines and Thomas, 1953) as equivalent fractions with respect to a fixed *CEC*. The Gapon convention is popular among soil scientists, and also uses equivalent fractions. It allows an easy calculation of exchangeable cations in heterovalent systems, and forms the basis for calculating the amount of exchangeable Na^+ from the Sodium Adsorption Ratio (*SAR*, the ratio of Na^+ over $Ca^{2+} + Mg^{2+}$ in water), which is an important parameter for estimating irrigation water-quality (Section 5.7). However, the Gapon equation does not perform well when several heterovalent cations are present and the Gaines-Thomas or the Vanselow convention should in general be preferred (Bolt, 1982; Evangelou and Phillips, 1988). In this book we use mostly the Gaines-Thomas convention.

EXAMPLE 5.3. *Exchange coefficients for different conventions as function of normality of solution*
Wiklander (1957) has determined the molar Ca/K-ratios on several clay minerals and a synthetic resin in equilibrium with a solution that contains equivalent amounts of Ca^{2+} and K^+ in Cl^- solutions of different normality. His results are presented below:

	CEC, meq/100 g	0.1 N	0.01 N	0.001 N	10^{-4} N
Kaolinite	2.3	–	1.8	5.0	11.1
Illite	16.2	1.1	3.4	8.1	12.3
Smectite	81.0	1.5	(10.0)	22.1	38.8
Resin	250.	3.27	10.8	36.0	89.9

Calculate the equivalent and molar fraction of Ca^{2+} and K^+ on the clay, and also the exchange coefficients according to the conventions of Gaines and Thomas, Vanselow, and Gapon. Correct solute concentrations with activity coefficients.

ANSWER
Before starting, it is noted that a 10-fold dilution of the solution gives an increase of the $\sqrt{[Ca^{2+}]}/[K^+]$ ratio of $\sqrt{10} = 3.2$. We expect therefore that each 10-fold dilution will increase the β_{Ca}/β_K ratio by about a factor of 3.2, and we observe that the resin (a synthetic ion exchanger) behaves quite nicely according to this relation. The clay minerals show a somewhat deviating behavior, but the same trend is followed. We are therefore confident that the mass action equation can be applied, and that the calculation of the exchange constant is meaningful. The calculations will be done here for smectite only, leaving the other clay minerals as an exercise.
    From the given molar ratio on the clay, we calculate the molar fractions in combination with:

$$\beta_{Ca}^M + \beta_K^M = 1$$

e.g. for smectite in 0.1N is $\beta_{Ca}^M = 1.5 \, \beta_K^M$, and $\beta_K^M = 1/2.5 = 0.40$, $\beta_{Ca}^M = 0.60$.
    The equivalent fractions are obtained with the help of Equation (5.18):

$$\beta_K^M = \frac{meq_{K-X}}{meq_{K-X} + (meq_{Ca-X_2})/2} = \frac{(meq_{K-X})/CEC}{(meq_{K-X})/CEC + (meq_{Ca-X_2})/(2 \cdot CEC)} = \frac{\beta_K}{\beta_K + \beta_{Ca}/2}$$

and, also in this case is:

$$\beta_{Ca} + \beta_K = 1$$

which gives $\beta_K = \beta_K^M/(2 - \beta_K^M)$.
    For 0.1N we obtain $\beta_K = 0.25$, and $\beta_{Ca} = 0.75$.
    The molal concentrations in solution are for the 0.1N solution $m_{K^+} = 0.05$, $m_{Ca^{2+}} = 0.025$, $m_{Cl^-} = 0.1$ mol/kg H_2O. This gives $I = \frac{1}{2} \sum m_i z_i^2 = 0.125$, and $\gamma_1 = 0.77$ and $\gamma_2 = 0.36$, and the solute activities are $[K^+] = 0.0385$ and $[Ca^{2+}] = 0.009$.

The exchange coefficients are now calculated for the different conventions from:

Gaines & Thomas

$$K_{Ca\backslash K} = \frac{\beta_{Ca}^{0.5} \, [K^+]}{\beta_K [Ca^{2+}]^{0.5}}$$

Vanselow

$$K_{Ca\backslash K}^V = \frac{(\beta_{Ca}^M)^{0.5} \, [K^+]}{\beta_K^M \, [Ca^{2+}]^{0.5}}$$

Gapon

$$K_{Ca\backslash K}^G = \frac{\beta_{Ca} \, [K^+]}{\beta_K \, [Ca^{2+}]^{0.5}}$$

and give:

	0.1 N	0.01 N	0.001 N	10^{-4} N
$K_{Ca\backslash K}$	1.39	(2.29)	1.47	0.79
$K_{Ca\backslash K}^V$	0.78	(1.17)	0.74	0.40
$K_{Ca\backslash K}^G$	1.20	(2.24)	1.45	0.79

It is apparent that the Ca\K selectivity of smectite varies with total solute concentrations, and we should keep in mind that the 'constant' of the mass action equation is actually a variable in the exchange equations.

5.3.1 *Values for exchange coefficients*

The value for the equilibrium coefficient of Na\Ca exchange as written in Equation (5.19) is around 0.4 (Gaines and Thomas convention). Other cations have approximate exchange coefficients listed in Table 5.5. The table gives coefficients rather than constants, since the numeric values are not constant (Example 5.3), but depend on type of exchanger in the soil, and also on water composition. This is implicit in the fact that the exchanger does not behave as 'ideal', and corrections with activity coefficients as applied for ions in water, are not yet generally known for the exchanger surface. A range is given where many measurements in different soils and for different clay minerals are available.

The exchange coefficients are expressed with Na^+ as the reference cation. Exchange coefficients among other cation-pairs are obtained by combining two reactions. For example, the exchange reaction among Al^{3+} and Ca^{2+} is obtained from:

$$Na^+ + \tfrac{1}{2} Ca\text{-}X_2 \leftrightarrow Na\text{-}X + \tfrac{1}{2} Ca^{2+} \qquad K_{Na\backslash Ca} = 0.4$$

and

$$Na^+ + \tfrac{1}{3} Al\text{-}X_3 \leftrightarrow Na\text{-}X + \tfrac{1}{3} Al^{3+} \qquad K_{Na\backslash Al} = 0.6$$

Subtracting the two reactions, and dividing the two exchange coefficients gives:

$$\tfrac{1}{3} Al^{3+} + \tfrac{1}{2} Ca\text{-}X_2 \leftrightarrow \tfrac{1}{3} Al\text{-}X_3 + \tfrac{1}{2} Ca^{2+} \qquad K_{Al\backslash Ca} = 0.7$$

The coefficients for the Gapon and Vanselow convention are identical with the Gaines-Thomas values in the case of homovalent exchange. For heterovalent exchange of divalent ions and Na^+ the coefficients can be obtained from:

Table 5.5. Values for exchange coefficients with respect to Na^+ (Gaines-Thomas convention, i.e. equivalent fractions are used for exchangeable cations). Based partly on a compilation by Bruggenwert and Kamphorst, 1982.

Equation: $Na^+ + 1/i\,I\text{-}X_i \leftrightarrow Na\text{-}X + 1/i\,I^{i+}$ with $K_{Na\backslash I} = \dfrac{[Na\text{-}X]\,[I^{i+}]^{1/i}}{[I\text{-}X_i]^{1/i}\,[Na^+]} = \dfrac{\beta_{Na}\,[I^{i+}]^{1/i}}{\beta_I^{1/i}\,[Na^+]}$

Ion I^+	$K_{Na\backslash I}$	Ion I^{2+}	$K_{Na\backslash I}$	Ion I^{3+}	$K_{Na\backslash I}$
Li^+	1.2 (0.95-1.2)	Mg^{2+}	0.50 (0.4-0.6)	Al^{3+}	0.6 (0.5-0.9)
K^+	0.20 (0.15-0.25)	Ca^{2+}	0.40 (0.3-0.6)	Fe^{3+}	?
NH_4^+	0.25 (0.2-0.3)	Sr^{2+}	0.35 (0.3-0.6)		
Rb^+	0.10	Ba^{2+}	0.35 (0.2-0.5)		
Cs^+	0.08	Mn^{2+}	0.55		
		Fe^{2+}	0.6		
		Co^{2+}	0.6		
		Ni^{2+}	0.5		
		Cu^{2+}	0.5		
		Zn^{2+}	0.4 (0.3-0.6)		
		Cd^{2+}	0.4 (0.3-0.6)		
		Pb^{2+}	0.3		

$$K^V_{Na\backslash I^{2+}} = \frac{2\,K_{Na\backslash I^{2+}}}{\sqrt{1 + \beta_{Na}}}$$

and

$$K^G_{Na\backslash I^{2+}} = \frac{K_{Na\backslash I^{2+}}}{\sqrt{1 - \beta_{Na}}}$$

The selectivity for cations follows, in general for sediments and soils, the lyotropic series. Cations with the same charge are held more strongly when their hydration number is smaller, i.e. when the hydration shell of water molecules that stick to the ion is smaller. A similar sequence is observed with synthetic, strongly acid ion exchange resins (Helfferich, 1959; Riemann and Walton, 1970). It is of interest to note that weakly acid resins exhibit just the reverse selectivity in the high pH range when the resin is fully deprotonated. This may indicate that strong acid cation exchangers are, in general, the more important in the natural environment.

5.3.2 *Calculation of exchanger composition*

It is not difficult to calculate the fractions of exchangeable cations β on a sediment which are in equilibrium with a given water composition, and with values of the equilibrium coefficients known. All the exchangeable fractions are expressed in the fraction β_I of one ion I^{i+}, known solute activities and equilibrium constants with Equation (5.20). Thus, for a general set of ions $I^{i+}, J^{j+}, K^{k+}, ..$ we have:

$$1/i\,I^{i+} + 1/j\,J\text{-}X_j \leftrightarrow 1/i\,I\text{-}X_i + 1/j\,J^{j+}$$

which gives:

$$\beta_J = \frac{\beta_I^{j/i} \cdot K_{J \backslash I}^j \cdot [J^{j+}]}{[I^{i+}]^{j/i}}$$ (5.21a)

and also β_K, β_L, .., are all expressed as a function of β_I. All fractions are introduced in:

$$\beta_I + \beta_J + \beta_K + .. = 1$$ (5.21b)

to give an equation with one unknown, β_I. When only monovalent and divalent ions are present, a quadratic equation results that can be solved (Example 5.4).

EXAMPLE 5.4: *Cation exchange complex in equilibrium with coastal dune groundwater*
Calculate the exchangeable cations in dune sand with a *CEC* of 1 meq/100g, in equilibrium with dunewater, $Na^+ = 1$, $Mg^{2+} = 0.5$, $Ca^{2+} = 2$ mmol/l. Assume that activity is numerically equal to concentration in mol/l, use the equilibrium constants from Table 5.5.

ANSWER
It is advantageous to express the exchangeable fractions of divalent cations in the fraction of the monovalent Na^+ (This gives immediately a quadratic equation). Thus:

$$\beta_{Mg} = \frac{\beta_{Na}^2 \cdot [Mg^{2+}]}{K_{Na \backslash Mg}^2 \cdot [Na^+]^2}$$

$$\beta_{Ca} = \frac{\beta_{Na}^2 \cdot [Ca^{2+}]}{K_{Na \backslash Ca}^2 \cdot [Na^+]^2}$$

$$\beta_{Mg} + \beta_{Ca} + \beta_{Na} = 1$$

Therefore:

$$\beta_{Na}^2 \cdot \left(\frac{[Mg^{2+}]}{K_{Na \backslash Mg}^2 \cdot [Na^+]^2} + \frac{[Ca^{2+}]}{K_{Na \backslash Ca}^2 \cdot [Na^+]^2} \right) + \beta_{Na} - 1 = 0$$

Substitution of the constants from Table 5.5, and analytical concentrations for solute activities yields:

$$\beta_{Na}^2 \cdot \{0.5 \cdot 10^{-3} / (0.5 \cdot 10^{-3})^2 + 2 \cdot 10^{-3} / (0.4 \cdot 10^{-3})^2\} + \beta_{Na} - 1 = 0$$

which gives $\beta_{Na} = 0.00827$, and subsequently, from back substitution, $\beta_{Mg} = 0.137$ and $\beta_{Ca} = 0.855$. We assume that the factor $10\rho_b/\varepsilon = 60$ can recalculate the concentrations from meq/100 g to meq/l pore water (cf. Example 5.1). Thus, the exchangeable cations amount to $meq_{Na-X} = \beta_{Na} \cdot CEC \cdot 60 = 0.5$ meq/l, $meq_{Mg-X_2} = 8.2$, and $meq_{Ca-X_2} = 51.3$ meq/l.

Use of the Vanselow-convention gives mole fractions. These can be recalculated to equivalent fractions with (cf. Problem 5.14)

$$\beta_I = \frac{\beta_I^M \cdot i}{\beta_I^M \cdot i + \beta_J^M \cdot j + \beta_K^M \cdot k + ...}$$ (5.22)

The Gapon convention allows for a particularly easy calculation of the exchangeable fractions. All fractions are linearly related according to:

$$\beta_J = \frac{\beta_I \cdot K^G_{J\backslash I} \cdot [J^{j+}]^{1/j}}{[I^{i+}]^{1/i}}$$

When all fractions are entered in the sum $\Sigma\beta = 1$, a little rearrangement gives:

$$\beta_I = \frac{[I^{i+}]^{1/i}}{\sum\limits_{J=I,J,K,\ldots} K^G_{J\backslash I} \cdot [J^{j+}]^{1/j}} \tag{5.23}$$

(Note that $K_{I\backslash I} = 1$ for all conventions). The latter equation is also known as the *multicomponent Langmuir equation*.

5.3.3 *Distribution coefficients*

The selectivity coefficient indicates the relative tendency of two elements to become adsorbed. If one of the elements is present in trace concentrations, a small change in concentration of the trace element does not influence the exchangeable fractions of the other, major ions. If, moreover, the concentrations of the major ions remain constant, the concentration of the exchangeable (adsorbed) trace element shows a linear relationship with solute concentration. The ratio among the two is termed a distribution coefficient K'_d:

$$K'_d = \frac{\text{mmol per g solid}}{\text{mmol per ml water}} \tag{5.24}$$

The distribution coefficient has the dimension (ml/g); the chemical unit should disappear, since the same unit (mmol, mg, meq, etc.) is used for solute as well as adsorbed concentrations. It is a lumped parameter that gives a direct link between the total, analytical concentrations in the solution and in the solid, and it can be convenient for keeping an overview. However, we may use the exchange concept to calculate a priori values for distribution coefficients.

If a heavy metal M^{m+} is in exchange equilibrium with Ca^{2+}, we have:

$$1/m\,M^{m+} + \frac{1}{2}\,Ca\text{-}X_2 \leftrightarrow 1/m\,M\text{-}X_m + \frac{1}{2}\,Ca^{2+}$$

with

$$K_{M\backslash Ca} = \frac{[M\text{-}X_m]^{1/m}\,[Ca^{2+}]^{0.5}}{[Ca\text{-}X_2]^{0.5}\,[M^{m+}]^{1/m}}$$

from which we obtain:

$$\frac{[M\text{-}X_m]}{[M^{m+}]} = \frac{\beta_M}{[M^{m+}]} = \left(K_{M\backslash Ca}\frac{\beta^{0.5}_{Ca}}{[Ca^{2+}]^{0.5}}\right)^m$$

The equivalent fraction β_M is multiplied with *CEC*, and divided by 100 to give exchangeable $M\text{-}X_m$ in meq/g; the activity $[M^{m+}]$ is multiplied with charge m to give (numerically) aqueous concentration in meq/ml. We thus have the distribution coefficient in ml/g:

$$K'_d = \frac{CEC/100}{m}\left(K_{M\backslash Ca}\frac{\beta^{0.5}_{Ca}}{[Ca^{2+}]^{0.5}}\right)^m \tag{5.25}$$

Note that m is here the charge of metal ion M^{m+}.

In fresh water $\beta_{Ca} \approx 1$ (cf. Example 5.4), the Ca^{2+} concentrations can be estimated for given CO_2 pressure as shown in Chapter 4, and with a further estimate of *CEC*, the distribution-coefficient can be obtained (Example 5.5 and 5.6). The distribution coefficient can also be related to relative transport velocities as will be shown in Chapter 9.

EXAMPLE 5.5: *Estimate the distribution coefficient of* Zn^{2+}
Porewater in a quartz sediment is fresh, with $m_{Ca^{2+}} = 2$ mmol/l. The sediment has 0.5% clay (< 2 μm) and 0.1% organic carbon, and porosity $\varepsilon = 0.2$. Estimate the distribution coefficient for Zn^{2+} at trace quantities, and express the distribution in dimensionless units.

ANSWER
Estimate *CEC* with Equation (5.7): $CEC = 0.7 \cdot 0.5 + 3.5 \cdot 0.1 = 0.7$ meq/100g. The exchange coefficient $K_{Zn\backslash Ca} = K_{Na\backslash Ca} / K_{Na\backslash Zn} = 1$ (from Table 5.5). With Equation (5.25) the distribution coefficient is: $K'_d = 0.007/2 \cdot (1 \cdot 1/\sqrt{2 \cdot 10^{-3}})^2 = 1.75$ ml/g. Bulk density $\rho_b = 2.65 \cdot (1 - \varepsilon) = 2.1$ g/cm^3 (cf. Example 5.1). Hence the dimensionless distribution coefficient is $1.75 \cdot \rho_b/\varepsilon = 18.6$.

Major elements are normally neglected in the calculation of the distribution coefficient from experimental data, but they can be incorporated by using the multicomponent formulae of the previous section. When the charge of the trace element and major ions is equal, a simple linear equation is obtained. For example for Sr^{2+}, as a trace element, with respect to major ions Mg^{2+} and Ca^{2+}:

$$\beta_{Sr} = \frac{[Sr^{2+}]}{[Sr^{2+}] + K^2_{Ca\backslash Sr} \cdot [Ca^{2+}] + K^2_{Mg\backslash Sr} \cdot [Mg^{2+}]} \qquad (5.26)$$

where the equilibrium constants $K_{I\backslash Sr}$ are defined with respect to Na^+ in Table 5.5.

EXAMPLE 5.6: *Distribution coefficient for* Sr^{2+}
Let us consider the applicability of the multicomponent relation (5.26) on data provided by Johnston et al. (1985) who carefully determined Sr distribution coefficients for sediments near a radioactive waste site. The distribution coefficients were obtained for sands and tills at varying Sr^{2+} concentrations in groundwater and synthetic solutions. A typical result is shown in Figure 5.11 for two samples of weathered till. Synthetic groundwater composition, and *CEC* of the two samples are given in Table 5.6.

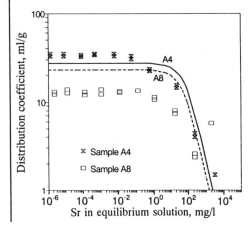

Figure 5.11. Strontium distribution curves for two soil samples. Experimental data from Johnston et al., 1985.

Table 5.6. Water compositions and sediment *CEC* in an experiment for determining the Sr^{2+} distribution coefficient (Johnston et al., 1985).

Sample	*CEC* meq/100 g	Ca^{2+} mmol/l	Mg^{2+} mmol/l	Na^+ mmol/l	K^+ mmol/l	Sr^{2+} mmol/l
A4	5.3	.75	.80	.09	.08	$1 \cdot 10^{-8}$ to 50
A8	3.4	.75	.33	.3	.43	$1 \cdot 10^{-8}$ to 50

The groundwater used for sample A4 contains mainly Ca^{2+} and Mg^{2+} in solution, which allows the experimental results to be modeled as a cation exchange reaction with respect to only the divalent ions (Mg^{2+} and Ca^{2+}). The distribution coefficient is in our exchange model:

$$K'_d = \frac{CEC/100}{2} \frac{\beta_{Sr}}{m_{Sr^{2+}}} \qquad (5.27)$$

where $m_{Sr^{2+}}$ is molality of Sr^{2+} (mol/kg H_2O). We obtain the exchange coefficients $K^2_{Ca\backslash Sr} = 0.77$, and $K^2_{Mg\backslash Sr} = 0.49$ from Table 5.5. We furthermore assume $[I] = \gamma_2 \cdot m_I$ (i.e. neglect complexes in solution) and find on combination of Equations (5.26) and (5.27):

$$K'_d = \frac{CEC/200}{C_{Sr}/87600 + 0.77 \cdot m_{Ca^{2+}} + 0.49 \cdot m_{Mg^{2+}}}$$

where C_{Sr} is concentration of Sr^{2+} in mg/l (we use mg/l to allow for a direct comparison with experimental data from Johnston et al., 1985). The conditions for sample A4 thus lead to:

$$K'_d(A4) = 0.0265/(C_{Sr}/87600 + 9.7 \cdot 10^{-4})$$

and for sample A8:

$$K'_d(A8) = 0.017/(C_{Sr}/87600 + 7.4 \cdot 10^{-3})$$

Lines according to these equations are plotted in Figure 5.11.

EXAMPLE 5.7: *Exchange coefficient of Cd^{2+} vs. Na^+ on montmorillonite*
Garcia-Miragaya and Page (1976) determined adsorption percentages of Cd^{2+} from $NaClO_4$ solutions of different normality on 2.05 meq/l montmorillonite. A selection of their data is presented in the table.

$NaClO_4$ N	Cd_{total} µM	adsCd %	$NaClO_4$ N	Cd_{total} µM	adsCd %
0.01	0.40	97.1	0.05	0.40	73.6
	0.80	96.3		0.80	69.2

Determine the exchange coefficient $K_{Na\backslash Cd}$ for the Gaines and Thomas convention.

ANSWER
Inspection of the data shows that the percentage of Cd^{2+} which is adsorbed (adsCd) appears to be independent of the amount of Cd^{2+}, which is added to the solution. In other words, the ratio of sorbed Cd^{2+} over solute Cd^{2+} is independent of Cd_{total}, and the distribution coefficient is a constant. The concentration of Na^+ has a marked effect, however, and we expect that cation exchange is operative. The reaction is:

$$Na^+ + 0.5Cd\text{-}X_2 \rightarrow Na\text{-}X + 0.5Cd^{2+}$$

with

$$K_{Na\backslash Cd} = \frac{\beta_{Na}}{\gamma_1 m_{Na^+}} \frac{(\gamma_2 \, m_{Cd^{2+}})^{0.5}}{\beta_{Cd}^{0.5}}$$

We have X = exchange capacity of the montmorillonite, 2.05 mM,

m_{Na^+} = solution normality N,

$\beta_{Cd} \cdot X/2$ = adsCd $\cdot$ $Cd_{total}/100$ or β_{Cd} = adsCd $\cdot$ $Cd_{total}/(50X)$,

$m_{Cd^{2+}}$ = $Cd_{total}(1 - adsCd/100)$,

and assume $\beta_{Na} = 1$. This gives us:

$$K_{Na\backslash Cd} = \frac{1}{\gamma_1 N} \frac{(\gamma_2 \cdot (1 - adsCd/100))^{0.5}}{(adsCd/50X)^{0.5}}$$

and we obtain:

for N = 0.01: $\gamma_2 = 0.67$; $\gamma_1 = 0.90$; $K_{Na\backslash Cd} = 0.50$ and 0.57

0.05: $\gamma_2 = 0.46$; $\gamma_1 = 0.82$; $K_{Na\backslash Cd} = 0.32$ and 0.35

5.4 CHROMATOGRAPHY OF CATION EXCHANGE

Cation exchange among water and solid aquifer material is important when ion concentrations vary along a flowline because of non-steady state situations. Most of the theoretical and experimental work on cation exchange has been confined to systems with only two cations. However, natural situations are never limited to only two ions, and we have to study the *multicomponent* effects of cation exchange during transport. The principles of ion exchange are now routinely applied in the chemical laboratory for separation and analysis of ions from a complex mixture with ion chromatography. Details of the calculations of multicomponent transport are postponed to Chapters 9, 10 and 11, but it is appropriate to give examples already here, as these may wet your appetite to try and understand more of this important and fascinating subject.

The chemical analyst uses a column packed with beads of an ion exchanger. Pores within and between the beads are filled with fluid, and the total amount of fluid in the column is termed the column pore volume, or shortly, the pore volume (V_0, m^3). The column is flushed with different solutions, and as a result of cation exchange a pattern of concentrations develops at the end of the column. Cations which are strongly selected by the exchanger will displace other ions from the exchanger, and be transported at a relatively low velocity. When the initial solution in the column contains ions that have a lower selectivity than the cation in the injected solution, it is termed *displacement chromatography*. The ions which are adsorbed on the exchanger leave the column in an ordered manner with this technique. First comes the ion which exhibits lowest selectivity, than the next selected, etc., and finally comes the cation which is injected into the column. If, on the other hand, the cation in the incoming solution has a lower affinity for the exchanger, it is termed *elution chromatography*. In this case, the concentrations at the outlet of the column change more gradually. Sometimes the entering fluid has a total ion concentration which differs from the resident solution. This will induce a salinity front, or an anion jump when one pore volume has been displaced from the column.

5.4.1 *Field examples of refreshening*

Displacement chromatography applies to the field situation where fresh water with mainly

Ca^{2+} displaces seawater with mainly Na^+ and Mg^{2+}. Ca^{2+} is favored by natural ion exchangers over Mg^{2+} and Na^+. It is possible to get some idea of the timescales of flow and cation exchange, if we simplify fresh water to a displacing solution which contains Ca^{2+} and HCO_3^- only. Example 5.8 shows how such a calculation is performed.

EXAMPLE 5.8: *Flushing of an exchange complex*
Cultivated fields on islands in the Dutch tidal flat, the Wadden sea, often show brackish patches in an otherwise fresh groundwater area. The brackish water is a remnant of transgressions during the last century. They occur where clay layers at the surface obstruct the downward percolation of fresh water. Estimate the time of refreshening and flushing of the exchange complex for the situation depicted in Figure 5.12. *CEC* of the sediment is 1 meq/100g, bulk density $\rho_b = 2.0$ g/cm^3, and porosity $\varepsilon = 0.3$. The salt water has seawater composition, the fresh water is pure Ca^{2+} and HCO_3^- water in equilibrium with $P_{CO_2} = 0.01$ atm.

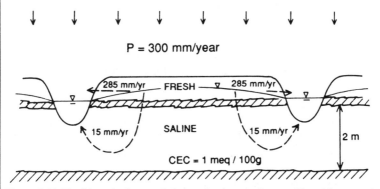

Figure 5.12. Flushing of salt water below a clay layer: a 2 m aquifer with a porosity $\varepsilon = 0.3$, contains 0.6 m water. The precipitation surplus is 300 mm/yr, of which 285 mm/yr flows directly into the ditches, while 15 mm/yr percolates downward into the aquifer.

ANSWER
First we calculate the years needed to flush the pore volume with the precipitation surplus of 15 mm/yr. The high anion concentrations are flushed after one pore volume, and the concentrations drop from those in seawater to the fresh HCO_3^- level. Then the flushing of the exchange complex begins. The amounts of exchangeable cations on the sediment are so great that the fresh water equilibrates initially with the original seawater exchange complex. This continues until all Na^+ has been flushed. Next comes Mg^{2+} which in turn is displaced by Ca^{2+}. The timelength is in all cases the ratio of the amount of exchangeable cation over the amount of freshwater Ca^{2+}, multiplied with the timelength for one pore volume flush.
    The aquifer contains initially 0.6 m salt water, which is displaced in 40 years with 15 mm/yr. The amount of exchangeable cations in the sediment can be calculated on a pore water basis: 2 g sediment has a volume of 1 cm^3, and contains 0.02 meq exchangeable cations, and 0.3 ml water. A sediment column of $2 \times 1 \times 1$ m^3 contains $2 \times 2 \cdot 10^6 = 4 \cdot 10^6$ g sediment, with $4 \cdot 10^4$ meq exchangeable cations, and 0.6 m^3 water. Percolation gives $0.015 \times 1 \times 1$ m$^3 = 0.015$ m^3/yr. The fraction of exchangeable Na^+ in equilibrium with seawater is calculated as in Example 5.4:

$$\beta_{Na}^2 \cdot \{[Ca^{2+}]/ (K_{Na\backslash Ca} \cdot [Na^+])^2 + [Mg^{2+}]/ (K_{Na\backslash Mg} \cdot [Na^+])^2\} + \beta_{Na} - 1 = 0$$

Assume that activity equals molal concentration, i.e. $[Na^+] = 0.485$, $[Mg^{2+}] = 0.055$, and $[Ca^{2+}] = 0.011$. We use the exchange constants from Table 5.5, i.e. $K_{Na\backslash Ca} = 0.4$, and $K_{Na\backslash Mg} = 0.5$, and find $\beta_{Na} = 0.583$, and subsequently $\beta_{Mg} = 0.318$, and $\beta_{Ca} = 0.099$. The sediment column contains $0.583 \times 4 \cdot 10^4$ meq Na-X.

Fresh water in equilibrium with calcite at $P_{CO_2} = 0.01$ atm contains: $Ca^{2+} = 10^{-2.1} \cdot (P_{CO_2})^{1/3} = 1.7$ mmol/l (cf. Chapter 4). The percolating volume thus brings $0.015 \times 1700 \times 2 = 51$ meq Ca^{2+} each year into the soil. The time needed to flush all Na-X is then $23316 / 51 = 457$ years. We find similarly the time for Mg^{2+}, if we assume that Mg-X$_2$ flushing starts only after all exchangeable Na-X has been removed (cf. Example 5.9), as 249 years. Summarizing then, we find as time for the removal of salt water, exchangeable Na-X, resp. Mg-X$_2$:

	Years for removal
Salt (high Cl⁻)	40
Na⁺ (bad soil structure)	457
Mg²⁺ (hydrological tracer)	249
Total:	746

This gives some idea of the time scales involved. Figure 5.13a shows how water compositions change in time at the end of a flowline. We will discuss the bad soil strucure associated with NaHCO$_3$ type water in later sections of this chapter, but we may note already that with reclamation of saline soils, gypsum is used to increase the Ca^{2+} concentration in percolating water and speed up the removal of Na⁺. In these (simple) calculations, effects of dispersion and geological heterogeneities were neglected. These tend to hasten initial breakthrough of Ca^{2+}, but to slow down the complete removal of Na⁺ and Mg^{2+}.

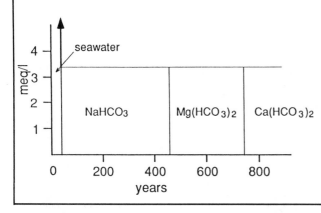

Figure 5.13a. Water compositions at the end of a flowline as seawater is displaced by fresh water.

Figure 5.13a shows how water compositions change in time at the end of a flowline in the aquifer of Example 5.8. We note that it is also simple to calculate flushing times in the following way, considering that $CEC = 66.7$ meq/l pore water. Of this, 58.3% is initially Na-X, or 38.9 meq/l. Fresh water contains $1.7 \times 2 = 3.4$ meq Ca^{2+}/l. To flush Na-X requires $38.9/3.4 = 11.4$ pore volumes of fresh water: Hence $11.4 \times 40 = 457$ years are necessary. Here, we simply calculate time with respect to flushing of one pore volume through the column. If the flowline is twice as long, twice as much time is needed for all transitions or fronts. We have, in other words, found the *characteristic velocity* $v = dx/dt$ of a front. We can illustrate this in a time-distance diagram, in which the transitions appear as straight lines (v is a constant for each front, Figure 5.13b).

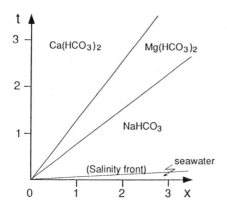

Figure 5.13b. A time-distance diagram illustrating the propagation of the fronts calculated in Example 5.8.

EXAMPLE 5.9: *Water composition after a salinity front*
We have assumed in Example 5.8 that only Na^+ is present in water, as soon as the seawater anion concentrations have been flushed. We must calculate the exact composition of $NaHCO_3$ water (with 3.4 meq HCO_3^-/l) which has lost Ca^{2+}, and has equilibrated with the seawater exchange complex.

ANSWER
The exchangeable fractions in equilibrium with seawater are $\beta_{Na} = 0.583$, $\beta_{Mg} = 0.318$, $\beta_{Ca} = 0.099$. The anion concentration is 3.4 meq HCO_3^-/l, which limits the total meq of cations to 3.4 meq/l. Hence:

$$m_{Na^+} + 2\,m_{Mg^{2+}} + 2\,m_{Ca^{2+}} = 0.0034 \qquad (5.28)$$

We furthermore have from exchange equilibrium

$$[Ca^{2+}] = \frac{\beta_{Ca} \cdot (K_{Na\backslash Ca} \cdot [Na^+])^2}{\beta_{Na}^2} \qquad (5.29)$$

and

$$[Mg^{2+}] = \frac{\beta_{Mg} \cdot (K_{Na\backslash Mg} \cdot [Na^+])^2}{\beta_{Na}^2} \qquad (5.30)$$

If we assume that $[I^{i+}] = m_I$, we obtain on combining Equations (5.28), (5.29) and (5.30):

$$m_{Na^+}^2 \cdot \left(\frac{2\beta_{Ca} K_{Na\backslash Ca}^2}{\beta_{Na}^2} + \frac{2\,\beta_{Mg} K_{Na\backslash Mg}^2}{\beta_{Na}^2} \right) + m_{Na^+} - 0.0034 = 0$$

This quadratic equation can be readily solved with the fractions β_I given above (from Example 5.8), and with $K_{Na\backslash Ca} = 0.4$, and $K_{Na\backslash Mg} = 0.5$, to give $m_{Na^+} = 3.39 \cdot 10^{-3}$ mol/l. Further back-substitution in Equation (5.29) and (5.30) gives $m_{Mg^{2+}} = 2 \cdot 10^{-6}$, and $m_{Ca^{2+}} = 0.5 \cdot 10^{-6}$ mol/l. We see that our assumption, that Na^+ is the only cation after the salinity jump, is indeed justified.

The classical field example where the chromatographic pattern of Example 5.8 has been observed is the injection experiment by Valocchi et al. (1981). Fresh water from a sewage plant with tertiary treatment (i.e. giving near drinking water quality) was injected into a brackish-water aquifer, shown in outline in Figure 5.14. The brackish water contained high amounts of Na^+ and Mg^{2+}, and the fresh water had Ca^{2+} as the major cation; the cation exchange capacity of the silty aquifer sediment was fairly high with 10 meq/100 g. The injection produced a clear chromatographic pattern in observation wells, of which an

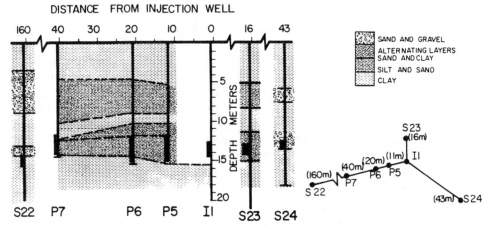

Figure 5.14. Aquifer outline showing injection well and observation wells (from Valocchi et al., 1981. Copyright by the Am. Geophys. Union).

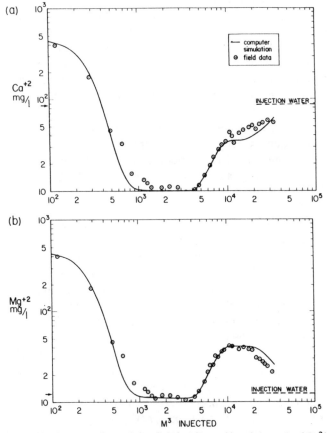

Figure 5.15. Comparison of simulated and actual breakthrough of Ca^{2+} and Mg^{2+} in boring S23 of Figure 5.14 during a field injection of fresh water into a brackish aquifer. Drawn line gives the simulated water qualities (from Valocchi et al., 1981. Copyright by the Am. Geophys. Union).

example is shown in Figure 5.15 for well S23. Only the concentrations of Mg^{2+} and Ca^{2+} have been presented by Valocchi et al., but the chromatographic pattern is already displayed by these two ions.

Figure 5.15 shows that the Ca^{2+} and Mg^{2+} concentrations start to decrease in the observation well after about 200 m^3 have been injected. Subsequently these ions attain concentrations which are even lower than originally present in the injection water, and the concentrations remain low until about 5000 m^3 have been injected. This stage corresponds to the dilution step after the salinity front, where Na^+ becomes the dominant cation in the diluted solution that is in equilibrium with the salt water exchange complex (cf. Example 5.9, Problem 5.7). After about 5000 m^3 have been injected, the Ca^{2+} and Mg^{2+} concentrations increase again. Ca^{2+} increases towards the concentration in the injection water, but Mg^{2+} surpasses that concentration quite markedly. This stage corresponds to the displacement of Mg^{2+} from the cation exchange complex by Ca^{2+}.

The fronts arrive gradually at the observation well because dispersion smears out concentration steps (as will be discussed in Chapter 9). If we neglect dispersion for the present, we can calculate the lengths of the $NaHCO_3$ stage, assuming that this length is determined by the amount of Na^+ that must be removed from the exchange complex, and the concentration increase of Na^+ in solution after the salinity front compared with the concentration in the injection water (cf. Problem 5.7). The quality changes in fig. 5.15 have been modeled with a numerical model using exchange constants obtained from batch experiments with the same aquifer material, and with obviously very good results.

The injection experiment by Valocchi et al. (1981) displayed the cation exchange phenomena especially clear, as result of the high cation exchange capacity of the aquifer, and, also, because the conditions of the aquifer before injection are accurately known. This is not always true in field studies where refreshening and salinization may have occurred repeatedly in the same aquifer in the past. A reversal of flow does not restore the original quality conditions, because the cation exchange processes are non-linear. The chromatographic pattern may therefore become somewhat blurred. However, refreshening quality patterns have been observed, for example in studies by Lawrence et al. (1976), Chapelle and Knobel (1983), Stuyfzand (1985, 1993, cf. Figure 5.2), Appelo et al. (1989), and Beekman (1991).

A refreshening pattern should display, near the fresh/salt front, a $NaHCO_3$ water type (or a high Na/Cl-ratio), which on the fresh water side merges into a $MgHCO_3$ water quality, and finally into the normal fresh $CaHCO_3$ water quality. These successive fronts have indeed been observed in the Dutch subsoil. Figure 5.16 shows water qualities along a West to East transect near Hoorn in The Netherlands (Beekman, 1991). Fresh water infiltrates at point 0 in this transect, and flows away through a ca. 100 m thick aquifer. The aquifer has been salinized during Holocene transgressions. Polder development in the Middle Ages initiated the flow of groundwater towards the drained areas, and started the infiltration of fresh water in relatively higher grounds. The process must have been slow in the beginning because floodgates were employed in the early days, and these permitted only limited quantities of water to drain. The use of windmills in the 16th century, and the advent of steam power still later, enabled the drainage of deeper lakes with higher rates of upward seepage. The flow pattern in the aquifer will have changed as new polders were established and the regional drainage pattern was adjusted. As a consequence, the water qualities may not be directly connected with present-day flow patterns. For example, the freshening pattern in Figure 5.16 is particularly clear in the central-east section of the profile. However,

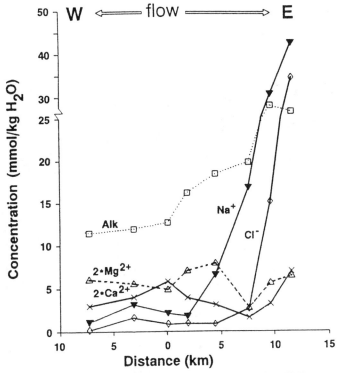

Figure 5.16. Chromatographic pattern along a flowline in a W-E transect through the subsoil of Hoorn, The Netherlands. Water infiltrates at point 0, and CaHCO₃ changes successively into MgHCO₃ and NaHCO₃ water quality types (Beekman, 1991).

the indicated freshening has now stopped, and must be a relict of earlier conditions since most groundwater flow is now directed towards the west and north of the infiltration area.

In some Pleistocene aquifers with a very coarse sediment texture, the high NaHCO₃ type water with a high Na/Cl ratio is less clearly present, although a freshening takes place according to the flow pattern. This absence may well be due to the low selectivity of these quartz rich sediments for Na^+. The equilibrium coefficient $K_{Na\backslash Ca}$ has been found to be as low as 0.05 in some of these sediments (Griffioen, 1992), which greatly reduces the amounts of exchangeable Na-X on the sediment. The NaHCO₃ type water (or high Na/Cl-type) is more explicitly present in the aquifers studied by Lawrence et al. (1976), and Chapelle and Knobel (1983). In these studies, however, only a very small relative increase of Mg^{2+} was observed, both in a glauconite-rich sediment which has substantial cation exchange capacity (Chapelle and Knobel, 1983), and almost none at all in a limestone aquifer as illustrated in Figure 5.17 (Lawrence et al., 1976; Edmunds and Walton, 1983). The profile of the limestone aquifer is from the work by Edmunds and Walton, and displays the enhanced Na/Cl ratio at the salinity front quite clearly. However, the small Mg^{2+} increase occurs towards the more saline end of the profile, which is the wrong side for a freshening front. The increase has been attributed to incongruent dissolution of carbonates by Edmunds and Walton (1983), and this may well be the appropriate explanation here, even though the exact reaction remains somewhat unclear.

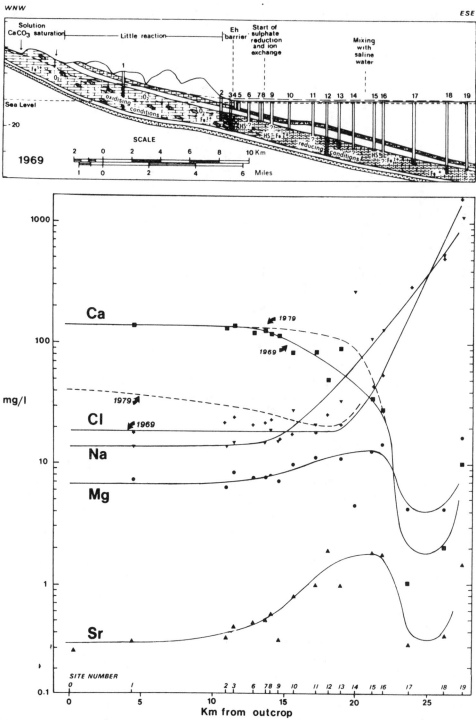

Figure 5.17. Hydrogeological cross section through the Lincolnshire Limestone, England, with hydro-geochemical trends. *Solid line* (and data points) refer to 1969. The aquifer has been resampled in 1979, and deviations are indicated by the *dotted line* (from Edmunds and Walton, 1983).

5.4.2 *Quality patterns with salinization*

The development of a chromatographic pattern depends on the amounts of exchangeable cations, and their amounts in solution: the ratio of the two gives the velocity of a front as shown in Example 5.8. This ratio is small, and the succession of fronts will be relatively rapid, when the aquifer has a low cation exchange capacity, or when concentrations in water are high. The high concentrations in the case of seawater intrusion, therefore, hasten the transitions of water qualities, and reduce the lateral extent of a water type in an aquifer. This may be the reason that a chromatographic sequence of salinization has not been described from observations in an aquifer (as far as we know), although the $CaCl_2$ water types are an ubiquitous indicator of salt water upconing in coastal areas (Section 5.1). Information on the sequence of compositions with seawater intrusions must therefore be obtained from column experiments, such as described by Appelo et al., 1990.

Beekman and Appelo (1990) have performed column experiments, displacing fresh water with seawater diluted 1:1 with distilled water. Results of a seawater intrusion experiment are shown in Figure 5.18. The large initial increase of Ca^{2+} as the seawater front arrives at the outlet is quite clear, and leads to the expected $CaCl_2$ water type. There is at the same time a conspicuous increase of Mg^{2+}, which is a result of the initial anion-jump. The total solute concentrations increase after the salinity front, while the cations are still in equilibrium with the original (in this case fresh) water quality. Such an increase implies that the concentration ratios of differently charged cations must change, i.e. the *concentration effect* of cation exchange applies again. The ratio of the divalent over the monovalent cations must increase when total concentrations increase (just the opposite of the dilution effect discussed in Example 5.8 and 5.9). It can be seen in Figure 5.18 that Ca^{2+} and Mg^{2+} increase initially congruent, in line with the equal concentration of both ions in fresh water. When adsorbed Ca^{2+} is exhausted, the behavior of the two ions diverges: Ca^{2+} decreases and Mg^{2+} increases towards the level in the salt water. The point of divergence arrives rapidly, because the amount of adsorbed Ca^{2+} is small in comparison with the concentration of solute Ca^{2+} which must be maintained after the anion jump (cf. Problem 5.8). The

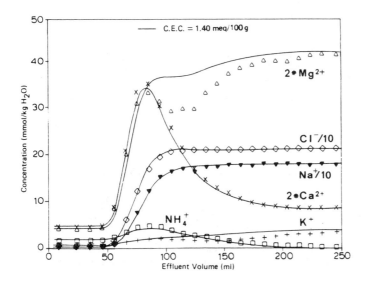

Figure 5.18. Column experiment with once diluted seawater displacing fresh water (from Beekman and Appelo, 1990).

decrease of Ca^{2+} therefore starts already before Cl^- has reached the final salt water concentration.

Cation exchange is a relatively fast reaction, and it is often assumed that equilibrium between adsorbed and solute ions is maintained. An equilibrium model can be applied, when flow velocities are less than ca. 100 m/yr (we will come back to this point in Chapter 11). The good results of equilibrium modeling, shown in Figure 5.15 and 5.18, do indeed confirm that the so-called *Local Equilibrium Assumption* is valid. The ease of a numerical model is of course that any situation within the capacities of the model can be described as a substitute for the painstaking field and laboratory experiments. Some more results of the numerical model used for the column experiment in Figure 5.18 are shown in Figure 5.19 and 5.20. (The model will be discussed in Chapter 10). Figure 5.19 gives the modeled profile of ion concentrations when salt water intrudes a fresh water aquifer (we 'look' along a flowline, or we 'sample' the column at several points at one time, instead of looking at changes in effluent as in Figure 5.18). Equilibrium with calcite has been imposed in the model, and this acts as a constraint on the increase of Ca^{2+} concentrations. The precipitation of calcite is manifest in the decrease of HCO_3^-. (Note that this decrease is not immediately evident in field samples, as discussed in Section 5.1). The general pattern is otherwise similar to the column experiment shown in Figure 5.18.

When however, seawater diffuses into a stagnant fresh groundwater, part of the exchanged ions from the sediment diffuse away at the same rate as the displacing ions arrive. The difference in diffusion velocity between a conservative element and an exchangeable element is partly determined by the *CEC*. When the *CEC* is not too large, the effects of cation exchange tend to be reduced, and are much less than when seawater intrudes into the aquifer by flow. This means that the resulting water composition will appear rather to be a simple mixture of fresh and salt water. Figure 5.20 shows the result of a simulation of such diffusive salt water intrusion.

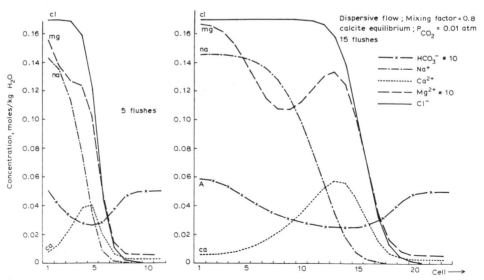

Figure 5.19. Simulation of sea water intrusion in a fresh water aquifer with a mixing cell geochemical model. Horizontal axis ('cell') represents distance of intrusion (from Appelo and Willemsen, 1987).

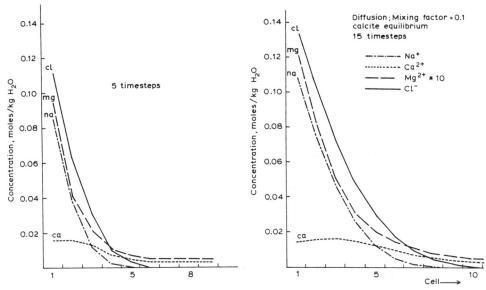

Figure 5.20. Simulation of salt water intrusion by diffusion (salt water concentrations are as used in Figure 5.19) (from Appelo and Willemsen, 1987).

5.5 SPECIFIC ADSORPTION

It was noted before, that dissociation and association of protons produces a pH-dependent charge on oxides and hydroxides. The protons are attached to the surface of these minerals in a much more selective way than would follow from the normal exchange coefficients in Table 5.5. Such selective adsorption is also found for other ions at trace concentrations, especially for heavy metals. An acid synthetic ion exchanger such as DOWEX-50 has about equal selectivity for H^+ and Na^+, but the selectivity of weakly acid ion exchangers for H^+ over Na^+ may easily be 10^6 or more. Similar selectivity for H^+ is displayed by oxides and hydroxides.

The selectivity, and the acid dissociation constant, can be found from a titration curve with NaOH or HCl. Let us start simply, and use NaOH to titrate water to which HCl has been added. Electroneutrality then requires that:

$$m_{H^+} + m_{Na^+} = m_{OH^-} + m_{Cl^-} \qquad (5.31)$$

If we neglect activity corrections, etc., we may equate activity [i] with molal concentration m_i, and from the dissociation constant derive $m_{OH^-} = 10^{-14}/m_{H^+}$. The relation between [$H^+$] or m_{H^+} and added base m_{Na^+} is then:

$$m_{Na^+} = 10^{-14}/m_{H^+} - m_{H^+} + m_{Cl^-} \qquad (5.32)$$

Upon adding an amphoteric solute AH (i.e. a substance which can act as acid and as base), two more species, AH_2^+ and A^- must be included in the electroneutrality equation. The relation between m_{Na^+} and m_{H^+} is now:

$$m_{Na^+} = 10^{-14}/m_{H^+} - m_{H^+} + m_{Cl^-} + m_{A^-} - m_{AH_2^+} \qquad (5.33)$$

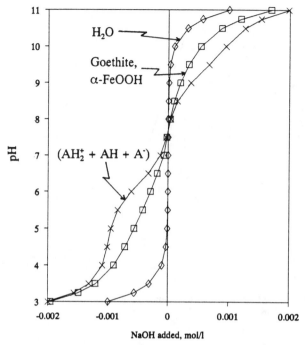

Figure 5.21. Titration curves for pure water, for water and 1 mmol/l of an amphoteric acid AH, and for goethite. The amphoteric acid AH and goethite have both $\log K_{a1} = -6.4$ and $\log K_{a2} = -9.25$, as intrinsic acidity constants. The constant-capacitance model has been used to calculate the goethite titration curve. Background electrolyte is 0.01 N NaCl.

The species are interrelated as:

$$AH_2^+ \leftrightarrow AH + H^+ \qquad K_{a1} \tag{5.34}$$

and

$$AH \leftrightarrow A^- + H^+ \qquad K_{a2} \tag{5.35}$$

where K_{a1} and K_{a2} are acid dissociation constants. We can calculate the fractions of AH_2^+, AH and A^- as a function of $[H^+]$ with the fractionation formula (Chapter 4, Section 4.2.1). Multiplying the fractions with total A_t $(= AH_2^+ + AH + A^-)$ gives actual concentrations which are entered in Equation (5.33), and the relation between m_{Na^+} and m_{H^+} can be calculated. The resulting titration curves, which depict the change in pH as NaOH is added, are drawn in Figure 5.21 for a system with water only, and with an amphoteric solute. For comparison, the titration curve of goethite (α-FeOOH) is drawn with an equal amount of acid neutralizing capacity as of AH.

The most conspicuous difference between the titration curves of AH and goethite is the much more gradual variation of pH in the goethite suspension as compared with the AH-solution. This gradual change may be explained as due to the presence of a variety of acid sites on the goethite surface, at different crystal sites, and with different K_a's (Van Riemsdijk et al., 1986). An alternative explanation is that the goethite surface has only two acid dissociation constants, just as AH. However, a fundamental difference with AH is that the acid groups on goethite are positioned close together on the surface. If one group has dissociated, one might expect increased difficulty for the neighbor groups to dissociate since more negative charge is available to hold the remaining H^+ more vigorously. This is different from a solution of AH in which an individual molecule is not affected by dissociation of other AH molecules. Thus, if a portion of $\equiv FeOH$ on the goethite surface has dissociated into $\equiv FeO^-$ and H^+, the apparent dissociation constant K_{a2} for the remaining $\equiv FeOH$ decreases as pH increases. Likewise for the association of $\equiv FeOH$ with H^+: as more $\equiv FeOH_2^+$ has formed, is the attachment of more H^+ more difficult, and also in this case, will the apparent K_{a1} decrease as pH increases.

5.5.1 *Surface complexation and electrostatic models*

The phenomenon is considered a result of the electric potential which develops on the solid surface when $\equiv SOH_2^+$ or $\equiv SO^-$ forms. The Gibbs free energy of the surface (desorption) reaction is a combination of a chemical and a Coulombic term:

$$\Delta G_{des} = \Delta G_{chem} + \Delta G_{coul} \tag{5.36}$$

The Coulomb part is the difference in energy of the state where one mole of an ion resides at the surface where the potential is Ψ_0, and the state where that ion resides in the bulk of the solution with potential $\Psi = 0$ (Volt). For desorption it amounts to:

$$\Delta G_{coul} = \Delta G_{\Psi=0} - \Delta G_{\Psi=\Psi_0} = zF(0 - \Psi_0) = -zF\Psi_0 \tag{5.37}$$

where z is the charge of the ion, and F is the Faraday constant ($96485 \ C/mol$). The chemical term gives an *intrinsic* dissociation constant K_{int} according to the standard relation (Chapter 3):

$$RT \ln K_{int} = -\Delta G_{chem}$$

The *apparent* dissociation constant K_a includes in fact the Coulomb term:

$$RT \ln K_a = -\Delta G_{chem} - \Delta G_{coul} = RT \ln K_{int} + zF\Psi_0 \tag{5.38}$$

or:

$$\log K_a = \log K_{int} + \frac{zF\Psi_0}{RT \ln 10} \tag{5.39}$$

For example for dissociation of a surface proton:

$$\equiv SOH_2^+ \leftrightarrow \equiv SOH + H^+ \tag{5.40}$$

the apparent dissociation constant is:

$$K_{a1} = \frac{[\equiv SOH][H^+]}{[\equiv SOH_2^+]} \tag{5.41}$$

The symbol K_{a1} is used to indicate the first of two possible deprotonation steps of the surface. The apparent constant for dissociation of $\equiv SOH$ into $\equiv SO^-$ is denoted as K_{a2}.

The difficulty is again to relate surface concentrations and activities. The surface activity includes the Coulomb term, but this term is not included in the standard state. The tacit assumption in the definition of the standard state, as discussed by Sposito (1983, 1984), is that the surface complex has an activity of one when fully covering the surface, but in a charge-free environment. This places the difficulty at the way Ψ_0 at the surface should be measured or estimated. Several models have been proposed, which are at least adequate to explain the observations of laboratory experiments, and we will discuss one of them.

The *constant capacitance* model (Schindler and Stumm, 1987) assumes a linear relation among surface charge and potential:

$$\Psi_0 = \sigma/\kappa\varepsilon \qquad (5.42)$$

where σ is the specific surface charge in C/m^2, and $\kappa\varepsilon$ is the specific capacitance in F/m^2.

When a surface species $\equiv SOH_2^+$ is being formed, the surface charge due to this species increases, and becomes:

$$\sigma = F\{\equiv SOH_2^+\}/A_s \qquad (5.43)$$

where A_s is the specific surface in m^2/kg, and $\{\equiv SOH_2^+\}$ is expressed as moles per kg adsorbing solid.

In the constant capacitance model the surface potential, from Equation (5.42), increases concomitantly, and

$$\Psi_0 = \frac{F\{\equiv SOH_2^+\}}{\kappa\varepsilon A_s} \qquad (5.44)$$

We can combine Equations (5.44) and (5.39), and obtain for the first dissociation constant, with $z = 1$ for H^+:

$$\log K_{a1} = \log K_{int} + \frac{F^2}{\kappa\varepsilon A_s RT \ln 10} \cdot \{\equiv SOH_2^+\} \qquad (5.45)$$

Part of the factor before $\{\equiv SOH_2^+\}$ is customarily lumped together as:

$$\alpha_1 = \frac{F^2}{\kappa\varepsilon A_s RT}$$

so that we write Equation (5.45) more simple as:

$$\log K_{a1} = \log K_{int} + \frac{\alpha_1}{\ln 10} \cdot \{\equiv SOH_2^+\} \qquad (5.46)$$

and similar for K_{a2}:

$$\log K_{a2} = \log K_{int} - \frac{\alpha_2}{\ln 10} \cdot \{\equiv SO^-\} \qquad (5.47)$$

The factors α_1 and α_2 can be obtained from plots of the surface charge (i.e. of $\equiv SOH_2^+$, or $\equiv SO^-$) vs. the apparent $\log K_a$, and these plots also provide an estimate of K_{int} (cf. Stumm and Morgan, 1981). The factors α_1 and α_2 should be identical for K_{a1} and K_{a2}, as follows from the derivation, but this is often not the case (Schindler and Stumm, 1987). In calculating the goethite titration curve of Figure 5.21, it was assumed that $\alpha_1 = \alpha_2$, and $\alpha_1/\ln 10 = 2$.

The factor α_1 and α_2 must be determined for each background electrolyte (an inert salt in solution), and the constant capacitance model can neither describe the effects of concentration changes of the background electrolyte. A background electrolyte such as NaCl has the

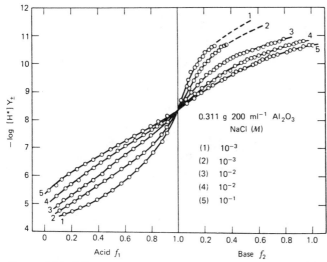

Figure 5.22. Effect of different concentrations of background electrolyte on titration curves of γ-Al$_2$O$_3$. f is equivalent fraction of titrant added (from Stumm and Morgan, 1981. Reprinted by permission of John Wiley & Sons, Inc.).

effect to equalize the slope of the titration curve over a large pH interval, as shown in Figure 5.22. This equalization indicates that the apparent dissociation constants increase, i.e. the effect of the electrostatic term is counteracted by the background electrolyte.

The electrolyte effect is most easily explained as due to an ion association of the background electrolyte with the charged surface. For example of Na$^+$ with $\equiv$SO$^-$:

$$\equiv\text{SONa} \leftrightarrow \text{Na}^+ + \equiv\text{SO}^- \qquad K_{\text{SONa}} \tag{5.48}$$

with

$$K_{\text{SONa}} = \frac{[\text{Na}^+]\,[\equiv\text{SO}^-]}{[\equiv\text{SONa}]} = \frac{[\text{Na}^+]\,\beta_{\equiv\text{SO}^-}}{\beta_{\equiv\text{SONa}}} \tag{5.49}$$

where $\beta_{\equiv\text{S}..}$ is the fraction of the surface species.

Now, the apparent acid dissociation constant is calculated from:

$$K_{a2} = \frac{[\text{H}^+]\cdot f}{(1-f)} = \frac{[\text{H}^+]\cdot f}{\beta_{\equiv\text{SOH}}} \tag{5.50}$$

where f is the fraction which has dissociated. The value of f is obtained from a titration experiment, and amounts to the sum:

$$f = \beta_{\equiv\text{SO}^-} + \beta_{\equiv\text{SONa}} \tag{5.51}$$

Combination with Equation (5.49) gives:

$$f = \beta_{\equiv\text{SO}^-}\cdot(1 + [\text{Na}^+]/K_{\text{SONa}}) \tag{5.52}$$

We introduce this relation in Equation (5.50). The quotient $[\text{H}^+]\cdot\beta_{\equiv\text{SO}^-}/\beta_{\equiv\text{SOH}}$ has an apparent constant defined by Equation (5.47), so that

$$\log K_{a2} = \log K_{int} - \frac{\alpha_2}{\ln 10} \cdot \{\equiv SO^-\} + \log \left(1 + \frac{[Na^+]}{K_{SONa}}\right) \qquad (5.53)$$

This equation suggests that K_{a2} decreases with increasing dissociation of the surface groups (increasing pH), and increases when the salt concentration increases. If the dissociation constant K_{SONa} is small, the increase of K_{a2} would be about tenfold for every tenfold increase in concentration. This is indeed observed in titrations of polymere solutions (Marinsky, 1987). Titrations of humic and fulvic acids, which are of interest in our context, follow a similar trend, as illustrated in Figure 5.23.

We assume that the electrostatic term for a fulvic acid may be approximated as

$$\frac{\alpha_2}{\ln 10} \cdot \{\equiv SO^-\} = \alpha_f \cdot f$$

so that Equation (5.53) can be rewritten as (we use the notation $-\log = p$):

$$pK_a = pK_{int} + \alpha_f \cdot f - \log \left(1 + \frac{[Na^+]}{K_{FNa}}\right) \qquad (5.54)$$

where the constants for the fulvic acid $\equiv FH$ follow the notation used earlier for a solid surface $\equiv SOH$. The plot (Figure 5.23) of pK_a vs. f shows an approximate linear increase with f, and also a decrease as concentration of $NaNO_3$ increases. The value of K_{FNa} can be estimated from such plots, with $pK_a = (pK_a)_1$ for $(m_{Na^+})_1$ and $pK_a = (pK_a)_2$ for $(m_{Na^+})_2$, using:

$$K_{FNa} + (m_{Na^+})_1 = 10^{(pK_a)_1 - (pK_a)_2} \cdot (K_{FNa} + (m_{Na^+})_2)$$

as follows from Equation (5.54). Lines in Figure 5.23 are drawn with the value of $pK_{int} = 2.8$, $\alpha_f = 3.7$, and $K_{FNa} = 0.03$.

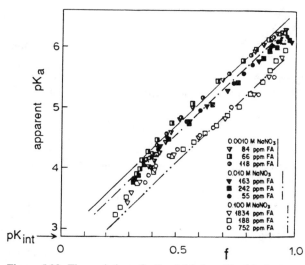

Figure 5.23. The variation of pK_a of fulvic acid with degree of dissociation at three different ionic strength levels. Ionic medium is $NaNO_3$. Lines are drawn according to the constant capacitance/ion association model. The fulvic acid is from Armadale Bh soil horizon. Experimental data after Marinsky (1987).

The electrostatic term which influences the association of ions of the background electrolyte with the solid surface was not included in the developed model. This term is contained in so-called *double layer models* (Dzombak and Morel, 1990), and *triple layer models* (Davis and Kent, 1990). Both models are succesful in explaining surface phenomena of oxides which have been found in laboratory experiments. A considerable problem with these models is that we do not yet know how the more complicated electrostatic terms can be estimated for soils and aquifer materials. We can only reiterate that the assumption of a constant exchange capacity, and neglect of the electrostatic term is adequate for describing observed exchange behavior of major elements, as has been illustrated in Sections 5.3 and 5.4.

5.5.2 *Specific adsorption of heavy metals*

The charge and the potential on oxide surfaces increases as the result of protonation or deprotonation of the surface oxygens. Protons are therefore called *potential determining ions* (*pdi*'s). The electrochemical behavior of simple salts such as Ag_2S or AgI has been extensively studied, and for these salts the ions forming the crystal lattice were found to operate as *pdi*'s. Oxides and clay minerals have, besides protons, a multitude of ions which can act as a *pdi*. Earth-alkaline ions, which are normally considered as inert background electrolytes for simple salt solids, may lead to a reversal of potential of the surface (Parks, 1990). The sign of the potential can be measured from the electrophoretic mobility of the suspended particles in an electric field maintained by two electrodes. An example is shown in Figure 5.24 (Fuerstenau et al., 1981). Rutile particles in a 3.3 µmol/l $Ca(NO_3)_2$ solution move towards the positive electrode at high pH, and their velocity increases as pH, and therefore the negative potential of the surface, is higher. However, the electrophoretic mobility diminishes when concentration of the background electrolyte increases, and it can even become inverted in the sense that movement is towards the negative electrode.

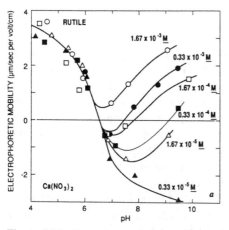 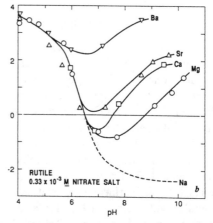

Figure 5.24. Dependence of the electrophoretic mobility of rutile particles on pH and electrolyte concentration of NO_3^- salts of alkaline-earth ions. (a) Mobility at various concentrations of $Ca(NO_3)_2$. (b) Mobility at an electrolyte concentration of 0.33 mmol/l of alkaline-earths (from Fuerstenau et al., 1981).

We can write a reaction for Ca^{2+} adsorption at the negative rutile surface as:

$$Ca^{2+} + \equiv TiO^- \rightarrow \equiv TiOCa^+ \tag{5.55}$$

which indicates a charge inversion at the surface. It can be seen in Figure 5.24 that the electrophoretic mobility is not affected in the low pH range, where the potential is positive as a result of protonation. Ca^{2+} is therefore a genuine background electrolyte at low pH. However, Figure 5.24b shows that the heavier earth-alkaline ions Sr^{2+} and Ba^{2+} have an influence which extends to lower pH values. These ions act also as *pdi*'s, and show specific adsorption just as the proton.

Many heavy metals show specific adsorption, and are adsorbed even on a surface that is positively charged through protonation. Figure 5.25 shows adsorption curves of a few heavy metals on rutile, and indicates that adsorption is essentially complete at pH < 6. Rutile has a Point of Zero Net Proton Charge at ca. pH = 6.8, which means that the (positive) heavy metals have become attached to a positive surface. We may suppose that the heavy metal (M^{m+}) competes with a proton for the surface site, and write:

$$\equiv SOH + M^{m+} \leftrightarrow \equiv SOM^{(m-1)+} + H^+ \qquad K_{SOM} \tag{5.56}$$

Many heavy metals have a charge of two in solution, and we could also consider the reaction as a true ion exchange reaction without charge effects on the solid, i.e. for M^{2+}:

$$2 \equiv SOH + M^{2+} \leftrightarrow (\equiv SO)_2 M + 2H^+ \qquad \beta_{(SO)_2 M} \tag{5.57}$$

where $\beta_{(SO)_2 M}$ indicates a complex formation constant. (This association with two surface sites is also termed a bidentate ligand formation).

If we calculate the ratio $\bar{M}/M$, where $\bar{M}$ is adsorbed, and M is the solute concentration of M, for a high and a low ratio, e.g. of $10/1$ and of $1/10$, we find for Equation (5.56):

$$\frac{[H^+]_{high}}{[H^+]_{low}} = \frac{1}{100} \cdot \frac{[\equiv SOH]_{high}}{[\equiv SOH]_{low}} \tag{5.56a}$$

and for Equation (5.57):

$$\frac{[H^+]_{high}}{[H^+]_{low}} = \frac{1}{10} \cdot \frac{[\equiv SOH]_{high}}{[\equiv SOH]_{low}} \tag{5.57a}$$

If $\bar{M}$ is present in trace quantities only, $[\equiv SOH]_{high} = [\equiv SOH]_{low}$, the increase in the adsorption ratio from $1/10$ to $10/1$ requires a pH increase of either 1 (for Equation 5.57) or of 2 (for Equation 5.56). The increase corresponds to an adsorption interval from 9.1 to 90.9%. Inspection of Figure 5.25 shows that Equation (5.56) is more adequate for the heavy metals Pb^{2+}, Cd^{2+} and Cu^{2+} on rutile because the adsorption isotherm covers 2 pH units over the indicated interval. However, other solids may give a steeper increase of adsorption with pH, especially when adsorption occurs at pH values near the pK_a's of the solid.

Ions which show specific adsorption can be discerned from ions which behave as a background electrolyte by the strength of the surface complexation reaction. The strength is indicated by the surface complexation constant. Values of K_{SOM} have been tabulated by Schindler and Stumm (1987) for a number of oxides with cations and anions, and a selection is reproduced here in Table 5.7. The table shows that earth-alkaline cations have up to 6 orders of magnitude smaller tendency to displace protons from Al_2O_3 and TiO_2 than heavy metals such as Pb^{2+}. The ions Ca^{2+} and Mg^{2+} are therefore true background electrolytes when the surface is neutral or positive, and become *pdi*'s only when the

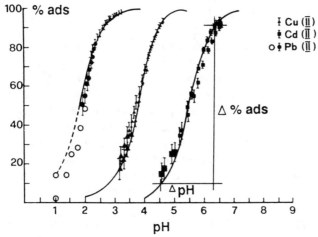

Figure 5.25. Adsorption of some divalent metal ions on TiO_2 (rutile). Comparison of pH values at 10 and 90% adsorption allows the reaction mechanism to be determined. Original data from Fürst (1976), cited by Schindler and Stumm (1987).

concentration is high enough to counteract the low K_{SOM}. Heavy metals such as Pb^{2+}, Cd^{2+}, etc., on the other hand, actively compete with protons for surface sites, and already do so at the low concentrations which are of environmental interest. The small value of K_{SOM} of background electrolyte ions explains why the *CEC* is not much different in electrolytes of freshwater ionic strength and say, seawater. The very large specificity of heavy metals for surfaces which can deprotonate also explains why pH is so often found to be the most important factor which determines the distribution coefficient of heavy metals in soils (Lexmond, 1980; Christensen, 1989; Workman and Lindsay, 1990). The ion exchange coefficients are generally about equal for heavy metals and earth-alkaline ions (Table 5.5), but the tendency of heavy metals to displace protons from oxide surfaces is more than a factor 1000 stronger (Table 5.7).

Table 5.7. Intrinsic exchange constants of cations and anions with H^+ and OH^-. Data from Schindler and Stumm, 1987.

Reaction: $M^{2+} + \equiv SOH \leftrightarrow \equiv SOM^+ + H^+$ K_{SOM}			Reaction: $A^{a-} + \equiv SOH \leftrightarrow \equiv SA + OH^-$ K_{SA}		
Solid	M^{2+}	$\log K_{SOM}$	Solid	A^{a-}	$\log K_{SA}$
γ–Al_2O_3	Mg^{2+}	-5.4	α–FeOOH	F^-	-4.8
	SO_4^{2-}	-5.8		SO_4^{2-}	-5.8
	Ca^{2+}	-6.1			
	Ba^{2+}	-6.6			
	Cu^{2+}	-2.1			
	Pb^{2+}	-2.2			
TiO_2	Mg^{2+}	-5.9			
	Co^{2+}	-4.3			
	Cu^{2+}	-1.4			
	Cd^{2+}	-3.3			
	Pb^{2+}	0.4			

5.6 THE GOUY-CHAPMAN THEORY OF THE DOUBLE LAYER

So far, we have considered equations which relate concentrations in solution to those on the exchanger surface. If we want to describe what is actually going on when a charged surface attracts or repels ions from a solution we must go down to the molecular level (Moore, 1972; Feynman et al., 1989). In our case, we could consider the cations and anions as point charges which can move in an electric field. A bulk solution in which the number of cations and anions are equal, has zero electric potential. A negatively charged surface has a negative potential that attracts positive charges and repels negative ones. However, diffusion will smear out concentration steps, and cause the transitions to become more gradual. The net effect (Figure 5.26) is a gradual change of concentrations in the socalled Gouy-Chapman diffuse double layer (*DDL*) which surrounds the charged surfaces in a solution.

Also, the electric potential will not change abruptly at the solid-solution interface, but gradually lessen as distance from the interface increases. The particles in the potential

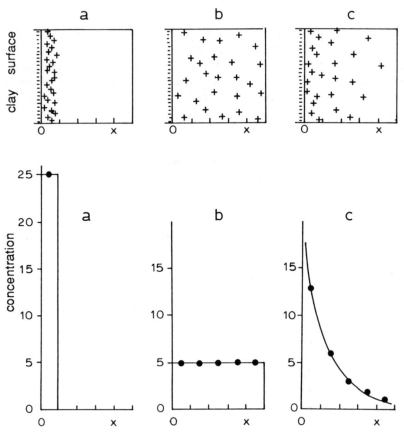

Figure 5.26. Distribution of counterions. (a) Minimal energy and maximal electrostatic attraction. (b) Maximal concentration distribution and entropy. (c) Compromise: the distribution in a diffuse double layer. Adapted from Bolt and Bruggenwert, 1978.

energy field are distributed according to Boltzmann's theorem:

$$n_{x1}/n_{x2} = \exp\left(-(U_{x1} - U_{x2})/kT\right) \tag{5.58}$$

where $U_{x1} - U_{x2}$ is the difference in potential energy at points $x1$ and $x2$, where the average number of particles is n_{x1} and n_{x2} (ions/m^3), k is the Boltzmann constant ($13.8 \cdot 10^{-24}$ J/K), and T is the absolute temperature (K).

The potential energy which is used here for a single particle, is similar to the Gibbs free energy for one mole of particles. In fact, if U_x and k are multiplied with Avogadro's number N_a ($= 6 \cdot 10^{23}$/mol), the relationship is for moles of particles i:

$$m_{i,x1}/m_{i,x2} = \exp\left(-(\Delta G_{i,x1} - \Delta G_{i,x2})/RT\right)$$

If only the electric potential changes, the difference in energy, $U_{x1} - U_{x2}$, in the double layer is related to the electric potential:

$$U_{x1} - U_{x2} = zq_e(\Psi_{x1} - \Psi_{x2}) \tag{5.59}$$

where z is charge of the ion, q_e is charge of the electron ($1.6 \cdot 10^{-19}$ C), and Ψ the potential (Volt).

In the free solution (infinite distance from the charged surface), $\Psi_\infty = 0$, and therefore for a positive ion with charge $+1$ at x:

$$n_{+,x}/n_{+,\infty} = \exp\left(-q_e\Psi_x/kT\right) \tag{5.60}$$

and for a negative ion with charge -1 at x:

$$n_{-,x}/n_{-,\infty} = \exp\left(q_e\Psi_x/kT\right) \tag{5.61}$$

In the bulk solution of two monovalent ions, is $n_{+,\infty} = n_{-,\infty}$, as required by electroneutrality. But near the charged surface, there is a charge density:

$$\rho = q_e(n_+ - n_-) = q_e \cdot n_\infty \left(e^{-q_e\Psi/kT} - e^{q_e\Psi/kT}\right) \tag{5.62}$$

and the electrical field changes as:

$$\frac{dE}{dx} = \frac{\rho}{\varepsilon} \tag{5.63}$$

where ε is the dielectric constant, or permittivity (of water $7.08 \cdot 10^{-10}$ F/m at 25°C).

A further relation connects the change of the potential with the work done as an electric charge is moved in this electrical field:

$$d\Psi = -\int_\infty^x E \cdot dx \tag{5.64}$$

which together with Equation (5.63) gives Poisson's equation:

$$\frac{d^2\Psi}{dx^2} = -\frac{\rho}{\varepsilon} \tag{5.65}$$

which in turn, can be combined with Equation (5.62):

$$\frac{d^2\Psi}{dx^2} = -\frac{q_e n_\infty}{\varepsilon}\left(e^{-q_e\Psi/kT} - e^{q_e\Psi/kT}\right) \tag{5.66}$$

The particle densities are a function of the potential. But, we only know the potential via a differential equation which contains the particle densities again. Integration with the right boundary conditions can provide us with a solution. The problem can be simplified even further when the potentials are small, since in that case the exponent can be approximated as:

$$\exp\left(\pm q_e \Psi / kT\right) = 1 \pm q_e \Psi / kT \tag{5.67}$$

Equation (5.66) then gives:

$$\frac{d^2\Psi}{dx^2} = \frac{2n_\infty q_e^2}{\varepsilon kT} \Psi \tag{5.68}$$

which we simplify with

$$\kappa^2 = 2n_\infty q_e^2 / (\varepsilon kT) \tag{5.69}$$

to

$$\frac{d^2\Psi}{dx^2} = \kappa^2 \Psi \tag{5.70}$$

Let us try to integrate Equation (5.70) so that we can calculate Ψ as a function of x. We must integrate twice, and will obtain two constants which we can determine from the boundary conditions. Thus, it can be readily verified by differentiation that a general solution is of the form:

$$\Psi = A \cdot \exp\left(-\kappa x\right) + B \cdot \exp\left(+\kappa x\right) \tag{5.71}$$

Two boundary conditions at $x = \infty$, and at $x = 0$, provide the values for A and B. The potential in the free solution, when $x = \infty$, is $\Psi_\infty = 0$, and this leads to $B = 0$. When $x = 0$, we must find $\Psi = \Psi_0$ and therefore $A = \Psi_0$. The solution for our problem is thus:

$$\Psi = \Psi_0 \cdot \exp\left(-\kappa x\right) \tag{5.72}$$

The potential decays exponentially from the charged surface, and so will the extent over which the influence of a diffuse double layer can be noted in different concentrations of cations and anions. If we go back to Equations (5.60) and (5.61), we may calculate the densities of positive and negative point charges as a function of distance from the surface, and hence obtain an idea of the concentrations of the cations and anions. An example is provided in Figure 5.27. When the concentration in the bulk solution (n_∞) increases, κ increases, and the potential decays faster.

5.6.1 *Thickness of the double layer*

The units of κ^2 are m^{-2}, as can be verified in Equation (5.69). The parameter $1/\kappa$ is termed the Debije length; it also appears in the Debije-Hückel theory for calculating ion activity coefficients. The ionic strength

$$I = \frac{1}{2} \sum m_i z_i^2 \tag{5.73}$$

can be incorporated in the Debije length, and gives:

$$1/\kappa = \sqrt{\frac{\varepsilon \cdot RT}{2(N_a q_e)^2} \cdot \frac{1}{1000 \, I}} \tag{5.74}$$

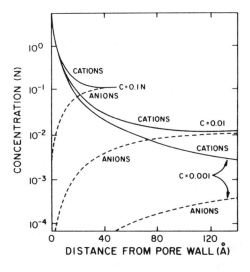

Figure 5.27. Distribution of cations and anions in the double layer on a negatively charged surface at 3 concentration levels in the free solution (from Nielsen et al., 1986. Copyright by the Am. Geophys. Union).

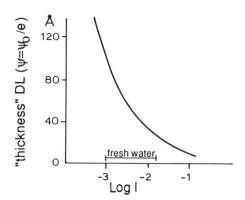

Figure 5.28. The extension of a double layer ('thickness') as a function of ionic strength.

with $\varepsilon = 7.08 \cdot 10^{-10}$ F/m, $R = 8.314$ J/K.mol, $T = 298$ K, $N_a = 6 \cdot 10^{23}$/mol, $q_e = 1.6 \cdot 10^{-19}$ C, and I expressed in mol/l we obtain:

$$1/\kappa = 3.09/\sqrt{I} \quad \text{(Å units)} \tag{5.74a}$$

It is customary to consider the point where $\Psi = \Psi_0/e$, i.e. where $x = 1/\kappa$, as thickness of the *DDL*. The influence of the ionic strength on the thickness of the *DDL* is shown in Figure 5.28.

5.6.2 *Practical aspects of double layer theory*

The thickness of the double layer (Figure 5.28) is important, since it shows how far a negative potential from a clay mineral extends in solution. Clay particles are repelled from each other when the negative potentials from opposite clay particles overlap. A suspension of clay particles is stable if the double layer is so large that the clay particles repel each other: the colloidal solution remains peptized. A small double layer on the other hand, may induce flocculation, i.e. the suspension is separated in a clear supernatant solution, and a flocculated clay. The double layer thickness increases with decreasing ionic strength. If, for example, Ca^{2+} is replaced by Na^+ at equal normality, ionic strength would decrease by a factor of 2, and the Debije length increases by 1.4. Thus, swelling of clay in the soil may be associated with low ionic strength and monovalent cations in the soil solution. Such swelling of clay can induce transport of clay particles, and this may, in turn, lead to a clogging of pores, and to a decrease of the hydraulic conductivity. Figure 5.29 illustrates this decrease as a function of SAR, i.e. the $Na/\sqrt{Ca}$ ratio in solution, and of total concentration (normality).

The effect may be even more outspoken when fresh water displaces seawater in an aquifer. Initially, in seawater, the concentrations are so high that the clay will remain flocculated, irrespective of the high Na^+ concentration in seawater. Dilution by for example, rainwater, will decrease the ionic strength, and may induce clay movement. Since initially the cations in solution are buffered to a high extent by cations from the exchange complex, there will be an additional effect from the Na^+ which is released from the sediment exchanger in return for Ca^{2+}. The effect is visible in experiments by Goldenberg (1985) who mixed dune sands with smectite clay and found a marked decrease of hydraulic conductivity as seawater is replaced by fresh water. In later experiments he could actually observe the migration of the clay particles (Goldenberg and Mandel, 1988). Such clay transport also influences other hydraulic properties of an aquifer, such as porosity and dispersivity (Mehnert and Jennings, 1985; Beekman and Appelo, 1990), and it can clog the wells when fresh water is injected into brackish water aquifers (Brown and Signor, 1974).

The extent of the double layer determines whether pores are electrostatically 'free'. Pores which are so small that the double layers overlap (or become hatched at lower moisture contents), are completely filled with a negative potential. Anions are then unable to pass the pores; but, cations are hindered as well since the outflowing solution must remain neutral. A consequence is that only watermolecules and uncharged species (H_4SiO_4, uncharged complexes) can pass the pore. This process is termed *hyperfiltration* or *membrane filtration*. From Figure 5.28 it follows that the double layer extends to about $0.01~\mu m$ at an ionic strength = 0.001. Some effect may therefore be expected in clay soils (particles and pores $\ll 1~\mu m$), or compacted clays. It is of interest that the process is self-regulating. The double layer will shrink as concentrations and ionic strength increase, and a residual brine forms, thereby allowing the passage of more and more solute ions. The

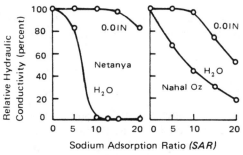

Figure 5.29. Hydraulic conductivity of a sandy loam (Netanya) and a silty loam (Nahal Oz) soil as a function of the SAR and the concentration of the leaching solutions (from Shainberg and Oster, 1978).

double layer approaches the size of a single layer of ions if concentrations are higher than 1 mol/l, and hyperfiltration would be reduced in that case. The process of membrane filtration has nevertheless been invoked in explaining anomalous ion ratios in deep brines (Kharaka and Berry, 1973; Demir, 1988), and it may even induce calcite precipitation (Fritz and Eady, 1985). A warning in relation to sampling of turbid waters is also appropriate here. Hyperfiltration may take place when water samples are filtered over membrane filters (poresize 0.45 μm, or sometimes even 0.1 μm) at too high a pressure, and against a too large resistance of the filtered solids.

A related phenomenon takes place when water is squeezed from compacting sediments or clay suspensions. The pressure filtrate may have *higher* ion concentrations than the original pore water (Bolt, 1961), and it is likewise known that the concentrations in the remaining pore water decrease when water is pressed out of the suspension (Von Engel-hardt and Gaida, 1963). These effects are difficult to trace in stratigraphic sequences in the field, because other water-rock interaction processes may be more dominating. However, the effects are also of importance when sediment samples are centrifuged or squeezed to obtain interstitial water for chemical analysis (De Lange, 1984; Saager et al., 1990). The effects are related to compaction of the diffuse double layer, and can be quantitatively described with the Donnan theory of suspensions (Appelo, 1977). Two examples are shown in Figure 5.30 and 5.31. Figure 5.30 gives the composition of water squeezed from a Na-illite suspension as a function of V, the relative volume of water that remains in the suspension. Initially the salt concentration is 0.96 mmol Na^+/l, and clay content amounts to 42 meq CEC/l. The figure shows that the salt concentration in the pressure filtrate increases sharply when about half of the water has been pressed from the suspension. Another example, in Figure 5.31, shows the concentrations in the pore solutions of a clay as function of the void ratio, i.e. the volume of water over the volume of solids. Three different salt concentrations in the initial pore water, varying from 0.157 to 1.07 eq/l, show all a marked decrease in concentration when water is squeezed from the clay suspension. The effect is quite notable, due to the high CEC of 0.67 or 0.72 eq/l in the initial suspension.

Another aspect of the relative anion deficit in the DDL is *anion exclusion*. Whenever pores are so small that double layers overlap, or when waterfilms become, in unsaturated soils, thinner than the DDL, the anions are hesitant to enter the pores with a negative

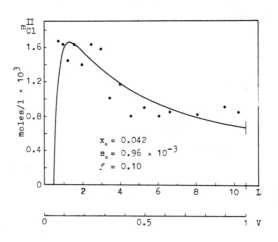

Figure 5.30. Concentrations in pressure filtrate from a Na-illite suspension. Data points from Bolt, 1961, line calculated by Appelo, 1977. See text for explanation of axes.

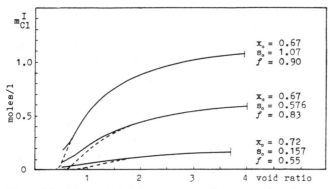

Figure 5.31. Concentrations in pore water of a compacting clay suspension. Solid line indicates data from Von Engelhardt and Gaida, 1963, dotted line (shown where deviating from the measured data) is calculated by Appelo, 1977.

potential. The effect is that the path of anions is confined to part of the porespace only, and this is easily measured in column experiments. An anion, affected by exclusion, will show an earlier breakthrough than water (found in, for example, tritium); the faster transport may easily be as much as 10% in soils (Bond et al., 1982; James and Rubin, 1986), or even 50% in compacted clay (Demir, 1988). A field experiment by Gvirtzmann and Gorelick (1991) shows similar effects.

We have seen (Figure 5.27) that the concentration of the anions will decrease towards a negatively charged surface. The relative deficit of anions (or more generally, of the co-ions which have the same valence as the adsorbing solid) balances part of the charge of the solid, and is thus rightly termed *negative adsorption* (Van Olphen, 1977). Negative adsorption increases with higher salt concentrations on solids with a fixed charge, such as clay minerals. A higher anion deficit means that less cations are adsorbed, and the cation exchange capacity may be expected to decrease when solute concentrations increase. This effect can amount to about 5% lower apparent *CEC* of soils in seawater compared with fresh water (Bolt and Bruggenwert, 1978). Since other adsorption mechanisms may play a role, such as the competition with protons, or the adsorption of complexes (Sposito et al. 1983; Griffioen and Appelo, 1993), the net effect is very difficult to detect experimentally in a sediment or soil if only batch exchange experiments are performed.

5.6.3 *Potentials in a double layer*

During the discussion of the effects of an electric potential on ion adsorption in Section 5.5.1, it was noted that we need the potential to account for effects of exchange and adsorption. Now that we have solved Poisson's equation you may ask whether we are able to give more precise calculations. If the charge density of the solid is known, as with clay minerals with a structural charge σ, we know the electrical field at the surface:

$$E_{x=0} = \sigma/\varepsilon \qquad (5.75)$$

E is also the gradient of Ψ, so that by differentiation of Equation (5.72) we obtain:

$$E_{x=0} = -(\partial\Psi/\partial x)_{x=0} = \kappa\Psi_0 \qquad (5.76)$$

Combination with Equation (5.75) gives the constant capacitance model:

$$\Psi_0 = \sigma/\kappa\varepsilon \tag{5.42}$$

If we calculate Ψ_0 for a charge of -0.05 C/m^2 (cf. Example 5.2), and for ionic strength = 0.1 mol/l we find $\Psi_0 = -0.069$ V. This value is based on a simplified assumption used for solving Poisson's equation; a stricter calculation for the conditions mentioned gives $\Psi_0 = -0.039$ V (cf. Van Olphen, 1977). Let us calculate the difference in $[H^+]_0$ at the surface with $[H^+]_\infty$ in the free solution (Equation 5.60):

$$\log [H^+]/[H^+]_0 = -F\Psi_0/(\ln 10 \cdot RT) \tag{5.77}$$

which gives:

$$pH_0 - pH_\infty \approx 16.9\,\Psi_0 \tag{5.78}$$

If $\Psi_0 = -0.069$ V, the pH at the surface is 1.2 units lower than in bulk solution. Related to the lower pH is the lower hydrolysis of complexes at a clay surface. For example the hydrolysis of Al,

$$Al(OH)_2^+ + 2\,H^+ \leftrightarrow Al^{3+} + 2\,H_2O$$

is shifted to the right. When Al^{3+} is exchanged by another cation, and enters the free solution, it will tend to hydrolyze, take up OH^- and precipitate as Al-hydroxide. The corresponding drop in pH is always apparent in the uniformly lower pH values measured in 1N KCl, as compared with pH measured in pure water.

The lower pH at negatively charged surfaces is clearly demonstrated in pH measurements of sediments, but it may be that the surface pH, as calculated from *DDL* theory, is too low. The theory given so far, assumes that the ions are point charges which can approach the solid surface to zero distance. In reality the ions are surrounded by a layer of water molecules. In the inner-sphere complexes this layer of water molecules is removed; in the outer-sphere complexes it is still present. However, even in the inner-sphere complexes, the cation still has a finite size which prevents it from approaching the solid surface to zero distance. The Stern double layer theory takes the finite distance of approach into account, and it can moreover incorporate chemical interaction terms by introduction of a specific adsorption potential (Van Olphen, 1977). It is, therefore, a more comprehensive model which can be used to describe adsorption and ion exchange. However, the application of the Stern model (or the closely related triple layer model) has been mainly confined to description of laboratory experiments with oxides (Westall and Hohl, 1980; Davis and Kent, 1990), and application to a heterogeneous soil has only recently been performed (Sposito et al., 1988; Zachara et al., 1989; Zhang and Sparks, 1990).

5.7 IRRIGATION WATER QUALITY

Soils with a high concentration of Na^+ in solution are notorious for having a bad physical structure. The permeability is reduced (Figure 5.29), and heavy agricultural machinery is not supported. The effects are related to a high degree of swelling of the clays, which in turn is caused by extending double layers around the clay particles. Diffuse double layers increase in thickness when monovalent ions make up a high proportion of the exchangeable cations. Problems with the soil structure arise when Na^+ makes up about 15% of the exchangeable cations, and ionic strength is lower than ca. 0.015 (i.e. fresh water). With Na^+, Mg^{2+}, and Ca^{2+} as the dominant cations in natural water, the fraction of exchangeable Na^+ is a function of the ratio of Na^+ over the square root of Ca^{2+} and Mg^{2+} concentrations. The fraction of Na^+ can be calculated with the Gaines-Thomas convention (Example 5.4):

$$\beta_{Na}^2 \cdot \{[Ca^{2+}]/(K_{Na\backslash Ca} \cdot [Na^+])^2 + [Mg^{2+}]/(K_{Na\backslash Mg} \cdot [Na^+])^2\} + \beta_{Na} - 1 = 0$$

However, in irrigation studies the Gapon convention is more popular, and it offers an easier relationship between exchanger fraction and solute concentrations. From the reaction:

$$Na^+ + Ca_{0.5}\text{-}X \leftrightarrow Na\text{-}X + \tfrac{1}{2} Ca^{2+} \tag{5.19a}$$

we have:

$$K_{Na\backslash Ca}^G = \frac{[Na\text{-}X]\,[Ca^{2+}]^{0.5}}{[Ca_{0.5}\text{-}X]\,[Na^+]} = \frac{\beta_{Na}\,[Ca^{2+}]^{0.5}}{\beta_{Ca}\,[Na^+]} \tag{5.20a}$$

or

$$\frac{\beta_{Na}}{\beta_{Ca}} = K_{Na\backslash Ca}^G \cdot \frac{[Na^+]}{[Ca^{2+}]^{0.5}} \tag{5.79}$$

The value of the Gapon constant $K_{Na\backslash Ca}^G = 0.5$. A further simplification is to assume that Mg^{2+} and Ca^{2+} are equally selected by the soil, i.e. $K_{Na\backslash Ca}^G = K_{Na\backslash Mg}^G$. This means that we can add the Ca^{2+} and Mg^{2+} concentrations together. Since furthermore $\beta_{Mg+Ca} = 1 - \beta_{Na}$ in a soil with only Na^+, Mg^{2+} and Ca^{2+}, we obtain:

$$\frac{\beta_{Na}}{1 - \beta_{Na}} = K_{Na\backslash Ca}^G \cdot \frac{[Na^+]}{\sqrt{[Ca^{2+}] + [Mg^{2+}]}} \tag{5.80}$$

where the quotient $\beta_{Na}/(1 - \beta_{Na})$ is termed the *exchangeable sodium ratio (ESR)*. The *ESR* is thus related to the activity of Na^+ over the square root of the sum of Ca^{2+} and Mg^{2+} activities.

EXAMPLE 5.10: *Calculation of the exchangeable sodium ratio (ESR)*
Calculate the ratio of $Na/\sqrt{Ca}$ in water where the limit of 15% Na^+ in the exchange complex is reached.

ANSWER
Equation (5.80) gives:

$$\frac{0.15}{1 - 0.15} = 0.5 \cdot \frac{[Na^+]}{\sqrt{[Ca^{2+}]}}$$

and the limiting ratio $Na/\sqrt{Ca}$ is 0.35 $(mol/l)^{.5}$ (assuming concentrations equal to activities).

Soil scientists use the *sodium adsorption ratio (SAR)* of water as a measure of *ESR*, to estimate the suitability of irrigation water. In the formula for *SAR*, concentrations in mmol/l are used, and the Mg^{2+} concentration is added to the Ca^{2+} concentration. With this convention Equation (5.80) becomes:

$$\frac{\beta_{Na}}{1 - \beta_{Na}} = 0.5 \cdot \frac{1000}{\sqrt{1000}} \cdot \frac{m_{Na^+}}{\sqrt{m_{Ca^{2+}} + m_{Mg^{2+}}}} \tag{5.80a}$$

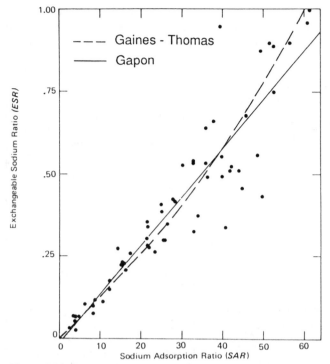

Figure 5.32. Relationship between *SAR* and *ESR* in a number of soils; full line gives the empirical relationship (agrees with Gapon convention), broken line gives Gaines-Thomas convention with $K_{Na\backslash Ca} = 0.37$. Adapted from US Soil Salinity Lab., 1954.

where m_i is here the concentration of *i* in mmol/l. In this equation is

$$SAR = \frac{m_{Na^+}}{\sqrt{m_{Ca^{2+}} + m_{Mg^{2+}}}}$$

so that Equation (5.80) may be written as:

$$ESR = 0.0158 \cdot SAR$$

The critical $ESR = 0.15$, is reached when $SAR = 10$. The equation obtained from exchange theory can be compared with the empirical relationship (US Soil Salinity Lab, 1954):

$$ESR = -0.013 + 0.015 \, SAR \tag{5.81}$$

which is the relationship found in a number of soils (Figure 5.32). The figure also includes a plot of *ESR* calculated with the Gaines-Thomas convention, with $K_{Na\backslash Ca} = 0.37$. The difference between the two equations is only marginal in this simple system with two cations, and use of the easy Gapon convention is here quite acceptable.

The quality of irrigation water is not only determined by *SAR* in water at the time of application, since the ratio of Na^+ over the square root of $Ca^{2+} + Mg^{2+}$ concentrations in water may change as a result of reactions in the soil. Most important is the concentrating effect of evapotranspiration, and the precipitation of calcite, and both processes increase *SAR*. Evapotranspiration has an effect because of the square root relationship for the Ca^{2+}

concentration in the formula for *SAR*, and the precipitation of calcite has an effect because it decreases the Ca^{2+} concentration in water. An adjusted *SAR* can be calculated in which calcite precipitation is taken into account (Ayers & Westcot, 1985).

EXAMPLE 5.11: *Calculation of SAR adjusted for calcite precipitation*
An analysis of groundwater provided the following results:

pH	Na^+	K^+	Mg^{2+}	Ca^{2+}		Cl^-	HCO_3^-	SO_4^{2-}
8.5	62	0.1	9.0	81		168	183	9.0 mg/l

- Is the analysis reliable?
- Is the water quality suitable for irrigation when irrigation return flow is 10%?

ANSWER
Recalculating the analysis in meq/l gives:

Na^+	K^+	Mg^{2+}	Ca^{2+}	$\Sigma+$		Cl^-	HCO_3^-	SO_4^{2-}	$\Sigma-$
2.7	0.002	0.74	4.05	7.49		-4.74	-3.00	-0.19	-7.93

The difference is -0.44, or 3% of Σ(cations-anions), and the analysis is good. *SAR* of the water is 1.7, and it offers excellent irrigation quality. However, concentrations increase upon evapotranspiration of 90% of the water, and this gives, in itself, an increase of *SAR*. One can expect moreover, that calcite precipitates, so that the Ca^{2+} concentration might perhaps decrease despite the concentration increase due to evapotranspiration. The Ca^{2+} concentration which is in equilibrium with calcite, can be calculated with the equations derived in Chapter 4, and it is easiest to assume a constant CO_2 pressure:

$$Ca^{2+} + 2HCO_3^- \leftrightarrow CaCO_3 + CO_2 + H_2O \qquad K = 10^{5.9}$$

With a CO_2 pressure of 0.01 atm, then at equilibrium with calcite:

$$[Ca^{2+}][HCO_3^-]^2 = 10^{-7.9} \qquad (5.82)$$

Note that $[HCO_3^-]$ as used here, is the actually measured concentration which may have been influenced by reactions other than calcite dissolution: we cannot assume $m_{HCO_3^-} = 2m_{Ca^{2+}}$, as was done in Chapter 4.

Neglecting activity coefficients and complexes, allows to use concentrations instead of activities. When the ions are 10 times concentrated through evapotranspiration, $m_{Ca^{2+}} = 20.2$ mmol/l $= 10^{-1.69}$ mol/l, and $m_{HCO_3^-} = 30$ mmol/l $= 10^{-1.52}$ mol/l. This gives:

$$(0.02)(0.03)^2 = 10^{-4.74}$$

An amount Δ will precipitate, which can be calculated with Equation (5.82):

$$(0.02 - \Delta)(0.03 - 2\Delta)^2 = 10^{-7.9}$$

Trial and error gives $\Delta = 0.0143$ mol/l, and the final concentrations as:

Na^+	Mg^{2+}	Ca^{2+}	HCO_3^-	
27	5.7	3.7	1.4	mmol/l

In the resulting water is $SAR = 27/\sqrt{(5.7 + 3.7)} = 8.8$, which is still on the safe side.

It may be noted once more, that water types with more equivalents Ca^{2+} than HCO_3^- offer relatively safe irrigation water qualities from the *SAR* point of view.

5.8 FINAL REMARKS

So far, we considered cation exchange mainly in relation to salt/fresh water displacements in aquifers, and salinization of soils. Here, the theory is adequate for application. Various aspects of ion exchange come about in relation to other subjects. For example, the relative decrease of Cl^- in ecosystem balances in the Western USA has been attributed to a possible adsorption of Cl^- on kaolinite in Chapter 2. Acidification and release of Al^{3+} from Al-hydroxides in the soil is buffered by cation exchange: Al^{3+} is taken up by the soil exchanger and base cations such as Ca^{2+} and Mg^{2+} are released. It has been a long agricultural practice to apply K^+ as a fertilizer. The throughput to groundwater lags behind NO_3^- (cf. Figure 2.15), because K^+ is taken up by the clay minerals.

There are difficulties associated with these other aspects of the solid-solution interface. Anion adsorption is not normally analyzed and, in fact, negative adsorption on negatively charged surface is often considered more important. However, positively charged surfaces undoubtedly exist in the natural assemblage of minerals and amorphous substances which make up the soil, and these may act as anion exchangers. Exchange and adsorption of Al^{3+} is a particularly difficult subject to study. At a pH of 5 to 8, the solubility of $Al(OH)_3$ is low, and it is difficult to discern between adsorption and precipitation in this pH range which is of particular importance for us. The adsorption of K^+ may be irreversible, in the sense that release of K^+ and its exchange for another cation is so slow, that uptake of K^+ might better be termed 'illitization', i.e. reconstitution of a mineral with 'illite' or 'muscovite' properties and only limited cation exchange ability.

These processes also affect the cation exchange capacity which is measured by displacing the adsorbed cations with a high loading of some cation (Scheffer and Schachtschabel, 1970; Page et al. 1982). The high normality of the solution which is used, has effect on the exchanging properties of the solid surfaces: extra protons may be desorbed, the double layer is compressed, specific adsorption of complexes may play a role. It is perhaps really surprising that use of average selectivity coefficients can give such an adequate description of the complex reactions on the solid surface, as we have found in this chapter.

PROBLEMS

5.1. Calculate the structural charge per gram of illite, $K_{0.7}[Si_{3.4}Al_{0.6}]^{iv}[Al_{1.9}Fe^{2+}_{0.1}]^{vi}O_{10}(OH)_2$, as well as the charge density. The $\vdash O \dashv$ layers are stacked 20 units deep; the unit cell size is as of smectite.

5.2. Do the same for vermiculite, $Mg_{0.4}[Si_{3.8}Al_{0.2}]^{iv}[Mg_{0.6}Al_{1.4}]^{vi}O_{10}(OH)_2$; unit cell size is as of smectite, and the layers are completely separated.

5.3. Repeat the calculation of Example 5.4 with all concentrations in water 10 times higher (note how the relative proportion of exchangeable Na-X changes). Also for seawater with concentrations from Table 5.2.

5.4. Determine the exchange coefficient $K_{Na\backslash Cd}$ on smectite from the following data of Garcia-Miragaya and Page (1976). The amount of smectite is 2.05 meq/l.

NaClO$_4$	Cd$_{total}$	adsCd
N	μM	%
0.03	0.40	91.1
	0.80	87.8

5.5. In the dunes of the island of Ameland two boreholes (A and B) have been drilled. Water from three filters (1 – 3), and the sea (4) has the following composition (concentrations in mmol/l):

	1	2	3	4
pH	6.7	6.8	7.0	8.2
Na$^+$	0.86	18.4	112.	485.
K$^+$	0.04	0.1	3.9	10.6
Mg^{2+}	0.09	0.45	10.	55.1
Ca^{2+}	3.0	0.9	31.3	10.7
Cl$^-$	1.0	3.0	200.	566.
HCO$_3^-$	6.1	18.0	0.3	2.4
SO$_4^{2-}$	0.02	–	–	29.3

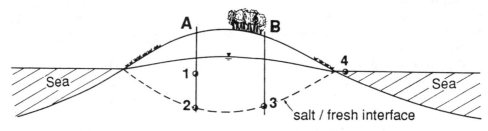

a. Explain the composition of these water samples.
b. Is the salt/freshwater interface moving upward or downward at A, B?.
c. Calculate the exchangeable fractions β_{Na}, β_K, β_{Mg} and β_{Ca} on the sediment with the Gaines-Thomas convention, for sample 1 and 4 (assume activity = total molality).
d. Recalculate the fractions of exchangeable cations in meq/l pore water, given $CEC = 1$ meq/100 g; $\rho_b = 1.8$ g/cm^3; $\varepsilon = 0.3$.

5.6. In the southern part of the Nile delta the composition of the groundwater is determined by:
  – composition of Nile water, which is used as irrigation water;
  – processes during infiltration of Nile water in the soil;
  – processes in the soil.
  a. What are the processes during infiltration? Data (concentrations in mmol/l):

	Na$^+$	K$^+$	Mg^{2+}	Ca^{2+}	HCO$_3^-$	Cl$^-$	SO$_4^{2-}$	pH
Nile water	.5	.1	.4	.7	2.2	.5	.1	
Groundwater	8.0	.1	.2	.7	2.7	4.5	1.0	8.1

b. In the northern part of the delta, near the sea, water is pumped up with a composition of:

	Na$^+$	K$^+$	Mg^{2+}	Ca^{2+}	HCO$_3^-$	Cl$^-$	SO$_4^{2-}$	
	180	2.5	35	15	7.5	270	1.5	(borehole 121)

– What processes influenced this composition?
– If more water is pumped up, will water become more brackish or more fresh?
c. Calculate *TDS* (Total Dissolved Solids) of the analysis of water in **b.**

5.7. Native groundwater in the injection test of Valocchi et al. (1981) had as composition $Na^+ = 86.5$, $Mg^{2+} = 18.2$ and $Ca^{2+} = 11.1$ mmol/l (Figure 5.15). Injected water has $Na^+ = 9.4$, $Mg^{2+} = 0.5$, $Ca^+ = 2.13$, $Cl^- = 14.66$ mmol/l. Selectivity coefficients were (Gaines and Thomas convention, assuming activity = molal concentration) $K_{Na\backslash Mg} = 0.54$ and $K_{Na\backslash Ca} = 0.41$. Sediment *CEC* = 750 meq/l pore water.
Calculate
a. The composition of the exchange complex in equilibrium with native groundwater
b. The composition of water after the salinity jump, in equilibrium with the original exchange complex (compare with Figure 5.15),
c. The composition of the exchange complex in equilibrium with injection water
d. The number of pore volumes that the exchange complex can maintain the high Na^+ concentration found under **b.**
i.e.

$$V = \frac{(m_{Na\text{-}X})_{5.7a} - (m_{Na\text{-}X})_{5.7c}}{(m_{Na^+})_{5.7b} - (m_{Na^+})_{inj.\ water}}$$

When the outcome is multiplied with $V_0 = 260$ m³, the pore volume for observation well S23, the rise of Mg^{2+} must be found: compare with Figure 5.15.

5.8. Fresh groundwater used in the column experiment shown in Figure 5.18 has $Na^+ = 5.6$, $Mg^{2+} = 1.9$, and $Ca^{2+} = 2.0$ mmol/l. Water is injected with 235 meq anions/l. The *CEC* of the sediment is 60 meq/l pore water. $K_{Na\backslash Mg} = 0.55$, $K_{Na\backslash Ca} = 0.35$ (Gaines and Thomas convention, solute activities = molal concentrations).
Calculate
a. The composition of the exchange complex,
b. The composition of water after the salinity jump in equilibrium with the original exchange complex,
c. The number of pore volumes that the exchange complex can maintain the high Ca^{2+} concentration found under **b.**

5.9. The plot of pK_{app} vs. f of a humic acid at 3 different concentrations of $Al(NO)_3$ is presented in Figure 5.33. Estimate pK_{int}, $\alpha/\ln 10$, and K_{h-Na}. Assume that Al^{3+} forms the complex:

$$\equiv hOAl^{2+} \leftrightarrow Al^{3+} + \equiv hO^- \qquad K_{h-Al}$$

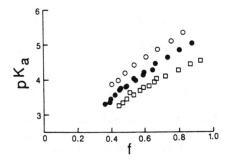

Figure 5.33. Apparent acid dissociation constant plotted against degree of dissociation at concentrations of 0.001, 0.01, and 0.1 mol/l of background electrolyte (from Tipping et al., 1988. Reprinted with permission from Pergamon Press PLC).

5.10. Derive the equation for the thickness of the double layer, Equation (5.74). Hint: derive (5.69) for ions with variable valence z, note that $kN_a = R$.

5.11. Calculate the potential at a smectite surface in a solution of 0.01 mol/l NaCl, and estimate the concentration of Na^+ and Cl^- at 0, 10, 50 and 100Å distance from the smectite surface. The smectite has a surface charge of 0.09 C/m^2.

5.12. Calculate *SAR* in the irrigation return flow water of Example 5.11 without taking calcite precipitation into account. Compare with *SAR* of the irrigation water.

5.13. Calculate the exchange coefficients for Ca\K exchange on illite and an ion exchange resin from the data of Wiklander (1955), as given in Example 5.3.

5.14. Derive a general formula to calculate equivalent fractions from molar fractions, and vice versa. Hint: use $\beta_I \cdot CEC = i \cdot \beta_I^M \cdot TEC$, and find 2 relations for TEC/CEC in which only β^M's or β's appear.

REFERENCES

Appelo, C.A.J., 1977, Chemistry of water expelled from compacting clay layers: a model based on Donnan equilibrium. *Chem. Geol.* 19, 91-98.

Appelo, C.A.J., 1979, Layer deformation and crystal energy of micas and related minerals. II. Deformation of the coordination units. *Am. Mineral.* 64, 424-431.

Appelo, C.A.J., Ponten, J.F. and Beekman, H.E., 1989, Natural ion-chromatography during fresh-/sea water displacements in aquifers: a hydrogeochemical model of the past. In D.L. Miles (ed.), *Water-Rock Interaction: Proc. 6th Water-Rock Interaction Symp., Malvern*. Balkema, Rotterdam, 23-28.

Appelo, C.A.J. and Willemsen, A., 1987, Geochemical calculations and observations on salt water intrusions. I. a combined geochemical/mixing cell model. *J. Hydrol.* 94, 313-330.

Appelo, C.A.J., Willemsen, A., Beekman, H.E. and Griffioen, J., 1990, Geochemical calculations and observations on salt water intrusions. II. Validation of a geochemical model with laboratory experiments. *J. Hydrol.* 120, 225-250.

Ayers, R.S. and Westcot, D.W., 1985, Water quality for agriculture. FAO Irr. Drain. Pap. 29.

Back, W., 1966, Hydrochemical facies and groundwater flow patterns in northern part of Atlantic coastal plain. US Geol. Surv. Prof. Paper 498-A, 42 pp.

Beekman, H.E., 1991, *Ion chromatography of fresh and salt water intrusions*. Ph.D. thesis, Free University, Amsterdam, 198 pp.

Beekman, H.E. and Appelo, C.A.J., 1990, Ion chromatography of fresh- and salt-water displacement: laboratory experiments and multicomponent transport modelling. *J. Contam. Hydrol.* 7, 21-37.

Bolt, G.H., 1961, The pressure filtrate of colloidal suspensions, II. Experimental data on homoionic clays. *Kolloid-Z.* 175, 144-150.

Bolt, G.H. (ed.), 1982, *Soil chemistry, B. Physico-chemical models*. Elsevier, Amsterdam, 527 pp.

Bolt, G.H. and Bruggenwert, M.G.M., 1978 (eds), *Soil chemistry, A. Basic elements*. Elsevier, Amsterdam, 281 pp.

Bond, W.J., Gardiner, B.N. and Smiles, D.E., 1982, Constant flux absorption of a tritiated calcium chloride solution by a clay soil with anion exclusion. *Soil Sci. Soc. Am. J.* 46, 1133-1137.

Breeuwsma, A., Wösten, J.H.M., Vleeshouwer, J.J., Van Slobbe, A.M. and Bouma, J. 1986, Derivation of land qualities to assess environmental problems from soil surveys. *Soil Sci. Soc. Am. J.* 50, 186-190.

Brown, R.F. and Signor, D.C., 1974, Artificial recharge – State of the art. *Ground Water* 12, 152-160.

Bruggenwert, M.G.M. and Kamphorst, A., 1982, Survey of experimental information on cation exchange in soil systems. In G.H. Bolt (ed.), *Soil Chemistry, B. Physico-chemical models*. Elsevier, Amsterdam, 141-203.

Chapelle, F.H., 1983, Groundwater geochemistry and calcite cementation of the Aquia aquifer in Southern Maryland. *Water Resour. Res.* 19, 545-558.

Chapelle, F.H. and Knobel, L.L., 1983, Aqueous geochemistry and the exchangeable cation composition of glauconite in the Aquia aquifer, Maryland. *Ground Water* 21, 343-352.

Christensen, T.H., 1989, Cadmium soil sorption at low concentrations. VIII. Correlation with soil parameters. *Water Air Soil Pollution 44*, 71-82.

Davis, J.A. and Kent, D.B., 1990, Surface complexation modeling in aqueous geochemistry. In M.F. Hochella, Jr. and A.F. White (eds), *Mineral-water interface geochemistry*. Reviews in Mineralogy 23, Mineral. Soc. Am., 177-260.

De Lange, G.J., 1984, Shipboard pressure-filtration system for interstitial water extraction. *Mededel. Rijks Geol. Dienst* (Haarlem, Netherlands) 38, 209-214.

Demir, I., 1988, The interrelation of hydraulic and electrical conductivities, streaming potential and salt fitration during the flow of chloride brines through a smectite layer at elevated pressures. *J. Hydrol.* 98, 31-52.

Dzombak, D.A. and Morel, F.M.M., 1990, *Surface complexation modeling: hydrous ferric oxide*. Wiley & Sons, New York, 393 pp.

Edmunds, W.M. and Walton, N.R.G., 1983, The Lincolnshire limestone – Hydrogeochemical evolution over a ten-year period. *J. Hydrol.* 61, 201-211.

Evangelou, V.P. and Phillips, R.E., 1988, Comparison between the Gapon and Vanselow exchange selectivity coefficients. *Soil Sci. Soc. Am. J.* 52, 379-382.

Feynman, R., Leighton, R.B. and Sands, M.L., 1989, *The Feynman lectures on physics*, vol. 2, California Inst. Technol.

Fritz, S.J. and Eady, C.D., 1985, Hyperfiltration-induced precipitation of calcite. *Geochim. Cosmochim. Acta* 49, 761-768.

Fuerstenau, D.W., Manmohan, D. and Raghavan, S., 1981, The adsorption of alkaline-earth metal ions at the rutile/aqueous interface. In P.H. Tewari (ed.), *Adsorption from aqueous solutions*. Plenum Press, New York, 93-117.

Gaines, G.L. and Thomas, H.C., 1953, Adsorption studies on clay minerals. II. A formulation of the thermodynamics of exchange adsorption. *J. Chem. Phys.* 21, 714-718.

Gapon, E.N., 1933, Theory of exchange adsorption. V. *(in Russian). J. Gen. Chem. (USSR)* 3, 667-669 (*Chem. Abstr.* 28, 4516, 1934).

Garcia-Miragaya, J. and Page, A.L., 1976, Influence of ionic strength and inorganic complex formation on the sorption of trace amounts of Cd by montmorillonite. *Soil Sci. Soc. Am. J.* 40, 658-663.

Goldenberg, L.C., 1985, Decrease of hydraulic conductivity in sand at the interface between seawater and dilute clay suspensions. *J. Hydrol.* 78, 183-199.

Goldenberg, L.C. and Mandel, S., 1988, Some processes in the sea-water/fresh-water interface, as influenced by the presence of gases. *Natuurwet. Tijdschr.* 70, 288-299 (*SWIM Conf. Ghent, Belgium*).

Griffioen, J., 1992, *Cation-exchange and carbonate chemistry in aquifers following groundwater flow*. Ph.D. thesis, Free University, Amsterdam, 182 pp.

Griffioen, J. and Appelo, C.A.J., 1993, Adsorption of Ca^{2+} and $CaHCO_3^+$ by exchangers in Ca-H-Cl-CO_2 systems. *Soil Sci. Soc. Am. J.* 57, 716-722.

Grim, R.E., 1968, *Clay mineralogy*. McGraw-Hill, New York, 596 pp.

Gvirtzman, H. and Gorelick, S.M., 1991, Dispersion and advection in unsaturated porous media enhanced by anion exclusion. *Nature* 352, 793-795.

Helfferich, F., 1959, *Ionenaustauscher*, Band 1: Grundlagen, Struktur, Herstellung. Verlag Chemie, Weinheim, 520 pp.

Helfferich, F., 1962, *Ion exchange*. McGraw-Hill, New York.

James, R.V. and Rubin, J., 1986, Transport of chloride ion in a water-unsaturated soil exhibiting anion-exclusion. *Soil Sci. Soc. Am. J.* 50, 1142-1149.

Johnston, H.M., Gillham, R.W. and Cherry, J.A., 1985, Distribution coeffcients for strontium and cesium in overburden at a storage area for low-level radioactive waste. *Can. Geotechn. J.* 22, 6-16.

Kharaka, Y.K. and Berry, F.A.F., 1973, Simultaneous flow of water and solutes through geological membranes. I. Experimental investigation. *Geochim. Cosmochim. Acta* 37, 2577-2603.

Lawrence, A.R., Lloyd, J.W. and Marsh, J.M., 1976, Hydrochemistry and groundwater mixing in part of the Lincolnshire limestone aquifer, England. *Ground Water* 14, 12-20.

Lexmond, T.M., 1980, The effect of soil pH on copper toxicity to forage maize growth under field conditions. *Neth. J. Agr. Sci.* 28, 164-183.

Marinsky, J.A., 1987, A two-phase model for the interpretation of proton and metal ion interaction with charged polyelectrolyte gels and their linear analogs. In W. Stumm (ed.), *Aquatic surface chemistry.* Wiley & Sons, New York, 49-81.

Mehnert, E. and Jennings, A.A., 1985, The effect of salinity-dependent hydraulic conductivity on saltwater intrusion episodes. *J. Hydrol.* 80, 283-297.

Moore, W.J., 1972, *Physical chemistry.* 5th ed. Longman, London, 977 pp.

Nativ, R., Issar, A. and Rutledge, J., 1983, Chemical composition of rainwater and floodwaters in the Negev desert, Israel. *J. Hydrol.* 62, 201-223.

Nielsen, D.R., Van Genuchten, M.Th. and Biggar, J.W., 1986, Water flow and solute transport processes in the saturated zone. *Water Resour. Res.* 22, 89S-108S.

Nkedi-Kizza, P., Rao, P.S.C., Jessup, R.E. and Davidson, J.M., 1982, Ion exchange and diffusive mass transfer during miscible displacement through an aggregated oxisol. *Soil Sci. Soc. Am. J.* 46, 471-476.

Parks, G.A., 1967, Surface chemistry of oxides in aqueous systems. In Stumm, W. (ed.), *Equilibrium concepts in aqueous systems.* Adv. Chem. Ser. 67, Am. Chem. Soc., Washington, 121-160.

Parks, G.A., 1990, Surface energy and adsorption at mineral-water interfaces: an introduction. In M.F. Hochella, Jr. and A.F. White (eds), *Mineral-water interface geochemistry.* Reviews in Mineralogy 23, Mineral. Soc. Am., 133-175.

Paterson, R., 1970, *An introduction to ion exchange.* Heyden & Son, London, 109 pp.

Pauling, L., 1960, *The nature of the chemical bond.* Cornell Univ. Press, Ithaca, NY, 644 pp.

Riemann, W. and Walton, H.F., 1970, *Ion exchange in analytical chemistry.* Pergamon Press, Oxford.

Saager, P.M., Sweerts, J.-P. and Ellermeijer, H.J., 1990, A simple pore-water sampler for coarse, sandy sediments of low porosity. *Limnol. Oceanogr.* 35, 747-751.

Scheffer, F. and Schachtschabel, P., 1970, *Lehrbuch der Bodenkunde.* Enke, Stuttgart, 448 pp.

Schindler, P.W. and Stumm, W., 1987, The surface chemistry of oxides, hydroxides and oxide minerals. In W. Stumm (ed.), *Aquatic surface chemistry.* Wiley & Sons, New York, 83-110.

Shainberg, I. and Oster, J.D., 1978, Quality of irrigation water. Int. Irr. Inf. Cent. Bet Dagan, Israel.

Sposito, G., 1983, On the surface complexation model of the oxide-aqueous solution interface. *J. Colloid Interface Sci.* 91, 329-340.

Sposito, G., 1984, *The surface chemistry of soils.* Oxford Univ. Press, New York, 234 pp.

Sposito, G., Holtzclaw, K.M., Charlet, L., Jouany, C. and Page, A.L., 1983, Na-Ca and Na-Mg exchange in Wyoming bentonite in ClO_4^- and Cl^- background ionic media. *Soil Sci. Soc. Am. J.* 47, 51-56.

Sposito, G., De Wit, J.C.M. and Neal, R.H., 1988, Selenite adsorption on alluvial soils: III Chemical modeling. *Soil Sci. Soc. Am. J.* 52, 947-950.

Stumm, W. and Morgan, J.J., 1981, *Aquatic chemistry.* 2nd ed. Wiley & Sons, New York, 780 pp.

Stuyfzand, P.J., 1985, Hydrochemistry and hydrology of the dune area between Egmond and Wijk aan Zee (in Dutch). KIWA, SWE, 85.012, Nieuwegein.

Stuyfzand, P.J., 1993, *Hydrochemistry and hydrology of the coastal dune area of the Western Netherlands.* Ph.D. Thesis, Free University Amsterdam, 366 pp.

Tipping, E., Backes, C.A. and Hurley, M.A., 1988, The complexation of protons, aluminium and calcium by aquatic humic substances: a model incorporating binding-site heterogeneity and macro-ionic effects. *Water Res.* 22, 597-611.

US Soil Salinity Lab. Staff., 1954, Diagnosis and improvement of saline and alkali soils. *USDA Handb.* 60, US Gov. Print. Office, Washington DC.

Valocchi, A.J., Street, R.L. and Roberts, P.V., 1981, Transport of ion-exchanging solutes in groundwater: chromatographic theory and field simulation. *Water Resour. Res.* 17, 1517-1527.

Van Olphen, H., 1977, *An introduction to clay colloid chemistry.* 2nd ed. Wiley & Sons, New York, 318 pp.

Van Riemsdijk, W., Bolt, G.H., Koopal, L.K. and Blaakmeer, J., 1986, Electrolyte adsorption on heterogeneous surfaces. *J. Colloid Interface Sci.* 109, 219-228.

Vanselow, A.P., 1932, Equilibria of the base-exchange reactions of bentonites, permutites, soil colloids and zeolites. *Soil Sci.* 33, 95-113.

Van Stempvoort, D.R., Reardon, E.J. and Fritz, P., 1990, Fractionation of sulfur and oxygen isotopes in sulfate by soil sorption. *Geochim. Cosmochim. Acta* 54, 2817-2826.

Von Engelhardt, W. and Gaida, K.H., 1963, Concentration changes of pore solutions during the compaction of clay sediments. *J. Sedim. Petrol.* 33, 919-930.

Westall, J. and Hohl, H., 1980, A comparison of electrostatic models for the oxide/solution interface. *Adv. Colloid Interface Sci.* 12, 265-294.

Wiklander, L., 1957, Cation and anion exchange phenomena. In F.E. Bear (ed.), *Chemistry of the soil.* Reinhold Publ. Cy, New York, 107-148.

Workman, S.M. and Lindsay, W.L., 1990, Estimating divalent cadmium activities measured in arid-zone soils using competitive chelation. *Soil Sci. Soc. Am. J.* 54, 987-993.

Zachara, J.M., Ainsworth, C.C., Cowan, C.E. and Resch, C.T., 1989, Adsorption of chromate by subsurface soil horizons. *Soil Sci. Soc. Am. J.* 53, 418-428.

Zhang, P.C. and Sparks, D.L., 1990, Kinetics and mechanisms of sulfate adsorption/desorption on goethite using pressure jump relaxation. *Soil Sci. Soc. Am. J.* 54, 1266-1273.

CHAPTER 6

Silicate weathering

The effect of dissolution of Ca-carbonate on groundwater chemistry is directly displayed by significant increases of dissolved calcium and carbonate in water (Figure 4.1). For weathering of silicate minerals, the resulting changes in water chemistry are less apparent because dissolution of most silicate minerals is a very slow process. Still, weathering of silicate minerals is estimated to contribute about 45% to the total dissolved load of the world's rivers (Stumm and Wollast, 1990). In sediments free of carbonate minerals, silicate weathering is, furthermore, the most important buffer mechanism against acidification of soil- and groundwater.

6.1 WEATHERING PROCESSES

Traditionally, weathering of detrital silicate minerals has been studied in recent or fossil soils. An example of the results of such a study is shown in Figure 6.1. Soils may be exposed to chemical weathering over thousands of years and the variation in mineralogical composition which develops as a function of depth and time, displays very slow degradation processes of minerals.

The parent granodiorite rock in Figure 6.1 consists of mainly quartz, K-feldspar (microcline), plagioclase (oligoclase), biotite and amphibole (hornblende). Moving upward in the sequence, one observes that first plagioclase, biotite and the amphibole disappear, while quartz and K-feldspar are apparently more resistant to weathering. The successive disappearance of different silicate minerals reflects their differences in dissolution rate. This overall kinetic control on the distribution of primary silicate minerals during weathering was recognized as early as 1938 by Goldich who designed the empirical weathering sequence shown in Figure 6.2. It displays olivine and Ca-plagioclase as the most easily weathered minerals, and quartz as the mineral most resistant to weathering.

The second important observation to be made from Figure 6.1 is that new, secondary minerals, like clays (illite, kaolinite and montmorillonite; see Section 5.2.1 for a general description of clay minerals) and Fe-oxides are formed during the weathering process. These are the insoluble remnants which form during *incongruent* dissolution of silicate minerals. Incongruent dissolution means that the ratio of the elements which appear in solution differs from that in the dissolving mineral. Weathering reactions for some common primary minerals are listed in Table 6.1, where the clay mineral kaolinite is used as an example of a weathering product.

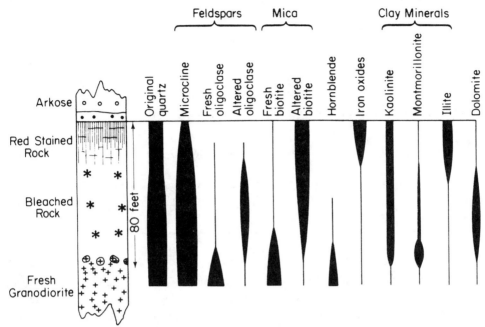

Figure 6.1. The mineralogical composition of a fossil soil developed on a granodiorite (Blatt, Middleton and Murray, 1980).

The formation of secondary products is due to the insolubility of Al-compounds. The reactions in Table 6.1 are therefore written so that aluminum remains conserved in the solid phase. The effect of silicate weathering on the water chemistry is primarily the addition of cations and silica. All silicate weathering reactions are acid consuming, and therefore have a pH buffering effect. Under unpolluted conditions, carbonic acid is the most important source of protons and, as indicated by the last equation in Table 6.1, bicarbonate will be produced during weathering of silicates. Finally, iron that is present in silicate minerals may form Fe-oxide as an insoluble weathering product (Figure 6.1).

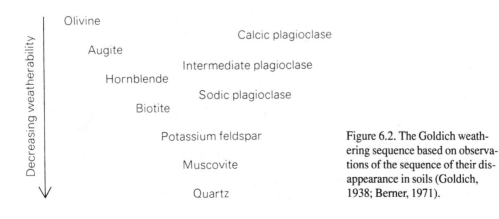

Figure 6.2. The Goldich weathering sequence based on observations of the sequence of their disappearance in soils (Goldich, 1938; Berner, 1971).

Table 6.1. Weathering reactions for different silicate minerals to the clay mineral kaolinite

$$2NaAlSi_3O_8 + 2H^+ + 9H_2O \rightarrow Al_2Si_2O_5(OH)_4 + 2Na^+ + 4H_4SiO_4$$
Albite   Kaolinite

$$2KAlSi_3O_8 + 2H^+ + 9H_2O \rightarrow Al_2Si_2O_5(OH)_4 + 2K^+ + 4H_4SiO_4$$
K-feldspar

$$CaAl_2Si_2O_8 + 2H^+ + H_2O \rightarrow Al_2Si_2O_5(OH)_4 + Ca^{2+}$$
Anorthite

$$[Ca_{1.15}MgAl_{.3}Si_{1.7}]O_6 + 4.3H^+ + 0.95H_2O \rightarrow 0.15Al_2Si_2O_5(OH)_4 + 1.15Ca^{2+} + Mg^{2+} + 1.4H_4SiO_4$$
Pyroxene

$$2K[Mg_2Fe][AlSi_3]O_{10}(OH)_2 + 10H^+ + .5O_2 + 7H_2O \rightarrow Al_2Si_2O_5(OH)_4 + 2K^+ + 4Mg^{2+} + 2Fe(OH)_3$$
Biotite $+ 4H_4SiO_4$

$$CO_2 + H_2O \rightarrow H^+ + HCO_3^-$$

Weathering reactions will also take place in silicate rocks or sandstones percolated by groundwater, although the mineralogical transformations here often are more difficult to detect than in soils. Some examples of groundwater compositions which may result from weathering of silicate minerals are shown in Figure 6.3. The high silica content in all groundwaters indicates active degradation of silicate minerals. The highest concentrations

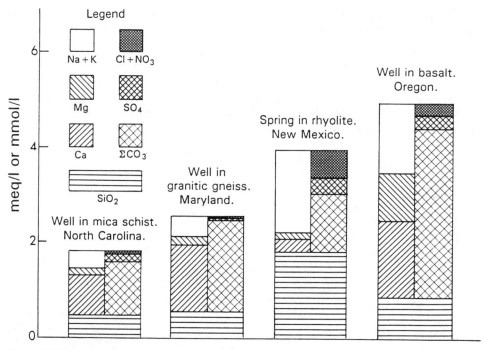

Figure 6.3. Examples of groundwater compositions in igneous and metamorphic rocks. Dissolved silica is expressed as mmol/l while charged ions are displayed as meq/l (modified from Hem (1985)).

are found in volcanic rocks (rhyolite and basalt) which contain more reactive material than rock types like mica schists or granite. Sodium contributes in all groundwaters significantly to the cations (K^+ which has been added to Na^+ in Figure 6.3 is in most cases a minor constituent) and its presence is not balanced by chloride as would be expected when seawater was the source of sodium. Sodium is mainly derived from weathering of Na-feldspar like albite or any member of the plagioclase solid solution series between albite and anorthite (Ca-feldspar).

Plagioclase weathering releases in addition Ca^{2+}, although also weathering of amphiboles, pyroxenes, etc., may contribute to the Ca^{2+} concentration. The increase in cation concentration is, as suggested by Table 6.1, accompanied by an increase in dissolved bicarbonate and it appears that in some cases even carbonate precipitation can be the result of silicate weathering as indicated by the presence of dolomite in the weathering profile (Figure 6.1). The total dissolved concentrations in groundwaters of silicate rocks (Figure 6.3) are normally low. This is due to the slow dissolution kinetics of most silicate minerals, but groundwater flow along fracture zones also restricts effective flushing of the bulk of the rock in many massive igneous rocks.

Figure 6.1 shows that different clay minerals such as montmorillonite and kaolinite may form as a silicate weathering product. In addition gibbsite ($Al(OH)_3$) can be a weathering product of silicates. Using albite as an example, the transformation into different weathering products can be described by the following equations. In the case of montmorillonite we assume the presence of Mg^{2+}, for example leached from pyroxenes, amphiboles or biotite:

$$3NaAlSi_3O_8 + Mg^{2+} + 4H_2O \rightarrow 2Na_{0.5}Al_{1.5}Mg_{0.5}\,Si_4O_{10}(OH)_2 + 2Na^+ + H_4SiO_4$$
albite montmorillonite (6.1)

$$2NaAlSi_3O_8 + 2H^+ + 9H_2O \rightarrow Al_2Si_2O_5(OH)_4 + 2Na^+ + 4H_4SiO_4$$
albite kaolinite (6.2)

$$NaAlSi_3O_8 + H^+ + 7H_2O \rightarrow Al(OH)_3 + Na^+ + 3H_4SiO_4$$
albite gibbsite (6.3)

The alteration of albite to montmorillonite consumes no acid, but with kaolinite and

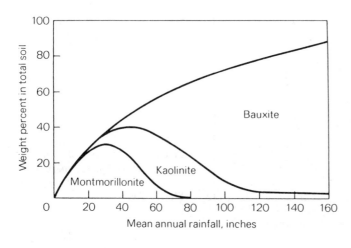

Figure 6.4. Weathering products on volcanic rocks on the island of Hawaii as a function of mean annual rainfall. Bauxite corresponds to Al-hydroxides and the contents of different minerals are plotted cumulatively as weight percent of the total soil (Berner, 1971).

gibbsite as weathering products, increasing amounts of protons are consumed. When albite alters to montmorillonite, 89% of the Si is preserved in the weathering product, decreasing to 33% for weathering to kaolinite and finally to 0% for gibbsite. The sequence from montmorillonite over kaolinite to gibbsite as weathering products accordingly represents different intensities of leaching, which is associated with the removal of increasing amounts of silica and cations. While the reactions are written above as alteration of albite, the clay minerals can also transform into each other.

The importance of leaching predicts that the type of weathering product depends on the hydrological conditions as well as on the rate of mineral weathering; montmorillonite is formed preferentially in relatively dry climates, where the flushing rate of the soil is low, and its formation is favored when rapidly dissolving material such as volcanic rock is available. Gibbsite, on the other hand, forms typically in tropical areas with intense rainfall and under well drained conditions. Here, gibbsite and other Al-hydroxides may form a thick weathering residue, termed *bauxite*, that constitutes the most important Al-ore.

A classic example of the effect of precipitation, and leaching intensity, on the composition of weathering residues is known from soils on Hawaii where bauxite, kaolinite and montmorillonite form on basalt at different altitudes on the volcanic slopes (Figure 6.4). Montmorillonite is found in areas with low rainfall and therefore relatively longer residence times of water in the soil, so that the concentrations of dissolved ions may become higher. At the other end of the scale, bauxite forms in high rainfall areas where leaching of

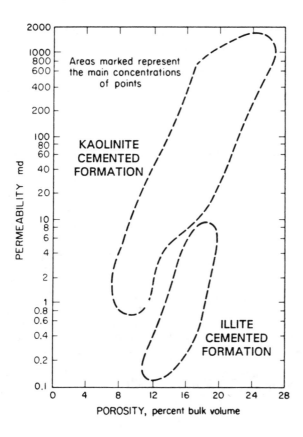

Figure 6.5. The effect of authigenic clay formation on the permeability of a sandstone (Blatt, Middleton and Murray, 1980).

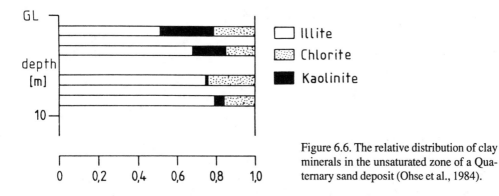

Figure 6.6. The relative distribution of clay minerals in the unsaturated zone of a Quaternary sand deposit (Ohse et al., 1984).

the soil is intensive and the residence time of water in the soil short, so that dissolved ion concentrations are low. Note also in Figure 6.4 that the amount of weathering products increases with the amount of rainfall.

Authigenic clay mineral formation due to the alteration of primary silicates has been studied extensively in sedimentary petrology, since the clay mineralogy may seriously affect the permeability of the rock. While kaolinite forms neat booklets that have only a moderate effect on the permeability, illite forms hairy aggregates that clog pore throats and reduce the permeability very significantly. This feature is well recognized in oil exploration and is illustrated in Figure 6.5, where a sandstone cemented with illite is shown to have a permeability that is at least a magnitude lower than a sandstone cemented with kaolinite for the same porosity.

The clay minerals that originate from silicate weathering are deposited mainly as coatings on clastic grains, but in some cases the clay can be washed down from the soil zone into the profile. The total amount of clay present in sand deposits amounts normally to less than a few percent and consists in most cases of a mixture of different clay minerals which occur closely intermingled with iron oxides. An example of the relative distribution of clay minerals in the unsaturated zone is shown in Figure 6.6. Note how in this figure the abundance of kaolinite, which represents a more extreme stage of weathering than illite and chlorite, decreases downward. This is as expected, since the upper part of the profile is subject to more intensive leaching than the lower part of the profile.

6.2 THE STABILITY OF WEATHERING PRODUCTS

The most extreme product of silicate weathering is gibbsite or Al-hydroxide, and we will use this mineral as a starting point for our discussion of the stability of weathering products. For gibbsite we may write the solubility product:

$$Al(OH)_{3\,gibbsite} \leftrightarrow Al^{3+} + 3OH^- \tag{6.4}$$

with the mass action equation

$$K_{gibbsite} = [Al^{3+}][OH^-]^3 = 10^{-32.64} \tag{6.5}$$

Equation (6.5) indicates that the activity of aluminum in water, $[Al^{3+}]$, will be very strongly dependent on the pH. Dissolved aluminum has, however, a distinct tendency to form

Table 6.2. The stability of dissolved Al-hydroxy complexes at 25°C (Ball et al., 1980).

Reaction	Log K	ΔH_r^0
$Al^{3+} + H_2O \leftrightarrow Al(OH)^{2+} + H^+$	-4.99	11.9
$Al^{3+} + 2H_2O \leftrightarrow Al(OH)_2^+ + 2H^+$	-10.13	22.0
$Al^{3+} + 4H_2O \leftrightarrow Al(OH)_4^- + 4H^+$	-22.05	44.1

hydroxy complexes which may increase the solubility of gibbsite significantly. The stability of the most important Al-hydroxy complexes is listed in Table 6.2.

The total amount of aluminum in solution consists of both complexed and uncomplexed aluminum and can be described by the mass balance equation:

$$\Sigma Al = m_{Al^{3+}} + m_{Al(OH)^{2+}} + m_{Al(OH)_2^+} + m_{Al(OH)_4^-} \tag{6.6}$$

In some cases complexes between aluminum and fluoride, sulfate and organic matter have to be considered (Driscoll, 1980).

The set of equations consisting of the mass action equations in Table 6.2, Equation (6.5), and the mass balance (6.6) can be solved by the methods described in Section 3.2, and can be used to display the solubility of total dissolved aluminum (Figure 6.7). It shows that in the low pH range (pH < 4) hydroxy complexes are insignificant. Their importance increases towards higher pH values and at pH > 7 the aluminate complex completely dominates the solubility of gibbsite. In the near neutral pH range, aluminum is highly insoluble, with total dissolved aluminum concentrations in the order of 1 μmol/l, close to the detection limit of standard analytical equipment. Only in the low pH range may

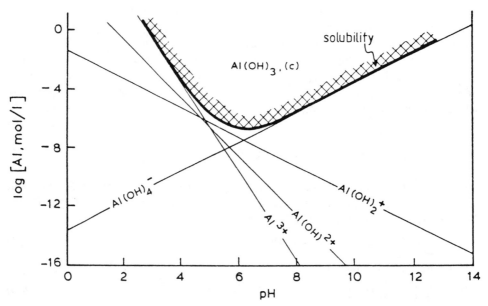

Figure 6.7. Solubility constraints on total dissolved aluminum by gibbsite (Al(OH)$_3$) and aqueous Al-hydroxy complexes.

significant amounts of aluminum come into solution, an aspect which will be discussed in Section 6.6 on groundwater acidification.

The next weathering product to be considered is kaolinite, for which we write the dissolution reaction:

$$Al_2Si_2O_5(OH)_4 + 6H^+ \leftrightarrow 2Al^{3+} + 2H_4SiO_4 + H_2O \tag{6.7}$$

The mass action equation, in logarithmic form, is:

$$\log K = 2 \log [Al^{3+}] + 2 \log [H_4SiO_4] + 6pH \tag{6.8}$$

Equation (6.8) allows us to display the stability of kaolinite (Figure 6.8), for an assumed constant value of total dissolved silica, in a diagram that is similar to Figure 6.7 and again includes the effect of Al-hydroxy complexes.

The overall shape of the stability field of kaolinite is similar to that for $Al(OH)_3$, and indicates that if equilibrium with kaolinite controls dissolved aluminum, its concentration becomes extremely low in the near neutral pH range. Some groundwater data have been included in Figure 6.8 and appear to confirm that weathering products like kaolinite or $Al(OH)_3$ control the aluminum concentration in groundwater. It is difficult to determine precisely which weathering product is in equilibrium with the groundwater. First of all, the stability of Al-hydroxides is highly variable. As shown in Table 6.3, amorphous $Al(OH)_3$ has a solubility product that is more than two orders of magnitudes larger then synthetic gibbsite. If synthetic gibbsite were included in Figure 6.8, its stability line would plot close to that of kaolinite.

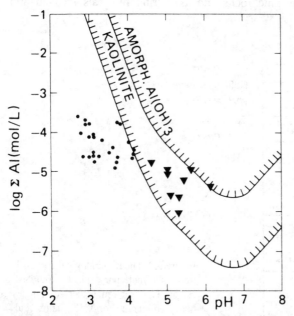

Figure 6.8. Solubility constraints on total dissolved aluminum by kaolinite and amorphous $Al(OH)_3$ including Al-hydroxy complexes. The concentration of total dissolved silica is 10^{-4} mol/l. Circles denote stream waters and triangles denote groundwaters (reprinted with permission from Nordstrom, 1982).

Table 6.3. The stability of some Al-containing weathering products at 25°C.

Mineral	Log K	Source
Synthetic gibbsite ($Al(OH)_3$)	-33.88	May et al. (1979)
Natural gibbsite ($Al(OH)_3$)	-32.64	Hem and Robertson (1967)
Amorphous ($Al(OH)_3$)	-31.59	Hayden and Ruben (1974)
Kaolinite ($Al_2Si_2O_5(OH)_4$)	$+7.44$	May et al. (1986)
Jurbanite ($AlOHSO_4$)	-17.8	Nordstrom (1982)

Additional uncertainty concerns the very slow dissolution kinetics of clay minerals which make it questionable whether true equilibrium is ever attained. This problem is encountered in the interpretation of field data, as well as in the determination of clay mineral stability by equilibration with water in the laboratory. May et al. (1986) found it necessary to equilibrate kaolinite with water for 1237 days in order to approach equilibrium while equilibrium with smectite (montmorillonite) was never attained in experiments that lasted as long. Despite these uncertainties there appears little doubt that the precipitation of secondary minerals keeps the aluminum concentration in groundwater in the near neutral pH range at a very low level, which confirms the essentially incongruent dissolution behavior of primary silicate minerals.

6.3 STABILITY OF PRIMARY SILICATES

Primary silicate minerals comprise feldspars, amphiboles, pyroxenes, micas, etc., that are present in the igneous and metamorphic rocks and as detrital minerals in sand and

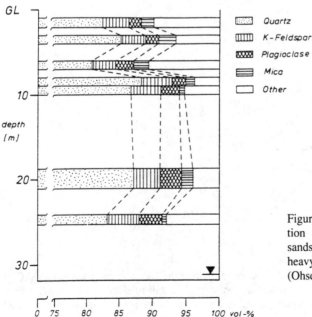

Figure 6.9. The mineralogical composition of Quaternary glacial outwash sands. The fraction 'other' comprises heavy minerals, organic matter etc. (Ohse et al., 1984).

sandstones. The mineralogical composition of a Quaternary sand deposit is illustrated in Figure 6.9, and shows that quartz is by far the dominant component of the sand.

Still, weatherable minerals like plagioclases, micas and K-feldspar are also present in significant amounts. The stability of these minerals in a groundwater system can be evaluated by calculating the saturation state of the groundwater for a given mineral. For example for albite, we may write the dissociation reaction:

$$NaAlSi_3O_8 + 4H^+ + 4H_2O \rightarrow Na^+ + Al^{3+} + 3H_4SiO_4 \qquad (6.9)$$
albite

As for other primary silicate weathering reactions (Table 6.1), the single arrow is used in the equation to indicate the irreversible character of the reaction. The mass action expression for Equation (6.9) is

$$\log IAP = \log [Na^+] + \log [Al^{3+}] + 3 \log [H_4SiO_4] + 4pH \qquad (6.10)$$

The *IAP* for a given mineral is easily calculated from a groundwater analysis, using the WATEQ program, and can be compared with the equilibrium value *K*. The results of such calculations carried out for groundwater in a quaternary sand deposit, are displayed in Figure 6.10. Anorthite and albite are used as end-members to describe the plagioclases in Figure 6.10, even though at least anorthite is not present as a separate mineral in the deposit (Figure 6.9). Strong subsaturation is observed for all three feldspars in the unsaturated zone and the upper part of the saturated zone, while the degree of subsaturation is less in the deeper part of the profile.

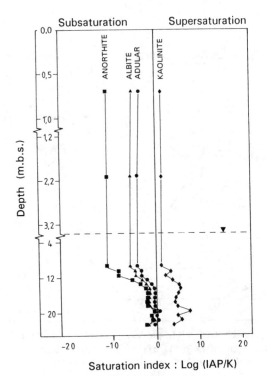

Figure 6.10. The saturation index of groundwater in the unsaturated and saturated zone of a Quaternary sandy aquifer for plagioclases (anorthite ($CaAl_2Si_2O_8$), albite ($NaAlSi_3O_8$) and adular ($KAlSi_3O_8$)) and the clay mineral kaolinite (Ohse et al., 1983).

Dissolution of plagioclases is thus, from an equilibrium point of view, expected to take place. The fact that subsaturation for plagioclases persists over most of the profile, stresses, however, the slow dissolution kinetics of feldspars. Note also the slight supersaturation of groundwater with respect to the weathering product kaolinite. Although the saturation state approach outlined above is useful, it also has several drawbacks. First of all, the aluminum concentration in groundwater is often below the detection limit as discussed in Section 6.2. Second, the number of variables, which may affect the saturation state, already amounts in the simple case of albite (6.10) to four. The stability of albite is therefore difficult to display in a diagram that includes all relevant parameters. To circumvent these problems, silicate stability diagrams have been developed which assume that all aluminum is preserved in the weathering product. An example of such a diagram for Ca-silicates is shown in Figure 6.11. It contains stability fields for the Ca-feldspar anorthite and its possible weathering products, gibbsite, kaolinite and Ca-montmorillonite expressed as a function of $\log([Ca^{2+}]/[H^+]^2)$ and $\log[H_4SiO_4]$.

The peculiar choice of the y-axis parameter is easily understood when we consider the reaction between anorthite and gibbsite:

$$CaAl_2Si_2O_8 + 2H^+ + 6H_2O \rightarrow 2Al(OH)_3 + Ca^{2+} + 2H_4SiO_4 \qquad (6.11)$$
anorthite   gibbsite

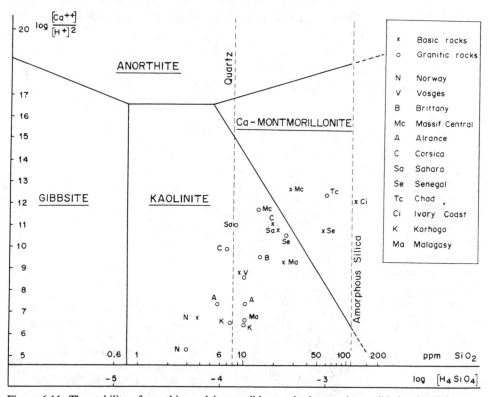

Figure 6.11. The stability of anorthite and its possible weathering products gibbsite, kaolinite and Ca-montmorillonite. Also included are the compositions of selected surface water samples from areas with crystalline rocks (reprinted with permission from Tardy, 1971).

Note that all aluminum released from anorthite is preserved in gibbsite, so that the relative stability between the two minerals only is controlled by dissolved silica, Ca^{2+} and pH. The mass action equation of (6.11) is:

$$\log K = \log [Ca^{2+}] + 2 \log [H_4SiO_4] - 2 \log [H^+] = 6.78 \tag{6.12}$$

Rearranging (6.12) yields:

$$\log K = \log ([Ca^{2+}]/ [H^+]^2) + 2 \log [H_4SiO_4] = 6.78 \tag{6.13}$$

Now (6.13) can be plotted conveniently as a straight line with slope -2 in the stability diagram (Figure 6.11) and the four variables have been reduced to two. In general, the expression on the y-axis has the form $\log([cation^{n+}]/ [H^+]^n)$ which basically reflects that the charge of released cations must be balanced by consumption of H^+. Similarly, the equilibrium between Ca-montmorillonite and kaolinite is described by the reaction:

$$2H^+ + 3Ca_{.33}[Si_{7.33}Al_{.67}] [Al_4]O_{20}(OH)_4 + 23H_2O \leftrightarrow 7Al_2Si_2O_5(OH)_4$$
$$+ 8H_4SiO_4 + Ca^{2+}$$

Ca-montmorillonite	kaolinite	(6.14)

and

$$\log K = \log ([Ca^{2+}]/ [H^+]^2) + 8 \log [H_4SiO_4] = - 15.7 \tag{6.15}$$

The boundary between kaolinite and gibbsite is described by the reaction:

$$Al_2Si_2O_5(OH)_4 + 5H_2O \leftrightarrow 2Al(OH)_3 + 2H_4SiO_4 \tag{6.16}$$
kaolinite $\qquad\qquad\qquad$ gibbsite

and

$$\log K = 2 \log [H_4SiO_4] = - 9.6 \tag{6.17}$$

Accordingly, a H_4SiO_4 activity of $10^{-4.8}$ indicates equilibrium between kaolinite and gibbsite and this phase boundary results in a straight line parallel to the y-axis in the stability diagram (Figure 6.11). Finally, in the reaction between anorthite and kaolinite all silica is preserved in the solid phase (Table 6.1), so that the boundary plots parallel to the $\log [H_4SiO_4]$ axis.

It should be remembered that silicate stability diagrams in which aluminum is retained in the solid phases, implicitly assume equilibrium with the depicted phases. For example, a water sample that plots in the kaolinite field in Figure 6.11 can show subsaturation for kaolinite when plotted in Figure 6.8, because of a low $[Al^{3+}]$. A better statement would be that according to Figure 6.11, kaolinite is more likely to be a stable mineral for this water composition than for example gibbsite.

Figure 6.11 also includes the stability lines for quartz and amorphous silica. For both substances, the solubility is described by the reaction:

$$SiO_{2(s)} + 2H_2O \leftrightarrow H_4SiO_4 \tag{6.18}$$

H_4SiO_4 remains undissociated at pH values below 9, and the solubility of $SiO_{2(s)}$ phases is given by the solubility product of (6.18):

$$K = [H_4SiO_4] \tag{6.19}$$

For quartz, $K = 10^{-4}$ at 25°C, but quartz has extremely sluggish reaction kinetics and therefore rarely appears to control dissolved silica concentrations. As in other cases, where the most stable mineral reacts very slowly, a range of less stable forms of $SiO_{2(s)}$ exists which comprises amorphous $SiO_{2(s)}$ or *opal*, (often found in marine sediments as diatom tests), *chalcedony* (cryptocrystalline quartz), *cristobalite*, *tridymite* and *opal-CT* (disordered cristobalite-tridymite) (Williams et al., 1985). The most soluble phase is amorphous $SiO_{2(s)}$ which has a solubility product of about $10^{-2.7}$ (25°C) and places the upper constraint on dissolved silica concentrations.

The composition of surface waters from different crystalline massifs in Europe and Africa plot dominantly in the stability field of kaolinite (Figure 6.11). Basic rocks (basalts etc.) are more reactive than granitic rocks and will yield higher total dissolved solid concentrations (Figure 6.3). Accordingly, waters from basic rocks plot in general closer to the stability field of montmorillonite than those derived from granitic rocks. The amount of leaching also affects the water composition. Therefore, water from a granitic area in rainy Norway plots way down in the kaolinite field while a sample from a granitic area in arid Chad plots in the montmorillonite field.

Figure 6.11 considers only Ca-feldspar and its weathering products. For a more complete picture, sodium and potassium feldspars should also be considered (corresponding diagrams can be found in Tardy (1971)). However, the consideration of pure end-member minerals is a crude simplification of the real world, where solid solutions are more abundant than pure minerals. It has been noted already, that plagioclases consist of a series of solid solutions of albite and anorthite and the variability in composition of clay minerals such as montmorillonite is even greater. Uncertainties concerning the stability and dissolution/precipitation behavior of such solid solutions, as well as of pure end-members, may affect the stability fields considerably, and caution should be exercised in the conclusions drawn from the diagrams. The second point of concern is again the slow reaction kinetics of

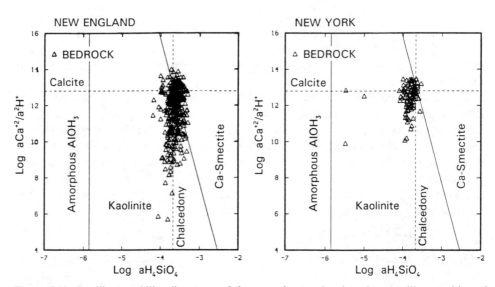

Figure 6.12. Ca-silicate stability diagrams and the groundwater chemistry in crystalline granitic and metamorphic rocks of New England and shales with limestones and dolomite in the state of New York (reprinted with permission from Rogers, 1989).

silicate minerals and it seems questionable whether true equilibrium is ever attained. Tardy (1971) reports that many waters which plot in the kaolinite stability field are derived from springs where montmorillonite is found in the watershed. Still, the value of silicate stability diagrams is that they enable a quick overview of stability relations between minerals and water in highly complicated systems.

An example of the application of silicate stability diagrams in groundwater geochemistry is presented in Figure 6.12, which compares the groundwater chemistry of an area with crystalline, granitic and metamorphic rocks in New England with an area containing shales, limestones and dolomite in the state of New York. In both cases, most of the data groups within the stability field of kaolinite and at the boundary with Ca-smectite (montmorillonite). However, the presence of carbonate minerals in New York-state aquifers influences the data distribution greatly. The variability of water compositions is very small and equilibrium with calcite appears to dominate the water chemistry. On the other hand, in the crystalline rocks of New England, a wide range of water compositions is found, which indicates that no single mineral phase controls the water composition and probably also reflects the slow reaction kinetics of silicate minerals in combination with differences in residence time of the water in the aquifer.

6.4 THE MASS BALANCE APPROACH TO WEATHERING

The problems encountered in treating silicate-water reactions as equilibrium systems has stimulated the use of other approaches. One of the best alternatives is the use of mass balance calculations which relate changes in water chemistry to the dissolution or precipitation of minerals and basically have the character of bookkeeping. For reactions between mineral and water we could write the general reaction:

$$\text{reactant phase} \rightarrow \text{weathering residue} + \text{dissolved ions} \qquad (6.20)$$

For the congruent dissolution of calcite in water containing carbonic acid the dissolution reaction is:

$$H_2CO_3 + CaCO_3 \rightarrow Ca^{2+} + 2HCO_3^- \qquad (6.21)$$

Water that percolates through a soil with calcite should accordingly be enriched with two moles of HCO_3^- for each mole of Ca^{2+}. For the incongruent dissolution of albite to kaolinite

$$2NaAlSi_3O_8 + 2H_2CO_3 + 9H_2O \rightarrow Al_2Si_2O_5(OH)_4 + 2Na^+ + 2HCO_3^- + 4H_4SiO_4$$
$$\text{albite} \qquad\qquad\qquad\qquad \text{kaolinite} \qquad\qquad\qquad (6.22)$$

the release of H_4SiO_4 is twice the release of HCO_3^- or Na^+.

Exactly the same can be done for the other reactions listed in Table 6.1. The results are displayed graphically in Figure 6.13 and show the expected water composition for each reaction. Note that a given water composition is not necessarily unique for a specific mineral, as is illustrated by the fact that dissolution of calcite (6.21) and weathering of anorthite to kaolinite yield the same water compositions (Figure 6.13). Rocks consist normally of mixtures of minerals and in some cases it is possible to reconstruct the contributions of different weathering reactions to the water composition.

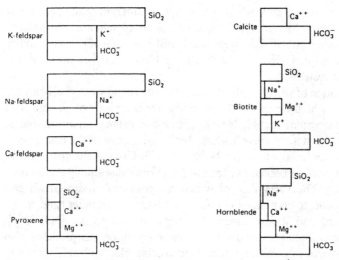

Figure 6.13. The composition of waters resulting from alteration of different silicate minerals to kaolinite in the presence of carbonic acid, according to the reactions listed in Table 6.1 (Garrels and Mackenzie, 1971).

EXAMPLE 6.1: *Mass balance and the water chemistry of the Sierra Nevada (USA)*
In a classic paper Garrels and Mackenzie (1967) related the groundwater chemistry in a granitic area, Sierra Nevada (USA), to silicate weathering reactions by the use of mass balance calculations. The approach used by Garrels and Mackenzie was stepwise substraction of different weathering reactions.

The chemistry of snow and ephemeral spring waters is listed in Table 6.4. The composition of snow was subtracted from the spring water composition to obtain the contribution by rock weathering.

Table 6.4. The composition of snow and ephemeral spring water (Garrels and Mackenzie, 1967). Rock weathering indicates the concentration in spring water minus the concentration in snow. Concentrations are given in mmol/l.

	Snow	Spring	Rock weathering
Na^+	0.024	0.134	0.110
Ca^{2+}	0.01	0.078	0.068
Mg^{2+}	0.007	0.029	0.022
K^+	0.008	0.028	0.020
HCO_3^-	0.018	0.328	0.310
Si	0.003	0.273	0.270
pH		6.2	

The granites of the Sierra Nevada contain the primary minerals plagioclase, biotite, quartz and K-feldspar and the weathering product kaolinite. Quartz is little reactive and is left out of consideration. The intuitive expectation is that weathering of the primary minerals to kaolinite (Table 6.1) may explain the spring water composition. The average composition of the minerals found in the area are listed in Table 6.5.

Table 6.5. The composition of some minerals in the granitic rocks in the Sierra Nevada (Garrels and Mackenzie, 1967).

Plagioclase	$Na_{0.62}Ca_{0.38}Al_{1.38}Si_{2.62}O_8$
Biotite	$KMg_3AlSi_3O_{10}(OH)_2$
K-Feldspar	$KAlSi_3O_8$
Kaolinite	$Al_2Si_2O_5(OH)_4$

The calculations proceed as follows. First all Na^+ is attributed to weathering of plagioclase. This requires the alteration of $0.110/0.62 = 0.177$ mmol/l plagioclase to kaolinite:

$$0.177 \, Na_{0.62}Ca_{0.38}Al_{1.38}Si_{2.62}O_8 + 0.246 \, CO_2 + 0.367 \, H_2O \rightarrow 0.122 \, Al_2Si_2O_5(OH)_4$$
$$+ 0.110 \, Na^+ + 0.068 \, Ca^{2+} + 0.246 \, HCO_3^- + 0.220 \, SiO_2$$

Subtracting the contribution of plagioclase weathering (Table 6.6) from the water chemistry shows that it also accounts for all Ca^{2+}, as well as most of the HCO_3^- and SiO_2. The next step is to attribute all Mg^{2+} to biotite weathering and $0.022/3 = 0.0073$ mmol/l biotite is altered to kaolinite:

$$0.0073 \, KMg_3AlSi_3O_{10}(OH)_2 + 0.051 \, CO_2 + 0.026 \, H_2O \rightarrow 0.0037 \, Al_2Si_2O_5(OH)_4$$
$$+ 0.0073 \, K^+ + 0.022 \, Mg^{2+} + 0.051 \, HCO_3^- + 0.015 \, SiO_2$$

Table 6.6. The contributions of weathering different silicates to the composition of ephemeral spring waters in the Sierra Nevada (mmol/l).

	Rock weathering	Minus Plagioclase	Minus Biotite	Minus K-feldspar
Na^+	0.110	0.000	0.000	0.000
Ca^{2+}	0.068	0.000	0.000	0.000
Mg^{2+}	0.022	0.022	0.000	0.000
K^+	0.020	0.020	0.013	0.000
HCO_3^-	0.310	0.064	0.013	0.000
Si	0.270	0.050	0.035	0.009

This removes most of the HCO_3^- as well as some of the K^+. The final step is to attribute the remainder of K^+ to K-feldspar weathering and accordingly 0.013 mmol/l K-feldspar must alter to kaolinite:

$$0.013 \, KAlSi_3O_8 + 0.013 \, CO_2 + 0.0195 \, H_2O \rightarrow 0.0065 \, Al_2Si_2O_5(OH)_4 + 0.013 \, K^+$$
$$+ 0.013 \, HCO_3^- + 0.026 \, SiO_2$$

What is left is only a minor amount of SiO_2 and the calculation shows that the three weathering reactions may explain the water chemistry surprisingly well. According to this reaction scheme, the dominant reaction is the transformation of plagioclase into kaolinite under the consumption of carbonic acid, while minor amounts of biotite and K-feldspar dissolve.

Basically, mass balance calculations consist of solving sets of linear equations and for this purpose standard computer programs, or the specially designed programs BALANCE (Parkhurst et al., 1982) or NETPATH (Plummer et al., 1991) can be used. Both programs can handle mass balances including redox reactions, isotope balances and simple mixing of end-member waters. To illustrate mass balance calculations in a more general and formal way, we treat in Example 6.2 the problem of ephemeral spring waters in the Sierra Nevada, once more using the program BALANCE.

EXAMPLE 6.2: *Mass balance calculation of the contribution of rock weathering to the composition of ephemeral spring waters in the Sierra Nevada with the computer code BALANCE*
BALANCE uses a mass balance equation of the form:

$$\text{initial solution} + \text{reactant phases} \rightarrow \text{final solution} + \text{product phases} \qquad (6.23)$$

And the mass balance equations on elements is:

$$\Delta m_{T,k} = m_{T,k\text{(final)}} - m_{T,k\text{(initial)}} = \sum_{p=1}^{P} \alpha_p b_{p,k} \qquad (6.24)$$

For each element $k = 1$ to J.

Here $\Delta m_{T,k}$ is the change in total molality of the k-th element between final and initial solution, while J is the number of elements included in the calculation. P is the number of reactant and product phases in Reaction (6.23), α_p is the calculated mass transfer of the p-th phase and $b_{p,k}$ denotes the stoichiometric coefficient of the k-th element in the p-th phase. BALANCE requires that the total number of phases is equal to the total number of elements in order to be able to solve the set of linear equations.

For the Sierra Nevada spring waters, the number of elements, excluding H and O, in Table 6.4 is six. Four minerals are listed in Table 6.5 and since the number of phases should equal the number of elements ($P = J$), we need two additional phases. An obvious choice is the uptake/release of CO_2 by the solution, since H_2CO_3 is an important weathering agent. The choice of the last phase is less obvious, but dissolution/precipitation of calcite could be used as a possible Ca^{2+} source or sink. An additional constraint is that all aluminum released from primary minerals is preserved in the weathering products. Therefore, we also consider a mass balance on aluminum, which in practice can be done by considering zero concentrations of aluminum in rain and spring water and including gibbsite as an additional phase. The mass balances equations to be solved are now:

$$\Delta m_{T,Na} = 0.134 - 0.024 = 0.62 \alpha_{\text{plagioclase}}$$

$$\Delta m_{T,Ca} = 0.78 - 0.01 = 0.38 \alpha_{\text{plagioclase}} + 1\alpha_{\text{calcite}}$$

$$\Delta m_{T,Mg} = 0.029 - 0.007 = 3\alpha_{\text{biotite}}$$

$$\Delta m_{T,K} = 0.028 - 0.008 = 1\alpha_{\text{biotite}} + 1\alpha_{\text{K-feldspar}}$$

$$\Delta m_{T,C} = 0.328 - 0.018 = 1\alpha_{\text{calcite}} + 1\alpha_{\text{CO2-gas}}$$

$$\Delta m_{T,Si} = 0.273 - 0.003 = 2.62 \alpha_{\text{plagioclase}} + 3\alpha_{\text{biotite}} + 3\alpha_{\text{K-feldspar}} + 2\alpha_{\text{kaolinite}}$$

$$\Delta m_{T,Al} = 0.000 - 0.000 = 1.38 \alpha_{\text{plagioclase}} + 1\alpha_{\text{biotite}} + 1\alpha_{\text{K-feldspar}} + 2\alpha_{\text{kaolinite}} + 1\alpha_{\text{gibbsite}}$$

Output of BALANCE (actually its interactive successor BALNINPT) is shown below. The first part lists the concentrations in spring waters and snow corresponding to Table 6.4. The second part contains the stoichiometric composition of different minerals. Plagioclase and biotite have been defined specially for this case while the remainder is from the database. The positive (+) sign for plagioclase, biotite and K-feldspar indicates that only dissolution is allowed, while the negative sign (−) for kaolinite and gibbsite indicates that only precipitation is allowed. For the other phases both dissolution (in-gassing) or precipitation (degassing) is allowed.

```
BALNINPT output for Sierra Nevada ephemeral springs (Garrels
and Mackenzie, 1967)

PART I
  1 AL      .000      .000      .000
  2 NA      .134      .024      .000
  3 CA      .078      .010      .000
  4 MG      .029      .007      .000
  5 K       .028      .008      .000
  6 C       .328      .018      .000
  7 SI      .273      .003      .000
```

```
PART II
   1 plagiocl F+   NA   .620   CA   .380   AL  1.380   SI  2.620
   2 biotite  F+   K   1.000   MG  3.000   AL  1.000   SI  3.000
   3 KAOLINIT F-   AL  2.000   SI  2.000        .000        .000
   4 CO2 GAS  F    C   1.000   RS  4.000        .000        .000
   5 CALCITE  F    CA  1.000   C   1.000   RS  4.000        .000
   6 K-FELDSP F+   K   1.000   AL  1.000   SI  3.000        .000
   7 GIBBSITE F-   AL  1.000        .000        .000        .000

PART III
   plagiocl  + F          .1774
   biotite   + F          .0073
   KAOLINIT  - F         -.1274
   CO2 GAS     F          .3094
   CALCITE     F          .0006
   K-FELDSP  + F          .0127
   GIBBSITE  - F         -.0100


 1 MODELS WERE TESTED.
 1 MODELS WERE FOUND WHICH SATISFIED THE CONSTRAINTS.
```

Part III lists the calculated mass transfer (α) for each mineral, where a positive value indicates dissolution and a negative value precipitation. Comparison with the calculations by hand in Example 6.1 fortunately shows a very good agreement. Again, the most important reaction is the alteration of plagioclase to kaolinite, while only small amounts of biotite and K-feldspar dissolve. Calcite and gibbsite, the phases that were not considered in manual calculations, dissolve or precipitate only in very small amounts.

Mass balance calculations of this type are often a useful approach in identifying possible reactions that may explain differences in water chemistry along a flowpath, not only in the case of silicate weathering, but also for carbonates, redox reactions etc. Often they form the first step in elucidating geochemical processes before attempting the more complex approach of pathway modeling as described in Chapter 10. One should, however, keep the following limitations in mind:

1. The solution of the mass balance equations is not necessarily unique. Different choices of phases may lead to equally consistent reaction schemes. In other words, a mass balance calculation does not prove that the reactions take place.

2. There are no thermodynamic constraints on mass balance calculations. The mass balance calculation may predict impossible reactions, like precipitation of plagioclase at low temperature, or gibbsite precipitation when kaolinite is the stable phase.

3. Mass balance calculations do not consider what is kinetically consistent. For example they could predict that quartz is dissolving, but not plagioclase, even though this would be kinetically quite unreasonable.

4. Mass balance calculations assume steady state; water samples along a flow path are usually taken at the same time and differences in water chemistry are assumed to be solely due to reactions with minerals and not, for example, to temporal variations in the composition of water entering the system.

5. Mass balances assume a homogeneous reaction between the points of analysis. This condition is hard to test in aquifer systems and certainly questionable in soil systems unless mass balances are carried out for the different horizons.

In other words, use your common geochemical sense in the use and interpretation of mass balance calculations. In the Sierra Nevada example, one could plot the water chemistry of the spring in the silicate stability diagram given in Figure 6.11 in order to check whether kaolinite is stable relative to gibbsite (do this yourself, neglecting differences between activities and molar concentrations). Also the preferential dissolution of plagioclase compared to K-feldspar as observed in Examples 6.1 and 6.2 appears kinetically reasonable.

Finally, note that the rate of mineral dissolution or precipitation can be derived from mass balance calculations when the travelled distance and the groundwater flow rate are known. A number of examples of mass balance calculations are discussed in Drever (1988) and we will return to the subject in Section 6.6.

6.5 KINETICS OF SILICATE WEATHERING

The importance of dissolution kinetics for weathering of primary silicate minerals, such as feldspars, amphiboles and pyroxenes, was already recognized on a qualitative level by Goldich (1938) who constructed the weatherability series shown in Figure 6.2. Especially, due to increasing problems with acid rain, a large number of studies have been carried out over the last twenty years in order to obtain a better quantitative understanding of the dissolution kinetics of primary silicates. Typically such studies concerned laboratory dissolution experiments of pure minerals.

The interpretation of silicate dissolution kinetics from such experiments has evolved with the improvement of the experimental methods. In early studies, (Wollast, 1967; Busenberg and Clemency, 1976) batch reactors were used in which secondary reaction products such as clay minerals or Al-hydroxides could precipitate. These and insufficient preparation of the crystals (removing fines and super-reactive sites due to crushing, (cf. Holdren and Berner, 1979) affected the interpretation of experimental results. Generally, dissolution rates were found to decrease continuously with time. This was interpreted in terms of a rate limiting solid state diffusion process through a secondary or leached layer (Wollast, 1967; Stumm and Wollast, 1990) very similar to that used in steel corrosion. Such a mechanism predicts ever decreasing rates, so that mineral dissolution practically comes to a standstill after some period of time.

Subsequent work has shown that dissolution rates become constant when crystals are pretreated correctly and examples of such linear dissolution behavior are illustrated in

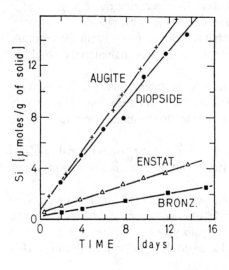

Figure 6.14. Linear dissolution kinetics expressed by the release of silica to the solution over time for pyroxenes and amphiboles at pH 6. Experiments with enstatite, augite and diopside were carried out at 50°C while bronzite was dissolved at 20°C and $P_{O_2} = 0$ (reprinted with permission from Schott and Berner, 1985).

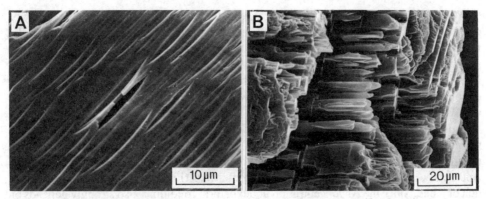

Figure 6.15. Etch pitting of amphiboles separated from a quaternary aquifer deposit. Grains were ultrasonically cleaned. A) An early stage of etch pitting with shallow lenses parallel to the c-axis. B) A terminal stage of etch pitting (Postma and Brockenhuus-Schack, 1987).

Figure 6.14 by the release of Si from several minerals. Linear dissolution kinetics are compatible with a mechanism where surface reaction controls the reaction rate rather than transport of reactants or products to and from the mineral surface (see Section 3.5.5).

A further indication for surface processes to be the rate limiting step in the overall dissolution process is the development of crystallographically controlled etch pits on the mineral surfaces which can be identified by SEM. Such etch pits have been observed both during experimental studies as well as on naturally weathered silicate grains found in soils or aquifer material. An example for amphibole crystals separated from a sandy deposit is illustrated in Figure 6.15.

A first impression of the time scales involved in silicate dissolution kinetics can be obtained by calculating the change in crystal size for different silicates using Equation (3.40). This has been done by Lasaga (1984) (Table 6.7) who calculated the lifetime of 1 mm crystals at pH 5 for various common minerals. Note that the range of lifetimes is very large; the Ca-feldspar (anorthite) dissolves about 700 times faster than Na-feldspar (albite) and about 4500 times faster than a K-feldspar. Qualitatively, the relative lifetimes for different minerals calculated from laboratory dissolution experiments, are in good agreement with the Goldich weathering sequence shown in Figure 6.2 .

Table 6.7. The mean lifetime in years of 1 mm crystals of various minerals at 25°C and pH 5 (Lasaga, 1984).

Mineral	Lifetime
Quartz	34,000,000
Muscovite	2,700,000
Forsterite	600,000
K-feldspar	520,000
Albite	80,000
Enstatite	8,800
Diopside	6,800
Nepheline	211
Anorthite	112

Work done with fluidized bed reactors, in which the composition of the solution can be controlled and kept below saturation with secondary reaction products, have revealed a much more detailed picture of silicate dissolution kinetics. Chou and Wollast (1985) studied albite dissolution with this technique and found an initial rapid phase in which Na^+ is exchanged for H^+, followed by a much slower dissolution process. Steady state dissolution rates for the slow process are shown in Figure 6.16 as a function of pH (Chou and Wollast, 1985).

Figure 6.16 shows that the albite dissolution rate is strongly pH dependent and has a minimum around pH 6-7. Chou and Wollast (1985) found furthermore that the dissolution rates of albite are also distinctly influenced by aluminum concentrations in the solution. Apparently, dissolved aluminum at low concentrations (Figure 6.17) inhibits the dissolution of albite. Since the Al concentration under natural conditions also strongly depends on pH (Figure 6.7), the two rate controlling parameters interact in a complex way. In addition,

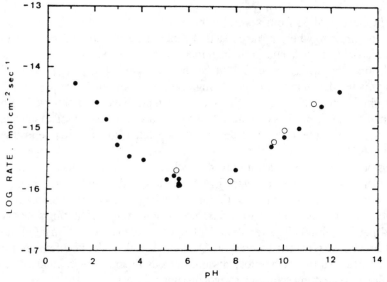

Figure 6.16. Dissolution rates of albite as a function of pH. (Chou and Wollast, 1985, reprinted by permission from American Journal of Science).

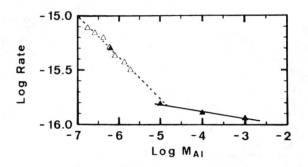

Figure 6.17. The effect of dissolved Al concentrations on the dissolution rate of albite at pH 3 (Chou and Wollast, 1985, reprinted by permission from American Journal of Science).

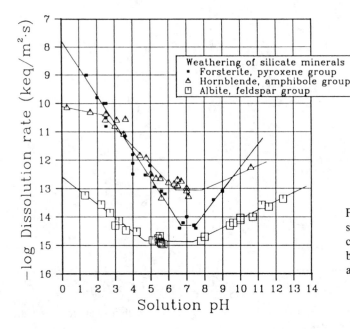

Figure 6.18. Experimental dissolution rates for different silicate minerals at 25°C (reprinted by permission from Sverdrup and Warfvinge, 1988).

the P_{CO_2} appears to affect the dissolution rate of feldspars and Sverdrup (1990) proposed a parallel dissolution mechanism involving carbonic acid. Clearly, feldspar dissolution is a highly complex process.

The dissolution kinetics of several other silicate minerals is compared with albite in Figure 6.18. The overall pattern of the rate distribution as a function of pH is for amphiboles and pyroxenes similar to that of albite. However, both the absolute rates and the rate dependency on the pH vary significantly. For example, in the near-neutral pH range the dissolution rate of hornblende is two orders of magnitude higher than that of albite. A comprehensive review of most studies on silicate dissolution kinetics can be found in Sverdrup (1990).

The description of silicate dissolution kinetics as displayed in Figures 6.16 to 6.18 is essentially empirical. Work is, however, in progress to obtain a more mechanistic insight in the processes that are happening at the silicate surface by using surface chemical techniques (Casey et al., 1989; Hellmann et al., 1990; Hochella and White, 1990) and by the application of surface complexation theory (Brady and Walther, 1989; Stumm and Wollast, 1990; Blum and Lasaga, 1991). At present there are few studies that apply silicate dissolution kinetics to field situations and these show generally that dissolution rates measured in the field are much slower than rates predicted from laboratory studies (Paces, 1983; Velbel, 1985; White and Peterson, 1990; Brantley, 1992). However, a succesful comparison of field and laboratory rates will be presented in Section 6.6.2.

6.6 GROUNDWATER ACIDIFICATION

The occurrence of acid groundwater is reported from an increasing number of localities in different parts of the industrialized world (e.g. Appelo et al., 1982; Hultberg and Wenblad,

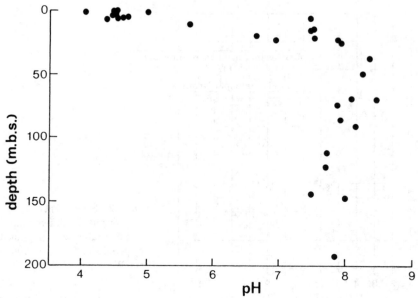

Figure 6.19. The pH of groundwater versus depth, collected from different wells in sandy aquifers of the Veluwe area, Netherlands (Appelo et al., 1982).

1980; Eriksson, 1981; Böttcher et al., 1985) where aquifer materials are free of carbonates. The threat of acidification to groundwater resources is illustrated in Figure 6.19, which shows a strong decrease of pH in the youngest groundwater of a carbonate-free sandy aquifer in the Veluwe area of the Netherlands.

A range of different processes may cause acidification of groundwater. First, natural acidification occurs through CO_2 production and root respiration in the soil by the overall reaction:

$$CH_2O + O_2 \rightarrow H_2O + CO_2 \qquad (6.25)$$

However, as pointed out in Section 4.2.2, the lower pH limit to be expected from this process is not less than about 4.6 even in the absence of any buffering processes within the aquifer. Thus, anthropogenic processes must have a significant effect on the acidification of groundwater.

The detrimental effect of acid rain on forests as well as on lake waters of northern Europe and North America is well documented (Wright and Henriksen, 1978; Likens et al., 1977; Ulrich et al., 1979; Drabløs and Tollan, 1980) and acid rain may be expected to affect groundwater reservoirs as well. The acid rain problem originates from fossil fuel combustion, which produces nitrous oxides and SO_2, that are subsequently oxidized in the atmosphere and precipitate as dilute sulfuric and nitric acid solutions (Berner and Berner, 1987; see also Chapter 2). Table 6.8 compares the composition of rainwater in 1938, before heavy industrialization, with that around 1980 in the Netherlands. A decrease in pH of almost one unit is evident and associated with significant increases of sulfate and nitrogen compounds. Evapotranspiration furthermore concentrates the acidity of the solution that enters the subsoil. In first approximation, the effect of evapotranspiration can be estimated by multiplying the measured rainwater concentrations with the evapotranspiration concen-

Table 6.8. The composition of 1938 rain and 1980 acid rain and the effect of evapotranspiration in the Netherlands (Appelo, 1985). Concentrations are in mmol/l. Evapotranspiration factor = 3 and $P_{CO_2} = 10^{-3.5}$, [1] = Na + K.

	1938 Rain	× 3	1980 Rain	× 3
pH	5.4	5.62	4.52	3.08
Na^+	0.07[1]	0.21[1]	0.073	0.219
K^+			0.004	0.012
Mg^{2+}	0.03	0.09	0.009	0.027
Ca^{2+}	0.04	0.12	0.014	0.042
NH_4^+	0.03	–	0.13	–
Cl^-	0.09	0.27	0.085	0.255
Alk	0.06	0.002	–	–
SO_4^{2-}	0.045	0.135	0.071	0.213
NO_3^-	–	0.09	0.061	0.573

tration factor (Chapter 2). A more sophisticated approach, using the program PHREEQE, includes oxidation of ammonia and equilibration with atmospheric CO_2 and will be discussed later in Section 10.3. The effect of evapotranspiration, calculated with PHREE-QE, is listed both for 1938 and 1980 rain in Table 6.8. The results show that 1938 rainwater does not become acid upon concentration. However, concentration of 1980 acid rain leads to a pH down to about 3 which is due to the oxidation of the much higher ammonium level in 1980 rainwater by the reaction:

$$NH_4^+ + 2O_2 \rightarrow NO_3^- + 2H^+ + H_2O \tag{6.26a}$$

The second source of anthropogenic acidification is the excessive use of ammonia and manure as fertilizers. The oxidation of ammonia by oxygen is here the main acidifying process in the soil.

$$NH_3 + 2O_2 \rightarrow NO_3^- + H^+ + H_2O \tag{6.26b}$$

When plants in the soil consume all the produced nitrate, the proton production due to the oxidation process, is balanced by the HCO_3^- production of the denitrification process (Chapter 7):

$$5CH_2O + 4NO_3^- \rightarrow 2N_2 + 4HCO_3^- + CO_2 + 3H_2O \tag{7.78}$$

However, the ubiquitous presence of nitrate in aquifers is ample proof of the opposite and nitrification of ammonia must be considered as an important acidifying process. The net effect of nitrification on the pH of recharge water is largely determined by the amount of lime which is applied together with fertilizers to the soil.

The third major man-made acidifying process is the oxidation of pyrite (FeS_2). Pyrite is found, at least in small quantities, in most reducing sediments and a lowering of the groundwater table, caused by production wells, may result in the oxidation of pyrite. This process is discussed in more detail in Chapter 7, but for our present purposes it can be described by the overall reaction:

$$2FeS_2 + 15/2O_2 + 5H_2O \rightarrow 2FeOOH + 4SO_4^{2-} + 8H^+ \tag{6.27}$$

This reaction is one of the most strongly acid-producing reactions in nature.

We conclude from the above discussion that acidification of groundwater cannot be attributed to one single cause, but acid rain is certainly the most serious one since it affects large areas of forests and uncultivated lands.

6.6.1 *Buffering processes in aquifers*

Acid water that is introduced into soils and aquifer systems is expected, at least in part, to be neutralized by reactions with soil and aquifer materials and the task at hand is to establish a reasonable understanding of buffering processes. The first qualitative assessment is to consider the relation between the geology and the suspectibility for acidification. Obviously, rocks which contain carbonate minerals are unlikely to develop acid groundwater because of the fast dissolution kinetics of carbonates (Chapter 4). In a regional survey of groundwater in the UK, Edmunds and Kinniburgh (1986) pointed out that low alkalinity groundwaters are most vulnerable for acidification and decreasing alkalinity values over time are indeed a good warning for groundwater acidification in the future (Hultberg and Wenblad, 1980). For carbonate free rocks, Edmunds and Kinniburgh (1986) listed granites, acid igneous rocks and clean quartz sandstones, which mainly contain slowly dissolving silicate minerals, as most susceptible to acidification, while basic and ultrabasic rocks, that are dominated by more quickly dissolving silicates, were found at the other end of the scale.

A serious case of groundwater acidification in a carbonate free sandy aquifer below a coniferous forest in Germany is illustrated in Figure 6.20. The elevated sulfate concentrations which are found throughout the profile indicate the influence of acid rain. Acid groundwater, with pH values slightly above 4, is only found in the upper part of the profile which indicates that proton neutralizing processes must be in operation. The upper layers of groundwater, with a low pH, also contain high aluminum concentrations and suggest that buffering processes are related to processes involving aluminum.

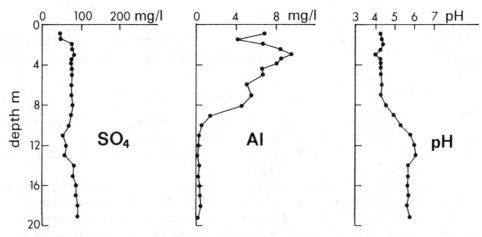

Figure 6.20. Groundwater acidification in the saturated zone of a sandy aquifer below a coniferous forest in Germany. The displayed values represent the average of four samples taken during a two year period (modified from Böttcher et al., 1985).

There are three dominant processes that may have such a pH buffering effect:
1. Dissolution of $Al(OH)_3$ and perhaps of $AlOHSO_4$.
2. Ion exchange of aluminum and possibly also of protons.
3. Weathering of primary silicates.

When acid water enters the soil or aquifer it may react with gibbsite and other forms of $Al(OH)_3$ or clay minerals, that have accumulated over time as weathering products of primary minerals. These weathering products display relatively fast dissolution kinetics and may bring considerable amounts of aluminum in solution. The buffering effect of such reactions is illustrated by the reaction

$$Al(OH)_{3 \text{ gibbsite}} \leftrightarrow Al^{3+} + 3OH^- \tag{6.28}$$

EXAMPLE 6.3: *Acid groundwater formation and gibbsite buffering*
In the Veluwe (The Netherlands), acid groundwater develops from acid rain. Groundwater is potentially even more acid than rain because acidity is concentrated by evapotranspiration. pH rises when gibbsite dissolves in the acid groundwater. This example shows how to calculate buffering and dissolution reactions when acid rain becomes concentrated groundwater.
  Rainwater at Deelen (Veluwe), has the following composition:

pH	Na^+	K^+	Mg^{2+}	Ca^{2+}	NH_4^+		Cl^-	HCO_3^-	SO_4^{2-}	NO_3^-	
4.32	70	4	9	15	120		78	0	74	64	(μmol/l)

  a. This water is concentrated 4 times by evapotranspiration and becomes groundwater. All NH_4^+ is oxidized to NO_3^-. Calculate the composition.
  b. $Al(OH)_3$ (gibbsite) dissolves in the soil. Calculate pH and Al-concentration. (Consider only Al^{3+}, neglect other Al-complexes)

$$Al(OH)_3 \leftrightarrow Al^{3+} + 3OH^- \qquad K = 10^{-32.0}$$

ANSWER
ad a. First calculate the composition in μeq/l:

H^+	Na^+	K^+	Mg^{2+}	Ca^{2+}	NH_4^+	$\Sigma+$	Cl^-	HCO_3^-	SO_4^{2-}	NO_3^-	$\Sigma-$
48	70	4	18	30	120	290	78	0	148	64	290

Concentrate 4 times:

H^+	Na^+	K^+	Mg^{2+}	Ca^{2+}	NH_4^+		Cl^-	HCO_3^-	SO_4^{2-}	NO_3^-	
192	280	16	72	120	480		312	0	592	256	

Oxidize all NH_4^+ (assume O_2 is in excess): $NH_4^+ + 2O_2 \rightarrow NO_3^- + 2H^+ + H_2O$

H^+	Na^+	K^+	Mg^{2+}	Ca^{2+}	NH_4^+		Cl^-	HCO_3^-	SO_4^{2-}	NO_3^-	
1152	280	16	72	120	0		312	0	592	736	

which gives pH = 2.9.

ad b. When gibbsite dissolves, H^+ is consumed:

$$Al(OH)_3 \leftrightarrow Al^{3+} + 3OH^- \qquad K = 10^{-32}$$
$$+3H^+ + 3OH^- \leftrightarrow 3H_2O \qquad K = (10^{14})^3 = 10^{42}$$

+ $\overline{\phantom{Al(OH)_3 + 3H^+ \leftrightarrow Al^{3+} + 3H_2O}}$ Π $\overline{\phantom{K = 10^{10}}}$

$$Al(OH)_3 + 3H^+ \leftrightarrow Al^{3+} + 3H_2O \qquad K = 10^{10}$$

Hence: $[Al^{3+}]/[H^+]^3 = 10^{10}$

Each Al^{3+} consumes three H^+ so that:

$$[Al^{3+}] = ([H^+]_i - [H^+]_e)/3$$

Combining the last two equations and substituting the data of concentrated, acid rain:

$$((1152 \cdot 10^{-6} - [H^+]_e)/3) / [H^+]_e^3 = 10^{10}$$

By trial and error we find $[H^+]_e = 10^{-4.48}$ (pH = 4.48), and $[Al^{3+}] = 0.37$ mmol/l.

As was discussed in Section 6.2 (Figure 6.7) the solubility of gibbsite and of poorer crystalline forms of $Al(OH)_3$ appears to set an upper limit to the Al^{3+} activity in solution. However, Figure 6.8 also showed that these phases seem not to control the Al^{3+} activity in solution when the pH is below 4.5. Van Breemen (1973) proposed a hypothetical $AlOHSO_4$ phase as a possible solubility constraint at pH values below about 4.5. Subsequently Nordstrom (1982) proposed that this hypothetical phase could be the mineral jurbanite $(AlOHSO_4 \cdot 5H_2O)$. Neglecting crystal water, the dissolution reaction for jurbanite is:

$$AlOHSO_4 \leftrightarrow Al^{3+} + OH^- + SO_4^{2-} \qquad (6.29)$$

The saturation state of groundwater with respect to gibbsite and jurbanite in the unsaturated zone of a sandy sediment is shown in Figure 6.21.

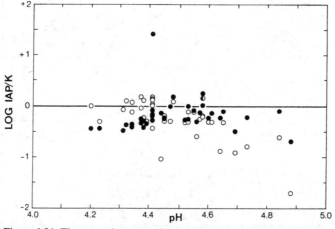

Figure 6.21. The saturation state of groundwater for gibbsite and jurbanite versus pH in the unsaturated zone of a sandy sediment in Denmark. Closed and open symbols indicate the saturation states for respectively gibbsite and jurbanite; $K_{jurbanite} = 10^{-17.8}$ and $K_{gibbsite} = 10^{-34.2}$ (Hansen and Postma, in prep.).

The results show that the groundwater with a pH below 4.5 is subsaturated for gibbsite but at pH values above 4.5, equilibrium with gibbsite tends to be approached. For jurbanite, the reverse trend is found, which could indicate an equilibrium control by jurbanite in the low pH range. However, possible solubility control by jurbanite is still highly controversial (Nordstrom and Ball, 1986; Neal et al., 1986; Förster, 1986) and the mineral jurbanite has never been identified in sediments or soils.

Aluminum brought into solution by dissolution of weathering products may interact with the cation exchange complex of the sediment. As discussed in Chapter 5, the pool of adsorbed cations, even in sediments with *CEC* values below 1 meq/100 g, is compared to the dissolved cations very large. Replacement of base cations (Ca^{2+}, Mg^{2+}, Na^+ and K^+) on the cation exchange complex by Al^{3+} can be summarized as a reaction with Ca^{2+} that in most cases is the dominant cation on the exchange complex:

$$1/3Al^{3+} + 1/2Ca\text{-}X_2 \leftrightarrow 1/3Al\text{-}X_3 + 1/2Ca^{2+} \tag{6.30}$$

The Al^{3+} released by dissolution of gibbsite adsorbs on the exchange complex according to (6.30), and allows additional gibbsite to dissolve and thereby to increase pH, until equilibrium between gibbsite, the exchange complex and the solution has been established. Such calculations are rapidly carried out by computer programs, as will be illustrated in Chapter 10. An interesting side effect of the role of aluminum adsorption as buffering

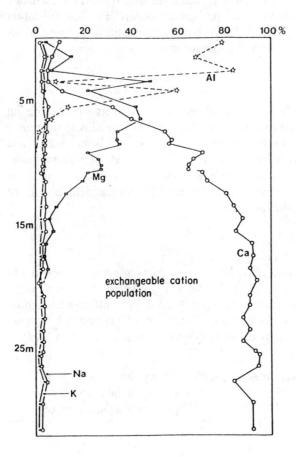

Figure 6.22. The composition of the exchange complex in the unsaturated zone of a sandy deposit in Germany (reproduced by permission of the Geological Society from Dahmke et al., 1986, Near surface geochemical processes in Quaternary sediments, J. Geol. Soc. 143).

reaction is that major variations in Na^+ contents of the groundwater which occur seasonally in the unsaturated zone (Chapter 2) may have a significant effect on the pH, due to the competition of Na^+ with other cations for positions on the exchange complex (Moss and Edmunds, 1989; Hansen and Postma, in prep.). Strong increases in Na^+ concentrations may desorb Al^{3+} from the exchange complex and by subsequent gibbsite precipitation lower the pH.

The distribution of aluminum on the exchange complex in the unsaturated zone is illustrated in Figure 6.22. In the uppermost part of the profile, Al^{3+} replaces Ca^{2+} as the dominant ion on the exchange complex, with Mg^{2+} as an intermediate. With time and increased acidification load, it is expected that the Al^{3+} adsorption front moves slowly downward. The coupled dissolution of gibbsite and adsorption of Al^{3+} on the exchange complex has a strongly delaying effect on the downward progression of acid groundwater. It is, however, important to realize that the exchange complex acts as a storage for acidity rather than as true acid neutralization. If a remediation scheme is planned for de-acidification, for example by applying lime to the soil (Warfvinge, 1988) the whole acidification process is reversed: addition of base induces precipitation of gibbsite, and thereby lowers the Al^{3+} activity in solution, which causes again desorption of Al^{3+} from the exchange complex and exchange by Ca^{2+}. The amount of base must therefore be sufficient to precipitate all adsorbed Al^{3+} as gibbsite before pH can increase to near neutral values. At the same time, a high concentration of Ca^{2+}, which exchanges with Al^{3+} on the exchange complex, may cause an acid pulse when the Al^{3+} precipitates as $Al(OH)_3$. Deacidification by artificial means must therefore proceed slowly, with concentrations of base cations which do not exceed the natural solute concentrations too much.

6.6.2 Field weathering rates

The only permanent neutralization of acid input in groundwater reservoirs is by weathering of primary silicate minerals. As discussed before, weathering of primary silicate minerals is controlled by their dissolution kinetics. Weathering rates of primary silicates under field conditions are difficult to measure and very little information is available from groundwater reservoirs. However, great efforts have been made to obtain reliable silicate weathering rates in soil acidification studies (Sverdrup, 1990; Rosén, 1991). In this field, a range of different approaches has been used which include:

1. Mass balance calculations
2. Historical rates
3. Experimental laboratory rates

Mass balance studies of small watersheds have been carried out at a number of localities to study the effect of acid rain (e.g. Likens et al., 1977; Paces, 1983; Velbel, 1985; Katz et al., 1985). Mass balance equations for watersheds are formulated slightly different from those discussed in Section 6.4 for a groundwater flow path. For the rate of weathering in a whole watershed we may write the general mass balance equation for base cations (Ca^{2+}, Mg^{2+}, K^+, Na^+) (Drever, 1988):

$$[\text{solutes from weathering}] = [\text{solutes in runoff}] + [\text{change in exchange pool}]$$
$$+ [\text{change in biomass}]$$
$$- [\text{solutes from atmosphere}]$$

$$(6.31)$$

Table 6.9. Yearly element balances in the Hubbard Brook area, New Hampshire, USA (Likens et al., 1977)

Component	Chemical element							
	Ca	Mg	Na	K	N	S	P	Cl
	Standing stock (kg/ha)							
Aboveground biomass	383	36	1.6	155	351	42	34	*
Belowground biomass	101	13	3.8	63	181	17	53	*
Forest floor	372	38	3.6	66	1256	124	78	*
	Annual flux (kg/ha.yr)							
Bulk precipitation input	2.2	0.6	1.6	0.9	6.5	12.7	0.04	6.2
Gaseous or aerosol input	*	*	*	*	14.2	6.1	*	?
Weathering release	21.1	3.5	5.8	7.1	0	0.8	?	*
Streamwater output								
Dissolved substances	13.7	3.1	7.2	1.9	3.9	17.6	0.01	4.6
Particulate matter	0.2	0.2	0.2	0.5	0.1	<0.1	0.01	*
Vegetation uptake	62.2	9.3	34.8	64.3	79.6[a]	24.5[a]	8.9	*
Litter fall	40.7	5.9	0.1	18.3	54.2	5.8	4.0	*
Root litter	3.2	0.5	0.01	2.1	6.2	0.6	1.7	*
Throughfall and stemflow	6.7	2.0	0.3	30.1	9.3	21.0	0.7	4.4
Root exudates	3.5	0.2	34.2	8.0	0.9	1.9	0.2	1.8
Net mineralization	42.4	6.1	0.1	20.1	69.6	5.7	?	?
Aboveground biomass accretion	5.4	0.4	0.03	4.3	4.8	0.8	0.9	*
Belowground biomass accretion	2.7	0.3	0.12	1.5	4.2	0.4	1.4	*
Forest floor accretion	1.4	0.2	0.02	0.3	7.7	0.8	0.5	*

*Small, unmeasured. [a]Root uptake

The results of such a mass balance is shown for the Hubbard brook catchment area (Table 6.9) and illustrates some of the problems in obtaining weathering rates from mass balances. First of all the weathering rate is very small compared to the amount of elements fixed in the biomass and one may also note that the rate of annual element cycling by the biomass is much larger than the release by weathering. Accordingly small changes in total biomass, which can be difficult to measure, may yield significant apparent variations in the weathering rate.

The same problem applies to the pool of exchangeable cations which, under the stress of acid input, typically is not in a steady state. The change in composition of the exchange pool is also difficult to analyze and in most cases the weathering rates obtained by the mass balance approach include the rate of depletion of the pool of exchangeable cations.

An additional example of weathering rates obtained from mass balances is shown in Table 6.10 for the Gårdsjön area in Sweden. The Gårdsjön area is located on a granitic bedrock with shallow podzolic soils. It has been extensively studied for lake and soil acidification and will be used here to compare different approaches for estimating weathering rates (Sverdrup and Warfvinge, 1988). The mass balance considers biomass cation accumulation, but considers the pool of exchangeable cations to be at steady state. The latter assumption is based on the observation that brooks draining the soils have shown a constant pH of 4.2 during the last 15 years. An important source of uncertainty in the mass balance is atmospheric deposition which is mainly due to problems in estimating dry deposition. It is interesting to note that the weathering rates for Ca^{2+} and Mg^{2+} which are

Table 6.10. The weathering rates (keq/ha.yr) calculated from mass balances in three watersheds (F1-F3) in the Gårdsjön area, Sweden (Sverdrup, 1990).

	F1	F2	F3
Ca^{2+}			
Deposition	0.18-0.43	0.18-0.43	0.18-0.43
Outflow	0.44	0.43	0.53
Net bioaccumulation	0.2	0.18	0.09
Weathering rate	0.46-0.21	0.43-0.18	0.44-0.19
Mg^{2+}			
Deposition	0.44-0.50	0.44-0.50	0.44-0.50
Outflow	0.85	0.67	0.67
Net bioaccumulation	0.05	0.044	0.022
Weathering rate	0.46-0.42	0.27-0.21	0.25-0.19
K^+			
Deposition	0.06-0.16	0.06-0.16	0.06-0.16
Outflow	0.12	0.08	0.11
Net bioaccumulation	0.05	0.045	0.02
Weathering rate	0.11-0.02	0.07-0.0	0.07-0.0
Na^+			
Deposition	2.0-2.2	2.0-2.2	2.0-2.2
Outflow	2.0-2.2	2.0-2.2	2.0-2.2
Weathering rate	0.02	0.02	0.02
Sum	0.67-1.05	0.41-0.79	0.4-0.78

derived from easily weatherable minerals, Ca-rich plagioclases, pyroxenes and amphiboles are significantly higher than for K^+ and Na^+ which originate from slowly dissolving minerals, Na-rich feldspars and K-feldspar.

Some of the problems in estimating the weathering rate from mass balance calculations can be solved by the use of the stable strontium isotopes ^{87}Sr and ^{86}Sr (Åberg et al., 1989; Jacks et al., 1989). The $^{87}Sr/^{86}Sr$ ratio in atmospheric deposition differs from that in bedrock so that the contribution by weathering can be estimated from the difference between the $^{87}Sr/^{86}Sr$ ratio in deposition and in runoff. This approach assumes isotopic equilibration between soil solution, exchange complex and runoff. The weathering rate is calculated from the relation:

$$R_w = P_{bc} \frac{q - a}{s - q} \tag{6.32}$$

Where R_w is the base cation weathering rate, P_{bc} the base cation deposition rate, a the $^{87}Sr/^{86}Sr$ ratio in the deposition, q the $^{87}Sr/^{86}Sr$ ratio in the runoff and s the $^{87}Sr/^{86}Sr$ ratio in the mineral matrix. The combination of weathering rates obtained with the Sr-isotope methods, with those estimated from traditional mass balances also allows a separation between the weathering rate and depletion of the cation exchange complex (Åberg et al., 1989).

In the second approach, weathering rates are estimated from the change in chemical composition through a soil which has developed over a known period of time (Bain et al.,

1991; Olsson and Melkerud, 1991; Sverdrup, 1990). Here an inert tracer is needed for which the content of quartz or zircon is often used since these minerals dissolve extremely slowly. Considering the quartz content as a conservative component, the weathering rate can be calculated from the equation (Sverdrup and Warfvinge, 1991):

$$R_w = \left(\frac{1}{\Delta t}\right) \left(\frac{X_{S,Q}}{X_{B,Q}} X_{B,i} - X_{S,i}\right) \frac{m_{soil,i}}{M_{W,i}} n_{BC,i} \tag{6.33}$$

where $X_{B,Q}$ = the fraction of quartz in the parent rock; $X_{B,i}$ = the fraction of mineral i in the parent rock; $X_{S,Q}$ = the fraction of quartz in the soil; $X_{S,i}$ = the fraction of mineral i in the soil; $m_{soil,\,i}$ = the amount of mineral i in the soil (kg/ha); $M_{W,i}$ = molecular weight of mineral i (kg/kmol); $n_{BC,i}$ = equivalent base cations per formula unit mineral; Δt = the considered time interval.

The weathering rates calculated from the compositional difference between the top and the bottom of the soil profile, with a total magnitude of 0.8 m, in the F3 catchment of Gårdsjön are listed in Table 6.11. The results show a total weathering rate which is considerably lower than calculated from the mass balance approach (Table 6.10). There are several problems with the historical approach. First it represents an average weathering rate over a period of about 12,000 years and weathering rates may decrease with soil age (Bain et al., 1991). Furthermore the pH of the soil solution has previously probably been considerably higher than at present, which in part could explain the low historical weathering rates.

The third and most rigorous approach is the kinetic dissolution model (Sverdrup and Warfvinge, 1988; Sverdrup, 1990). In this approach, the distribution of minerals in the soil has to be determined, their surface areas estimated as well as the composition of the exchange complex. Using kinetic rate laws determined in the laboratory, the model calculates the dissolution of different minerals considering the effects of the soil solution composition, available surface areas and the temperature (see Sverdrup and Warfvinge, 1988 for details). The total weathering rate is given by the equation:

$$R_w = \sum_{i=1}^{horizons} \theta_i \sum_{j=1}^{minerals} r_j \cdot A_{w_{ij}} \cdot 3600 \cdot 365 \cdot 24 \tag{6.34}$$

where R_w is the total weathering rate for the soil profile, and θ_i, the degree of moisture saturation. The second term describes the sum of weathering rates for different minerals,

Table 6.11. Weathering rates, R_w (keq/ha.yr) calculated from compositional differences between the top and bottom of the soil in the F3 catchment in Gårdsjön (Sverdrup and Warfvinge, 1988). R_w is the weathering rate expressed as release of Ca, Mg, Na and K.

Mineral	Bottom of profile composition R_w	%	Ions released
Microcline	0.03	16	K^+
Plagioclase	0.04	16	Na^+, Ca^{2+}
Hornblende	0.03	1.0	Mg^{2+}, Ca^{2+}
Epidote	0.04	2.0	Ca^{2+}
Biotite	0.02	1.0	Mg^{2+}, K^+
Sum	0.16		

Table 6.12. The total production of cations (keq/ha.yr)by silicate weathering as predicted by the kinetic model in the F3 watershed at Gårdsjön (Sverdrup and Warfvinge, 1988).

Minerals	Soil cation production			Average rate
	0-0.3 m	0.3-0.6 m	0.6-0.9 m	
Microcline	0.02-0.04	0.04-0.08	0-0.01	0.03-0.05
Plagioclase	0.12-0.17	0.13-0.28	0.10-0.14	0.12-0.21
Hornblende	0.20-0.34	0.15-0.20	0.06-0.08	0.17-0.26
Epidote	0.14-0.16	0.10-0.15	0.06-0.08	0.12-0.15
Sum	0.48-0.71	0.42-0.71	0.22-0.31	0.44-0.67

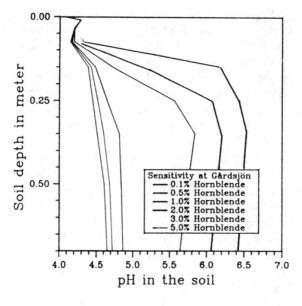

Figure 6.23. The sensitivity of the pH to the hypothetical content of hornblende at Gårdsjön as predicted by the kinetic dissolution model (reprinted by permission from Sverdrup, 1990).

with r_j, the weathering rate for mineral j, $A_{w_{ij}}$ total exposed surface areas of mineral j, while the last term recalculates the rates from seconds to years. The rate of mineral dissolution is described by the equation:

$$r = k_{H^+} \frac{[H^+]^n}{[M]^x[Al^{3+}]^y} + \frac{k_{H_2O}}{[Al^{3+}]^u} + k_{CO_2} \cdot P_{CO_2}^m + k_{org}\,[org]^{0.5} \tag{6.35}$$

In this rate equation the first term describes the effect of pH and solution composition on the rate, the second term the rate contribution due to hydrolysis and aluminum activity, the third term describes the effect of CO_2 and the last term the effect of organic acids. k_i indicates the rate coefficients for the different processes, n, x, y, u and m the apparent reaction orders to be determined experimentally and the subscript M the base cations.

The results (Table 6.12) indicate that hornblende, epidote and plagioclase are the major contributors to silicate weathering while microcline and other silicates play a minor role. This is particularly notable since hornblende and epidote constitute at most 2% of the mineral content of the soil, while the content of microcline and plagioclase range between

12-16%. The important role of even minor amounts of hornblende and epidote is obviously due to their fast dissolution kinetics. The effect of the hornblende content of the soil is illustrated in Figure 6.23 and shows that a few percent of hornblende in the soil is sufficient to prevent acidification.

The results of the kinetic model for the Gårdsjön F3 catchment can be compared with the results from mass balance calculations for the same catchment listed in Table 6.10 and the estimates for the total weathering rate obtained by the two different methods agree surprisingly well. The good correspondence between the weathering rates calculated from mass balances and the kinetic dissolution model is encouraging, but amazing considering the number of approximations and uncertainties involved in both methods.

PROBLEMS

6.1. Which of the following minerals would you expect to be present in a thoroughly weathered sediment: olivine, K-feldspar, quartz, gibbsite or smectite.

6.2. The pH and aluminum speciation (in molality) of a groundwater sample is shown below.

pH	Al^{3+}	$AlOH^{2+}$	$Al(OH)_2^+$
4.53	$6.0 \cdot 10^{-5}$	$7.7 \cdot 10^{-6}$	$6.0 \cdot 10^{-7}$

What is the pH of the water entering the reservoir when gibbsite dissolution has been the only buffering process. Neglect differences between activity and molal concentrations.

6.3. Derive the relationship between pH and $[SO_4^{2-}]$ at simultaneous equilibrium between gibbsite and jurbanite.

6.4. Listed below are the compositions of ephemeral and perennial springs in the Sierra Nevada (Garrels and Mackenzie, 1967). Perennial springs are considered to represent a longer residence time in the reservoir and the difference between perennial and ephemeral spring is considered to be due to rock weathering. Concentrations are given as mmol/l

	Perennial spring	Ephemeral spring	Rock weathering
Na^+	0.259	0.134	0.125
Ca^{2+}	0.260	0.078	0.182
Mg^{2+}	0.071	0.029	0.042
K^+	0.040	0.028	0.012
HCO_3^-	0.895	0.328	0.567
Si	0.410	0.273	0.137
SO_4^{2-}	0.025	0.010	0.015
Cl^-	0.030	0.014	0.016
pH	6.8	6.2	

The minerals presumed to be present are:

Halite	NaCl
Gypsum	$CaSO_4$
Plagioclase	$Na_{0.62}Ca_{0.38}Al_{1.38}Si_{2.62}O_8$
Biotite	$KMg_3AlSi_3O_{10}(OH)_2$
Kaolinite	$Al_2Si_2O_5(OH)_4$
Ca-smectite	$Ca_{0.17}Al_{2.33}Si_{3.67}O_{10}(OH)_2$
Calcite	$CaCO_3$

a. Perform a mass balance calculation to explain the change in composition in terms of dissolution and precipitation of the listed minerals. Hints: Use the change in Al^{3+} concentration (which is...). Assign all Cl^- and SO_4^{2-} to respectively halite and gypsum. Do not include HCO_3^- and K^+ in the mass balance.

b. Plot the water composition of the perennial spring into the stability diagram in Figure 6.11 and evaluate whether the chosen weathering products are reasonable.

c. Are the calculated mineral transfers kinetically consistent?

REFERENCES

Åberg, G., Jacks, G. and Hamilton, P.J., 1989, Weathering rates and $^{87}Sr/^{86}Sr$ ratios; an isotopic approach. *J. Hydrol.* 109, 65-78.

Appelo, C.A.J., 1985, CAC, computer aided chemistry, or the evaluation of groundwater quality with a geochemical computer model (in Dutch). *H₂O* 26, 557-562.

Appelo, C.A.J., Krajenbrink, G.J.W., van Ree, C.C.D.F. and Vasak, L., 1982, *Controls on groundwater quality in the NW Veluwe catchment* (in Dutch). Soil Protection Series 11, Staatsuitgeverij, Den Haag, 140 pp.

Bain, D.C., Mellor, A., Robertson, M.S.E. and Buckland, S.T., 1991, Variations in weathering processes and rates with time in a chronosequence of soils from Glen Feshie, Scotland. In Selinus, O., (ed.), *Proc. 2nd Int. Symp. Env. Geochemistry, Uppsala.*

Ball, J.W., Nordstrom, D.K. and Jenne, E.A., 1980, Additional and revised thermodynamical data and computer code for WATEQ2 – a computerized chemical model for trace and major element speciation and mineral equilibria of natural waters. US Geol. Surv. Water Supply Inv. 78-116.

Berner, E.K. and Berner, R.A., 1987, *The global water cycle.* Prentice-Hall, New Jersey, 397 pp.

Berner, R.A., 1971, *Principles of chemical sedimentology.* McGraw-Hill, New York, 240 pp.

Blatt, H., Middleton, G. and Murray, R., 1980, *Origin of sedimentary rocks,* 2nd ed. Prentice-Hall, New Jersey, 782 pp.

Blum, A.E. and Lasaga, A.C., 1991, The role of surface speciation in the dissolution of albite. *Geochim. Cosmochim. Acta* 55, 2193-2201.

Böttcher J., Strebel, O. and Duynisveld, H.M., 1985, Vertikale Stoffkonzentrationsprofile im Grundwasser eines Lockergesteins-Aquifers und deren Interpretation (Beispiel Fuhrberger Feld). *Z. dt. geol. Ges.* 136, 543-552.

Brady, P.V. and Walther, J.V., 1989, Controls on silicate dissolution rates in neutral and basic pH solutions at 25°C. *Geochim. Cosmochim. Acta* 53, 2823-2830.

Brantley, S.L., 1992, Kinetics of dissolution and precipitation – Experimental and field results. In Y.K. Kharaka and A.S. Maest (eds), *Water-Rock Interaction: Proc. 7th Water Rock Interaction Symp., Utah, USA.* Balkema, Rotterdam, 1, 3-6.

Busenberg, E. and Clemency, C.V., 1976, The dissolution kinetics of feldspars at 25°C and 1 atm CO_2 partial pressure. *Geochim. Cosmochim. Acta* 40, 41-49.

Casey, W.H., Westrich, H.R., Arnold, G.W. and Banfield, J.F., 1989, The surface chemistry of dissolving labradorite feldspar. *Geochim. Cosmochim. Acta* 53, 821-832.

Chou, L. and Wollast, R., 1985, Steady-state kinetics and dissolution mechanisms of albite. *Am. J. Sci.* 285, 963-993.

Dahmke, A., Matthess, G., Pekdeger, A., Schenk, D. and Schulz, H.D., 1986, Near-surface geochemical processes in Quaternary sediments. *J. Geol. Soc.* 143, 667-672.

Drabløs, D. and Tollan, A. (eds), 1980, *Ecological impact of acid precipitation. Proc. Int. Conf., Sandefjord, Norway, March 11-14, 1980.*

Drever, J.I., 1988, *The geochemistry of natural waters.* 2nd ed., Prentice Hall, New Jersey, 437 pp.

Driscoll, C.T., 1980, Aqueous speciation of aluminium in the Adirondack region of New York State, USA. *Proc. Int. Conf., Sandefjord, Norway, March 11-14, 1980.* 214-215.

Edmunds, W.M. and Kinniburgh, D.G., 1986, The susceptibility of UK groundwaters to acidic deposition. *J. Geol. Soc.* 143, 707-720.

Eriksson, E., 1981, Aluminium in groundwater: possible solution equilibria. *Nordic Hydrol.* 12, 43-50.

Förster, R.A., 1986, A multicomponent transport model. *Geoderma* 38, 261-278.

Garrels, R.M. and Mackenzie, F.T., 1967, Origin of the chemical compositions of some springs and lakes. In W. Stumm (ed.), *Equilibrium concepts in natural water systems.* Adv. in Chem. Series 67, Am. Chem. Soc., 222-242.

Garrels, R.M. and Mackenzie, F.T., 1971, *Evolution of Sedimentary Rocks.* W.W. Norton and Company, Inc., New York, 397 pp.

Goldich, S.S., 1938, A study in rock-weathering. *J. Geol.* 46, 17-58.

Hayden, P.L. and Ruben, A.J., 1974, In A.J Ruben (ed.), *Aqueous-environmental chemistry of metals.* Ann Arbor, 317-381.

Hellmann, R., Eggleston, C.M., Hochella, M.F. and Crerar, D.A., 1990, The formation of leached layers on albite surfaces during dissolution under hydrothermal conditions. *Geochim. Cosmochim. Acta* 54, 1267-1281.

Hem, J.D., 1985, Study and Interpretation of the Chemical Characteristics of Natural Water. US Geol. Surv. Water Supply Paper 2254, 3rd ed., 264 pp.

Hem, J.D. and Roberson, C.E., 1967, Form and stability of aluminum-hydroxide complexes in dilute solution. U.S. Geol. Surv. Water Supply Paper, 1827A.

Hochella, M.F. and White, A.F., 1990, *Mineral-water interface geochemistry.* Reviews in Mineralogy, 23, Mineral. Soc. Am., 603 pp.

Holdren, G.R. and Berner, R.A., 1979, Mechanism of feldspar weathering. I. Experimental studies. *Geochim. Cosmochim. Acta* 43, 1161-1171.

Hultberg, H. and Wenblad, A., 1980, Acid groundwater in southwestern Sweden. In D. Drabløs and A. Tollan (eds), *Ecological impact of acid precipitation. Proc. Int. Conf., Sandefjord, Norway, March 11-14, 1980,* 220-221.

Jacks, G., Åberg, G. and Hamilton, P.J., 1989, Calcium budgets for catchments as interpreted by strontium isotopes. *Nordic Hydrol.* 20, 85-96.

Katz, B.G., Bricker, O.P. and Kennedy, M.M., 1985, Geochemical mass-balance relationships for selected ions in precipitation and stream water, Catoctin Mountains, Maryland. *Am. J. Sci.* 285, 931-962.

Lasaga, A.C., 1984, Chemical kinetics of water-rock interactions. *J. Geophys. Res.* 89, 4009-4025.

Likens, G.E., Bormann, F.H., Pierce, R.S., Eaton, J.S. and Johnson, N.M., 1977, *Biogeochemistry of a forested ecosystem.* Springer Verlag, New York, 146 pp.

May, H.M., Helmke, P.A. and Jackson, M.L., 1979, Gibbsite solubility and thermodynamic properties of hydroxy-aluminium ions in aqueous solution at 25°C. *Geochim. Cosmochim. Acta* 43, 861-868.

May, H.M., Kinniburgh, D.G., Helmke, P.A. and Jackson, M.L., 1986, Aqueous dissolution, solubilities and thermodynamic stabilities of common aluminosilicate clay minerals: Kaolinite and smectites. *Geochim. Cosmochim. Acta* 50, 1667-1677.

Moss, P.D. and Edmunds, W.M., 1989, Interstitial water-rock interaction in the unsaturated zone of a Permo-Triassic sandstone aquifer. In D.L. Miles (ed.), *Water-Rock Interaction. Proc. 6th water-rock interaction symp., Malvern, UK.* Balkema, Rotterdam, 495-499.

Neal, C., Smith, C.J., Walls, J. and Dunn, C.S., 1986, Major, minor and trace element mobility in the acidic upland forested catchment of the upper River Severn, Mid Wales. *J. Geol. Soc.* 143, 635-648.

Nordstrom, D.K., 1982, The effect of sulfate on aluminum concentrations in natural waters: some stability relations in the system Al_2O_3-SO_3-H_2O at 298 K. *Geochim. Cosmochim. Acta* 46, 681-692.

Nordstrom, D.K. and Ball, J.W., 1986, The geochemical behavior of aluminum in acidified surface waters. *Science* 232, 54-56.

Ohse, W., Mathess, G. and Pekdeger, A., 1983, Gleichgewichts- und Ungleichgewichtsbeziehungen zwischen Porenwassern und Sedimentgesteinen im Verwitterungsbereich. *Z. dt. geol. Ges.* 134, 345-361.

Ohse, W., Matthess, G., Pekdeger, A. and Schulz, H.D., 1984, Interaction water-silicate minerals in the unsaturated zone controlled by thermodynamic disequilibria. *Int. Ass. Hydrol. Sci.* 150, 31-40.

Olsson, M. and Melkerud, P.A., 1991, Weathering rates in mafic soil mineral material. In K. Rosén (ed.), *Chemical weathering under field conditions*. Rep. Forest Ec. Forest Soil 63, Swedish Univ. Agri. Sci., 63-78.

Paces, T., 1983, Rate constants of dissolution derived from the measurements of mass balance in hydrological catchments. *Geochim. Cosmochim. Acta* 47, 1855-1863.

Parkhurst, D.L., Plummer, L.N. and Thorstenson, D.C., 1982, BALANCE – a computer program for calculating mass transfer for geochemical reactions in ground water. U.S. Geol. Surv. Water-Resources Inv., 82-14, 29 pp.

Plummer, L.N., Prestemon, E.C. and Parkhurst, D.L., 1991, An interactive code (NETHPATH) for modeling net geochemical reactions along a flow path. US Geol. Surv. Water Res. Inv., 91-4078, 100 pp.

Postma D. and Brockenhuus-Schack, B.S., 1987, Diagenesis of iron in pro-glacial sand deposits of late- and post-Weichselian age. *J. Sed. Petrol.* 57, 1040-1053.

Rogers, R.J., 1989, Geochemical comparison of groundwater in areas of New England, New York, and Pennsylvania. *Ground Water* 27, 690-712.

Rosén, K. (ed.), 1991, *Chemical weathering under field conditions*. Rep. Forest Ec. Forest Soil 63, Swedish Univ. Agri. Sci., 185 pp.

Schott, J. and Berner, R.A., 1985, Dissolution mechanisms of pyroxenes and olivines during weathering. In J.I. Drever (ed.), *The chemistry of weathering*. Reidel Publ., Dordrecht, 35-53.

Stumm, W. and Wollast, R., 1990, Coordination chemistry of weathering. *Rev. Geophysics* 28, 53-69.

Sverdrup, H.U., 1990, *The kinetics of base cation release due to chemical weathering*. Lund Univ. Press., Lund, 246 pp.

Sverdrup, H.U. and Warfvinge, P., 1988, Weathering of primary silicate minerals in the natural soil environment in relation to a chemical weathering model. *Water Air Soil Poll.* 38, 387-408.

Sverdrup, H.U. and Warfvinge, P., 1991, On the geochemistry of chemical weathering. In K. Rosén (ed.), *Chemical weathering under field conditions*. Rep. Forest Ec. Forest Soil 63, Swedish Univ. Agri. Sci., 79-119.

Tardy, Y., 1971, Characterization of the principal weathering types by the geochemistry of waters from some European and African crystalline massifs. *Chem. Geol.* 7, 253-271.

Ulrich, B., Mayer, R. and Khanna, P.K., 1979, *Die Deposition von Luftverunreinigungen und ihre Auswirkungen in Waldökosystemen im Solling*. Forstl. Fak. Univ. Göttingen u. Nieders., Forstl. Ver. 58, 291 pp.

Van Breemen, N., 1973, Dissolved aluminium in acid sulfate soils and in acid mine waters. *Soil Sci. Soc. Am. Proc.* 37, 694-697.

Velbel, M.A., 1985, Geochemical mass balances and weathering rates in forested watersheds of the southern blue ridge. *Am. J. Sci.* 285, 904-930.

Warfvinge, P., 1988, *Modeling acidification mitigation in watersheds*. Ph.D. Thesis, Lund University, 180 pp.

White, A.F. and Peterson, M., 1990, Role of reactive-surface-area characterization in geochemical kinetic models. In D.C. Melchior and R.L. Basset (eds), *Chemical Modeling of Aqueous Systems II*. Adv. in Chem. Series, 416, Am. Chem. Soc., 461-475.

Williams, L.A., Parks, G.A. and Crerar, D.A., 1985, Silica Diagenesis, I. Solubility Controls. *J. Sed. Petrol.* 55, 301-311.

Wollast, R., 1967, Kinetics of the alteration of K-feldspar in buffered solutions at low temperature. *Geochim. Cosmochim. Acta* 31, 635-648.

Wright, R.F. and Henriksen, A., 1978, Chemistry of small Norwegian lakes, with special reference to acid precipitation. *Limnol. Oceanogr.* 23, 487-498.

CHAPTER 7

Redox processes

Reduction and oxidation processes exert an important control on the distribution of species like O_2, Fe^{2+}, H_2S, CH_4, etc. under natural conditions in groundwater. They also play a major role in aquifer pollution problems such as nitrate from fertilizers, leaching from landfills, acid mine drainage and the mobility of heavy metals.

Redox reactions implicate the electron transfer from one atom to another. Electron transfer reactions are often very slow, which indicates that in this field of study, apart from equilibrium chemistry, kinetics also play a significant role. Many reactions proceed only at significant rates when mediated by bacterial catalysis. An example is the reduction of sulfate by organic matter which occurs both in aquifers and in marine sediments. Purely inorganically the reaction is unmeasurably slow. However, bacterial catalysis by *Desulfovibrio* allows this reaction to proceed rapidly in natural environments.

Redox problems in aquifers often concern the addition of an oxidant, like O_2 or nitrate, to an aquifer system containing a reductant. On the other hand, the addition of a reductant, like downward leaching of dissolved organic matter (*DOC*) from soils or landfills can also be important. In the following, we first treat some basic redox theory, and subsequently discuss a number of important redox processes in aquifers.

7.1 BASIC THEORY

As an example of a redox process, we consider the reaction between Fe^{2+} and Mn^{4+} as it would take place in acid solution:

$$2Fe^{2+} + MnO_2 + 4H^+ \leftrightarrow 2Fe^{3+} + Mn^{2+} + 2H_2O \tag{7.1}$$

In this reaction two electrons are transferred from Fe(II) to reduce Mn(IV) in MnO_2. Ferrous iron acts as a *reductant* and reduces Mn(IV), while MnO_2 can be called the *oxidant* that oxidizes Fe^{2+}. An alternative terminology is to call Fe^{2+} the *electron donor*, and Mn(IV) the *electron acceptor*. Reaction (7.1) also demonstrates that redox reactions may have a significant pH effect. Some general rules for balancing redox reactions are summarized in Table 7.1.

Redox reactions are conveniently split up into two half reactions:

$$Fe^{2+} \leftrightarrow Fe^{3+} + e^- \tag{7.2}$$

and

Table 7.1. Guide to balancing redox equations.

1.	Write for each half reaction the oxidized and reduced species into the equation and balance the elements, except H and O at left and right
2.	Balance the number of oxygen atoms by adding H_2O
3.	Balance the number of protons by adding H^+
4.	Balance electroneutrality by adding electrons
5.	Subtract the two half reactions to obtain the complete redox reaction

$$Mn^{2+} + 2H_2O \leftrightarrow MnO_2 + 4H^+ + 2e^- \tag{7.3}$$

The separation of a redox reaction into two half reactions originates from electrochemistry, where the half reactions may occur on different electrodes. Half reactions may misleadingly suggest that electrons are found in a free state in solution. This is untrue, since electrons only can be exchanged and are not found in a 'free' state.

Reactions (7.1) to (7.3) can be written in their general form as

$$bB_{red} + cC_{ox} \rightarrow dD_{ox} + gG_{red} \tag{7.4}$$

and half reactions

$$bB_{red} \rightarrow dD_{ox} + ne^- \tag{7.5}$$

$$gG_{red} \rightarrow cC_{ox} + ne^- \tag{7.6}$$

In terms of Gibbs free energy (Chapter 3) we may write for Reaction (7.4)

$$\Delta G_r = \Delta G_r^0 + RT \ln \frac{[D_{ox}]^d [G_{red}]^g}{[B_{red}]^b [C_{ox}]^c} \tag{3.22}$$

The Gibbs free energy of a reaction can be related to the voltage developed by a redox reaction in a electrochemical cell by the relation

$$\Delta G = nFE \tag{7.7}$$

Where E is a potential (emf) in Volts, F the Faraday's constant (96.42 kJ/Volt gram equivalent) and n the number of electrons transferred in the reaction. Substitution of (7.7) into (3.22) produces the *Nernst* equation:

$$E = E^0 + \frac{RT}{nF} \ln \frac{[D_{ox}]^d [G_{red}]^g}{[B_{red}]^b [C_{ox}]^c} \tag{7.8}$$

Here E^0 is the standard potential (Volt), or the potential with all substances present at unit activity at 25°C and 1 atm, similar to ΔG_r^0. As before, R is the gas constant ($8.314 \cdot 10^{-3}$ kJ/deg.mol), and T the absolute temperature.

For oxidation of H_2, we write the half reaction

$$H_2 \leftrightarrow 2H^+ + 2e^- \tag{7.9}$$

By definition reaction (7.9) has $\Delta G_r^0 = 0$ at 25°C and 1 atm, and according to (7.7) also $E^0 = 0$ Volt. Substituting (7.9) for half reaction (7.6) in (7.8) gives:

$$E = E^0 + \frac{RT}{nF} \ln \frac{[D_{ox}]^d P_{H_2}}{[B_{red}]^b [H^+]^2} \tag{7.10}$$

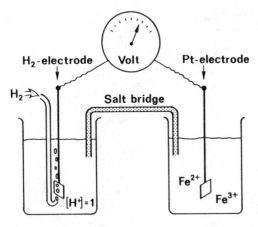

Figure 7.1. A schematic drawing of a redox cell.

The setup of such a redox cell is illustrated in Figure 7.1. At the left side the standard hydrogen electrode is shown, which consists of a Pt-electrode over which H_2 gas is bubbled in a solution of pH = 0, so that standard state conditions are fulfilled. In the compartment to the right, an inert Pt-electrode is placed in a solution containing Fe^{2+} and Fe^{3+}, corresponding to half reaction (7.2). The two electrodes are connected to a voltmeter and the electrical circuit is closed by a salt bridge. Under these conditions, both P_{H_2} and $[H^+]$ are equal to one in (7.10). When Fe^{2+} and Fe^{3+} are present in solution at unit activity, the voltmeter will registrate the E^0 of (7.2) but at other activity ratios of Fe^{3+} and Fe^{2+}, different E values are measured.

Since both P_{H_2} and $[H^+]$ in this setup always are equal to one, they are usually omitted from Equation (7.10) and indicated instead by adding the postscript h to E:

$$Eh = E^0 + \frac{RT}{nF} \ln \frac{[D_{ox}]^d}{[B_{red}]^b} \tag{7.11}$$

Note once more that both Eh and E^0 of any half reaction are expressed as the potential relative to the standard state H_2/H^+ reaction. The value of the standard potential E^0 of a half reaction indicates the tendency to release or to accept electrons and listings of standard potentials are useful to obtain a first overview over possible reactions. Standard potentials for a few reactions are listed in Table 7.2. Note that some confusion exists concerning sign conventions of redox reactions. In many texts half reactions are written as reduction reactions instead of oxidation reactions which are used here. With our convention, reducing agents have a more negative E^0 and oxidizing agents a more positive E^0.

If we apply the Nernst equation to Reaction (7.2), using E^0 from Table 7.2, we obtain:

$$Eh = E^0 + \frac{RT}{nF} \ln \frac{[Fe^{3+}]}{[Fe^{2+}]}$$

$$= 0.77 + \frac{(8.314 \cdot 10^{-3})(298.15)(2.303)}{96.42} \log \frac{[Fe^{3+}]}{[Fe^{2+}]}$$

$$= 0.77 + 0.059 \log \frac{[Fe^{3+}]}{[Fe^{2+}]} \tag{7.12}$$

Table 7.2. Standard potentials for a few reactions at 25°C, 1 atm.

Reaction	E^0, Volt
$Cr^{2+} \leftrightarrow Cr^{3+} + e^-$	-0.41
$H_2 \leftrightarrow 2H^+ + 2e^-$	0.00
$Cu^+ \leftrightarrow Cu^{2+} + e^-$	$+0.16$
$S^{2-} + 4H_2O \leftrightarrow SO_4^{2-} + 8H^+ + 8e^-$	$+0.16$
$As + 3H_2O \leftrightarrow H_3AsO_{3(aq)} + 3H^+ + 3e^-$	$+0.25$
$Cu \leftrightarrow Cu^+ + e^-$	$+0.52$
$H_3AsO_{3(aq)} + H_2O \leftrightarrow H_3AsO_{4(aq)} + 2H^+ + 2e^-$	$+0.56$
$Fe^{2+} \leftrightarrow Fe^{3+} + e^-$	$+0.77$
$Fe^{2+} + 3H_2O \leftrightarrow Fe(OH)_3 + 3H^+ + e^-$	$+0.98$
$2H_2O \leftrightarrow O_{2(g)} + 4H^+ + 4e^-$	$+1.23$
$Mn^{2+} + 2H_2O \leftrightarrow MnO_2 + 4H^+ + 2e^-$	$+1.23$

where the factor 2.303 converts natural to base ten logarithms. Similarly for Reaction (7.3)

$$Eh = E^0 + \frac{RT}{2 \cdot F} \ln \frac{[H^+]^4}{[Mn^{2+}]}$$

$$= 1.23 + \frac{(8.314 \cdot 10^{-3})(298.15)(2.303)}{2 \cdot 96.42} \log \frac{[H^+]^4}{[Mn^{2+}]}$$

$$= 1.23 + 0.03 \log \frac{[H^+]^4}{[Mn^{2+}]} \tag{7.13}$$

Note that in (7.13) neither MnO_2 nor H_2O appears, since they both by definition have unit activity. At equilibrium between the two half reactions, the *Eh* for both reactions should be the same. In other words, for a given *Eh* the distribution of all redox equilibria is fixed.

EXAMPLE 7.1: *Calculation of redox speciation with the Nernst equation*
A water sample contains $[Fe^{2+}] = 10^{-4.95}$ and $[Fe^{3+}] = 10^{-2.29}$ with pH = 3.5 at 25°C. (The activities were calculated following the procedures of Chapter 3). What would be the $[Mn^{2+}]$ if this water sample were in equilibrium with a sediment containing MnO_2?

ANSWER
First rewrite (7.12) as

$Eh = 0.77 + 0.059 (\log [Fe^{3+}] - \log [Fe^{2+}])$

Substitute iron activities

$Eh = 0.77 + 0.059 (-2.29 - (-4.95)) = 0.927$ V.

Next rewrite (7.13) as

$Eh = 1.23 + 0.03 (-4pH - \log [Mn^{2+}])$

$\log [Mn^{2+}] = (1/0.03)(1.23 - Eh - 0.12pH)$

substitute the given pH and the *Eh* calculated from the Fe^{3+}/Fe^{2+} couple

$\log [Mn^{2+}] = 33.3 (1.23 - 0.927 - 0.42) = -3.90$

$[Mn^{2+}] = 10^{-3.90}$

Thus for any lower $[Fe^{3+}]/[Fe^{2+}]$ ratio, or lower pH value, the $[Mn^{2+}]$ will increase and vice versa. The example illustrates that once the Eh is evaluated from one redox couple, the distribution of all other redox couples is fixed.

In order to obtain the E^0 for reaction (7.1), we simply subtract the E^0 of (7.13) from the E^0 of (7.12) (without multiplying the E^0 with the number of electrons transferred since E^0 already is expressed as potential per electron according to (7.7)).

$$E^0 = 0.77 - 1.23 = -0.46 \text{ Volt}$$

The negative voltage indicates that the reaction should proceed spontaneously to the right when all activities are equal to one. For redox reactions like (7.1), we return to the general form of the Nernst equation (7.8):

$$E = E^0 + \frac{RT}{2 \cdot F} \ln \frac{[Fe^{3+}]^2[Mn^{2+}]}{[Fe^{2+}]^2[H^+]^4}$$

$$= -0.46 + 0.03 \log \frac{[Fe^{3+}]^2[Mn^{2+}]}{[Fe^{2+}]^2[H^+]^4} \tag{7.14}$$

In this fashion, the Nernst equation can be used to express the distribution of any redox reaction at equilibrium.

As indicated by Equations (7.7) and (7.8), redox potentials can be related to Gibbs free energies. The practical significance of this relation is that it allows us to calculate the standard potential of any redox reaction directly from thermodynamic tables (see Section 3.3.1) since we may rewrite (7.7) for standard conditions as:

$$\Delta G_r^0 = nFE^0 \tag{7.15}$$

EXAMPLE 7.2: *Calculate E^0 from ΔG_r^0*

For reaction (7.2)

$Fe^{2+} \leftrightarrow Fe^{3+} + e^-$

$\Delta G_{f\,Fe^{2+}}^0 = -78.9 \text{ kJ/mol};$

$\Delta G_{f\,Fe^{3+}}^0 = -4.7 \text{ kJ/mol};$

$\Delta G_{f\,e^-}^0 = 0.0 \text{ kJ/mol (by definition)};$

$\Delta G_r^0 = \Delta G_{f\,Fe^{3+}}^0 + \Delta G_{f\,e^-}^0 - \Delta G_{f\,Fe^{2+}}^0 = -4.7 + 0 - (-78.9) = 74.2 \text{ kJ/mol}.$

Using (7.15)

$E^0 = \Delta G_r^0 / nF = 74.2/96.45 = 0.77 \text{ V}.$

7.1.1 *The significance of redox measurements*

Theoretically, the Eh determines the distribution of all redox equilibria in a similar way as the pH expresses the distribution of all acid-base equilibria. In contrast to pH, unfortunately Eh cannot be measured unambiguously in most natural waters.

Eh-measurements are made with an inert Pt-electrode against a standard electrode of a known potential (see Langmuir (1971) for detailed procedures). What we really want to know is the potential relative to the standard hydrogen electrode. However, since the latter is rather impractical to carry around in the field, a reference electrode of known potential is used and measured potentials are corrected accordingly

$$Eh = E_{meas} - E_{ref} \qquad (7.16)$$

For example, the potential of a calomel reference electrode ($KCl_{(sat)}$, $Hg_2Cl_{2(s)}$: $Hg_{(l)}$) is $E_{ref} = 244.4 \, mV \, (25°)$.

Although waters from oxidized environments generally yield higher *Eh* values than those from reduced environments, it has generally proved to be very difficult to obtain a meaningful quantitative interpretation in the sense of the Nernst equation. This is illustrated in Figure 7.2 were *Eh* values measured in the field are compared with those calculated with the Nernst equation from the analytic data for several half reactions. The results show disturbingly large variations between the two sets of data. For example for the important reaction which relates the O_2 content to the *Eh*,

$$2H_2O \leftrightarrow O_{2(g)} + 4H^+ + 4e^- \qquad (7.17)$$

the Pt-electrode is apparently unaffected by the O_2 concentration. The results for the other half reactions are not much better.

There are two reasons for the large discrepancies: lack of equilibrium between different redox couples in the same water sample (Lindberg and Runnells, 1984) and analytical difficulties in measuring with the Pt-electrode (Stumm and Morgan, 1981). The latter include lack of electroactivity at the Pt surface, as for O_2, mixing potentials, and poisoning

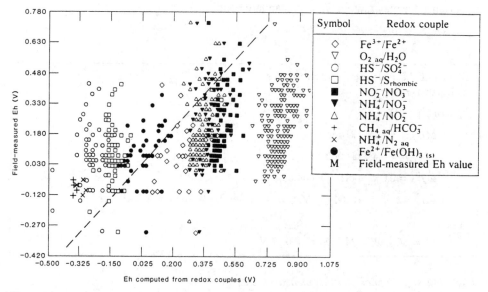

Figure 7.2. Comparison of groundwater field *Eh* measurements with potentials calculated for individual redox species (reproduced with permission, Lindberg and Runnells, 1984. Science, 225, 925-927 Copyright 1985 by the AAAS).

of the electrode. An example of poisoning is precipitation of FeOOH on the Pt surface when the electrode is immersed in an anoxic, Fe^{2+}-rich sample because of O_2 adsorbed on the electrode surface (Doyle, 1968).

A promising application of *Eh* measurements in field studies is in acid mine waters, where high concentrations of Fe^{2+} and Fe^{3+} appear to control the electrode potential (Nordstrom et al. 1979). Recently, encouraging results in the use of *Eh* measurements on the iron system at higher pH have been obtained by Macalady et al. (1990). There remains, however, little doubt concerning the conclusion that *Eh* measurements only should be interpreted quantitatively when there is a rigorous control on what is being measured.

7.1.2 *Redox reactions and the pe concept*

An alternative theoretical treatment of redox reactions simplifies the algebra considerably. In this approach the law of mass action is used on redox half reactions, for example, for Reaction (7.2):

$$K = \frac{[Fe^{3+}][e^-]}{[Fe^{2+}]} = 10^{-13.05} \tag{7.18}$$

where K is the equilibrium constant. In contrast to the Nernst equation, the electron activity appears here explicitly in the activity product. The electron activity should not be interpreted in terms of a concentration of electrons, since electrons are only exchanged, but rather as the tendency to release or accept electrons. In analogy to pH, one may define the parameter pe:

$$pe = -\log [e^-] \tag{7.19}$$

Just as for *Eh*, high positive values of pe indicate oxidizing conditions and low negative values reducing conditions. Rewriting (7.18) in logarithmic form yields the equation:

$$\log K = \log [Fe^{3+}] - pe - \log [Fe^{2+}] = -13.05 \tag{7.20}$$

In a similar fashion we may write for reaction (7.3)

$$\log K = -4pH - 2pe - \log [Mn^{2+}] = -41.52 \tag{7.21}$$

Equations (7.20) and (7.21) can be compared with the corresponding Nernst equations (7.12) and (7.13). The calculation of redox speciation, using the pe concept is demonstrated in the following Example 7.3 which is a direct analogue to Example 7.1, where the Nernst approach was used.

EXAMPLE 7.3: *Calculation of redox speciation using the pe concept*
A water sample contains $[Fe^{2+}] = 10^{-4.95}$ and $[Fe^{3+}] = 10^{-2.29}$ with pH = 3.5 at 25°C. (The activities were calculated following the procedures of Chapter 3). What would be the $[Mn^{2+}]$ if this water sample were in equilibrium with a sediment containing MnO_2?

ANSWER
First pe is calculated from (7.20) which is then substituted in (7.21). Rewriting (7.20) yields:

$$pe = -\log K + \log [Fe^{3+}] - \log [Fe^{2+}]$$

Substituting values gives:

$$pe = -(-13.05) + (-2.29) - (-4.95) = 15.71$$

Rewriting (7.21) yields:

$$\log [Mn^{2+}] = -\log K - 4pH - 2pe$$

Substituting log K, pH and the pe calculated above yields

$$\log [Mn^{2+}] = -(-41.52) - 4(3.5) - 2(15.71) = -3.90$$

Which is the same result as obtained in Example 7.1

EXAMPLE 7.4: *Calculation of K from thermodynamic data*
This example is an analogue to Example 7.2 where the Nernst equation was used. The mass action constant is related to ΔG_r^0 by Equation (3.24)

$$\Delta G_r^0 = -RT \ln K$$

At 25°C this is equal to

$$\Delta G_r^0 = -5.701 \log K$$

Following Example 7.2 the ΔG_r^0 of the Fe^{2+}/Fe^{3+} reaction is:

$$\Delta G_r^0 = \Delta G_{fFe^{3+}}^0 + \Delta G_{fe^-}^0 - \Delta G_{fFe^{2+}}^0 = -4.7 + 0 - (-78.9) = 74.2 \text{ kJ/mol}$$

Note again that $\Delta G_{fe^-}^0 = 0.0$ kJ/mol by definition. Substitution results in

$$\log K = 74.2/-5.701 = -13.02$$

Note the slight inconsistency with the K value in (7.20). Such inconsistencies are frequently found when using different sources of data.

There must of course be a simple relationship between Eh and pe, which can be found by combining (3.25), (7.7) and (7.8) written for a half reaction and which yields:

$$Eh = \frac{2.303 \, RT}{F} \, pe \tag{7.22}$$

At 25°C this is equal to

$$Eh = 0.059 \, pe \tag{7.23}$$

Both the Nernst equation and the pe concept are commonly used in the literature. The advantage of the pe concept is that the algebra of redox reactions becomes similar to other mass action expressions, allowing the same algorithm to be used in computer programs. The disadvantage of the use of pe is that it is a non-measurable quantity. For the sake of keeping calculations as simple as possible, we will in this book mainly use the pe concept.

7.2 REDOX DIAGRAMS

The number of dissolved redox and of mineral phases whose stability is influenced by redox conditions is overwhelming. Furthermore many redox reactions, as for example (7.3), are also strongly influenced by pH. In order to retain an overview in such complicated systems, redox diagrams are commonly used to express the stability of both dissolved species and minerals as a function of pe (or *Eh*) and pH. The prime force of such diagrams is that possible stable phases can be identified at a glance. Books like Garrels and Christ (1965) and Brookins (1988) that contain redox diagrams for many systems are very useful for this purpose. However, more detailed calculations are often needed to confirm critical aspects. The construction of redox diagrams, and their limitations, is discussed below. Further detailed guidelines for constructions of redox diagrams can be found in Garrels and Christ (1965), Stumm and Morgan (1981) and Drever (1988).

7.2.1 *Stability of water*

Although H_2O is not intuitively considered as a redox sensitive substance, it may take part in the following redox reactions

$$H_2O \leftrightarrow 2e^- + 2H^+ + 1/2O_{2(g)} \tag{7.24}$$

where O^{2-} in water is oxidized, and

$$1/2H_2 \leftrightarrow H^+ + e^- \tag{7.25}$$

where H^+ is being reduced. Very strong oxidants, that drive (7.24) to the right, cannot persist in natural environments because they react with H_2O. Likewise very strong reductants will reduce H_2O. Accordingly, the stability of water sets limits to the possible redox conditions to be encountered in natural environments. These can be quantified with the mass action equations of (7.24) and (7.25). For Reaction (7.24) it is:

$$\log K = 1/2 \log P_{O_2} - 2pe - 2pH = -41.55 \tag{7.26}$$

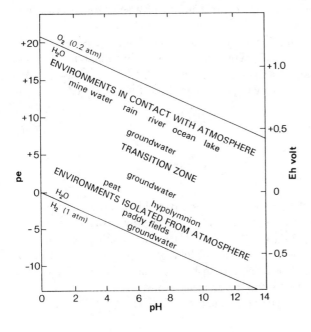

Figure 7.3. The stability of water and the ranges of pe- and pH-conditions in natural environments (modified from Garrels and Christ, 1965).

Substituting the atmospheric concentration of oxygen, $P_{O_2} = 0.2$ atm, leads to

$$pe = 20.60 - pH \tag{7.27}$$

Similarly for (7.25)

$$\log K = -pH - pe - 1/2 \log P_{H_2} \tag{7.28}$$

Here $\log K$ is zero by definition. To assign a value to P_{H_2} is more arbitrary, but the upper limit in surficial environments must be a value of one, which reduces (7.28) to

$$pe = -pH \tag{7.29}$$

Relationships (7.27) and (7.29) can be plotted in a pe/pH diagram (Figure 7.3) in order to display the range of redox conditions to be expected in natural environments.

Together with the upper and lower stability limits of water, ranges of pe/pH conditions encountered in natural environments are shown. Some caution is warranted since these ranges are based on *Eh* measurements in nature (Section 7.1.1). Groundwater environments are seen to cover a broad range from oxidizing to reducing environments. Note also that mine waters from sulfide mines may create environments with a extremely low pH, to which we later will devote a separate section.

7.2.2 *The stability of dissolved species and gases*

The great advantage of redox diagrams is that a quick overview is also obtained for less familiar systems. We illustrate this point using the arsenic system. Arsenic may substitute in pyrite (FeS_2) or be found as the separate mineral arsenopyrite (FeAsS). Locally, high arsenic concentrations in drinking water are a health hazard as it may cause black foot disease. This disease is associated with general lethargy of the patient; visible signs are a thickening and ultimate blackening of skin of face, hands and feet. The illness occurs in areas in Central Africa (Burkina Fasso) and in volcanic areas with arsenic fumaroles which are used for drinking water (a.o. in Japan).

Arsenic is found in natural waters in valences As(V) and As(III). We discuss here how to plot some of the stability fields of dissolved arsenic species in a stability diagram following Cherry et al. (1979). Both As(V) and As(III) form protolytes, which may release protons stepwise similar to carbonic acid. For As(V), the undissociated form is H_3AsO_4 and the second dissociation reaction is

$$H_2AsO_4^- \leftrightarrow HAsO_4^{2-} + H^+ \tag{7.30}$$

and the mass action equation

$$\log [HAsO_4^{2-}] - pH - \log [H_2AsO_4^-] = -6.9 \tag{7.31}$$

The boundary between the $H_2AsO_4^-$ and $HAsO_4^{2-}$ stability fields is drawn where their activities are the same. In that case (7.31) reduces to pH = 6.9; below this value $H_2AsO_4^-$ is the dominant form and at higher pH, $HAsO_4^{2-}$ predominates. The stability fields of both species are separated by a vertical line at pH 6.9 in a pe/pH diagram (Figure 7.4) since the boundary is independent of the pe. It is important to realize that in the stability field of $H_2AsO_4^-$, $HAsO_4^{2-}$ is also present, albeit at smaller concentrations.

In the same way we may write for the first dissociation of As(III):

$$H_3AsO_3 \leftrightarrow H_2AsO_3^- + H^+ \tag{7.32}$$

and

$$\log [H_2AsO_3^-] - pH - \log [H_3AsO_3] = -9.2 \tag{7.33}$$

Again a vertical line indicating equal activities of both species at pH 9.2 results in our pe/pH diagram (Figure 7.4).

There must of course also exist a boundary between the As(V) species $H_2AsO_4^-$ and the As(III) species H_3AsO_3 for which we may write the reaction:

$$H_3AsO_3 + H_2O \leftrightarrow H_2AsO_4^- + 3H^+ + 2e^- \tag{7.34}$$

and

$$\log [H_2AsO_4^-] - 3pH - 2pe - \log [H_3AsO_3] = -21.7 \tag{7.35}$$

And for equal activities of both species this reduces to

$$2pe = -3pH + 21.7$$

This boundary is dependent on both pe and pH and is also plotted in Figure 7.4.

The remaining boundaries between $HAsO_4^{2-}$ and H_3AsO_3, and between $HAsO_4^{2-}$ and H_2AsO_3 are found in exactly the same way. For the $HAsO_4^{2-}/H_3AsO_3$ boundary, the equations are:

$$H_3AsO_3 + H_2O \leftrightarrow HAsO_4^{2-} + 4H^+ + 2e^- \tag{7.36}$$

$$\log [HAsO_4^{2-}] - 4pH - 2pe - \log [H_3AsO_3] = -28.5 \tag{7.37}$$

$$2pe = -4pH + 28.5$$

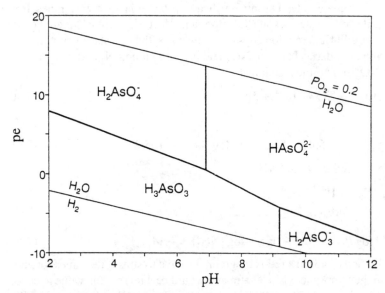

Figure 7.4. A partial pe-pH stability diagram for dissolved arsenic species. Boundaries indicate equal activities of both species.

And for the $HAsO_4^{2-}/H_2AsO_3^-$ boundary:

$$H_2AsO_3^- + H_2O \leftrightarrow HAsO_4^{2-} + 3H^+ + 2e^- \tag{7.38}$$

$$\log [HAsO_4^{2-}] - 3pH - 2pe - \log [H_2AsO_3^-] = -19.3 \tag{7.39}$$

$$2pe = -3pH + 19.3$$

Cherry et al. (1979) have suggested that separate analyses of dissolved As(V) and As(III) could serve as a useful redox indicator since this redox couple is positioned in the middle of the redox range normally found in groundwater.

Our next redox diagram concerns the nitrogen system. Nitrogen is an important component in the biogeochemical cycle, and is found in organic matter, as dissolved species and gases. In contrast, minerals containing nitrogen are generally very soluble and therefore rare in nature (but adsorption of NH_4^+ can be highly significant). The extensive use of fertilizers and their leaching from the soil has stimulated interest in the fate of nitrogen in aquifers. Section 7.5 will be devoted to this subject.

Nitrogen is found in nature in valences ranging from +5, in NO_3^-, to −3 in NH_4^+, and a reduction series can be written as:

$$NO_3^- \rightarrow NO_2^- \rightarrow N_{2(g)} \rightarrow NH_4^+$$

Intermediates between NO_2^- and $N_{2(g)}$, such as $NO_{(g)}$ and $N_2O_{(g)}$ are known to occur in aquifers, although rarely in significant amounts. Bacteria play an important role as catalysts in almost all nitrogen transformations in nature. In microbiology (Krumbein, 1983; Zehnder, 1988) the two important overall reactions are *denitrification* and *nitrification*. Denitrification concerns the reduction of nitrate to N_2, by bacteria, through a complicated pathway involving intermediates like nitrite. It should be noted that denitrification is not a reversible reaction; there are no bacteria which are able to oxidize N_2 to NO_3^-. Dissimilatory nitrate reduction to NH_4^+ is also possible in groundwater systems (Smith et al., 1991) but plays normally a subordinate role. During nitrification, bacteria oxidize amines from organic matter to nitrite and nitrate. Despite the important effect of microbial kinetics on the nitrogen system, the equilibrium relationships must always be the first starting point.

First, we define the boundaries between NO_3^- and N_2, and among N_2 and NH_4^+. For the boundary between NO_3^- and N_2 the reaction is:

$$1/2N_2 + 3H_2O \leftrightarrow NO_3^- + 6H^+ + 5e^- \tag{7.40}$$

with

$$\log K = \log [NO_3^-] - 6pH - 5pe - 1/2 \log P_{N_2} = -105.2 \tag{7.41}$$

And for the boundary between N_2 and NH_4^+:

$$NH_4^+ \leftrightarrow 1/2N_2 + 4H^+ + 3e^- \tag{7.42}$$

and

$$\log K = 1/2 \log P_{N_2} - 4pH - 3pe - \log [NH_4^+] = -14.0 \tag{7.43}$$

In order to plot (7.41) and (7.43) in a redox diagram, we have to substitute values for NO_3^-, N_2, NH_4^+. The P_{N_2} in the atmosphere is 0.77 atm and would be a reasonable value to chose. The choice of the activities of $[NO_3^-]$ and $[NH_4^+]$ is more arbitrary. An activity level of 10^{-3} for NO_3^- is commonly found in polluted groundwater and would therefore be reasonable.

Then for the sake of consistency, we may chose the same activity level for NH_4^+ even though activities in groundwater usually are lower. Substitution of these values simplifies (7.41) and (7.43) to:

$$6pH + 5pe = 102.3 \qquad (7.44)$$

and

$$4pH + 3pe = 16.94 \qquad (7.45)$$

These equations are displayed in Figure 7.5. The results show that NO_3^- is only stable near the upper stability limit of water while NH_4^+ becomes stable first near the lower stability limit of water. Above pH 9.2, $NH_{3(aq)}$ is stable relative to NH_4^+, and the pe/pH slope for the N_2/NH_3 boundary changes to -1, as indicated in the diagram.

In order to evaluate the stability of NO_2^-, which is intermediate in the redox sequence between NO_3^- and N_2, we consider the reaction:

$$NO_2^- + H_2O \leftrightarrow NO_3^- + 2H^+ + 2e^- \qquad (7.46)$$

and

$$\log K = \log [NO_3^-] - 2pH - 2pe - \log [NO_2^-] = -28.3 \qquad (7.47)$$

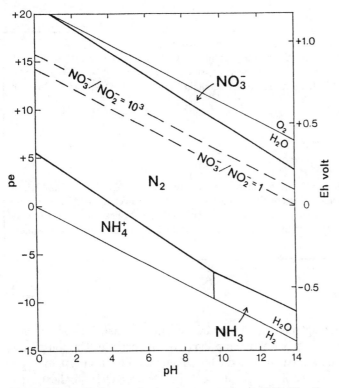

Figure 7.5. pe-pH diagram for the nitrogen system at 25°C. The diagram is valid under the conditions that $P_{N_2} = 0.77$ atm and the activities of dissolved species are 10^{-3} unless otherwise specified. Metastable boundaries are indicated by dashed lines.

Equation (7.47) can be drawn by assuming equal activities for NO_3^- and NO_2^-, in the same way as in the arsenic example. The resulting line for the NO_3^-/NO_2^- redox couple is in Figure 7.5 positioned well below the boundary of the NO_3^-/N_2 couple. Since N_2 is a more reduced form of nitrogen than NO_2^-, the conclusion must be that nitrite is an unstable species. One might argue that the assumption of equal activities of NO_3^- and NO_2^- is unreasonable, since the concentration of NO_2^- in groundwater normally is much lower than of NO_3^-.

Therefore we replot Equation (7.47) under the assumption that the activity of NO_3^- is a thousand times larger than of NO_2^-. Although the NO_3^-/NO_2^- boundary now moves closer to the NO_3^-/N_2 boundary, NO_2^- remains still an unstable species. First when the NO_3^-/NO_2^- ratio is increased to more than 10^7, would NO_2^- attain its own stability field. However, in that case the NO_2^- level becomes unmeasurably low. Therefore we may conclude that the presence of NO_2^- in groundwater is kinetically controlled as an intermediate product in the overall reduction of NO_3^- to N_2. In the same way it can be shown that intermediates like $NO_{(g)}$ and $N_2O_{(g)}$ are unstable. More generally, the example illustrates how unstable boundaries are identified during the construction of a redox diagram.

Note in Figure 7.5 that N_2-gas actually is thermodynamically unstable at the P_{O_2} of the earth's atmosphere. If all N_2 in the atmosphere were to oxidize to nitrate, the atmosphere would be become strongly depleted in oxygen and the proton production involved would acidify ocean waters to around pH 1.7 when mineral buffering reactions are disregarded. Fortunately, equilibrium thermodynamics do not apply here.

EXAMPLE 7.5: *Oxidation of the atmosphere's N_2 content to nitrate*
Thermodynamically, N_2 is unstable relative to nitrate at the P_{O_2} of the atmosphere. What would be the pH of ocean water if all the O_2 in the atmosphere were consumed by oxidation of N_2 to NO_3^- ? The O_2 content of the atmosphere is $3.7 \cdot 10^{19}$ mol. The oceans contain $13.7 \cdot 10^{23}$ g H_2O with an alkalinity of 2.3 meq/l.

ANSWER
For oxidation of N_2 to NO_3^- we may write

$$N_2 + 2.5O_2 + H_2O \rightarrow 2NO_3^- + 2H^+$$

Accordingly the $3.7 \cdot 10^{19}$ moles of O_2 present in the atmosphere could produce $3.0 \cdot 10^{19}$ mol H^+, or 21.6 mmoles H^+ per kg seawater. Subtracting the alkalinity content of seawater would leave us with 19 mmol/l H^+, corresponding to a pH of 1.7. Dissolution of carbonate and silicate minerals would in the long run probably buffer most of this acidity, but it would still leave us with an atmosphere deprived of oxygen.

7.2.3 *The stability of minerals*

The construction of redox diagrams which include minerals is analogous to those for dissolved species although the number of assumptions tends to increase. The construction of stability fields for minerals will be illustrated with a diagram for the iron system (Figure 7.6). The boundaries between dissolved species Fe^{2+}, Fe^{3+} and $Fe(OH)^{2+}$ indicate again equal activities, as in the first example. For example the boundary between Fe^{2+} and Fe^{3+} is described by Equation (7.20). Likewise, the stability of the aqueous complex $Fe(OH)^{2+}$ is described by:

$$Fe(OH)^{2+} \leftrightarrow Fe^{3+} + OH^- \tag{7.48}$$

$$K = \frac{[Fe^{3+}][OH^-]}{[Fe(OH)^{2+}]} = 10^{-11.6} \tag{7.49}$$

Which reduces in logarithmic form to:

$$\log [Fe^{3+}] - \log [Fe(OH)^{2+}] + pH = 2.4 \tag{7.50}$$

This relation is plotted for equal activities as a dashed vertical line at pH 2.4.

The stability field of amorphous iron oxyhydroxide, $Fe(OH)_3$, is under strongly oxidizing conditions determined by equilibria with dissolved ferric iron:

$$Fe(OH)_3 \leftrightarrow Fe^{3+} + 3OH^- \tag{7.51}$$

and

$$K = [Fe^{3+}][OH^-]^3 = 10^{-38.3} \tag{7.52}$$

Or in logarithmic form

$$\log [Fe^{3+}] + 3pH = 3.7 \tag{7.53}$$

In order to display (7.53) in the redox diagram, we have to assign a value to $[Fe^{3+}]$. The chosen value is arbitrary; normally a very small value (for example 10^{-6}) is selected to indicate that the concentration of the dissolved species becomes very small under natural conditions. However, the situation at hand is slightly more complex, since the aqueous complex $Fe(OH)^{2+}$ is also present in significant amounts. For total ferric iron in solution the mass balance is

$$\Sigma Fe(III) = m_{Fe^{3+}} + m_{Fe(OH)^{2+}} \tag{7.54}$$

The total concentration of dissolved ferric iron in equilibrium with $Fe(OH)_3$ can be found by substituting (7.49) and (7.52) into (7.54). To carry this out correctly, one must also know the activity coefficients of the dissolved ion. However, since the purpose of redox diagrams is to provide an overview of the system, differences between activities and molal concentrations are usually disregarded. In that case we obtain the equation:

$$\Sigma Fe(III) = \frac{10^{-38.3}}{[OH^-]^3} (1 + [OH^-] \cdot 10^{11.6}) \tag{7.55}$$

It has been solved for the values of $\Sigma Fe(III) = 10^{-3}$, 10^{-4} and $10^{-5.5}$, and the resulting boundaries are displayed in Figure 7.6. Note that the effect of the aqueous complex $Fe(OH)^{2+}$ is to increase the solubility of $Fe(OH)_3$.

The boundary between the Fe^{2+} and $Fe(OH)_3$ is described by the reaction

$$Fe^{2+} + 3H_2O \leftrightarrow Fe(OH)_3 + 3H^+ + e^- \tag{7.56}$$

where

$$\log K = -3pH - pe - \log [Fe^{2+}] = -16.8 \tag{7.57}$$

Again, in order to draw a stability line we have to chose a value for $[Fe^{2+}]$. In this case, a low value would be appropriate because the Fe^{2+} concentration usually is small in the presence

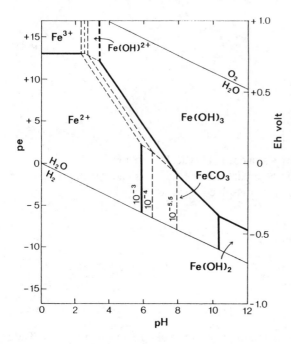

Figure 7.6. Stability relations in the system, Fe-H$_2$O-CO$_2$ at 25°C. $TIC = 10^{-2.5}$ mol/l. Solid/solution boundaries are specified for different Fe^{2+} activities. Heavy lines indicate 'realistic' boundaries that correspond to usual field conditions.

of Fe-oxides and the preferred boundary is indicated by a heavy line. Note also the very strong pH dependency of the boundary. The size of the stability field of Fe-oxides will vary significantly with the stability of the Fe-oxides (see Section 7.6). In Figure 7.6 we used the most soluble form which is amorphous Fe(OH)$_3$. When the least soluble Fe-oxide, hematite, is used in Figure 7.7, the stability field increases significantly.

The boundary between Fe^{2+} and siderite (FeCO$_3$) is described by the reaction:

$$FeCO_3 \leftrightarrow Fe^{2+} + CO_3^{2-} \tag{7.58}$$

and

$$K_{sid} = [Fe^{2+}][CO_3^{2-}] = 10^{-10.45} \qquad \text{at 25°C} \tag{7.59}$$

However, the [CO$_3^{2-}$] is related to the pH via the carbonic acid system. As discussed in detail in Chapter 4, the dissolved carbonate system can either be treated as a closed system (total dissolved carbonate is constant), or as an open system (P_{CO_2} is constant). The first case is elaborated below. The mass balance of dissolved carbonate is:

$$TIC = m_{H_2CO_3^*} + m_{HCO_3^-} + m_{CO_3^{2-}} \tag{7.60}$$

Again disregarding differences between activities and molal concentrations, K_1 and K_2 of carbonic acid (Section 4.2.1) can be substituted:

$$TIC = \frac{[H^+]^2[CO_3^{2-}]}{K_1 K_2} + \frac{[H^+][CO_3^{2-}]}{K_2} + [CO_3^{2-}] \tag{7.61}$$

Rearranging and substitution of (7.59) yields:

$$TIC = \frac{K_{sid}}{[Fe^{2+}]}\left[\frac{[H^+]^2}{K_1 K_2} + \frac{[H^+]}{K_2} + 1\right] \tag{7.62}$$

Finally filling in constants produces:

$$TIC = \frac{1}{[Fe^{2+}]}\left[[H^+]^2 \cdot 10^{6.39} + [H^+] \cdot 10^{0.09} + 10^{-10.45}\right] \qquad (7.63)$$

This equation has been solved for a *TIC* content of $10^{-2.5}$ and $[Fe^{2+}]$ of 10^{-3}, 10^{-4} and $10^{-5.5}$. The change in size of the siderite stability field with $[Fe^{2+}]$ is substantial and if the traditional boundary concentration of 10^{-6} were chosen, the siderite stability field would disappear altogether. The choice of a low equilibrium $[Fe^{2+}]$ is not reasonable in this case, since siderite in nature occurs in high $[Fe^{2+}]$ environments (Postma, 1981, 1982). There-fore, the boundaries with $[Fe^{2+}] = 10^{-3}$ and 10^{-4} are the most realistic ones.

The boundary between $FeCO_3$ and $Fe(OH)_3$ is found graphically as the intersect of the $Fe^{2+}/FeCO_3$ and $Fe^{2+}/Fe(OH)_3$ boundaries, or more formally by addition of (7.56) and (7.58), producing the reaction

$$FeCO_3 + 3H_2O \leftrightarrow Fe(OH)_3 + CO_3^{2-} + 3H^+ + e^- \qquad (7.64)$$

This reaction can again be combined with (7.61) to relate it to *TIC* and thereby to pH. The resulting boundary between $FeCO_3$ and solid $Fe(OH)_3$ is shown in Figure 7.6.

In summary, phase boundaries in redox diagrams have different meanings: between dissolved species they indicate a given activity ratio, between solid and dissolved species they indicate equilibrium for a specified solute activity, and if other components such as dissolved carbonate or sulfur species are involved, their concentrations should also be specified. Accordingly, figure captions of redox diagrams should be read carefully.

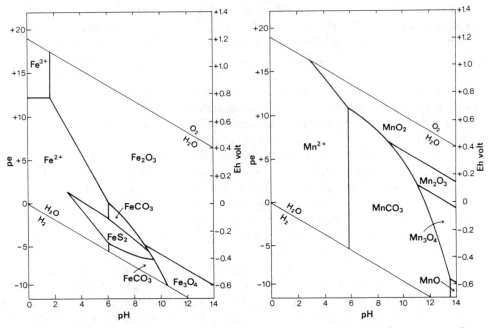

Figure 7.7. Stability relations for iron and manganese at 25°C, with $\Sigma S = 10^{-6}$ and $TIC = 10^0$ M. Solid-solution boundaries are drawn for $[Fe^{2+}] = 10^{-6}$ (modified from Krauskopf, 1979, reproduced with permission, copyright Mc Graw-Hill Inc., 1979).

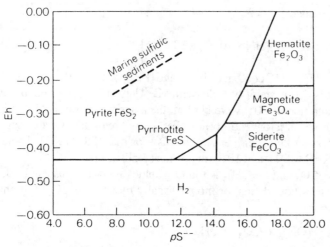

Figure 7.8. *Eh-pS^{2-} diagram where pS^{2-} = − log [S^{2-}]; pH = 7.37, P_{CO_2} = $10^{-2.4}$, 25°C (Berner, 1971, reproduced with permission, copyright Mc Graw-Hill Inc., 1971).

This point is stressed by comparison of Figure 7.6 with the often reproduced diagram for the iron system in Figure 7.7. We may note that the more stable Fe-oxide hematite has a considerably larger stability field than amorphous $Fe(OH)_3$, while magnetite, which was not considered in Figure 7.6 replaces $Fe(OH)_2$. Furthermore in Figure 7.7 also sulfur is included in the system and now a stability field for pyrite appears within that of siderite. Closer inspection of the figure caption shows that the diagram has been drawn for an unrealistically high *TIC* content of 1 mol/l, while ΣS is very low. These values were chosen in order to display both pyrite and siderite within the same diagram, since at more realistic *TIC* the siderite field would disappear altogether. Clearly, this diagram is unsuitable to evaluate the relative stability of pyrite and siderite. Pyrite is highly insoluble which means that as long as dissolved sulfide is available, it will precipitate all Fe^{2+}. Only when dissolved sulfide becomes exhausted may Fe^{2+} concentrations increase sufficiently to stabilize siderite. In order to elucidate this point, Berner (1971) modified the redox diagram of iron by replacing pH with pS^{2-} as the controlling parameter (Figure 7.8). It illustrates that siderite first becomes stable at extremely low dissolved sulfide concentrations, and also that magnetite can be a stable intermediate between siderite and hematite. Such modifications of redox diagrams may be useful for elucidating particular stability relations.

Despite these words of caution, the great value of redox diagrams is to obtain a quick overview, which is illustrated by comparing the diagrams for iron and manganese (Figure 7.7). First note that the presence of MnO_2 in sediments indicates strongly oxidizing conditions whereas hematite is stable over a much broader pe range. Also note that dissolved Mn^{2+} is stable over a wide range in contact with hematite, while inversely Fe^{2+} is unstable in contact with MnO_2. Furthermore, rhodochrosite is stable over a wide pe range, while the presence of siderite indicates strongly reducing conditions. Finally one may note that Mn-sulfide does not appear in the diagram, because it is much more soluble than Fe-sulfides, which corresponds with the extreme rarity of Mn-sulfide in recent environments. Thus a number of important conclusions can be obtained very easily from pe/pH diagrams, which otherwise would take many hours to calculate.

7.3 SEQUENCES OF REDOX REACTIONS AND REDOX ZONING

By evaluating pe/pH diagrams for elements of quantitative importance in groundwaters, sequences of reactions can be constructed which range from highly oxidized conditions to highly reducing conditions. Thus, O_2 reduces before nitrate (Fig. 7.5), which again is followed by reduction of Mn-oxides and by reduction of Fe-oxides (Fig. 7.7), etc. These sequences of reduction reactions are listed according to their redox potential at pH 7 in Figure 7.9.

Reduction reactions have to be coupled to oxidation processes which are listed in the

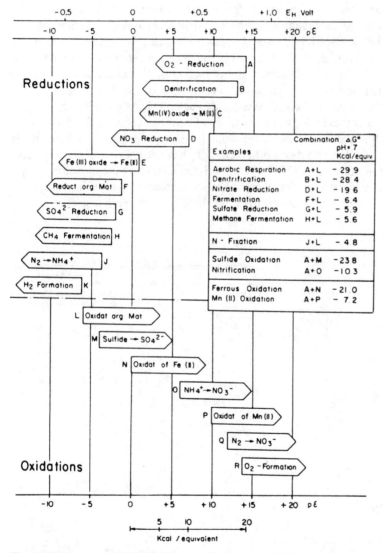

Figure 7.9. Sequences of important redox processes at pH 7 in natural systems (Stumm and Morgan, 1981, reproduced with permission of John Wiley & Sons, copyright 1981).

lower part of Figure 7.9. Examples of coupled reduction/oxidation reactions and their energy yield are displayed in the insert of Figure 7.9. Redox reactions with the standard potentials of their half reactions far apart have the greatest energy yield. It is generally observed in natural environments that redox processes proceed sequentially from the highest energy yield downward. Oxygenated water that enters an aquifer rich in organic matter will first be freed from its oxygen content and then from its nitrate content. Subsequently sulfate is reduced and finally methane may appear in the water. The reverse order is observed when organic-rich landfill leachate enters an oxic aquifer. Mn-oxides present in the sediment are the first to be reduced, followed by Fe-oxides. Then sulfate reduction takes place and the sequence ends again with methane producing conditions.

There are both equilibrium chemical and kinetic/microbial reasons why redox reactions follow sequences with a decreasing energy. If, for example, a mixture of Mn- and Fe-oxides is simultaneously reduced by organics (which is possible since both reactions yield a negative energy), the produced Fe^{2+} would reduce Mn-oxide by a fast reaction and precipitate as $Fe(OH)_3$ (Postma, 1985). The overall initial process would therefore become MnO_2 reduction, until MnO_2 is exhausted and reduction of Fe-oxides takes over. The sequence of redox reactions with a decreasing energy yield is also found for microbial mediated reactions. For example the sulfate reducing bacteria *Desulfovibrio* is an obligate anaerobe which first becomes active when anoxic conditions have been reached.

The effect of such sequences of redox reactions on water chemistry is schematically illustrated in Figure 7.10. In some cases, (O_2, NO_3^-, SO_4^{2-}), the *disappearance of a reactant* is monitored, while in other cases $(Mn^{2+}, Fe^{2+}, H_2S$ and $CH_4)$ it is the *appearance of a reaction product* that is registered. Note that the Fe^{2+} concentration approaches zero in the region dominated by sulfate reduction due to the insolubility of iron sulfides.

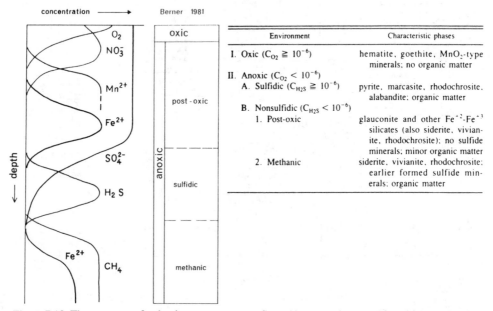

Figure 7.10. The sequence of reduction processes as reflected by groundwater composition. At the right is shown Berner's (1981a) classification of redox environments together with solids which are expected to form in each zone.

Due to problems with the interpretation of *Eh* measurements (Section 7.1.1), it has been proposed (Champ et al., 1979; Berner, 1981) to classify redox environments in terms of the presence or absence of indicative redox species (Figure 7.10). Berner (1981a) distinguishes first between *oxic* and *anoxic* environments, i.e. whether they contain measurable amounts of dissolved O_2 ($\geq 10^{-6}$ M). Then anoxic environments are subdivided into *post-oxic*, dominated by reduction of nitrate, Mn-oxide and Fe-oxide, *sulfidic*, where sulfate reduction occurs and finally the *methanic zone*. A further subdivision of the post-oxic zone into a *nitric* zone ($m_{NO_3^-} \geq 10^{-6}$ M, $m_{Fe^{2+}} \leq 10^{-6}$ M) and a *ferrous* zone ($m_{NO_3^-} \leq 10^{-6}$ M, $m_{Fe^{2+}} \geq 10^{-6}$ M) can be useful in groundwater environments. Not all zones need to be visible in a reduction sequence and in many cases the groundwater never passes the post-oxic state. In other cases, the sulfidic zone may follow the oxic zone directly, but the post-oxic zone is probably always an intermediate on a smaller scale. Figure 7.10 also shows some minerals which are expected to be found in each redox zone. Their occurrence can easily be understood by leafing through the redox diagrams on preceding pages.

An alternative way to characterize redox environments is to measure the H_2 concentration in groundwater, Lovley and Goodwin (1988), argue that H_2 is an important intermediate in most bacterial mediated redox reactions, so that steady state H_2 contents are related to the dominant redox reactions in a given environment. A compilation of the H_2 contents of a wide range of environments is presented in Figure 7.11, which shows indeed a clear relation between the dominant electron accepting reaction and the H_2 concentration.

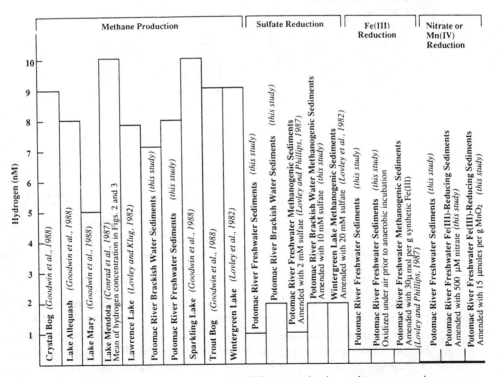

Figure 7.11. H_2 concentrations in sediments with different predominant electron accepting processes (reproduced with permission, Lovley and Goodwin, 1988).

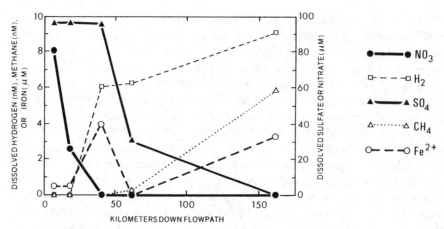

Figure 7.12. The distribution of redox species along the flow path of the Middendorf aquifer, South Carolina (modified from Lovley and Goodwin, 1988).

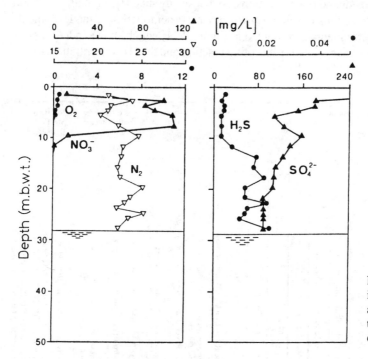

Figure 7.13. Redox zoning in a sandy Pleistocene aquifer affected by agricultural nitrate input (modified from Leuchs, 1988).

The same authors also measured H_2 contents and concentrations of other redox sensitive dissolved species along a flow path in an aquifer (Fig. 7.12). As groundwater travels through the aquifer over a considerable distance, it becomes increasingly more reduced, and the redox sensitive species change in concentration as predicted by Figure 7.10.

Redox zoning may also be apparent in a single boring, where in an homogeneous aquifer the change in concentration over depth represents different travel times through the aquifer upstream of the borehole. An example is presented in Figure 7.13. The oxic zone is here

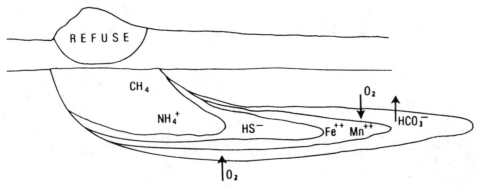

Figure 7.14. Schematic development of redox zones downstream of a landfill (modified from Baedecker and Back, 1979a).

nearly absent as the O_2 concentration in the top of the aquifer is already very low. The next 10 m of the sequence contain nitrate, indicating agricultural pollution, and belong to the post-oxic nitric zone. The zone of Fe-oxide reduction is apparently not well developed since the sulfidic zone with H_2S in the water and decreasing sulfate concentrations is found directly underneath.

A slightly different form of redox zoning often develops downstream of landfills (Figure 7.14) (Baedecker and Back, 1979a and b). Decomposition of organic rich material, domestic waste, sewage, discarded chemicals, produces an organic-rich leachate containing chemicals such as volatile fatty acids, fulvic acid-like compounds etc, which may enter an oxic aquifer. This is the reverse of the previous case, where oxic groundwater entered an anoxic aquifer. Downstream of a landfill, dissolved organic matter will first remove dissolved oxygen and nitrate upon mixing of leachate with groundwater. Large amounts of CO_2 are produced that often give rise to carbonate dissolution and to a strongly enhanced alkalinity (note that fatty acids from the leachate also may contribute considerably to titrated alkalinity). Subsequently, Mn and Fe-oxides in the sediments are reduced, releasing often large amounts of ferrous iron. Again the sequence proceeds through the sulfate reduction zone and terminates in the methane producing zone.

7.4 OXYGEN CONSUMPTION IN AQUIFERS

The atmosphere has a P_{O_2} of 0.2 atm, and using Henry's law it is easy to calculate for water in contact with the atmosphere a dissolved oxygen content of 8.26 mg/l (0.26 mmol/l) at 25°C (increasing to 12.8 mg/l at 5°C). Water that is air saturated with oxygen is continuously infiltrating through soils into aquifers and O_2 is therefore an important oxidant in aquifer systems.

Figure 7.15 shows the O_2 distribution in two sandy aquifers on quite different depth scales to demonstrate the variation in the O_2 reduction capacity of aquifer material. In the profile shown on the left, the oxygen concentration is already reduced to half that of air saturation at the groundwater table which is located at 1 m depth. O_2 decreases rapidly further downward, until total depletion at around 2 m depth. Reduction of O_2 starts already in the soil and is completed in the top of the saturated zone. It appears that when the water

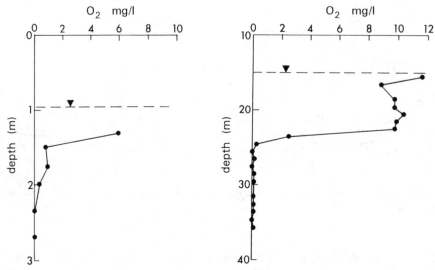

Figure 7.15. Examples of oxygen distribution in two sandy aquifers at Rodney (Canada) and Rabis Creek (Denmark). Note the different depth scales (modified from Miller et al., 1985 (left) and Postma et al., 1991 (right)).

table is close to the surface, downward transport of soil *DOC* may deplete the aquifer from oxygen (Starr, 1988; Starr and Gillham, 1989). Depletion of oxygen in the top layer of the aquifer by a similar mechanism has also been reported in sewage irrigation (Ronen et al. 1987).

In the other example (Figure 7.15, right), the unsaturated zone is 15 m thick and the upper 12 m of the saturated zone displays a nearly constant O_2 content that is close to air saturation. Evidently the upper 27 m of the sequence is unable to reduce any O_2 and below that depth oxygen is rapidly consumed, as the water enters a layer containing substances that can reduce oxygen. In the absence of reducing substances in the aquifer, oxygen saturated groundwater may travel a long way through aquifers. Winograd and Robertson (1982) reported groundwaters more than 10,000 years old, that had travelled up to 80 kilometers from their point of recharge and still were rich in oxygen. Generally, oxygen distributions are controlled by closed system conditions (similar to the case of CO_2 discussed in Chapter 4); i.e. below the water table gas exchange with the atmosphere ceases, and oxygen is gradually consumed along the flow path by reactions with reduced substances in the aquifer. However, in some cases (Rose and Long, 1988), oxygen contents are found to increase down the flow path due to addition of oxygen through the unsaturated zone.

Oxygen saturated groundwater may react with any reduced substance in the aquifer sediment, such as organic matter or Fe(II)-bearing minerals like pyrite or siderite. For the oxidation of organic matter we can write:

$$CH_2O + O_2 \rightarrow CO_2 + H_2O \qquad (7.65)$$

The extent to which this reaction will take place depends strongly on the reactivity of the available organic matter, which ranges from highly reactive peat over lignite to little reactive coal fragments. An additional important consequence of Reaction (7.65) is that CO_2 production within the aquifer may induce carbonate mineral dissolution (Thorstenson et al. 1979; Chapelle et al. 1987).

7.4.1 *Pyrite Oxidation*

The oxidation of pyrite and other sulfide minerals by oxygen, has a large environmental impact, and plays a key role in acid mine drainage, the formation of acid sulfate soils due to improved drainage by lowering the groundwater table, as a source of sulfate and iron in groundwater, as well as a source of heavy metals in the environment. It has been studied for almost a century but unclarified aspects still remain. Reviews can be found in the works of Lowson (1982) and Nordstrom (1982). The overall process is described by the reaction:

$$FeS_2 + 15/4O_2 + 7/2H_2O \rightarrow Fe(OH)_3 + 2SO_4^{2-} + 4H^+ \qquad (7.66)$$

It illustrates the strong generation of acid by pyrite oxidation. The complete oxidation process involves the oxidation of both the polysulfide S_2^{2-} and of Fe^{2+} as indicated in Reaction (7.66), but under natural conditions it often proceeds in two steps. The initial step is oxidation of the polysulfide to sulfate by O_2:

$$FeS_2 + 7/2O_2 + H_2O \rightarrow Fe^{2+} + 2SO_4^{2-} + 2H^+ \qquad (7.67)$$

Subsequently Fe^{2+} is oxidized by oxygen to Fe^{3+} which may precipitate as FeOOH, depending on pH.

$$Fe^{2+} + 1/4O_2 + H^+ \rightarrow Fe^{3+} + 1/2H_2O \qquad (7.68)$$

The energy yield of polysulfide oxidation is larger than of Fe^{2+} oxidation, and incomplete pyrite oxidation, because of an insufficient supply of electron acceptors, results in solutions rich in Fe^{2+} and SO_4^{2-}. Hydrogeologically, pyrite oxidation may occur in two settings with different results as illustrated in Figure 7.16.

In the first setting advective flow is controlling, and oxygenated groundwater enters a layer containing pyrite. Since air saturated groundwater contains about 0.33 mmol/l O_2, the maximum resulting increase in sulfate amounts to $(4/7) \cdot 0.33 = 0.19$ mmol/l SO_4^{2-} and $(2/7) \cdot 0.33 = 0.09$ mmol/l Fe^{2+} for incomplete pyrite oxidation according to Reaction (7.67) or $(8/15) \cdot 0.33 = 0.18$ mmol/l SO_4^{2-} in the case of complete pyrite oxidation according to Reaction (7.66). This situation is illustrated at the left in Figure 7.16 where the upper part of the saturated zone in a sandy aquifer has a constant O_2 content which corresponds approximately to air saturation, while the increase in SO_4^{2-} and Fe^{2+}, at the depth where O_2 disappears corresponds to that predicted by Reaction (7.67). pH remains unchanged because of sediment buffering.

In the other hydrogeological setting, pyritic swamp sediments are drained, which allows continuous O_2 diffusion from the surface into the pyritic layer via the unsaturated zone (Figure 7.16). In this case advective flow plays a minor role and the flux of oxygen is limited only by the thickness of the unsaturated zone according to Fick's first law of diffusion (see Chapter 9). The resulting sulfate and iron concentrations are huge and pH decreases to values close to 2. Dissolved sulfate and iron concentrations correspond to the stoichiometry of pyrite (Equation 7.67) although in an overall sense most iron was found to precipitate as FeOOH. The low pH values causes large increases in dissolved Al (Chapter 6) and a variety of secondary minerals may precipitate under such extreme pyrite oxidation

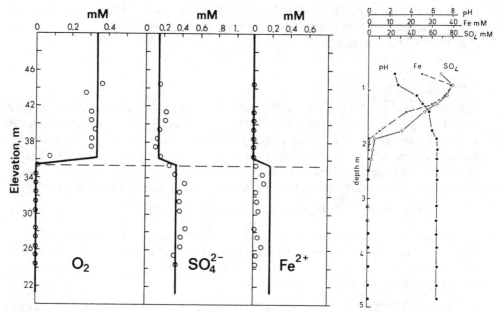

Figure 7.16. Pyrite oxidation by oxygen supplied by purely advective flow (left) and (right) by diffusive influx under stagnant water conditions. In the first case (from Postma et al., 1991) groundwater, air-saturated with O_2, is transported through a pyritic layer. To the right drained pyritic swamp sediments are oxidized by gaseous O_2 diffusion through the unsaturated zone into the pyritic layer (Postma 1983). Reproduced with permission.

conditions. These include gypsum, ferric hydroxysulfates like jarosite ($KFe_3(SO_4)_3 \cdot 9H_2O$) and several others (Van Breemen, 1976, Nordstrom, 1982; Postma, 1983). Generally these precipitates are relatively soluble and will disappear with time except for FeOOH.

In many cases intermediates between these two end-member situations of pyrite oxidation are found where both diffusive transport of O_2 and advective flow play a role. A typical setting is lowering of a groundwater table below a pyritic layer in the sediment.

A further example (Figure 7.17), shows acid drainage water that enters an aquifer. A plume of contaminants spreads through a sandy aquifer from sulfide mine tailings located at the left side of the dam. The plume contains in its center part more than 11,000 ppm sulfate, more than 5000 ppm dissolved iron, and a pH value of less than 4.8. Also aluminum, nickel and radium-226 are found at strongly enhanced concentrations (Morin and Cherry, 1986). Since the plume was found to migrate at a much slower rate than the groundwater velocity, neutralization and precipitation of sulfate and ferrous iron must occur at the margins of the plume. Morin and Cherry (1986) found equilibrium with gypsum throughout the plume and saturation to supersaturation for siderite. They proposed that calcite, which is present in small amounts in the sediments, dissolves. The resulting enhanced Ca^{2+} and dissolved carbonate concentrations apparently generate the simultaneous precipitation of gypsum and siderite, the latter possibly as solid solutions with calcite.

Sulfide mine tailings are a notorious source of heavy metal contamination for both streams and groundwater. Sulfide ores, contain apart from pyrite, a range of sulfide

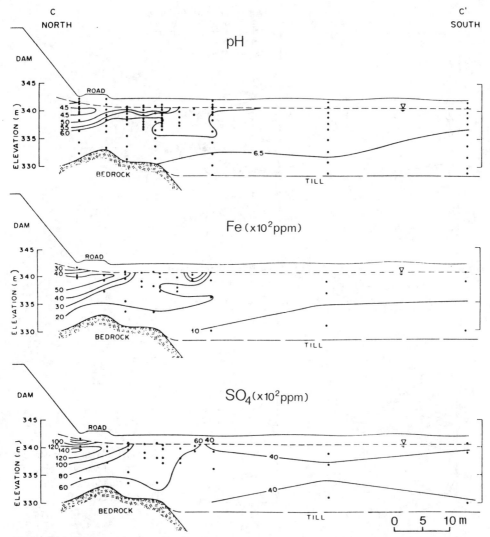

Figure 7.17. A plume of contaminants, emanating from sulfide mine tailings into a sandy aquifer. The tailings are located just north of the dam (modified from Dubrovsky et al., 1984).

minerals like sphalerite (ZnS), chalcopyrite ($CuFeS_2$) and arsenopyrite (FeAsS). These minerals provide an ample source of heavy metals for groundwater contamination as illustrated in Figure 7.18. The fate of the heavy metals appears in some cases to be controlled by the solubility of specific minerals such as anglesite ($PbSO_4$) or amorphous $Cr(OH)_3$, while for other elements (Co and Ni), adsorption and coprecipitation with Fe-oxyhydroxides seem to be controlling (Blowes and Jambor, 1990).

7.4.2 *Kinetics of pyrite oxidation*

Laboratory experiments of pyrite oxidation by O_2 shows this to be a slow process, which is at variance with the dramatic results of pyrite oxidation in the field as described above. This

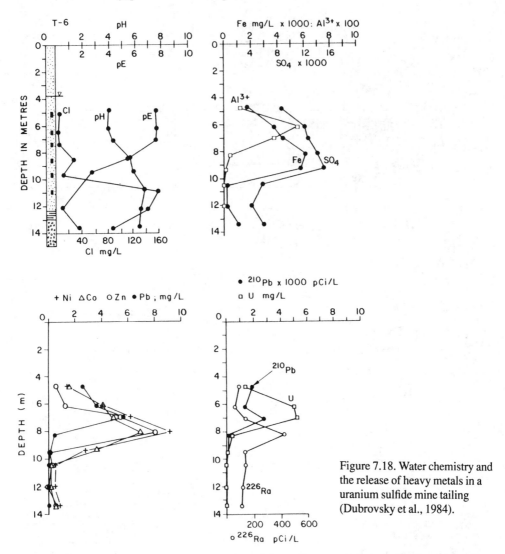

Figure 7.18. Water chemistry and the release of heavy metals in a uranium sulfide mine tailing (Dubrovsky et al., 1984).

discrepancy has induced extensive research into the kinetics of pyrite oxidation. The kinetic complexity is best illustrated by Figure 7.19, which gives an overview of possible intermediate sulfur species between sulfide (S^{2-}) and sulfate (SO_4^{2-}). Of these species, polysulfides (S_n^{2-}), sulfoxy anions like thiosulfate ($S_2O_3^{2-}$), polythionates ($S_nO_6^{2-}$), and sulfite (SO_3^{2-}) are sometimes observed in laboratory studies. In field studies, sulfate normally is the dominant oxidation product of pyrite oxidation. However, highly significant amounts of polysulfide, thiosulfate and polythionate have been reported from natural waters (Boulegue, 1977; Howarth et al., 1983) although they equally well may originate from dissolved sulfide oxidation as from pyrite oxidation.

In order to oxidize a S_2^{2-} group of pyrite to sulfate, 14 electrons have to be transferred. Since in each elementary reaction only one to two electrons can be transferred, a chain of reactions is needed among the sulfur species in Figure 7.19. The search is for the slowest or

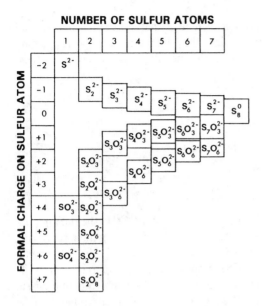

Figure 7.19. Sulfur species in solution as a function of their formal charge. The large number of intermediates between sulfide (S^{2-}) and sulfate (SO_4^{2-}) suggest a complicated oxidation mechanism (Willliamson and Rimstidt, 1990).

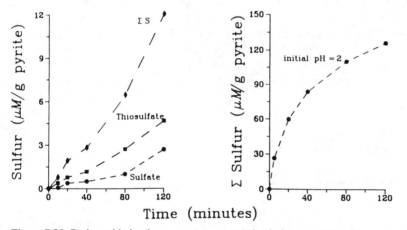

Figure 7.20. Pyrite oxidation by an oxygen saturated solution at pH 9 (left), and by ferric iron at pH 2 (right). Note the differences in scale on the y-axis (reproduced with permission from Moses et al., 1987).

rate limiting step, but both the pathway and the rate limiting step may vary for different conditions under which pyrite oxidation proceeds.

In experimental studies of pyrite oxidation by O_2 the production of sulfoxy anions is well documented at pH 6 and above. Figure 7.20 shows that thiosulfate is a major reaction product at pH 9, while sulfite, sulfate and polythionate are minor, but at pH 6, sulfate and polythionate become important products. The production of thiosulfate and polythionate during pyrite oxidation, indicates that the Fe-S bonds are broken before the S-S bonds and the release of decomposition products of sulfur from the pyrite surface must be rate controlling in the overall process. It is important to realize that sulfoxy anions are not expected to have a long life time in most aquifers since they are metastable species that are subject to a series of breakdown reactions. For example thiosulfate may become oxidized by O_2:

$$S_2O_3^{2-} + \tfrac{1}{2}O_2 \rightarrow SO_4^{2-} + S \tag{7.69}$$

or by FeOOH according to (Jørgensen, 1990):

$$S_2O_3^{2-} + 8FeOOH + 14H^+ \rightarrow 2SO_4^{2-} + 8\,Fe^{2+} + 11H_2O \tag{7.70}$$

or simply disproportionate (Jørgensen, 1990):

$$S_2O_3^{2-} + H_2O \rightarrow SO_4^{2-} + HS^- + H^+ \tag{7.71}$$

The effect of different parameters on the rate of pyrite oxidation by O_2 in the near neutral pH range, which corresponds to the conditions in most aquifers containing traces of pyrite, has been extensively studied. As expected, it was found that the rate is first order with respect to the surface area (Nicholson et al., 1988), emphasizing the importance of pyrite grain size in the sediment. The effect of pH in the near neutral range was found to be small (Moses et al., 1987), and microbial catalysis, although possible, does seem to increase the rate of pyrite oxidation substantially (Nicholson et al., 1988). However, the O_2 concentration has an important effect although its exact nature has been disputed (McKibben and Barnes, 1986; Nicholson et al., 1988). The relation between the O_2 concentration and the rate of pyrite oxidation at near neutral pH is shown in Figure 7.21. While at low O_2 concentrations a strong increase of the rate with increasing O_2 concentration is found, the effect decreases towards higher O_2 concentrations. This may indicate saturation of the number of sites on the pyrite surface which adsorb O_2. Accordingly, Nicholson et al. (1988)

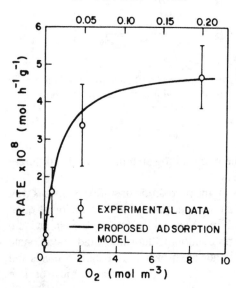

Figure 7.21. The rate of pyrite oxidation as a function of the oxygen concentration at near neutral pH (reproduced with permission from Nicholson et al., 1988).

described their data with a rate equation that is closely related to the Langmuir adsorption isotherm:

$$R = R_m KC / (1 + KC) \tag{7.72}$$

Here R is the rate of oxidation, R_m, the rate based on maximum saturation of all surface sites ($5.05 \cdot 10^{-8}$ mol/hour.g), K the adsorption constant for oxygen on pyrite (1.36 m^3/mol) and C the concentration of oxygen. Note that the right hand term, excluding R_m, is equal to a Langmuir adsorption isotherm (see Chapter 9). The rate controlling mechanism was considered to be the surface decomposition reaction.

The above experimental studies were all carried out on a time scale of hours to days. However, Nicholson et al. (1990) carried out experiments of pyrite oxidation at near neutral pH over periods of more than one year, and these are of special relevance for aquifer conditions. The results showed that reaction rates decreased significantly over time due to armoring of pyrite surfaces by FeOOH precipitates. This FeOOH layer of increasing thickness appears to act as a diffusion barrier for O_2, causing continuously decreasing pyrite oxidation rates.

The second mechanism for pyrite oxidation is by reaction with Fe^{3+} and is particularly important at low pH:

$$FeS_2 + 14Fe^{3+} + 8H_2O \rightarrow 15Fe^{2+} + 2SO_4^{2-} + 16H^+ \tag{7.73}$$

This reaction is illustrated experimentally in Figure 7.22 and the data shows weak increases in concentrations of Fe_{total} and S_{total} (mainly sulfate) over time which are due to pyrite decomposition. The difference between Fe_{total} and Fe^{2+} is equal to the ferric iron concentration which clearly is consumed and converted to ferrous iron in the course of the reaction. The rate increases with increasing ferric iron concentrations although the exact rate dependency is a matter of debate (McKibben and Barnes, 1986). Since the solubility of Fe^{3+} decreases as a third order function with increasing pH, pyrite oxidation by Fe^{3+} is

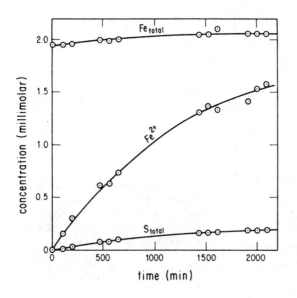

Figure 7.22. Experimental pyrite oxidation by ferric iron at pH 1.89 at 30°C (reproduced with permission from McKibben and Barnes, 1986).

particularly important at low pH, and is here responsible for the extreme effects of pyrite oxidation as shown in Figures 7.16 and 7.17.

A comparison of pyrite oxidation rates by O_2 and Fe^{3+} (Figure 7.20) shows that the latter process is about ten times faster than the first (note the differences in scale on the y-axis in Figure 7.20). However, as Figure 7.22 demonstrates, the process rapidly consumes all Fe^{3+} and pyrite oxidation would cease unless Fe^{3+} is replenished by some other process. This process is the oxidation of Fe^{2+} by oxygen

$$Fe^{2+} + 1/4O_2 + H^+ \rightarrow Fe^{3+} + 1/2H_2O \tag{7.74}$$

The kinetics of Fe^{2+} oxidation are displayed in Figure 7.23. The rate laws are given as inserts in the figure and show that at pH above about 4, the rate is first order with respect to P_{O_2} and $[Fe^{2+}]$ and second order with respect to $[OH^-]$; see Example 3.6 to appreciate the effect of pH on the oxidation rate of Fe^{2+}. Around pH 4 a change in mechanism occurs and below this value the Fe^{2+} oxidation rate becomes very slow and independent of the pH.

Above pH 4, the rate of Fe^{2+} oxidation increases rapidly with pH, but the produced Fe^{3+} is not available for pyrite oxidation due to the insolubility of FeOOH. Below pH 4.5 it appears that the rate of Fe^{2+} oxidation is considerably slower than of pyrite oxidation by Fe^{3+}. In other words, the oxidation of Fe^{2+} is apparently the rate limiting step in the overall process of pyrite oxidation (Singer and Stumm, 1970) and as a result, very little Fe^{3+} can be present at the site of pyrite oxidation. If the oxidation of Fe^{2+} at low pH is the rate limiting step for pyrite oxidation, then according to Figure 7.23, rates remain low for purely inorganic systems. However, iron oxidizing bacteria, such as *Thiobacillus ferrooxidans*, are able to increase the Fe^{2+} oxidation rate by up to five orders of magnitude which brings

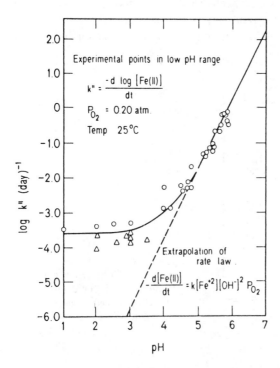

Figure 7.23. The oxidation rate of ferrous iron as a function of pH (reproduced with permission from Singer and Stumm, 1970. Science, 167, 1121-1123, Copyright 1970 by the AAAS).

the rate of Fe^{2+} oxidation within the same magnitude as that of pyrite oxidation by Fe^{3+} (Nordstrom, 1982). Thus bacterial catalysis is critical for the rate determining Fe^{3+}-regeneration step in pyrite oxidation at low pH.

7.5 NITRATE IN GROUNDWATER

Nitrate pollution of groundwater is an increasing problem in all European and North American countries (e.g. Gillham and Cherry, 1978; Obermann, 1982; Andersen and Kristiansen, 1984; Howard, 1985; Mariotti, 1986; Van Beek et al., 1988) and poses a major threat to drinking water supplies based on groundwater. The admissible nitrate concentration in drinking water (see Chapter 1) is 50 mg/l NO_3^- (corresponding to 11 mg/l NO_3-N or 0.8 mmol/l) and the recommended level is less than 25 mg/l NO_3^-. High nitrate concentrations in drinking water are believed to be a health hazard because they may cause methaemoglobinemia in human infants, a potentially fatal syndrome by which oxygen transport in the bloodstream is impaired.

Nitrate in groundwater is derived from various point and non-point sources, such as feed lots, septic tanks and oxidation of organically bound N in soils. However, the main cause for the increasing nitrate concentrations in shallow groundwater is the extensive application of fertilizers and manure in agriculture since the early sixties. The relation between land-use and nitrate pollution of aquifers is illustrated in Figure 7.24. It illustrates that very little nitrate is leached from the forest and heath areas while from arable land, plumes of nitrate containing waters spread through the aquifer. In order to evaluate the water resource quality of the future, it has to be clarified to what extent nitrate is transported as a conservative substance through the aquifer, and how chemical transformations within the aquifer may affect the nitrate concentrations.

7.5.1 *Processes of nitrate reduction*

Since nitrate does not form insoluble minerals that could precipitate, nor is adsorbed significantly under aquifer conditions, the only means for in situ nitrate removal from groundwater is by reduction. The effect of aquifer pollution by nitrate is to add besides

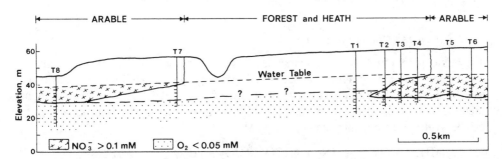

Figure 7.24. Nitrate pollution plumes emanating from agricultural fields into an unconfined sandy aquifer (Rabis Creek, Denmark). The groundwater flows from right to left. Numbers T1 through T8 refers to locations of multilevel samplers on which the plume distribution is based (reproduced with permission, Postma et al., 1991).

oxygen an additional electron acceptor to the system. Accordingly electron donors within the aquifer are required for nitrate to become reduced. Nitrate is primarily reduced to N_2 by overall reactions of the type:

$$2NO_3^- + 12H^+ + 10e^- \rightarrow N_2 + 6H_2O \tag{7.75}$$

This overall reaction comprises a transfer of five electrons per N atom and proceeds therefore through a complicated pathway with intermediates like NO_2^-, NO and N_2O. Reduction of NO_3^- to N_2 by typical components of reduced groundwater, such as Fe^{2+}, H_2S and CH_4, is thermodynamically favored although kinetic problems can be expected. However, reactions between reduced species dissolved in groundwater and nitrate are rarely quantitatively significant for nitrate reduction in aquifers, since the high nitrate concentrations introduced into the aquifer by fertilizers may far exceed the reducing capacity of dissolved species available by mixing. For substantial nitrate reduction in aquifers, reduction potential must be present within the sediments. Examples of solid phases that are common in aquifers and which thermodynamically may reduce nitrate, are organic matter, pyrite, and Fe(II)-silicates. The energy yield for reactions between nitrate and organic matter or pyrite can be evaluated from the following reactions:

$$5C + 2H_2O + 4NO_3^- \rightarrow 2N_2 + 4HCO_3^- + CO_2 \tag{7.76}$$
graphite

$$\Delta G_r^0 = -458.06 \text{ kJ/mol of } NO_3^-$$

$$2FeS_2 + 6NO_3^- + 2H_2O \rightarrow 3N_2 + 2FeOOH + 4SO_4^{2-} + 2H^+ \tag{7.77}$$
pyrite   goethite

$$\Delta G_r^0 = -415.84 \text{ kJ/mol of } NO_3^-$$

Graphite is used in (7.76) to represent the most stable form of organic matter. Other forms of organic matter would yield more energy. Note that the ΔG_r^0 values are normalized to nitrate. Equations (7.76) and (7.77) show that both organic matter and pyrite yield negative ΔG_r^0 values, and should therefore proceed to the right side. Since the energy yield of (7.76) is larger than of (7.77), nitrate reduction by organic matter should thermodynamically proceed before reduction by pyrite. However, as discussed below, the relative sequence of these two reactions is also strongly affected by reaction kinetics. In the case of Fe(II)-silicates, reaction kinetics are also expected to play a dominant role.

7.5.2 *Nitrate reduction by organic matter*

The reduction of nitrate by organic matter (*denitrification*) is well documented from soils and marine sediments. A number of studies (Obermann, 1982; Trudell et al, 1986; Morris et al., 1988; Starr, 1988; Smith and Duff, 1988) have shown that nitrate reduction by organic matter oxidation also can be important in aquifers. The process is bacterially catalyzed and can be written as an overall reaction:

$$5CH_2O + 4NO_3^- \rightarrow 2N_2 + 4HCO_3^- + CO_2 + 3H_2O \tag{7.78}$$

Intermediates of nitrate reduction, like nitrite, are sometimes found in groundwater in low concentrations and are then good indicators for ongoing nitrate reduction, but the process predominantly proceeds to the final product N_2. Bacterial nitrate reduction to ammonia is

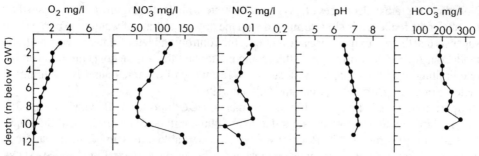

Figure 7.25. Nitrate reduction by organic matter oxidation. Average concentrations in the saturated zone at the Mussum waterworks, Germany (modified from Obermann, 1982).

also well known in microbiology (e.g. Zehnder, 1988), and was recently shown to be of some importance in aquifers (Smith et al. 1991), but the dominant product of nitrate reduction appears to be N_2.

The reduction of nitrate by oxidation of organic matter within the sediments has been described by Obermann (1982) (Figure 7.25). Note that the O_2 concentration in water at the groundwater table already has been depleted to less than half of air saturation. As predicted thermodynamically, O_2 is preferentially reduced compared to nitrate and the presence of low amounts of nitrite confirms that nitrate reduction is taking place. The oxidation of organic matter is reflected by increases of HCO_3^- and pH as predicted by Equation (7.78). It was shown that the decrease in nitrate (and O_2) between the water table and 8 m depth, is for about 60% balanced by the increase in total dissolved inorganic carbon, which strongly suggests that organic matter is the most important electron donor.

EXAMPLE 7.6: *Construct a redox balance for nitrate reduction by organic matter*
Based on the changes in concentrations between the water chemistry at the top of the saturated zone (Figure 7.25) and those at 8 m depth, the following redox balance can be constructed (modified from Obermann (1982). All concentrations are in mmol/l.

	Water table	8 m depth	Difference
NO_3^-	2.18	0.81	− 1.37
Ca^{2+}	3.37	3.24	− 0.13
HCO_3^-	3.11	4.42	+ 1.31
TIC	3.75	4.74	+ 0.99
O_2	0.11	0.03	− 0.08

If the decrease in nitrate is caused by oxidation of organic matter according to (7.78), it should be balanced by an increase in total dissolved inorganic carbon (*TIC*) of 5/4 · 1.37 = 1.71 mmol/l. The observed Δ*TIC* amounts to 0.99 mmol/l. Additional processes which could affect *TIC* are $CaCO_3$ precipitation induced by increasing pH and HCO_3^- (= 0.13 mmol/l) and also oxidation of organic matter by O_2 that produces an equivalent amount of *TIC* (= −0.08 mmol/l). Thus the corrected Δ*TIC* = 0.99 + 0.13 − 0.08 = 1.04 mmol/l. In other words 1.04/1.71 · 100% = 61% of the decrease in nitrate can be explained by organic matter oxidation while the remainder is probably due to variations in nitrate input. Sulfate concentrations in the profile are constant so that pyrite oxidation (see next section) is of no importance.

Note that the gradual decreases of oxygen and nitrate with depth (Figure 7.25) indicate that the rate of the reduction is slow compared with the downward rate of water transport which suggests that the reactivity of organic matter is the controlling factor for the extent of nitrate reduction. Although in principle both organic matter and the nitrate concentration can be rate limiting, it appears that in most cases, the reactivity of organic matter is the limiting factor (Smith and Duff, 1988; Starr and Gillham, 1989).

Apart from organic matter within the sediments, *DOC* transported downward from the soil also appears to generate nitrate reduction within aquifers. Starr (1988) and Starr and Gillham (1989) compared the Rodney site with a water table only 1 m below the surface, and the Alliston site with a deeper water table of 4 m below the surface (Figure 7.26). At the shallow site, *DOC* generated in the soil survives transport through the unsaturated zone and creates anoxic conditions in the aquifer. At the deep site most *DOC* becomes oxidized in the unsaturated zone. Since O_2 transport is much faster in the gas phase that is present in the unsaturated zone than in solution in the saturated zone, the residence time in the unsaturated zone determines the degree of aerobic degradation of *DOC*. Note that O_2 is absent already in the top of the saturated zone at Rodney, while O_2 levels close to air saturation are found down to more than 10 m at Alliston. Denitrification rates were measured in situ with the

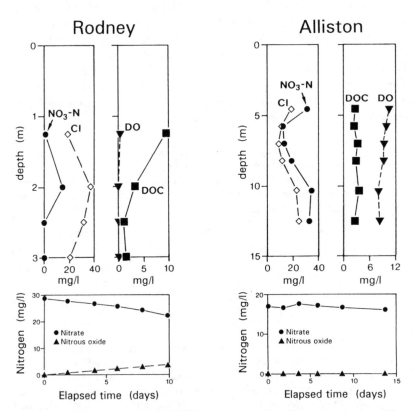

Figure 7.26. Nitrate reduction by *DOC* at Rodney (left) with a groundwater table at 1 m below the surface and at Alliston (right) with a water table at 4 m depth. The upper part shows field profiles while the lower part shows results of in situ denitrification measurements using the acetylene block technique (modified from Starr, 1988).

acetylene block technique, in which acetylene inhibits the conversion of N_2O to N_2, so that the amount of nitrate reduced can be recorded by the N_2O production. The results (Figure 7.26) showed, as expected, that denitrification was found at Rodney where higher *DOC* contents are present and was absent at Alliston were *DOC* is low even though organic carbon in the sediment is present.

In summary, it appears that nitrate reduction by organic matter is a common process in aquifers, while it is the reactivity of organic matter, which may vary from highly reactive recent material to low reactive lignite and coal, that controls the extent of nitrate reduction.

7.5.3 *Nitrate reduction by pyrite oxidation*

Nitrate reduction coupled with pyrite oxidation in aquifers has also been reported (Kölle et al., 1983; Strebel et al., 1985; Van Beek et al., 1988, 1989, Postma et al., 1991). This process involves the oxidation of both sulfur and Fe(II) and can be described by the reactions:

$$5FeS_2 + 14NO_3^- + 4H^+ \rightarrow 7N_2 + 5Fe^{2+} + 10SO_4^{2-} + 2H_2O \qquad (7.79)$$

and

$$10Fe^{2+} + 2NO_3^- + 14H_2O \rightarrow 10FeOOH + N_2 + 18H^+ \qquad (7.80)$$

As in the case of pyrite oxidation by oxygen, the energy of sulfide oxidation is larger than that of Fe(II) oxidation, so that incomplete pyrite oxidation yields Fe^{2+}-rich water. Inorganic oxidation of pyrite by nitrate seems not to be possible in anoxic systems (Postma, unpublished results), however, the sulfide in pyrite is apparently oxidized by *Thiobacillus denitrificans* (Kölle et al., 1987) while the Fe^{2+} subsequently is oxidized by nitrate with *Gallionella ferruginea* (Gouy et al., 1984). The additional presence of oxygen in field situations may furthermore generate complex reaction sequences.

Nitrate reduction by pyrite oxidation is illustrated in Figure 7.27 which shows the detailed water chemistry of the multilevel sampler T-2 in Figure 7.24 and is very similar to that described from the Fuhrberg aquifer by Böttcher et al. (1985) and Kölle et al. (1987). Water in the upper part of the sequence is derived from the forest and heath area and is free of nitrate, while water in the lower part is derived from arable land. The O_2 content is near constant and close to air saturation in the oxidized zone which indicates that no reduction of oxygen or nitrate takes place in the upper 12 m of the saturated zone, nor in the 15 m thick unsaturated zone. At the *redoxcline* oxygen and nitrate disappear abruptly and simultaneously, which indicates that the reduction process is fast compared to the downward water transport rate which amounts to about 0.75 m/yr. Oxidation of pyrite is reflected by increases in sulfate and Fe^{2+} and is in good agreement with the distribution of pyrite in the sediment. However, organic carbon is also present in the sediment and in fact in much larger amounts than pyrite. Since a slight increase in total dissolved inorganic carbon (*TIC*) is observed, the relative contributions of organic matter and pyrite oxidation have to be estimated. For this purpose, an electron balance across the redoxcline is needed which is complicated by historical variations in the groundwater content of nitrate and sulfate.

The expected reactions are listed in Table 7.3 and the relative contribution of each half reaction can be estimated by multiplying concentrations with the number of electrons transferred. The electron equivalents obtained in this way are of course only valid for the specified half reactions. The results are plotted cumulatively in Figure 7.28 where (c)

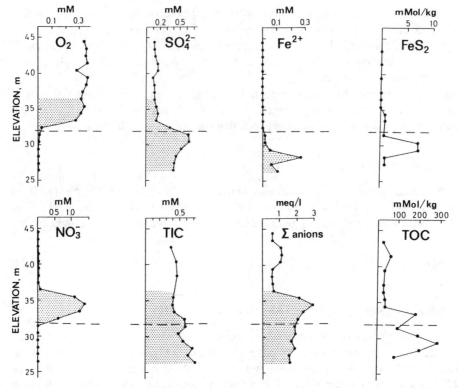

Figure 7.27. Pyrite oxidation by nitrate and oxygen in multilevel sampler T2 (for location see Figure 7.24) in the saturated zone of the Rabis Creek sandy aquifer. The dashed line indicates the depth where O_2 disappears and the shaded parts indicate nitrate contaminated water derived from arable land which is overlain by nitrate free water from a forrested area (modified from Postma et al., 1991).

Table 7.3. Electron equivalent for dissolved redox components which reflect their molar concentrations multiplied by the number of electrons transferred.

Reaction		Electron equivalents
$NO_3^- \rightarrow \frac{1}{2} N_2$	$+5e^-$	$5 \cdot m_{NO_3^-}$
$O_2 \rightarrow 2O^{2-}$	$+4e^-$	$4 \cdot m_{O_2}$
$CH_2O \rightarrow CO_2$	$-4e^-$	$4 \cdot TIC$
$S_{FeS_2} \rightarrow SO_4^{2-}$	$-7e^-$	$7 \cdot m_{SO_4^{2-}}$
$Fe_{FeS_2} \rightarrow FeOOH$	$-1e^-$	$0.5 \cdot (m_{SO_4^{2-}} - 2m_{Fe^{2+}})$

contains only water from arable land, while (a) and (b) contain water from the forested area on top of water from arable land. Figure 7.28 shows clearly that oxidation of pyrite-sulfur is the main electron donor for nitrate reduction, since both contributions of oxidation of organic matter and of pyrite-Fe(II) oxidation are minor. Of course this procedure is only permitted when solid carbonates do not dissolve or precipitate as is the case here. Note also that nitrate contamination in this aquifer increases the load of electron acceptor by a factor of five compared to input of only O_2. One may wonder in this case why organic matter which is present in much larger amounts than pyrite is not oxidized. The explanation is probably that the organic matter in these sediments consists of reworked Miocene lignite fragments, which have a very low reactivity so that the thermodynamically predicted sequence of processes is reversed. The same was found in the Fuhrberg aquifer (Kölle et al., 1987) where the sediments also contain both lignite and pyrite and again pyrite oxidation was found to be the dominant process.

A final example of nitrate reduction by pyrite oxidation is shown in Figure 7.29. Again, nitrate-free water derived from non-arable land is found on top of nitrate contaminated water from arable land. The increases in sulfate and Fe^{2+} below the depth where nitrate disappears indicates pyrite oxidation. An unexpected side effect of pyrite oxidation by

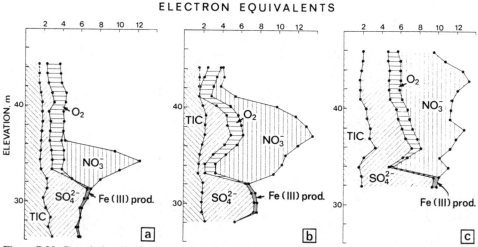

Figure 7.28. Cumulative distribution of electron equivalents in Rabbis Creek multisamplers T2 (a), T3 (b) and T5 (c) (see for location Figure 7.24). Electron equivalants are defined in Table 7.3 (reproduced with permission, Postma et al., 1991).

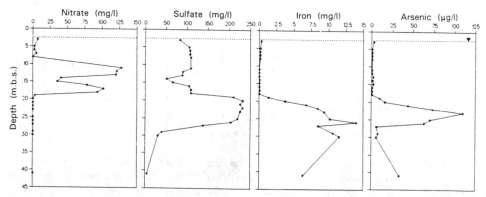

Figure 7.29. Nitrate reduction by pyrite oxidation in a sandy aquifer. Pyrite oxidation results in the release of arsenic from pyrite (modified from Van Beek et al., 1989).

nitrate are the high arsenic concentrations, well above the WHO limit of 0.05 mg/l, that are probably released from pyrite. It can be argued that arsenic would also be released during pyrite oxidation by oxygen in the absence of nitrate contamination. However, nitrate pollution at the level found here increases the amount of electron acceptor per liter by about a factor of five, which in principle increases the arsenic concentration by the same factor.

7.5.4 *Nitrate reduction by Fe(II)*

The third documented process of nitrate reduction is by reaction with Fe(II) according to the overall reaction

$$10Fe^{2+} + 2NO_3^- + 14H_2O \rightarrow 10FeOOH + N_2 + 18H^+ \tag{7.81}$$

Since dissolved Fe^{2+} concentrations in groundwater rarely exceed 0.1 mmol/l and nitrate

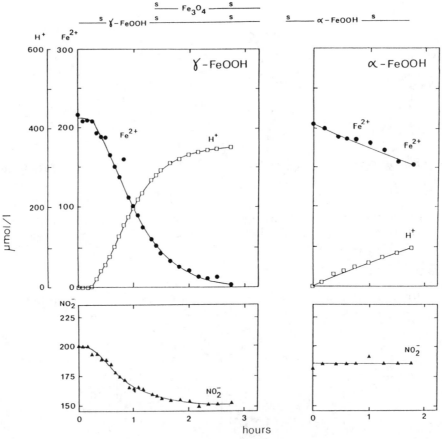

Figure 7.30. Reduction of nitrite by Fe^{2+} in the presence of FeOOH at pH 8.0, 25°C under anoxic conditions. Uppermost, identification of Fe-oxides by XRD is indicated. Note that the transformation of lepidocrocite (γ-FeOOH) to magnetite/greenrust (Fe_3O_4) induces nitrate reduction, while goethite (α-FeOOH) is unable to catalyze nitrite reduction (reproduced with permission, Sørensen and Thorling, 1991).

loads often amount to 1 mmol/l or more, it is clear that a source of Fe^{2+} in the sediment is required if this process should have some significance.

Such a source of Fe^{2+} could be pyrite oxidation as described in Section 7.4.1, dissolution of detrital Fe(II)-bearing minerals like magnetite, pyroxenes and amphiboles (Postma and Brockenhuus-Schack, 1987), or Fe(II) bearing clay minerals like nontronite and smectites. Reaction (7.81) is inorganically very slow but appears to be bacterially mediated (Gouy et al., 1984). Other reported catalytic effects are the presence of Cu^{2+} (Buresh and Moraghan, 1976) and the surface of freshly precipitated Fe-oxyhydroxide. Sørensen and Thorling (1991) mixed nitrite and Fe^{2+}solutions with suspensions of different forms of FeOOH and found that nitrite reduction was associated with the transformation of lepidocrocite to green rusts/magnetite which indicates that highly complex reaction kinetics take place in relation to the electron transfer (Figure 7.30). In this study, nitrite was used because nitrate was found to react much more slowly.

Postma (1990) studied the overall reaction between amphiboles and pyroxenes and nitrate in solution and found that nitrate reduction could take place at low rates ($\sim 4 \cdot 10^{-17}$ NO_3^- mol/cm^2.sec (25°C), but only under conditions where secondary reaction products, including FeOOH, precipitated on the detrital mineral surfaces in good agreement with the results of Sørensen and Thorling (1991). A rough estimate of the potential of nitrate reduction by Fe(II) silicates in a sandy aquifer, containing 1% Fe(II)-silicates yielded values in the order of 10^{-5} mol/l.yr NO_3^-, and thus requires long residence times to be significant. Field evidence of nitrate reduction by Fe(II) was presented by Vanek (1990).

7.6 IRON IN GROUNDWATER

Iron is a common constituent of anoxic groundwater. Redox diagrams (Figures 7.6 and 7.7) indicate that in the normal pH range of groundwater (pH = 5-8), dissolved iron is present as Fe^{2+}, since Fe^{3+} under these conditions is insoluble. During drinking water production, anoxic groundwater containing Fe^{2+} becomes aerated and since the oxidation of Fe^{2+} at near neutral pH is very fast (Section 3.5.2), Fe-oxyhydroxides are precipitated. Although iron in drinking water is non-poisonous, Fe-oxyhydroxides may clog distribution systems, stain clothing, ceramics, etc., and iron is therefore usually removed during water treatment. In simple hand pumping wells, slimy deposits from Fe^{2+} oxidizing bacteria can be a problem.

7.6.1 *Sources of iron in groundwater*

The main sources for Fe^{2+} in groundwater are the dissolution of Fe(II)-bearing minerals and the reduction of Fe-oxyhydroxides present in the sediments. Important Fe(II)-bearing minerals commonly present in aquifer materials comprise minerals like magnetite, ilmenite, pyrite, siderite, and Fe(II)-bearing silicates, like amphiboles, pyroxenes, olivine, biotite, glauconite and clay minerals such as smectites. Oxidative dissolution of pyrite as a source for Fe^{2+} in groundwater is an important process and has already been discussed extensively in Section 7.4.1. Also dissolution of siderite ($FeCO_3$) may yield high dissolved Fe^{2+} concentrations as discussed in more detail in the next section.

The dissolution of detrital Fe(II)-bearing silicates like amphiboles and pyroxenes is illustrated for the pyroxene bronzite in Figure 7.31. Bronzite dissolution rates (expressed as the $Si(OH)_4$ release) are observed to be considerably faster under anoxic than under oxic

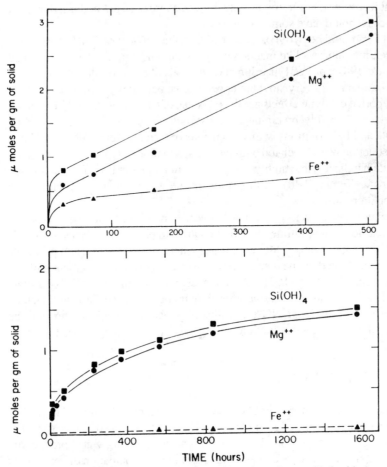

Figure 7.31. Dissolution kinetics of the pyroxene bronzite at pH 6 with (upper) anoxic ($P_{O_2} = 0$) conditions and (lower) oxic ($P_{O_2} = 0.2$) conditions. Note the differences in concentration and time scales (reproduced with permission from Schott and Berner, 1983).

conditions. Under anoxic conditions the release of dissolved ions becomes linear after the initial stage which indicates constant rates, while under oxic conditions the rate is constantly decreasing over time. Comparison of the release of Fe^{2+} in both cases, shows that while under anoxic conditions the Fe^{2+} concentration in solution increases, under oxic conditions little Fe^{2+} is released. In this case the mineral surface becomes covered by an inner layer of Fe(III)-silicate and an outer layer of FeOOH which inhibits the dissolution process by diffusion control (Schott and Berner, 1983; White, 1990). Mechanistic studies by White and Yee (1985) indicate that aqueous Fe^{3+} becomes reduced at the mineral surface to Fe^{2+}, thereby transforming the Fe(II) silicate matrix to a Fe(III) silicate matrix. In general, dissolution of Fe(II)-bearing silicates is much faster under anoxic conditions than under oxic conditions and in the latter case rates will decrease with exposure time. Magnetite, being a Fe(II,III)-oxide displays a similar behavior (Hochella and White, 1990).

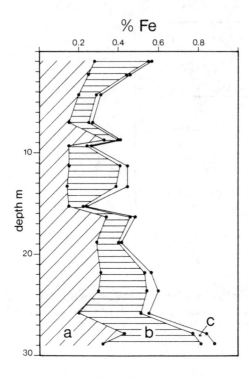

% Fe

Figure 7.32. Cumulative iron distribution (wt %) in a sandy aquifer based on separation by physical methods: a) light fraction with quartz and feldspar, b) weak magnetic fraction containing pyroxenes and amphiboles, c) strong magnetic fraction with magnetite and ilmenite (reproduced with permission from Postma and Brockenhuus-Schack, 1987).

Postma and Brockenhuus Schack (1987) found that amphiboles and pyroxenes in a sandy aquifer showed distinct dissolution features which indicate that such minerals may be an important source of Fe^{2+} in groundwater. However, the reported concentrations of Fe^{2+} were rather low (10-13 µmol/l), which emphasizes the importance of dissolution kinetics. The distribution of iron in the solid phase was analyzed (Figure 7.32) and suggested that an important pool of iron in aquifers is present in the amphibole-pyroxene fraction, while in this case the Fe(II, III) oxides magnetite and ilmenite are much less important. Note also that throughout the sediment, a considerable amount of iron is present as FeOOH coating on the surface of quartz and feldspar grains.

Reductive dissolution of FeOOH becomes important when an electron acceptor, such as dissolved organic matter, H_2S or methane enters a sediment containing FeOOH. *DOC* may originate from soils, organic rich sediment layers within the sequence, or from landfills. The pool of FeOOH in the sediment covers a broad range of minerals with both variable stability and variable kinetic properties. Common FeOOH minerals found in aquifers are ferrihydrite, goethite, lepidocrocite and hematite. In general the most unstable FeOOH minerals are also kinetically the most reactive. Identification in aquifer materials is difficult, and mostly empirical extraction methods, such as ammonium oxalate (Schwertmann, 1964), for poorly crystalline material, and dithionite-citrate-bicarbonate (DCB) (Mehra and Jackson, 1960), for total Fe-oxide contents are used. Postma et al. (1991) found that both fractions were abundant throughout a sandy aquifer in the anoxic as well as the post-oxic zone.

The mechanism of reductive dissolution of Fe-oxides has been investigated extensively by Banwart et al. (1989) and Hering and Stumm (1990). Results are summarized in Figure

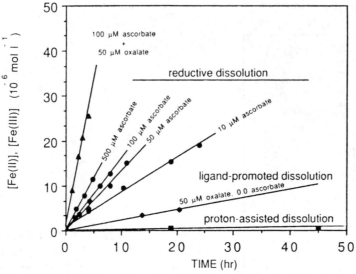

Figure 7.33. Dissolution of hematite at pH 3, by proton assisted dissolution, ligand (oxalate) promoted dissolution, reductant (ascorbate) promoted dissolution and combined ligand and reductant promoted dissolution (reproduced with permission from Banwart et al., 1989).

7.33. They show that hematite dissolution in the presence of protons only is very slow. However, organic ligands, here exemplified by oxalate, may accelerate the dissolution process considerably. A reductant like ascorbate facilitates dissolution by electron exchange with surface Fe(III) through an inner sphere complex. The reduction rate is found to increase with the ascorbate concentration in a fashion that can be described by a Langmuir isotherm (Banwart et al., 1989). The fastest dissolution is observed for the combination of both a ligand and a reductant. While these results describe a simplified laboratory setting, a range of natural organic compounds, such as phenols, tannic acid and cysteine are known to be able to reduce Fe-oxides. Furthermore, in aquifer systems bacterial catalysis is expected to play an important role. According to Lovley (1987), a variety of microorganisms is able to pair reduction of FeOOH to oxidation of organic matter and Lovley et al. (1990) identified acetate oxidizing, Fe(III) reducing bacteria in Cretaceous sediments at up to 200 m depth.

7.6.2 Solubility controls of iron in groundwater

The most obvious control on Fe^{2+} concentrations in groundwater is by oxidation to Fe^{3+} and precipitation as Fe-oxyhydroxides. This process occurs particularly in zones of groundwater discharge, near rivers etc., where Fe^{2+} rich groundwater comes in contact with atmospheric oxygen. As a result, bands of Fe-oxyhydroxides (and Mn-oxides) can form around the water table (Piispanen, 1985). Also within aquifers Fe-oxyhydroxides may control the solubility of dissolved Fe^{2+}, and the produced Fe-oxyhydroxides may apparently be transported along with the groundwater (Langmuir and Whittemore, 1971; Whittemore and Langmuir, 1975). The solubility of Fe-oxyhydroxides can be compared using the activity product:

$$IAP_{FeOOH} = [Fe^{3+}][OH^-]^3 \tag{7.82}$$

Equation (7.82) is valid for both Fe-oxides with the stoichiometry of hematite (α-Fe$_2$O$_3$) and of goethite (α-FeOOH). For goethite we may write:

$$FeOOH + H_2O \leftrightarrow Fe^{3+} + 3OH^- \tag{7.83}$$

and for hematite:

$$1/2Fe_2O_3 + 3/2H_2O \leftrightarrow Fe^{3+} + 3OH^- \tag{7.84}$$

As seen in Figure 7.34, Fe-oxyhydroxides cover a wide range of stability from freshly precipitated amorphous Fe(OH)$_3$ to the most stable phase hematite (for a review see Murray (1979)). With time, recrystallization of unstable phases towards more stable phases is to be expected, which essentially increases the stability field of Fe-oxyhydroxides in the redox diagram (Figures 7.6 and 7.7) and decreases the solubility of Fe^{2+}. In a field study, Langmuir and Whittemore (1971) and Whittemore and Langmuir (1975) found $-\log(IAP_{FeOOH})$ values ranging from 36 to 43 and they concluded that the groundwater was in equilibrium with Fe-oxyhydroxides. Since Fe^{2+} concentrations decreased concurrently with increasing $-\log(IAP_{FeOOH})$ values, they proposed aging of Fe-oxyhydroxides to be the controlling factor on dissolved iron concentrations.

At higher Fe^{2+} concentrations in groundwater, siderite (FeCO$_3$, $\log K_{sid} = -10.45$ at 25°C) may become an alternative solubility control. Saturation or slight supersaturation of groundwaters for siderite has been reported by several authors (Whittemore and Langmuir, 1975; Nesbitt, 1980; Margaritz and Luzier, 1985; Morin and Cherry, 1986). Siderite is a common constituent of sedimentary rocks, either in finely dispersed form or as concretions.

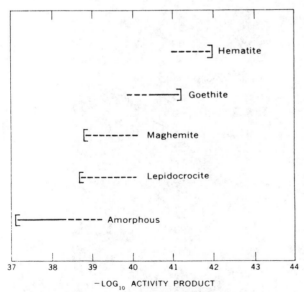

Figure 7.34. The stability ranges of common Fe-oxyhydroxides, hematite (α-Fe$_2$O$_3$), goethite (α-FeOOH), maghemite (γ-Fe$_2$O$_3$), lepidocrocite (γ-FeOOH) and amorphous Fe(OH)$_3$ which is equivalent to ferrihydrite (5Fe$_2$O$_3 \cdot$ 9H$_2$O). The activity products are expressed as $IAP = [Fe^{3+}][OH^-]^3$ and are given for 25°C and 1 atm (Langmuir, 1969).

Furthermore siderite commonly substitutes into calcite and dolomite. Direct evidence of siderite precipitation from aquifers is scarce, but Leuchs (1988) reported authigenic overgrowth of siderite on lignite in an anoxic aquifer and supersaturation of the groundwater for siderite. Siderite is commonly found in fresh water swamps (Postma, 1977, 1982) and lakes (Anthony, 1977), and is here often accompanied by the ferrous phosphate mineral vivianite ($Fe_3(PO_4)_2 \cdot 8H_2O$, log $K_{viv} = -36$). As was shown in Figure 7.8, siderite is only stable compared to pyrite at extremely low dissolved sulfide concentrations. In practice, this implies anoxic environments where sulfate is not available for sulfate reduction or where sulfate has become exhausted. Postma (1982) described a location with both fresh and brackish swamp sediments where brackish pore waters contained abundant sulfate and H_2S, and pyrite was forming. Under fresh water conditions the pore waters contained neither sulfate nor dissolved sulfide but were rich in Fe^{2+} and siderite formed. On a much larger scale, the same relation between environment of deposition and authigenic iron mineralogy was found in carboniferous sediments by Curtis (1967). SEM micrographs of some possible authigenic iron minerals are shown in Figure 7.35.

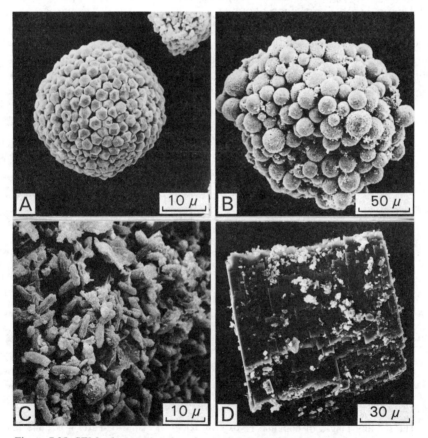

Figure 7.35. SEM micrographs of some common authigenic Fe(II) minerals. (A) Framboidal pyrite which are aggregates of pyrite crystals, (B) Polyframboidal pyrite, (C) Siderite and (D) A crystal fragment of vivianite (reproduced with permission from Postma, 1982).

7.7 SULFATE REDUCTION AND IRON SULFIDE FORMATION

Sulfate reduction by organic matter is catalyzed by bacteria of the genus *Desulfovibrio* (Jørgensen, 1982) according to the overall reaction:

$$2CH_2O + SO_4^{2-} \rightarrow 2HCO_3^- + H_2S \tag{7.85}$$

Here, CH_2O is used as a simplified representation of organic matter. The H_2S produced will for the major part react with Fe-oxides in the sediment and form iron sulfide minerals (see Section 7.7.1). H_2S gives a foul smell of rotting eggs and is also highly toxic, and for both reasons undesirable in drinking water. H_2S may also reoxidize to sulfate through a complex reaction sequence (see for review Morse et al., 1987). The importance of sulfate reduction in aquifer systems depends apart from the availability of organic matter, also on the supply of sulfate. Fresh waters are generally low in sulfate. However important local sources of sulfate can be dissolution of gypsum, oxidation of pyrite, mixing of fresh water with seawater, and also acid rain and fertilizers. Enhanced circulation around production wells apparently may stimulate sulfate reduction, releasing sulfate reducing bacteria to well waters and leading to clogging with iron sulfides (Van Beek and Van der Kooij, 1982).

Ongoing sulfate reduction can first of all be recognized by the presence of H_2S in groundwater (Figure 7.13). However if excess Fe-oxide is available in the sediment it may precipitate all H_2S, so that H_2S is not necessarily present. In aquifers influenced by mixing of fresh water and seawater, also the Cl/SO_4 ratio may indicate sulfate reduction. Seawater has a fixed Cl/SO_4 molar ratio of 19 so that significant higher values, in water with elevated Cl-concentrations, indicates sulfate reduction. Stable sulfur isotopes are also suitable to reveal ongoing sulfate reduction, since ^{32}S is preferentially consumed compared to ^{34}S during sulfate reduction. The ratio between the two stable isotopes is expressed as:

$$\delta^{34}S(\text{‰}) = \frac{^{34}S/^{32}S_{sample} - \,^{34}S/^{32}S_{standard}}{^{34}S/^{32}S_{standard}} \times 1000 \tag{7.86}$$

Since atmospheric $\delta^{34}S$ values are less than $+ 10$ ‰, the increases of $\delta^{34}S$ values with depth in association with decreasing sulfate concentrations (Figure 7.36) indicate sulfate reduction. Note that sulfate reduction only takes place in the anoxic zone (the low O_2

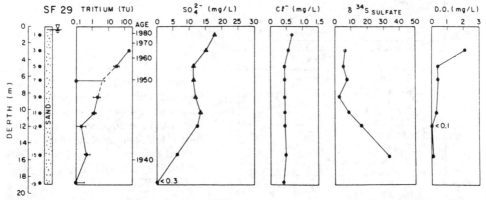

Figure 7.36. Sulfate reduction as indicated by $\delta^{34}S$ values in a sandy aquifer (Robertson et al., 1989).

contents below 5 m depth are possibly due to contamination). Finally, sulfate reduction in sediments can be measured directly using radiotracer methods (Jørgensen, 1978).

Sulfate reduction has extensively been studied in marine sediments where it plays an important role in the degradation of organic matter. Measured rates of sulfate reduction vary over orders of magnitude and the dominant control on the rates appears to be the reactivity of organic matter which can be described by the equation (Berner, 1980):

$$dG_m/dt = -kG_m \qquad (7.87)$$

Here G_m is the amount of metabolizable organic matter and k a first order decay constant. Since the reactivity of organic matter constantly decreases, G_m has to be subdivided in pools with decreasing reactivity, also called the multi-G-model (Westrich and Berner, 1984). In an alternative interpretation, Middelburg (1988) considered the G_m pool as a constant, in other words the total organic carbon content, while fitting k to an exponential function. For a time range of exposure of up to 10^6 years he found that k varied over 8 orders of magnitude. This very wide range emphasizes the variability of organic matter reactivity. A problem with the Middelburg model is that k becomes a function of time, rather then of organic matter reactivity. Recently Boudreau and Ruddick (1991) modeled organic matter reactivity using a gamma distribution function which considers an infinite number of pools of organic matter, each with a unique first order rate constant. Presently, there is little information concerning the reactivity of organic matter in aquifers, but a similar range in reactivity as in recent marine sediments can be expected.

The second control on the rate of sulfate reduction may be the sulfate concentration. Again studies of marine sediments have shown that the rate of sulfate reduction is zero order with respect to the sulfate concentration (Berner, 1981b) (Figure 3.10). In other

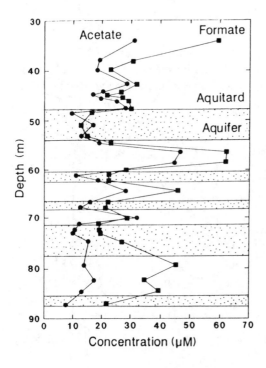

Figure 7.37. Concentrations of dissolved formate and acetate in pore waters of aquitard and aquifer sediments, indicating diffusion of dissolved organics into aquifer layers and consumption by sulfate reduction (McMahon and Chapelle, 1991, reprinted by permission from Nature, 349, 1509-1520, copyright Macmillan Magazins Ltd, 1979).

words, organic matter reactivity is limiting bacterial activity rather than sulfate availability. Only at sulfate concentrations below about 2 mmol/l, the rate of sulfate reduction appears to become first order with respect to the sulfate concentration (Bågander, 1977; Boudreau and Westrich, 1984).

Sulfate reduction may hydrogeologically occur in two settings. In the first case, organic matter dispersed through the aquifer material may induce sulfate reduction (Figure 7.36). In the second case, fine grained layers containing organic matter may release dissolved organics which subsequently diffuse into more permeable layers where it comes in contact with groundwater containing sulfate (Figure 7.37). Apparently, fermentation occurs in low permeable clayey aquitard layers that produce formate and acetate which then diffuse into sandy aquifer layers and are consumed by sulfate reduction.

7.7.1 *The formation of iron sulfides*

Pyrite (cubic FeS_2) and its polymorph marcasite (orthorhombic FeS_2) are the two most abundant iron sulfide minerals in ancient sedimentary rocks. Pyrite ($\Delta G_f^0 = -160.2$ kJ/mol) is more stable than marcasite ($\Delta G_f^0 = -158.4$ kJ/mol) and pyrite is also the form that is normally encountered in recent environments. In ancient deposits pyrite is particularly abundant in fine grained sediments, but it may also be present in sand-sized deposits. Pyrite occurs as disseminated single crystals, as framboids that are globular aggregates of crystals with a size of up to several hundreds microns (Love and Amschutz, 1966) (Figure 7.35), or as concretions. Marcasite is less stable, its formation is poorly understood, and it is often found as concretions in carbonate rocks. A recent review of the formation of iron sulfide minerals can be found in Morse et al. (1987).

Pyrite formation in recent marine sediments has been studied extensively. The overall pathway of formation is displayed in Figure 7.38, illustrating that in most cases it takes place as a two step process:

$$2FeOOH + 3HS^- \rightarrow 2FeS + S° + H_2O + 3OH^- \tag{7.88}$$

and

$$FeS + S° \rightarrow FeS_2 \tag{7.89}$$

In the first step, dissolved sulfide reacts with Fe-oxide, where part of the sulfide reduces Fe(III) and produces S°, while the remainder of dissolved sulfide precipitates as FeS. FeS is also called acid volatile sulfide (AVS) since it in contrast to pyrite it readily dissolves in HCl, and this property that forms the base of the analytical assay. FeS stains the sediment black and consists of amorphous FeS and extremely fine grained minerals like mackinawite ($Fe_{1-x}S$) and greigite (Fe_3S_4). FeS is less stable than pyrite but forms kinetically very fast and its formation is due to the sluggish precipitation kinetics of pyrite. The presence of FeS in aquifer sediments has been documented by Leuchs (1988) who also found the groundwater to be supersaturated with respect to FeS (log $K_{FeS} = -17.7$). In marine sediments, FeS is found to disappear with time (depth) as it transforms slowly to pyrite. The transformation of FeS to FeS_2 is actually an oxidation and can be written as an overall process as in Equation (7.89). In some cases, where dissolved sulfide concentrations remain very low, fast direct formation of pyrite seems also possible (Howarth, 1979; Howarth and Jørgensen, 1984).

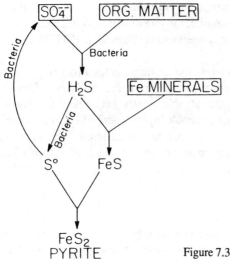

Figure 7.38. Pathways of pyrite formation (Berner, 1981b).

7.8 METHANE IN GROUNDWATER

The final stage in the reductive sequence, is methane formation. Methane is a common constituent of anoxic groundwater (Barker and Fritz, 1981; Leuchs, 1988; Grossman et al., 1989). It may degas from groundwater and accumulate in wells and buildings and so impose an explosion hazard. The origin of methane may either be *biogenic*, produced by microbial activity, or *thermocatalytic*. The latter takes place non-biologically at several kilometers depth in the earth and is often related to oil formation. The methane may migrate upwards into aquifers through fracture zones. Distinction between the two types of methane is usually done by stable carbon isotopes since thermocatalytic methane has $\delta^{13}C$ values greater than about −45 ‰, while biogenic methane has $\delta^{13}C$ values of less than −60 ‰ (Barker and Fritz, 1981). Biogenic methane from landfill sites is reported to have intermediate $\delta^{13}C$ values. Biogenic methane forms through a series of complex biogenic reactions (Vogels et al., 1989). In general, the predominant processes of methane formation are reduction of CO_2

$$CO_2 + 4H_2 \rightarrow 2H_2O + CH_4 \tag{7.90}$$

and fermentation of, for example, acetate

$$CH_3COOH \rightarrow CH_4 + CO_2 \tag{7.91}$$

As predicted from the theoretical redox sequence, methane containing groundwaters are usually low in sulfate. Although the exact mechanism has not been clarified, it appears that oxidation of methane also can be used as an energy source for sulfate reduction. This has been demonstrated particularly elegantly by Iversen and Jørgensen (1985) who used radiotracer methods to measure both sulfate reduction rates and methane oxidation rates in marine sediments. The interface between the methane producing and sulfate reducing zones was found to be sharp and showed coinciding maxima of the rates of methane oxidation and sulfate reduction.

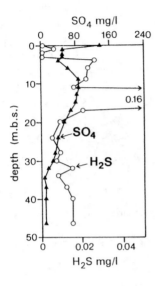
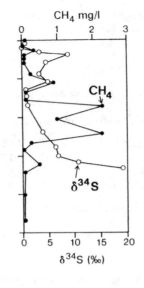

Figure 7.39. Sulfate reduction and methane formation in a sandy aquifer rich in organic remains (modified from Leuchs, 1988).

A groundwater profile through a sandy aquifer containing lignite fragments and other plant remains is presented in Figure 7.39 and shows both sulfate reduction and methane formation. The largest concentrations are found in the middle part of the profile, but smaller concentrations are also present in the upper part. Apparently, methane is here formed in micro environments and it diffuses afterwards in the bulk groundwater. This final example demonstrates that redox zoning can be found on a micro as well as a macro scale.

PROBLEMS

7.1a. List, using Figure 7.7, Fe and Mn minerals which are stable together. Do the same for combinations of dissolved species and minerals.

    b. Calculate and plot in Figure 7.6 the redox boundary between Fe^{2+} and hematite.

    c. Drainwater from the Haunstrup browncoal mine, in Denmark, showed the following variations over time:

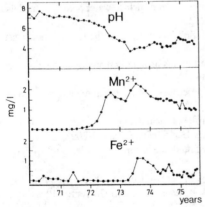

Find two explanations, in terms of pe/pH variations why Mn^{2+} is released earlier than Fe^{2+}.

d. An analysis of groundwater provided the following concentrations:

pH	NO_3^-	NO_2^-	
7.16	110	2.1	(mg/l)

Calculate the amount of Mn^{2+} in water, when the aquifer contains MnO_2.

7.2. Construct a pe-pH diagram for Pb^{2+} with the following data:

    1. $PbO_2 + 4H^+ + 2e^- \leftrightarrow 2H_2O + Pb^{2+}$ $\log K_1 = 49.2$
    2. $Pb^{2+} + 2e^- \leftrightarrow Pb$ $\log K_2 = -4.26$
    3. $PbO + 2H^+ \leftrightarrow Pb^{2+} + H_2O$ $\log K_3 = 12.7$
    4. $Pb^{2+} + 3H_2O \leftrightarrow Pb(OH)_3^- + 3H^+$ $\log K_4 = -28.1$

Use activity levels for Pb^{2+} and $Pb(OH)_3^-$ of 10^{-6} at the solid/solution boundaries.

7.3a. Calculate the amount of $MnCO_3$ (rhodochrosite) that can dissolve in water, if $P_{CO_2} = 10^{-1.5}$ atm

    $MnCO_3 \leftrightarrow Mn^{2+} + CO_3^{2-}$ $K = 10^{-9.3}$

b. Explain the terms 'congruent/incongruent' dissolution of a mineral.
c. Which type of dissolution is expected for rhodochrosite in a well-aerated soil: congruent dissolution or incongruent dissolution? Which mineral(s) will form? What is the effect on the P_{CO_2}?

7.4. Write a balanced equation for oxidation of organic material (CH_2O) by $KMnO_4$ (i.e. analysis of *COD*). Express the amount of $KMnO_4$ also as O_2.

7.5. Draw a pe/pH-diagram for sulfur-species. Consider $S_{(s)}$, $H_2S_{(aq)}$, HS^-, SO_4^{2-}. Activity of a species at a mineral boundary is 10^{-2}

    1. $H_2S_{(aq)} \leftrightarrow HS^- + H^+$ $K = 10^{-7}$
    2. $H_2S_{(aq)} + 4H_2O \leftrightarrow SO_4^{2-} + 10H^+ + 8e^-$ $K = 10^{-40.6}$
    3. $S_{(s)} + 4H_2O \leftrightarrow SO_4^{2-} + 8H^+ + 6e^-$ $K = 10^{-36.2}$

If S precipitates from spring water, what can be concluded about pH and H_2S-content?

REFERENCES

Andersen L.J. and Kristiansen, H., 1984, Nitrate in groundwater and surface water related to land use in the Karup basin, Denmark. *Environ. Geology* 5, 207-212.

Anthony, R.S., 1977, Iron-rich rhythmically laminated sediments in Lake of the Clouds, northeastern Minnesota. *Limnol. Oceanogr.* 22, 1, 45-54.

Baedecker, M.J. and Back, W., 1979a, Modern marine sediments as a natural analog to the chemically stressed environment of a landfill. *J. Hydrol.* 43, 393-414.

Baedecker, M.J. and Back, W., 1979b, Hydrogeological processes and chemical reactions at a landfill. *Ground Water* 17, 429-437.

Banwart, S., Davies., S. and Stumm, W., 1989, The role of oxalate in accelerating the reductive dissolution of hematite (α-Fe_2O_3) by ascorbate. *Colloids and Surfaces* 39, 303-309.

Barker, J.F. and Fritz, P., 1981, The occurrence and origin of methane in some groundwater flow systems. *Canadian J. Earth Sci.* 18, 1802-1816.

Berner, R.A., 1971, *Principles of chemical sedimentology*. McGraw-Hill, New York, 240 pp.

Berner, R.A., 1980, *Early Diagenesis – A Theoretical Approach*. Princeton University Press, Princeton, N.J., 241 pp.

Berner, R.A., 1981a, A new geochemical classification of sedimentary environments. *J. Sed. Petrol.* 51, 2, 359-365.

Berner, R.A., 1981b, Authigenic mineral formation resulting from organic matter decomposition in modern sediments. *Fortschr. Miner.* 59, 117-135.

Blowes, D.W. and Jambor, J.L., 1990, The pore-water chemistry and the mineralogy of the vadose zone of sulfidic tailings, Waite Amulet, Quebec. *Applied Geochem.* 5, 327-346.

Böttcher J., Strebel, O. and Duynisveld, H.M., 1985, Vertikale Stoffkonzentrationsprofile im Grundwasser eines Lockergesteins-Aquifers und deren Interpretation (Beispiel Fuhrberger Feld). *Z. dt. geol. Ges.* 136, 543-552.

Boudreau, B.P. and Ruddick, B.R., 1991, On a reactive continuum representation of organic matter diagenesis. *Am. J. Sci.* 291, 507-538.

Boudreau, B.P. and Westrich, J.T., 1984, The dependence of bacterical sulfate reduction on sulfate concentration in marine sediments. *Geochim. Cosmochim. Acta* 48, 2503-2516.

Boulegue, J., 1977, Equilibria in a sulfide rich water from Enghien-les-Bains, France. *Geochim. Cosmochim. Acta* 41, 1751-1758.

Brookins, D.G., 1988, *Eh-pH diagrams for geochemistry*. Springer Verlag, Berlin, 176 pp.

Buresh, R.J. and Moraghan, J.T., 1976, Chemical reduction of nitrate by ferrous iron. *J. Environ. Qual.* 5, 320-325.

Bågander, L.E., 1977, In situ studies of bacterial sulfate reduction at the sediment-water interface. In L.E. Bågander (ed.), *Sulfur fluxes at the sediment-water interface – an in situ study of closed systems, Eh and pH*. Microbial Geochem., II.

Champ, D.R., Gulens, J. and Jackson, R.E., 1979, Oxidation-reduction sequences in ground water flow systems. *Can. J. Earth Sci.* 16, 12-23.

Chapelle, F.H., Zelibor, J.L. Jr., Grimes, D.J. and Knobel, L.L., 1987, Bacteria in deep coastal plain sediments of Maryland: A possible source of CO_2 to groundwater. *Water Resour. Res.* 23, 1625-1632.

Cherry, J.A., Shaikh, A.U., Tallman, D.E. and Nicholson, R.V., 1979, Arsenic species as an indicator of redox conditions in groundwater. *J. Hydrol.* 43, 373-392.

Curtis, C.D., 1967, Diagenetic iron minerals in some British Carboniferous sediments. *Geochim. Cosmochim. Acta* 31, 2109-2123.

Doyle, R.W., 1968, The origin of the ferrous ion-ferric oxide Nernst potential in environments containing ferrous iron. *Am. J. Sci.* 266, 840-859.

Drever, J.I., 1988, *The geochemistry of natural waters*. 2nd ed. Prentice Hall, New Jersey, 437 pp.

Dubrovsky, N.M., Morin, K.A., Cherry, J.A. and Smyth, D.J.A., 1984, Uranium tailings acidification and subsurface contaminant migration in a sand aquifer. *Water Poll. Res. J. Canada* 19, 2, 55-89.

Garrels, R.M. and Christ, C.L., 1965, *Solutions, Minerals, and Equilibria*. Harper and Row, New York, 450 pp.

Gillham, R.W. and Cherry, J.A., 1978, Field evidence of denitrification in shallow groundwater flow systems. *Water Poll. Res. in Canada* 13, 53-71.

Gouy J., Bergé, P. and Labroue, L., 1984, *Gallionella ferruginea*, facteur de dénitrification dans les eaux pauvre en matière organique. *C. R. Acad. Sc. Paris* 298, 153-156.

Grossman, E.L., Coffman, B.K., Fritz, S.J. and Wada, H., 1989, Bacterial production of methane and its influence on ground-water chemistry in east-central Texas aquifer. *Geology* 17, 495-499.

Hering, J.G. and Stumm, W., 1990, Oxidation and reductive dissolution of minerals. In M.F. Hochella and A.F. White (eds), *Mineral water interface geochemistry*. Rev. in Mineral., Mineral. Soc. Am., 23, 427-465.

Hochella, M.F. and White, A.F. (eds), 1990, *Mineral water interface geochemistry*. Rev. in Mineral., Mineral., Soc. Am., 23, 603 pp.

Howard, K.W.F., 1985, Denitrification in a major limestone aquifer. *J. Hydrol.* 76, 265-280.

Howarth, R.W., 1979, Pyrite: Its rapid formation in a salt marsh and its importance in ecosystem metabolism. *Science* 203, 49-51.

Howarth, R.W., Giblin, A., Gale, J., Peterson, B.J. and Luther, G.W. III, 1983, Reduced sulfur compounds in the pore waters of a New England salt marsh. In R. Hallberg (ed.), *Environmental Biogeochemistry. Ecol. Bull.* 35, 135-152.

Howarth, R.W. and Jørgensen, B.B., 1984, Formation of ^{35}S-labelled elemental sulfur and pyrite in coastal marine sediments (Limfjorden and Kysing Fjord, Denmark) during short-term $^{35}SO_4^{2-}$ reduction measurements. *Geochim. Cosmochim. Acta* 48, 1807-1818.

Iversen, N. and Jørgensen, B.B., 1985, Anaerobic methane oxidation rates at the sulfate-methane transition in marine sediments from Kattegat and Skagerrak (Denmark). *Limnol. Oceanogr.* 30, 944-955.

Jørgensen, B.B., 1978, A comparison of methods for the quantification of bacterial sulfate reduction in coastal marine sediments I. Measurement with radiotracer techniques. *Geomicrobiol. J.* 1, 11-27.

Jørgensen, B.B., 1982, Ecology of the bacteria of the sulphur cycle with special reference to anoxic-oxic interface environments. *Phil. Trans. R. Soc. Lond.* B 298, 543-561.

Jørgensen, B.B., 1990, A thiosulfate shunt in the sulfur cycle of marine sediments. *Science* 249, 152-154.

Kölle W., Strebel, O. and Böttcher, J., 1987, Reduced sulfur compounds in sandy aquifers and their interactions with groundwater. *Int. Symp. Groundwater Monitoring, Dresden*, 12 pp.

Kölle W., Werner, P., Strebel, O. and Böttcher, J., 1983, Denitrifikation in einem reduzierenden Grundwasserleiter. *Vom Wasser* 61, 125-147.

Krauskopf, K.B., 1979, *Introduction to Geochemistry 2nd ed.* McGraw-Hill, New York, 617 pp.

Krumbein, W.E. (ed.), 1983, *Microbial Geochemistry*. Blackwell Scientific Publications, Oxford, 330 pp.

Langmuir, D., 1969, The Gibbs free energies of substances in the system $Fe-O_2-H_2O-CO_2$ at 25 °C. U.S. Geol. Surv. Prof. Paper, 650-B, B180-B184.

Langmuir, D., 1971, Eh-pH determination. In R.E Carver (ed.), *Procedures in Sedimentary Petrology*. Wiley & Sons, New York, 597-634.

Langmuir, D. and Whittemore, D.O., 1971, Variations in the stability of precipitated ferric oxyhydroxides. In J.D. Hem (ed.), *Non-equilibrium systems in natural water chemistry*. Adv. in Chem. Ser., 106, Am. Chem. Soc., 209-234.

Leuchs, W., 1988, *Vorkommen, Abfolge und Auswirkungen anoxischer redoxreaktionen in einem pleistozanen Porengrundwasserleiter*. Bes. Mitt. dt. Gewass. Jahrb. 52, Dusseldorf, 106 pp.

Lindberg, R.D. and Runnells, D.D., 1984, Ground water redox reactions: An analysis of equilibrium state applied to Eh measurements and geochemical modeling. *Science* 225, 925-927.

Love, L.G. and Amstutz, G.C., 1966, Review of microscopic pyrite from the Denovian Chattanooga Shale and Rammelsberg Banderz. *Fortschr. Miner.* 43, 273-309.

Lovley, D.R., 1987, Organic matter mineralization with the reduction of ferric iron: A review. *Geomicrobiol. J.* 5, 375-399.

Lovley, D.R., Chapelle, D.R. and Phillips, E.J.P., 1990, Fe(III)-reducing bacteria in deeply buried sediments of the atlantic coastal plain. *Geology* 18, 954-957.

Lovley, D.R. and Goodwin, S., 1988, Hydrogen concentrations as an indicator of the predominant terminal electron-accepting reactions in aquatic sediments. *Geochim. Cosmochim. Acta* 52, 2993-3003.

Lowson, R.T., 1982, Aqueous oxidation of pyrite by molecular oxygen. *Chem. Rev.* 82, 461-497.

Macalady, D.L., Langmuir, D., Grundl, T. and Elzerman, A., 1990, Use of model-generated Fe^{3+} ion activities to compute Eh and ferric oxyhydroxide solubilities in anaerobic systems. In D.C. Melchior and R.L. Basset (eds), *Chemical modeling of aqueous systems II*. ACS Symp. Ser. Am. Chem. Soc. 416, 350-367.

Magaritz, M. and Luzier, J.E., 1985, Water-rock interactions and seawater-freshwater mixing effects in the coastal dunes aquifer, Coos Bay, Oregon. *Geochim. Cosmochim. Acta* 49, 2515-2525.

Mariotti, A., 1986, La dénitrification dans les eaux souterraines, Principes et méthodes de son identification: une revue. *J. Hydrol.* 88, 1-23.

McKibben, M.A. and Barnes, H.L., 1986, Oxidation of pyrite in low temperature acidic solutions: Rate laws and surface textures. *Geochim. Cosmochim. Acta* 50, 1509-1520.

McMahon, P.B. and Chapelle, F.H., 1991, Microbial production of organic acids in aquitard sediments and it's role in aquifer geochemistry. *Nature* 349, 233-235.

Mehra, O.P. and Jackson, M.L., 1960, Iron oxide removal from soils and clays by a dithionite-citrate system buffered with sodium bicarbonate, *Clays Clay Minerals* 5, 317-327.

Middelburg, J.J., 1989, A simple rate model for organic matter decomposition in marine sediments. *Geochim. Cosmochim. Acta* 53, 1577-1581.

Miller, D.J., Gillham, R.W. and Starr, R.S., 1985, An in-situ method for determining rates of denitrification in groundwater. Institute for groundwater research. *Proc. Techn. Transfer Conf. No. 6, Toronto, Ontario, Dec. 11-12*, Part 3, 87-106.

Morin, K.A. and Cherry, J.A., 1986, Trace amounts of siderite near a uranium-tailings impoundment, Elliot Lake, Ontario, Canada, and its implication in controlling contaminant migration in a sand aquifer. *Chem. Geol.* 56, 117-134

Morse, J.W., Millero, F.J., Cornwell, J.C. and Rickard, D., 1987, The chemistry of the hydrogen sulfide and iron sulfide systems in natural waters. *Earth Sci. Rev.* 24, 1-42.

Morris, J.T., Whiting, G.J. and Chapelle, F.H., 1988, Potential denitrification rates in deep sediments from the southeastern coastal plain. *Environ. Sci. Technol.* 22, 832-836.

Moses, C.O., Nordstrom, D.K., Herman, J.S. and Mills, A.L., 1987, Aqueous pyrite oxidation by dissolved oxygen and ferric iron. *Geochim. Cosmochim. Acta* 51, 1561-1572.

Murray, J.W., 1979, Iron oxides. In R.G. Burns (ed.), *Marine Minerals*. Reviews in Mineralogy. Miner. Soc. Am.,6, 47-98.

Nesbitt, H.W., 1980, Characterization of mineral-formation water interactions in carboniferous sandstones and shales of the Illinois sedimentary basin. *Am. J. Sci.* 280, 607-630.

Nicholson, R.V., Gillham, R.W. and Reardon, E.J., 1988, Pyrite oxidation in carbonate-buffered solution: 1. Experimental kinetics. *Geochim. Cosmochim. Acta* 52, 1077-1085.

Nicholson, R.V., Gillham, R.W. and Reardon, E.J., 1990, Pyrite oxidation in carbonate-buffered solution: 2. Rate control by oxide coatings. *Geochim. Cosmochim. Acta* 54, 395-402.

Nordstrom, D.K., 1982, Aqueous pyrite oxidation and the consequent formation of secondary iron minerals. In Acid Sulphate Weathering (ed. D.K.Nordstrom). pp. 37-56. Soil Sci. Soc. Amer., Spec. Publ. No 10.

Nordstrom, D.K., Jenne, E.A. and Ball, J.W., 1979, Redox equilibria of iron in acid mine waters. In Jenne, E.A. (ed.), *Chemical Modeling in Aqueous Systems*. ACS Symp. Ser., 93, Am. Chem. Soc., 49-79.

Obermann, P., 1982, *Hydrochemische/hydromechanische Untersuchungen zum Stoffgehalt von Grundwasser bei landwirtschaftlicher Nutzung*. Bes. Mitt. dt. Gewäss. Jahrb. 42 pp.

Piisphanen, R., 1985, Geochemistry of groundwater iron precipitation in the light of an occurrence in glaciofluvial material near Hameenjarvi, Finland. *Geol. Foren. Stockh. Forh.* 107, 143-152.

Postma, D., 1977, The occurrence and chemical composition of recent Fe-rich mixed carbonates in a river bog. *J. Sed. Petrology* 47, 1089-1098.

Postma, D., 1981, Formation of siderite and vivianite and the pore-water composition of a recent bog sediment in Denmark. *Chem. Geology* 31, 255-244.

Postma, D., 1982, Pyrite and siderite formation in brackish and freshwater swamp sediments. *Am. J. Sci.* 282, 1151-1183.

Postma, D., 1983, Pyrite and siderite oxidation in swamp sediments. *J. Soil. Sci.* 34, 163-182.

Postma, D., 1985, Concentration of Mn and separation from Fe in sediments. I. Kinetics and stoichiometry of the reaction between birnessite and dissolved Fe(II) at 10°C. *Geochim. Cosmochim. Acta* 49, 1023-1033.

Postma, D., 1990, Kinetics of nitrate reduction by detrital Fe(II)-silicates. *Geochim. Cosmochim. Acta* 54, 903-908.

Postma, D., Boesen, C., Kristiansen, H. and Larsen, F., 1991, Nitrate reduction in an unconfined sandy aquifer: Water chemistry, reduction processes, and geochemical modeling. *Water Resour. Res.* 27, 2027-2045.

Postma, D. and Brockenhuus-Schack, B.S., 1987, Diagenesis of iron in proglacial sand deposits of late- and post-Weichselian age. *J. Sed. Petrology* 57, 1040-1053.

Robertson, W. D., Cherry, J. A. and Schiff, S. L., 1989, Atmospheric sulfur deposition 1950-1985 inferred from sulfate in groundwater. *Water Resour. Res.* 25, 1111-1123.

Ronen, D., Magaritz, M., Almon, E. and Amiel, A. J., 1987, Anthropogenic anoxification ('eutrophication') of the water table region of a deep aquifer. *Water Resour. Res.* 23, 1554-1560.

Rose, S. and Long, A., 1988, Dissolved oxygen systematics in the Tucson Basin aquifer. *Water Resour. Res.* 24, 127-136.

Schott, J. and Berner, R. A., 1983, X-ray photoelectron studies of the mechanism of iron silicate dissolution during weathering. *Geochim. Cosmochim. Acta* 47, 2233-2240.

Schwertmann, U., 1964, Differenzierung der Eisenoxide des Bodens durch Ektraktion mit Ammonium oxalat-lösung, *Z. Pflanzenernähr. Bodenk.* 105, 194-202.

Singer, P.C. and Stumm, W., 1970, Acid mine drainage: The rate limiting step. *Science* 167, 1121-1123.

Smith, R. L. and Duff, J. H., 1988, Denitrification in a sand and gravel aquifer. *Appl. Environ. Microbiol.* 54, 1071-1078.

Smith, R. L., Howes, B. L. and Duff, J. H., 1991, Denitrification in nitrate-contaminated groundwater: occurrence in steep vertical geochemical gradients. *Geochim. Cosmochim. Acta* 55, 1815-1825.

Sørensen, J. and Thorling, L., 1991, Stimulation by lepidocrocite (γ-FeOOH) of Fe(II)-dependent nitrite reduction, *Geochim. Cosmochim. Acta* 55, 1289-1294.

Starr, R.C., 1988, *An investigation of the role of labile organic carbon in denitrification in shallow sandy aquifers.* Ph. D. thesis, Univ. Waterloo, 148 pp.

Starr, R.C. and Gillham, R.W., 1989, Controls on denitrification in shallow unconfined aquifers. In H. E. Kobus and W. Kinzelbach (eds), *Contaminant Transport in Groundwater.* Balkema, Rotterdam, 51-56.

Strebel O., Böttcher, J. and Kölle, W., 1985, Stoffbilanzen im Grundwasser eines Einzugsgebietes als Hilfsmittel bei Klärung und Prognose von Grundwasserqualitätsproblemen (Beispiel Fuhrberger Feld). *Z. dt. geol. Ges.* 136, 533-541.

Stumm, W. and Morgan, J.J., 1981, *Aquatic chemistry.* 2nd ed. John Wiley & Sons, New York, 780 pp.

Thorstenson, D.C., Fisher, D.W. and Croft, M.G., 1979, The geochemistry of the Fox Hill-basal Hell Creek aquifer in the southwestern North Dakota and northwestern South Dakota. *Water Resour. Res.* 15, 1479-1498.

Trudell, M.R., Gillham, R.W. and Cherry, J.A., 1986, An in-situ study of the occurrence and rate of denitrification in a shallow unconfined sand aquifer. *J. Hydrol.* 86, 251-268.

Van Beek, C.G.E.M., Boukes, H., van Rijsbergen, D. and Straatman, R., 1988, The threat to the Netherlands waterworks by nitrate in the abstracted groundwater, as demonstrated on the well field Vierlingsbeek. *Wat. Supply* 6, 313-318.

Van Beek, C.G.E.M., Hettinga, F.A.M. and Straatman, R., 1989, The effects of manure spreading and acid deposition upon groundwater quality in Vierlingsbeek, the Netherlands. *IAHS Publ.*, 185, 155-162.

Van Beek, C.G.E.M. and Van der Kooij, D., 1982, Sulfate-reducing bacteria in ground water from clogging and nonclogging shallow wells in the Netherlands river region. *Ground Water* 20, 298-302.

Van Breemen, N., 1976, Genesis and solution chemistry of acid sulfate soils in Thailand. *Agric. Res. Rep.*, 263 pp.

Vanek, V., 1990, In situ treatment of iron-rich groundwater by the addition of nitrate. Rep. Lunds Universitet, 33 pp.

Vogels, G.D., Keltjens, J.T. and van der Drift, C., 1988, Biochemistry of methane production. In A.J.B. Zehnder (ed.), *Biology of anaerobic microorganisms.* Wiley & Sons, New York, 707-770.

Westrich, J.T. and Berner, R.A., 1984, The role of sedimentary organic matter in bacterial sulfate reduction: The *G* model tested. *Limnol. Oceanogr.* 29, 236-249.

White, A.F., 1990, Heterogeneous electrochemical reactions associated with oxidation of ferrous oxide and silicate surface. In M.F. Hochella and A.F. White (eds), *Mineral water interface geochemistry.* Rev. in Mineral. Mineral. Soc. Am., 23, 447-509.

White, A.F. and Yee, A., 1985, Aqueous oxidation-reduction kinetics associated with coupled electron-cation transfer from iron-containing silicates at 25°C. *Geochim. Cosmochim. Acta* 49, 1263-1275.

Whittemore, D.O. and Langmuir, D., 1975, The solubility of ferric oxyhydroxides in natural waters. *Ground Water* 13, 360-365.

Williamson, M.A. and Rimstidt, J.D., 1990, Thermodynamic and kinetic controls on the aqueous oxidation of sulfide minerals. 2nd Goldschmidt Conf. Abstract volume, Baltimore, p 91.

Winograd, I.J. and Robertson, F.N., 1982, Deep oxygenated ground water: Anomaly or common occurrence. *Science* 216, 1227-1230.

Zehnder, A.J.B. (ed), 1988, *Biology of anaerobic microorganisms*. Wiley & Sons, New York, 872 pp.

CHAPTER 8

Salt water, mixing and discharge/quality relations

This chapter is about mixing of waters. Mixing was used in previous chapters as the starting point for deciphering reactions and processes which had influenced the water composition. For example, the composition of rain water could be compared with that of sea water, diluted ($\approx$ mixed) with distilled water, and the comparison provided clues concerning other sources of rain water constituents. Conservative mixing of sea water and fresh water to the observed Cl^- concentration in groundwater was used to study cation exchange and other reactions during sea water intrusion (Chapter 5). Mixing of water types is important when spring water qualities are used to calculate complex silicate weathering reactions (Chapter 6). The chemical reactions which take place when two different waters are mixed is the subject of this chapter. We start with salt water (where we may consider the loss of water molecules during evaporation as mixing with negative quantities of distilled water), and describe the compositional changes as salts precipitate from solution. Mix-plots are often used to describe the processes in deep subsurface brines. Precipitation or dissolution may occur as a result of differences in CO_2 pressures, oxidation/reduction potential, etc. in the waters that mix. Mixing is quite obvious in surface waters, and changing quality with runoff may be used to discern runoff sources and contributing areas.

8.1 SALT WATER

In many areas of the world, pumping of groundwater for drinking and irrigation purposes is accompanied by undesirable upconing of salt water. The mining and oil industries are also faced with salt waters as a nuisance which requires specific precautions. In other areas, salt water is actively sought for therapeutic purposes (balneology, Heilbäder, Spas). Whether the salt is beneficial, or undesirable, it is important to know the source of salt. There are principally six processes which may increase the salt content of water to such extent that it is appropriately termed 'salt water'. These are:
- evaporation and concentration/precipitation of dissolved salts;
- dissolution of salts;
- mixing with recent sea water;
- mixing with old water (connate from buried marine sediments);
- volcanic exhalations;
- hyperfiltration.

It is often possible to use chemical criteria to distinguish between sources of the salts. Let us start with chemical changes during evaporation.

8.1.1 *Evaporative concentration*

Evaporation may be considered as the reverse of diluting a solution with distilled water: evaporation removes water molecules and concentrates the solutes. The increase in concentrations can lead to precipitation as soon as the solubility product of a solid is surpassed. Salts which precipitate from seawater upon evaporation are illustrated in Figure 8.1. The salts precipitate in an ordered sequence, starting with the least soluble solid (Ca-carbonate), and ending with the most soluble solids (Mg-salts).

The increase in concentration of individual ions can also be limited by precipitation of the salts. When the ions are initially present in water in the same stoichiometry as in the salts, the concentrations are limited simply by the solubility of the salt. For example, when dolomite precipitates from water in which originally $m_{Ca^{2+}} = m_{Mg^{2+}} = \frac{1}{4}m_{HCO_3^-}$, and concentrations are restricted by dolomite solubility, the ratios between the solute ions remain the same when the solubility limit has been reached. In this case there is actually no change in composition of the solution despite continuing evaporation and precipitation.

Natural waters do not contain ions in the stoichiometry that corresponds to the composition of simple salts. Whenever the ratio of the solute ion concentrations is different from the ratio in the precipitate a general principle applies when a solid of fixed composition precipitates during evaporation. Solutes with a concentration-ratio that is *higher* in the solution than in the precipitate, *increase* in concentration, and solutes with a *lower* concentration-ratio *decrease* in concentration. This principle can be demonstrated as in Example 8.1.

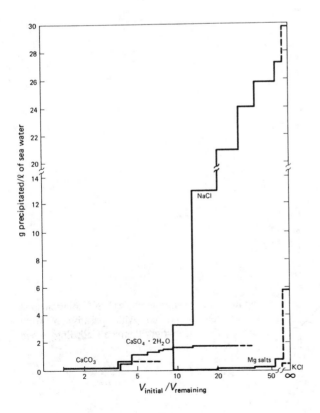

Figure 8.1. The sequence of salts which precipitate from evaporating sea water. Reprinted with permission from Drever, 1988 Copyright Prentice Hall, Englewood Cliffs, New Jersey.

EXAMPLE 8.1: *Concentration of a CaSO₄ solution (After Drever, 1988)*
A solution with $Ca^{2+} = 7$ and $SO_4^{2-} = 3.5$ mmol/l evaporates. Calculate the solution composition when gypsum precipitates.

ANSWER
Precipitation of gypsum requires a saturation index *SI* > 0 for the reaction:

$$Ca^{2+} + SO_4^{2-} + 2\,H_2O \rightarrow CaSO_4\cdot 2H_2O \qquad K_{gyps} = [Ca^{2+}]\,[SO_4^{2-}] = 10^{-4.6}.$$

When the solution is *n* times concentrated, *y* moles of gypsum precipitate. Hence:

$$m_{Ca^{2+}} = n \cdot (7\cdot 10^{-3} - y) \qquad m_{SO_4^{2-}} = n \cdot (3.5\cdot 10^{-3} - y)$$

or assuming activity numerically equal to concentration in mol/l:

$$K_{gyps} = 10^{-4.6} = n^2 \cdot (7\cdot 10^{-3} - y) \cdot (3.5\cdot 10^{-3} - y)$$

Thus *y* can be calculated as a function of *n*. For example for *n* = 3:

$$10^{-4.6} = 9 \cdot (7\cdot 10^{-3} - y) \cdot (3.5\cdot 10^{-3} - y)$$

$$y = 2.83\cdot 10^{-3}\,\text{mol/l}$$

Figure 8.2 shows how the solution composition of Example 8.1 changes with different concentration factors. We note that in the starting solution the ratio of $m_{Ca^{2+}}/m_{SO_4^{2-}} = 2$, which is twice that in gypsum. The Ca^{2+}/SO_4^{2-} ratio in solution further increases as evaporation proceeds, as illustrated in Figure 8.2.

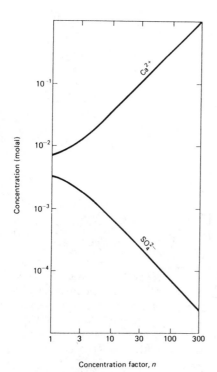

Figure 8.2. Concentrations in a CaSO₄ solution upon evaporation. Reprinted with permission from Drever, 1988. Copyright Prentice Hall, Englewood cliffs, New Jersey.

Common precipitates which occur when fresh water evaporates, are:

Calcite:

$$Ca^{2+} + 2\,HCO_3^- \rightarrow CaCO_3 + CO_2 + H_2O$$

'Sepiolite'

$$Mg^{2+} + 3\,H_4SiO_4^0 + 2\,HCO_3^- \rightarrow MgSi_3O_6(OH)_2 + 2\,CO_2 + 6\,H_2O$$

Gypsum:

$$Ca^{2+} + SO_4^{2-} + 2\,H_2O \rightarrow CaSO_4 \cdot 2H_2O$$

Chert (colloidal silica, chalcedony):

$$H_4SiO_4^0 \rightarrow SiO_2 + 2\,H_2O$$

The salts have been ordered here in their common sequence of precipitation from fresh water (cf. also Figure 2.14). At later stages Na-containing salts can precipitate, such as halite (NaCl), or trona (NaHCO$_3$·Na$_2$CO$_3$·2H$_2$O). The precipitation of sepiolite is not well documented, although it is known that water readily loses Mg^{2+} upon evaporation. The loss of Mg^{2+} might also occur through precipitation in dolomite, or a clay mineral (chlorite). In many instances an inverse relationship has been found between the Si and Mg^{2+} concentrations, so that precipitation in a silicate (sepiolite or clay mineral) appears to be probable. Indeed, Mg^{2+} seems to be of primary importance for the formation of clay minerals at low temperature (Harder, 1976).

Application of the ratio principle to evaporating water suggests that the composition of salt water and brines is a function of the concentration ratios of the ions in the starting solution. This principle has been worked out by Hardie and Eugster (1970) in an evolutionary sequence for evaporating waters, shown in Figure 8.3. The first mineral to precipitate is

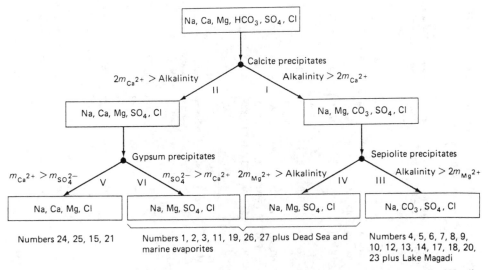

Figure 8.3. Some possible pathways for evaporation of natural waters according to the model of Hardie and Eugster (1970).

Table 8.1. The composition of brines from saline lakes in North America, compiled by Drever, 1988. The numbers refer to the Hardie-Eugster model in Figure 8.3. Concentrations in mg/l. tr = trace.

	1 Kamloops Lake No. 7, B.C.	2 Basque Lake No. 1, B.C.	3 Hot Lake Wash.	4 Lenore Lake Wash.	5 Soap Lake Wash.	6 Harney Lake Ore.	7 Summer Lake Ore.	8 Alkali Valley Ore.	9 Abert Lake Ore.	10 Surprise Valley Calif.	11 Great Salt Lake Utah	12 Honey Lake Calif.	13 Pyramid Lake Nev.	14 Winnemucca Lake, Nev.	15 Carson Sink Nev.	16 Rhodes Marsh Nev.
SiO_2	tr	–		22	101	31	103	542	645	36	48	55	1.4	11	19	142
Ca^{2+}			640	3	3.9	7	tr	–	–	11	241	–	10	4.2	261	17
Mg^{2+}	34,900	42,400	22,838	20	23	tr	tr	–	–	31	7,200	–	113	4.5	129	0.5
Na^+	10,900	13,660	7,337	5,360 }	12,500	8,826	6,567	117,000	119,000	4,090	83,600	18,300	1,630	4,350 }	56,800	3,680
K^+		1,570	891			336	264	8,850	3,890	11	4,070	1,630	134		3,240	102
HCO_3^-	2,400	3,020	6,296	6,090	12,270	4,425 }	5,916 }	2,510	–	1,410	251	5,490	1,390	744	322	23
CO_3^{2-}		–	–	3,020	5,130			91,400	60,300	664	–	8,020	–	128	–	648
SO_4^{2-}	160,800	195,710	103,680	2,180	6,020	1,929	695	46,300	9,230	900	16,400	12,100	264	918	786	2,590
Cl^-	200	1,690	1,668	1,360	4,680	6,804	3,039	45,700	115,000	4,110	140,000	9,680	1,960	5,290	88,900	3,070
Total	209,000	258,000	143,000	18,000	40,700	22,383	16,633	314,000	309,000	10,600	254,000	52,900	5,510	11,100	152,000	10,400
pH								10.1	9.8	9.2	7.4	9.7		8.7	7.8	9.5

	17 Mono Lake, Calif.	18 Deep Springs Lake, Calif.	19 Saline Valley, Calif.	20 Owens Lake, Calif.	21 Death Valley, Calif.	22 Searles Lake, Calif.	23 Soda Lake, Calif.	24 Bristol Dry Lake, Calif.	25 Cadiz Lake, Calif.	26 Danby Lake, Calif.	27 Salton Sea, Calif.
SiO_2	14		36	299	–	–	–	–	–	–	20.8
Ca^{2+}	4.5	3.1	286	43	–	–	–	43,296	4,504	325	505
Mg^{2+}	34	1.2	552	21	150	–	–	1,061	412	108	581
Na^+	21,500	111,000	103,000	81,398	109,318	110,000	114,213	57,365	22,603	137,580	6,249
K^+	1,170	19,500	4,830	3,462	4,043	26,000	tr	3,294	1,038	–	112
HCO_3^-	5,410	9,360	614	52,463 }	–	27,100	12,053	–	–	tr	232
CO_3^{2-}	10,300	22,000	–					–	–	–	–
SO_4^{2-}	7,380	57,100	22,900	21,220	44,356	46,000	52,026	223	280	13,397	4,139
Cl^-	13,500	119,000	150,000	53,040	140,196	121,000	124,618	172,933	44,764	119,789	9,033
Total	56,600	335,000	282,360	213,700	299,500	336,000	305,137	279,150	73,600	271,200	20,900
pH	9.6		7.35								

calcite, and depending on the ratio of Ca^{2+}/alkalinity, there will be enrichment of either Ca^{2+} or HCO_3^- in solution. The next step is determined by gypsum or sepiolite precipitation, and it is again the ratio of the ions in solution that determines whether concentrations increase or decrease. This scheme offers a simple and adequate classification for surface brines. The numbers in Figure 8.3 correspond with brine analyses given in Table 8.1 from saline lakes in North America. Similar brine compositions have been found in Australia (Jankowski and Jacobson, 1989). It can be seen that in many of these brines there is indeed a remarkable absence of one of the major ions.

Activity calculations in salt waters are generally more complex than in waters which have concentrations less than sea water. The difficulty is that activity coefficients deviate from Debije-Hückel theory. Computer programs which incorporate more empirical correction terms (Pitzer equations, e.g. Pitzer, 1987) have been developed (Van Gaans, 1990; PHRQPITZ, Plummer and Parkhurst, 1990), and also specific programs have been made to calculate the composition changes in evaporating brines (Harvie and Weare, 1980). An example of such a calculation for evaporating sea water is shown in Figure 8.4. The chemical formula of the complex salts which are indicated in this figure are given in Table

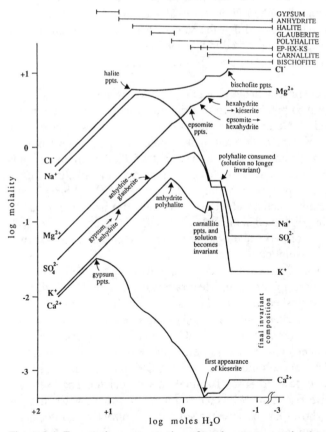

Figure 8.4. Evaporative concentration of modern seawater simulated using the computer program of Harvie and Weare (1980). Progressive changes in major ion concentrations are indicated as a function of moles of water that remain of 1 dm^3 of seawater. (Reproduced with permission from Hardie, 1991. © 1991 by Annual Reviews Inc.).

Table 8.2. Chemical composition of some of the common minerals of evaporites (from Hardie, 1991).

Anhydrite	$CaSO_4$	Kainite	$MgSO_4 \cdot KCl \cdot 11/4H_2O$
Antarcticite	$CaCl_2 \cdot 6H_2O$	Kieserite	$MgSO_4 \cdot H_2O$
Aragonite	$CaCO_3$	Langbeinite	$2MgSO_4 \cdot K_2SO_4$
Bassanite	$CaSO_4 \cdot 1/2H_2O$	Mirabilite	$Na_2SO_4 \cdot 10H_2O$
Bischofite	$MgCl_2 \cdot 6H_2O$	Nahcolite	$NaHCO_3$
Bloedite	$Na_2SO_4 \cdot MgSO_4 \cdot 6H_2O$	Natron	$Na_2CO_3 \cdot 10H_2O$
Calcite	$CaCO_3$	Polyhalite	$2CaSO_4 \cdot MgSO_4 \cdot K_2SO_4 \cdot 2H_2O$
Carnallite	$MgCl_2 \cdot KCl \cdot 6H_2O$	Rinneite	$FeCl_2 \cdot NaCl \cdot 3KCl$
Dolomite	$CaMg(CO_3)_2$	Shortite	$2CaCO_3 \cdot Na_2CO_3$
Epsomite	$MgSO_4 \cdot 7H_2O$	Sylvite	KCl
Glauberite	$CaSO_4 \cdot Na_2SO_4$	Tachyhydrite	$CaCl_2 \cdot 2MgCl_2 \cdot 12H_2O$
Gypsum	$CaSO_4 \cdot 2H_2O$	Thenardite	Na_2SO_4
Halite	$NaCl$	Thermonatrite	$Na_2CO_3 \cdot H_2O$
Hexahydrite	$MgSO_4 \cdot 6H_2O$	Trona	$NaHCO_3 \cdot Na_2CO_3 \cdot 2H_2O$

8.2. The equilibrium calculations predict rather complex transitions and recrystallizations. The recrystallization of gypsum into anhydrite is, for example, an effect of diminishing water activity in the brine. It may be noted that the salt industry is able to separate the different salts by letting the evaporation proceed in a sequential series of basins; the final result can be halite of a purity similar to that obtained from refinement using a salt deposit in a geological sequence.

We can now draw some general conclusions about irrigation water quality as well. Calcite is often the first mineral to precipitate from irrigation water. When alkalinity is larger than the sum of Ca^{2+} and Mg^{2+}, the concentrations of these cations will decrease in the return flow-part of irrigation water. The return flow-percentage must be carefully guarded to prevent problems associated with sodication (Chapter 5). If the cations Ca^{2+} and Mg^{2+} are initially present at higher equivalent concentrations than the alkalinity, these cations increase in concentration upon evaporation, and the $Na^+/(Ca^{2+} + Mg^{2+})^{1/2}$ ratio (*SAR*) will not change dramatically. Such a water type is favorable for irrigation. It can be found in areas where gypsum has dissolved in the water, or when acid anions (SO_4^{2-} or NO_3^- from acid rain, or from oxidation of pyrite and organic matter) are present in excess over HCO_3^-, and have caused calcite dissolution.

Imagine a rain event in an arid area, where salts are present in the surface soils. The salts will dissolve in water, but often precipitate again when rain water evaporates. With more intense showers the salts may be flushed to deeper layers, where evapotranspiration is no longer active. Such a process of repeated dissolution and precipitation will enrich the ions from rapidly dissolving salts in water. Alkali salts ($NaCl$, Na-carbonates) dissolve more rapidly than Ca^{2+} salts (dolomite, gypsum, calcite, etc.) and groundwater can become alkaline, with high Na^+ and CO_3^{2-} concentrations. As soon as CO_3^{2-} concentrations are high, Ca^{2+} concentrations must remain low, because of saturation with calcite. The low Ca^{2+} concentrations may in turn give rise to high F^- concentrations in groundwater in areas where fluorine is present in the host rock, because the solubility control by fluorite, CaF_2, now allows increased concentrations of the anion (Chapter 3). A few cycles of salt dissolution and precipitation with low Ca^{2+} concentrations in solution may also enrich F^- in the precipitating salts. Redissolution of these salts can subsequently lead to high F^- concentrations in shallow groundwater.

8.1.2 *Subsurface brines*

Waters that are more concentrated than seawater, so-called *brines*, are ubiquitous in the deeper earth. An example of Na^+ vs. Cl^- concentrations in pore waters from Jurassic, Cretaceous, and Tertiary formations in the Gulf Coast area is shown in Figure 8.5. The solid line in the figure indicates the trend of an evaporating seawater; concentrations of Na^+ and Cl^- increase concomitantly until saturation with halite (NaCl) occurs. Then, the Cl^- concentration continues to increase with increasing evaporation, while the Na^+ concentration decreases since the Cl^-/Na^+ ratio in the original water is greater than 1 (compare with Figure 8.4). Also indicated in Figure 8.5 is the mixing line for water in which halite has dissolved, which is very close to the seawater evaporation line. The samples follow these conservative trend-lines. However, the brines have also much higher Ca^{2+}/Cl^- ratios than expected for evaporating seawater, which indicates that extensive water-rock interactions have taken place. Since these processes may have changed the Na^+/Cl^- ratio as well, caution is warranted for too simple interpretations based on evaporation only (Collins, 1975). There is also a hydrological argument which pleads against simple evaporation as the major process explaining the salt content of deep groundwaters. Evaporating brines are a sign of water flow towards the surface (upward seepage). Under these hydrological conditions there is no water left which can be sampled later on as a relict, and only the precipitated salts are a witness of the arid conditions. Groundwater brines with concentrations higher than seawater have likely derived their higher concentrations from dissolution of minerals or salts.

The importance of evaporation for deep salt waters can also be judged from isotopic compositions of the water molecules. Most brines in deeper layers have an isotopic composition that points to a local rainwater source (Clayton et al., 1966; Kharaka and Carothers, 1986). Figure 8.6 shows a plot of δ^2H (= δD) versus $\delta^{18}O$ of brines from sedimentary basins in America. Both these isotopes of the water molecule are expressed relative to the composition of seawater, which therefore has δ^2H and $\delta^{18}O = 0$. The data for the brines from each basin fall on a line that can be extrapolated towards the meteoric water line, suggesting a single common source of the waters from local rainwater. The oxygen isotope shift is considered to be a result of isotope exchange between oxygen in water and in rock. Hydrogen is not a common constituent of rock forming minerals (except in the

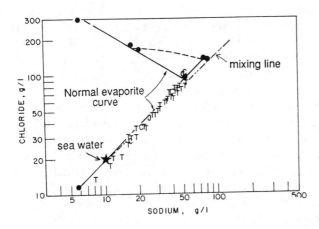

Figure 8.5. Sodium versus chloride concentrations for some formation waters Tertiary (T), Cretaceous (C), and Jurassic (J) sediments, compared with evaporating sea water (modified from Collins, 1975).

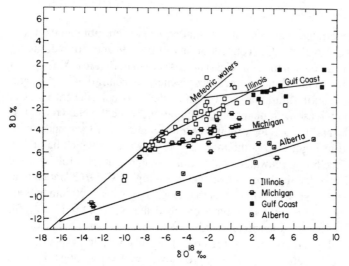

Figure 8.6. Isotopic compositions of brines in sedimentary basins in North America. The 'meteoric waters' lines is the locus of values for meteoric waters throughout the world (from Clayton et al., 1966. Copyright by the American Geophysical Union).

hydroxyl groups of clay minerals and hydroxides), and δ^2H shows no exchange, or at least not as pronounced a change as oxygen. The Gulf Coast samples are closest to a composition expected for connate seawater, without much evaporation.

The chloride in the brines must therefore have come from dissolution of salts. In a sedimentary basin abundant amounts of salts are available for dissolution and to increase the solute concentration. However, brines in igneous rocks in the Canadian Shield have a similar composition, but a source for Cl^- is not as easily available there (Frape and Fritz, 1987). The deep salt waters exhibit here an increase of Ca^{2+} which parallels the increase of Cl^-, as shown in Figure 8.7. The increasing Ca^{2+} concentration is a true indicator of water rock interaction, for which different explanations have been proposed. Dolomitization of calcite may be a possibility, and is suggested by relative losses of Mg^{2+} in the brines (Collins, 1975). Dissolution of gypsum or anhydrite and concomitant reduction of SO_4^{2-} can also give rise to high Ca^{2+} concentrations (De Boer, pers. comm.). Another explanation suggests salt sieving (also termed membrane filtration, or hyperfiltration, cf. Chapter 5) as the most likely cause for the relatively high Ca^{2+} concentration (Kharaka and Berry, 1973; Graf, 1982). The increases in deep brines in the Canadian shield are attributed to anorthite reactions (Frape and Fritz, 1987). (Anorthite is the Ca-feldspar; it may alter to kaolinite under release of Ca^{2+} into solution, or to albite, the Na-feldspar, while taking up Na^+).

The principle that a high concentration of a cation forbids the presence of the anion with which it can form a salt, precludes the presence of much CO_3^{2-} in the high Ca^{2+} brines. This, in turn, indicates a low CO_2 pressure, which may also be an important master variable. Many hydrothermal systems in New Zealand, Kenya and Europe are outgassing CO_2, and in these cases the Ca^{2+} concentrations are low. A high CO_2 pressure in hydrothermal areas is often related to calcination of carbonate rocks at a temperature above 500°C. Since carbonate rocks are ubiquitous on earth, it could be that temperature dictates the relative abundance of Ca^{2+} in the formation waters. The high Ca^{2+} concentrations observed in saline

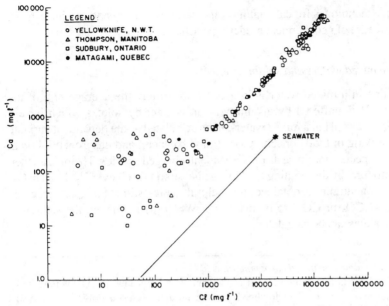

Figure 8.7. Ca/Cl concentrations in subsurface brines of the Canadian Shield (Reprinted with permission from Frape and Fritz, 1987. Copyright Geological Association of Canada).

waters as depicted in Figure 8.7 are confined to areas (sedimentary basins, kratons) with a low temperature troughout.

Lastly, we note that the relatively low Mg^{2+} concentrations in the brines, considered previously a sign of dolomitization, can also be a result of the formation of Mg-silicates. This is a response to more abundant supply of $H_4SiO_4^0$ derived from quartz, as this mineral becomes more soluble when temperature increases. It is, again, an example of inverse relationships among solutes and environmental conditions which are regulating water compositions.

8.2 MIXING OF DIFFERENT WATER TYPES

Mixing of different water types can induce reactions which lead to a water composition that is different from the conservative mixture. Examples are the change of pH as result of buffering reactions by carbonate species (which can give undersaturation with respect to calcite), the common ion effect which gives often supersaturation with respect to minerals, and the oxidation/reduction reactions which may occur in production wells, and are well known for their associated clogging effects. Mixing of waters in springs or wells with long screens can often be unraveled by repeated sampling under varying hydrological conditions (Mazor, 1983, 1992). The tools for calculations of mixing effects are the computer programs NETPATH (Plummer et al., 1991) or its predecessor BALANCE (Parkhurst et al., 1982) and PHREEQE (Parkhurst et al., 1980). NETPATH can compute mixing ratios of two end members as well as reactions for observed water compositions, PHREEQE can predict the theoretical effects of mixing and reactions. In this section we will once more discuss a few simple techniques for calculating water compositions due to mixing effects, before turning to the ease of geochemical modeling in Chapter 10.

8.2.1 *pH change in mixed waters and solubility effects*

The H^+ concentration in a mixed water is in general not simply the average m_{H^+} of the component waters. pH is buffered by alkalinity, or in general by solutes which have a pH-dependent charge. The pH in a mixed water can be calculated as the dependent variable after conservative mixing of total concentrations (mass balance), and electroneutrality. In most waters the CO_2 species dominate as pH-dependent, charged species. The pH can then be calculated in a mixture, by distributing total inorganic carbon (*TIC*) over H_2CO_3, HCO_3^-, and CO_3^{2-} until electroneutrality is obtained. It is helpful to pre-estimate the pH, since this often allows either H_2CO_3 or CO_3^{2-} to be discarded. When pH is known, the saturation index for solid carbonates can be calculated.

EXAMPLE 8.2 *pH calculation in mixed carbonate water*

Consider the two flowlines in a Dolomite massif from Example 4.3. In flowline A, dolomite was dissolved to saturation at $P_{CO_2} = 10^{-3.5}$, in flowline B passing through forest soil, dolomite dissolved to saturation at $P_{CO_2} = 10^{-2}$ atm. Now, the two flowlines mix 1:1 in a spring. Calculate composition of the spring, and the saturation index for dolomite.

Flowline	A	B	Mixture 1:1
Ca^{2+}	0.33	1.04	0.68 mmol/l
Mg^{2+}	0.33	1.04	0.68 mmol/l
HCO_3^-	1.31	4.16	2.73 mmol/l
pH	8.42	7.42	?
TIC	1.33	4.65	2.99 mmol/l
EC	131	416	273 µS/cm

ANSWER

When pH in the mixture is below 8, we can assume that only $H_2CO_3^*$ and HCO_3^- contribute to *TIC*. Hence:

$$m_{H_2CO_3^*} = TIC - m_{HCO_3^-} = 2.99 - 2.73 = 0.26$$

and pH is calculated from (assuming $m_{HCO_3^-} = [HCO_3^-]$):

$$10^{-6.3} = \frac{[H^+][HCO_3^-]}{[H_2CO_3^*]}$$

resulting in pH = 7.32.

With this pH, $[CO_3^{2-}]$ can be calculated from:

$$10^{-10.3} = \frac{[H^+][CO_3^{2-}]}{[HCO_3^-]}$$

and

$$SI_{dol} = \log([Ca^{2+}][Mg^{2+}][CO_3^{2-}]^2) - (-16.9) = -0.5$$

Example 8.2 shows that a mixed water of two flowlines with different P_{CO_2} becomes subsaturated with respect to solid carbonate, even though water of the two flowlines is at equilibrium. The large concentration of H_2CO_3 in water of the flowline with the higher P_{CO_2} acts as a buffer (and there is a small additional effect arising from the non-linear relation between activity coefficients and concentration). The effect of mixing on the solubility of a solid can be calculated with the mass action equation, using the relationships discussed in Chapter 4. For the reaction

$$CaCO_3 + H_2CO_3^* \leftrightarrow Ca^{2+} + 2\,HCO_3^- \tag{8.1}$$

the mass action expression is:

$$K_{(8.1)} = \frac{[Ca^{2+}]\,[HCO_3^-]^2}{[H_2CO_3^*]}$$

If we apply the electroneutrality relationship $2m_{Ca^{2+}} = m_{HCO_3^-}$, we obtain:

$$m_{Ca^{2+}} = (0.25\,K_{(8.1)})^{1/3} \cdot (m_{H_2CO_3^*})^{1/3} \tag{8.2}$$

Mixing of two waters, of which the compositions follow this parabolic relationship of Ca^{2+} and $H_2CO_3^*$ concentrations, will always result in *sub*saturation (cf. Problem 8.2). The renewed aggressiveness in the mixed water may cause calcite dissolution and possibly explain cave formation. The effect has been called *Mischungskorrosion* in the German literature (Bögli, 1978). It is illustrated in Figure 8.8, where groundwater with log $P_{CO_2} = -1.3$ is mixed with groundwater with log $P_{CO_2} = -3$, resp. -2.3. Also shown is the amount of $CaCO_3$ which would dissolve in the mixed water until equilibrium is re-established. Mischungskorrosion even operates when seawater (which is supersaturated with respect to calcite, and has a low CO_2 pressure) is mixed in certain proportions with fresh groundwater with a higher CO_2 pressure (Plummer, 1975). Sanford and Konikow

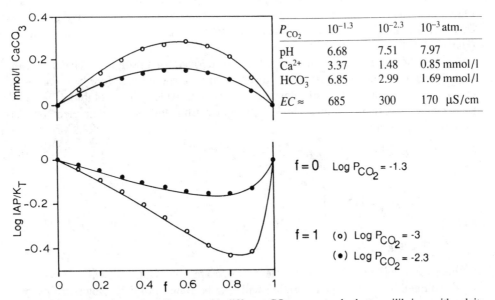

P_{CO_2}	$10^{-1.3}$	$10^{-2.3}$	10^{-3} atm.
pH	6.68	7.51	7.97
Ca^{2+}	3.37	1.48	0.85 mmol/l
HCO_3^-	6.85	2.99	1.69 mmol/l
$EC \approx$	685	300	170 µS/cm

$f = 0$ Log $P_{CO_2} = -1.3$

$f = 1$ (o) Log $P_{CO_2} = -3$

   (●) Log $P_{CO_2} = -2.3$

Figure 8.8. Mixing of two groundwaters with different CO_2 pressures, both at equilibrium with calcite, leads to undersaturation with respect to calcite. (Calculations with PHREEQE, cf. Chapter 10).

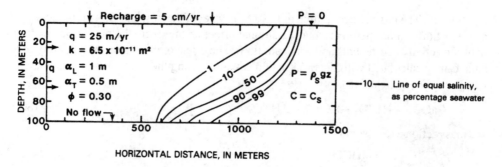

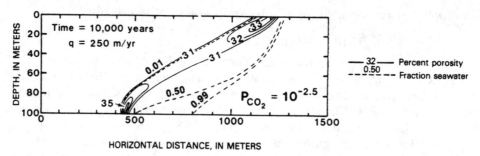

Figure 8.9. Development of porosity in the mixing zone of fresh and salt water. Calculations based on a hydraulic model (above), and PHREEQE (from Sanford and Konikow, 1989. Copyright by the American Geophysical Union).

(1989) have calculated the increase of porosity due to Mischungskorrosion, as shown in Figure 8.9. Such mixing of fresh and seawater near the coast might explain the abundant cave formation which is observed in many calcareous coasts, e.g. in the Mexican Yucatan (Hanshaw & Back, 1985).

A parabolic relationship similar to (8.2) can be derived for all solids which dissolve due to carbonic acid, and subsaturation as result of mixing should therefore be common for many minerals (carbonates, silicates and oxides/hydroxides). The solubility of simple salts, on the other hand, follows a hyperbolic relationship among the component ions. For example for gypsum:

$$CaSO_4 \cdot 2H_2O \leftrightarrow Ca^{2+} + SO_4^{2-} + 2\,H_2O \tag{8.3}$$

with

$$K_{(8.3)} = [Ca^{2+}]\,[SO_4^{2-}]$$

and

$$m_{Ca^{2+}} \approx K_{(8.3)}/m_{SO_4^{2-}}$$

In this case, a mixing of different waters, which are both saturated, will always give a *super*saturated mixture. Supersaturation will also be found in mixed waters, when other acids besides carbonic acid play a role, and the CO_2 pressure of the more acid water is lowest (Example 8.3).

EXAMPLE 8.3: *Seepage of acid groundwater in a river with neutral pH*
In the Veluwe (The Netherlands), acid groundwater develops from acid rain (Example 6.3). We calculated a composition:

pH	Na$^+$	K$^+$	Ca^{2+}	Al^{3+}	Cl$^-$	HCO$_3^-$	
4.48	280	16	120	370	312	0	(µmol/l)

while *TIC*-content of the groundwater is 100 µmol/l. This groundwater discharges in a rivulet that springs up in meadows fertilized with calcite. The surface water composition before mixing is:

pH	Na$^+$	K$^+$	Ca^{2+}	Al^{3+}	Cl$^-$	HCO$_3^-$	
6.40	590	240	820	0	700	1040	(µmol/l)

Calculate the composition after mixing, at a brook/groundwater ratio of 4:1, assuming that equilibrium with Al(OH)$_3$ is maintained.

ANSWER
After mixing at given ratio, the composition of the rivulet is:

H$^+$	Na$^+$	K$^+$	Ca^{2+}	Al^{3+}	Cl$^-$	HCO$_3^-$	H$_2$CO$_3$	
7	528	195	680	74	622	832	681	(µmol/l)

This water is oversaturated with respect to Al(OH)$_3$:

$$[Al^{3+}]/[H^+]^3 = (7.4 \cdot 10^{-5})/(7 \cdot 10^{-6})^3 = 2.1 \cdot 10^{11} > 10^{10}$$

so that Al(OH)$_3$ will precipitate, releasing H$^+$. The H$^+$ is taken up by HCO$_3^-$ (pH-buffering):

$$
\begin{array}{llr}
Al^{3+} + 3\ H_2O & \leftrightarrow Al(OH)_3 + 3\ H^+ & K = 10^{-10} \\
3H^+ + 3\ HCO_3^- & \leftrightarrow 3\ H_2CO_3 & K = (10^{6.3})^3 \\
\hline
Al^{3+} + 3\ HCO_3^- + 3\ H_2O \leftrightarrow Al(OH)_3 + 3\ H_2CO_3 & & K = 10^{8.9} \qquad (8.4)
\end{array}
$$

If all Al^{3+} precipitates, then

$$m_{HCO_3^-end} = m_{HCO_3^-begin} - 3\ m_{Al^{3+}} = 0.832 - 0.222 = 0.610\ mmol/l$$

and:

$$m_{H_2CO_3^*end} = m_{H_2CO_3^*begin} + 3\ m_{Al^{3+}} = 0.681 + 0.222 = 0.903\ mmol/l$$

Insert in (8.4):

$$[H_2CO_3^*]^3/([Al^{3+}] \cdot [HCO_3^-]^3) = 10^{8.9}$$

and find:

$$m_{Al^{3+}\ end} = 10^{-8.4}\ mol/l \qquad \text{(All Al}^{3+}\text{ indeed precipitates)}$$

and finally from the HCO$_3^-$/ H$_2$CO$_3^*$ equilibrium: pH = 6.13.

8.2.2 *The common ion effect in mixed waters*

Gypsum (CaSO$_4 \cdot$2H$_2$O) has a higher solubility than calcite (CaCO$_3$). Mixing of water in equilibrium with gypsum, with water in equilibrium with calcite, may result in precipitation

Table 8.3. Mixing of calcite-water with gypsum-water (10°C). Hand calculations as well as calculations with PHREEQE in which complexes etc. are included, are presented. Concentrations in mmol/l.

	Hand-calculations			PHREEQE-calculations		
	calcite	gypsum	mix 1:1	calcite	gypsum	mix
pH	7.3	7.0	7.3	7.32	7.29	7.28
P_{CO_2}	10^{-2}	0	$10^{-2.5}$	10^{-2}	0	$10^{-2.3}$
Ca^{2+}	1.6	4.42	3.01	2.13	13.92	8.03
HCO_3^-	3.2	0	1.6	4.26	0	2.14
SO_4^{2-}	0	4.42	2.21	0	13.92	6.96
TIC	3.52	0	1.76	4.79	0	2.40
SI_{cc}	0	$-\infty$	-0.02	0	$-\infty$	0

of calcite. This is called the *common ion effect*. Ca^{2+} is a common ion in the two minerals with different solubilities, and mixing may increase the Ca^{2+} concentration above the solubility limits of the least soluble mineral. The solubility of the two minerals (salts) must differ at least by a factor of 3 to compensate for dilution of the anion (in this case the carbonate ion). At temperatures below 25°C, gypsum and calcite solubilities do not differ enough to give appreciable oversaturation in mixed water. Higher temperatures are needed to produce oversaturation, and this occurs especially when the fraction of gypsum-water in the mixture is between 0.3 and 0.5.

Take as an example a calcite water of Section 4.2.3 at $P_{CO_2} = 10^{-2.0}$ atm, and mix it with gypsum water of Example 3.2, at zero P_{CO_2}. Table 8.3 gives results of simple hand-calculations as well as calculations with PHREEQE, which include activity corrections and aqueous complexes.

Note how the inclusion of complexes and activity coefficients gives much higher Ca^{2+} and SO_4^{2-} concentrations, especially in water in equilibrium with gypsum. The total Ca^{2+} concentration is more than 3 times larger in gypsum-water than in calcite-water, but the mixture is still not oversaturated with respect to calcite because complexes with SO_4^{2-} lower the Ca^{2+} activity.

A better example to illustrate the common ion effect is to mix water in equilibrium with fluorite (CaF_2) with gypsum water. In many areas in the world high concentrations of fluoride in water occur when the Ca^{2+} concentrations are low, as discussed in Chapter 3. Example 8.4 shows how F^- concentrations are lowered when gypsum water is mixed with high fluoride water.

EXAMPLE 8.4: *Mixing of gypsum-water with high F^- containing water*
Water from Lake Baringo (Kenya) contains 0.18 mmol Ca^{2+}/l, and 1 mmol F^-/l from volcanic exhalations in the Rift Valley. Calculate the saturation index with respect to fluorite, before and after mixing 1:1 with gypsum water.

ANSWER
Neglecting activity coefficients, etc., gives for Lake Baringo water:

$$SI_{fluorite} = \log \{0.18 \cdot 10^{-3} \cdot (10^{-3})^2\} - (-10.6) = 0.86$$

(which is already supersaturated).
  Mixing 1:1 with gypsum water from Table 8.2 gives:

Ca^{2+}	SO_4^{2-}	F^-	
2.3	2.21	0.5	(mmol/l)

and
$$SI_{fluorite} = \log\{2.3 \cdot 10^{-3} \cdot (0.5 \cdot 10^{-3})^2\} - (-10.6) = 1.36$$
which is even more supersaturated.

It is sometimes possible to separate contributions to river flow by considering the ions which are generated in specific areas of the drainage basin. An example for a limestone/dolomite/marls area is as follows:

EXAMPLE 8.5: *Contributions of rock types to total flow*
A drainage basin in Northern Italy is underlain by calcite, dolomite and gypsum containing marls. The water composition at the outlet is:

pH	Ca^{2+}	Mg^{2+}	HCO_3^-	SO_4^{2-}	
8.05	3.5	0.1	2.7	2.2	(mmol/l)

Calculate the contribution of each of the rock types to the water composition, assuming:
  – water from gypsum beds is saturated with respect to gypsum;
  – water from dolomite rock is saturated with respect to dolomite, with $P_{CO_2} = 10^{-2}$ atm.

ANSWER
Water from gypsum beds has, if only gypsum dissolves in water until saturation $m_{Ca^{2+}} = m_{SO_4^{2-}} = 4.42$ mmol/l. In the river, $m_{SO_4^{2-}} = 2.2$ mmol/l, a two-fold dilution. The contribution of water from gypsum is therefore 50% of total flow.
  Water from dolomite has, if only dolomite dissolves in water until saturation (use formulas from Chapter 4), $256(m_{Mg^{2+}})^6/(P_{CO_2})^2 = 10^{-12.0}$, and $m_{Mg^{2+}} = 0.85$ mmol/l. In the river, $m_{Mg^{2+}} = 0.1$ mmol/l, which means an 8-fold dilution of the pure dolomite water. Contribution of dolomite water is therefore 12%. There remains a contribution of 38% of runoff from areas with calcite as a mineral.

It is also possible to use a least-squares procedure for estimating end-member concentrations if more components are measured in the river water than the number of end-member waters which make-up the mixture (Christopherson et al., 1990).

8.2.3 *Mixing of reduced water and oxygenated water*

Clogging of well screens can be a problem in groundwater abstraction. A case which often occurs in phreatic aquifers is caused by the combined pumping from an oxygen containing upper groundwater layer and from a reduced, Fe^{2+}-containing deeper groundwater. When both water types mix, $Fe(OH)_3$ precipitates to clog the gravel pack and well screen. Figure 8.10 shows the situation.

Mixing takes place already in the gravel pack, and precipitation of $Fe(OH)_3$ occurs as indicated by the piezometric readings shown in Figure 8.10. The oxidation of Fe^{2+} is

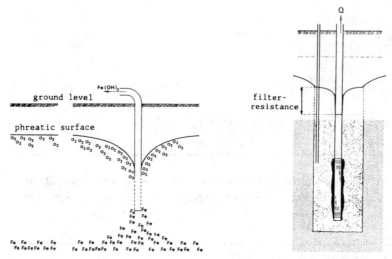

Figure 8.10. (left) Mixing of O_2 rich water with Fe^{2+} containing water in a water well, leads to precipitation of $Fe(OH)_3$. (right) Precipitation occurs in the gravel pack and onto the well screen (Van Beek and Brandes, 1977).

accelerated by iron bacteria, which have an excellent ability to remain fixed on their support even at high water flow velocities. The piezometric level will decrease as filter resistance builds up. Oxygenated water then enters the well at deeper and deeper levels, clogging also the deeper sections of the well. A clogged well can be regenerated mechanically by brushing and jetting at increased flow velocities through sections of the well, and/or chemically by injecting acid HCl-solutions. However, the problem will reappear when pumping is resumed. A possibly more permanent solution is to abstract water separately from different levels in the well, or to use different wells with separate flowlines for shallow and deep groundwater as illustrated in Figure 8.11.

Another type of clogging may occur at the border of the gravel pack and aquifer sand in anaerobic aquifers. As a result of the higher flow velocities near abstraction wells, there is a larger food supply for microorganisms in the form of dissolved organic carbon or SO_4^{2-}. Sulfate reducing bacteria such as *Desulfovibrio* sp. may start to grow on the higher supply. These bacteria are not able to withstand the high flow velocities in the gravel pack, but can survive at the outside. (This in contrast to the iron oxidizing bacteria *Gallionella* and *Leptothrix* sp.). Such well clogging outside the gravel pack is remedied by letting HOCl seep downward from the top of the gravel pack, while maintaining a small injection flow in the well which drives the hypochlorite solution into the aquifer.

Oxidation/reduction reactions in the subsoil can also be used advantageously as a water treatment technique. The common occurrence of Fe^{2+} in reduced groundwater necessitates a treatment before water is distributed as drinking water. It is possible to remove a large part of the iron already in the aquifer by injecting volumes of oxygenated water in the pumping well. This oxygenated water induces a precipitation of iron oxyhydroxide. The precipitation of iron hydroxide occurs mainly in the aquifer where ferrous iron is adsorbed to the sediment, and not in the gravel pack. When pumping is resumed, and the native groundwater reappears, it has an appreciably lower Fe^{2+} concentration for an extended period of

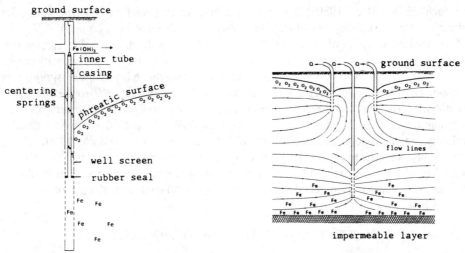

Figure 8.11. Abstraction at different levels in the aquifer can prevent well clogging resulting from mixing of incompatible water types. (Van Beek and Brandes, 1977).

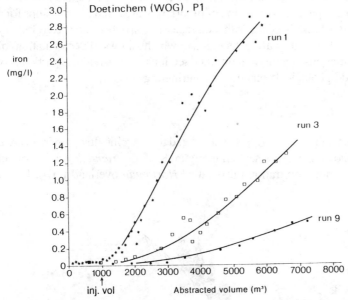

Figure 8.12. Iron concentration in groundwater as a function of pumped volume and number of injections of oxygenated water. With each run ca. 1000 m^3 of oxygenated water has been injected (from Van Beek, 1980a).

time. The efficiency of the system increases with each cycle of injection of oxygenated water, i.e. larger volumes of native groundwater can be pumped with lower Fe^{2+} concentration, as shown in Figure 8.12. The loss of ferrous iron has been attributed to sorption of Fe^{2+} onto the precipitated iron hydroxides, which increase in volume and reactive surface area with each run (Van Beek, 1980a). The technique functions especially well when the aquifer sediment is free of sulfides, and when pH is above 7.0 (Van Beek, pers. comm.).

The oxidative capacity of the injected water is limited by the amount of oxygen which can dissolve at atmospheric pressure in water, which is ca. 0.3 mmol/l. The capacity can be enhanced by using NO_3^- at a high concentration. However, NO_3^- is more sluggish in oxidizing Fe^{2+} than O_2, and it requires either microbiological mediation (Brons et al., 1991) or catalytic surfaces (Postma, 1990). An experimental injection of $NaNO_3$ solution in an esker aquifer in Sweden, indicated that more than 60% of the NO_3^- was consumed by oxidizing organic matter, and only about 40% by oxidizing Fe^{2+} (Vanek, 1990). These experiments are directed towards removal of Fe^{2+} from groundwater, but the reverse possibility, of reducing high NO_3^- concentrations as pollutants from excess fertilizers by injecting NO_3^- water into a reduced aquifer were also found applicable (Riha, 1980).

8.3 CHANGING WATER QUALITY WITH CHANGES IN DISCHARGE

Variations in discharge of springs or rivers are normally accompanied by changes in concentrations of solutes. A higher discharge resulting from rainfall or snow/ice-melt normally dilutes dissolved elements, but some elements may be enriched in the rapid flow component, and thus show an increase in concentration. The changing chemistry of runoff can be used as a tracer to decipher the generation of runoff in an area, and relationships for a year-round and for individual runoff-peaks with their recession have been applied (Gregory and Walling, 1973). A chemical routing technique, which allows the calculation of contributions (and the variations therein) to individual sections of the stream, is particularly useful for tracing the hydrogeological behavior of contributing areas.

8.3.1 *Runoff models*

The rapid flow component of peak runoff has been related to a specific flowpath in an area. Possible flowpaths are shown in a cross-section in Figure 8.13. *Overland flow* is the direct runoff of precipitation without infiltration (also called *Hortonian* overland flow). Direct

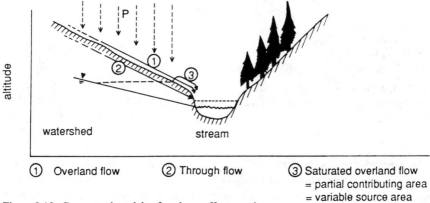

Figure 8.13. Conceptual models of peak runoff generation.

runoff occurs when the rainfall intensity is higher than the infiltration capacity of the soil. Horton (1945) observed this type of flow in badlands, where the surface had a very low infiltration capacity. Low infiltration capacities are also common for roads and other disturbances of the soil structure, and these are the rapid runoff generating areas. In vegetated natural areas (forests) this type of flow is an exception, since a loose, thoroughly rooted and burrowed top soil has a good infiltration capacity. To explain runoff peaks in forested areas, other mechanisms were found more applicable.

Rapid throughflow of precipitation parallel to the surface slope can be such a mechanism. Rapid throughflow has been observed in lysimeter trenches. Throughflow occurs when soil horizons act as flow barriers which hinder a direct downward percolation of precipitation. It is not necessarily saturated flow via a perched watertable. In peaty soils a special type of throughflow (called *piping*) occurs along preferential flowpaths. These have been traced in the Plynlimon catchments in Wales by actually following the gurgling of rapid water flow through the pipes with a stethoscope (Gilman and Newson, 1980; Jones, 1987).

Another type of rapid runoff generation is *saturated overland flow*. The underlying principle is a rapid rise of the water table with rain, caused by percolation of rainwater, or an increase of hydraulic head that increases groundwater discharge in the areas near to the river. Where the water table reaches the surface, rain flows as surface runoff towards the river and contributes to the runoff peak.

8.3.2 *Mixing models*

Each different mechanism of rapid flow generation may be expected to contribute with a different water quality to the river flow. The simplest technique of water quality based separation of runoff is to assume constant concentrations for the different components. For two components:

$$C_t Q_t = C_g Q_g + C_s Q_s \tag{8.5}$$

and

$$Q_t = Q_g + Q_s \tag{8.6}$$

where C is the concentration of a solute (mmol/l), Q is discharge (l/s), and subscripts g, s, t indicate groundwater, direct, and total runoff components.

The ratio of two runoff components is then:

$$\frac{Q_s}{Q_g} = \frac{C_g - C_t}{C_t - C_s} \tag{8.7}$$

Classical separation using this relationship has been performed by Pinder and Jones (1969), for a stream draining a karstic carbonate terrain. They calculated a groundwater discharge based on Mg^{2+} and Ca^{2+} concentrations in river runoff, as shown in Figure 8.14. The concentrations of ions in the groundwater component were assumed constant, and estimated from baseflow values. The direct runoff concentrations were estimated by Pinder and Jones to be identical to river concentrations in the upper reach of the stream, where concentrations showed minor change when discharge increased. However, this assumption may not be generally applicable, and appreciable changes in concentrations of the rapid flow component have been found to exist (Nakamura, 1971; Pilgrim et al., 1978).

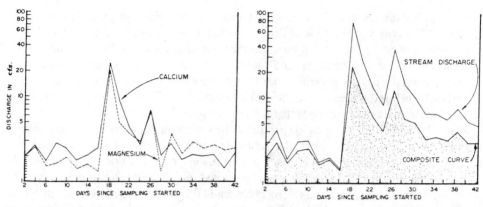

Figure 8.14. Groundwater component of total runoff, calculated using Ca^{2+} and Mg^{2+} concentrations in river water. The composite curve is estimated as an average using other elements as well (from Pinder and Jones, 1969. Copyright by the American Geophysical Union).

Similar separation of flow components was claimed by Kunkle (1965), Toler (1965), Sklash and Farvolden (1979), Herrmann and Stichler (1980), Van de Griend and Arwert (1983), Hooper et al. (1990). Separation with isotopes can be considered least ambiguous, and it appears that the groundwater component (or resident water component) during peak flow accounts for more than 80% of the water which is discharged (Sklash and Farvolden, 1979; Herrmann and Stichler, 1980).

An alternative to calculating the runoff components, is to use a hydrological model to study the changes in water chemistry when runoff varies. Reworking of (8.5) and (8.6) gives:

$$C_t = \frac{C_g + C_s \cdot Q_s/Q_g}{1 + Q_s/Q_g} \tag{8.8}$$

We may try to simplify this equation with a concept of the hydrology of the catchment. Johnson et al. (1969) assumed that Q_s/Q_g is directly proportional to stream discharge:

$$Q_s/Q_g = \beta \cdot Q_t$$

and defined the difference between groundwater and rapid runoff concentrations as

$$C_d = C_g - C_s$$

Inserting these relations in (8.8) gives:

$$C_t = \frac{C_d}{1 + \beta \cdot Q_t} + C_s \tag{8.9}$$

Johnson et al. (1969) plotted concentrations in stream discharge for the Hubbard Brook catchments against the factor $1/(1 + \beta \cdot Q_t)$, to obtain $C_d = C_g - C_s$ from the slope, and C_s as the intercept. A few of their plots are presented in Figure 8.15. It appeared that some elements (Na^+, Si, Mg^{2+} and Ca^{2+}) were diluted with increasing stream discharge, while others (Al^{3+}) increased with stream discharge. NO_3^- and K^+ showed a seasonal behavior, concentrations being consistently low during summer, but showing an increase with discharge during the winter period.

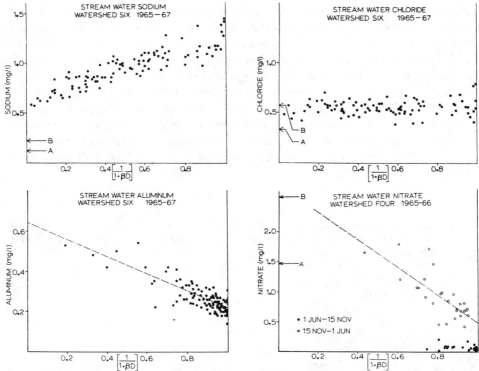

Figure 8.15. Variation of concentrations in Hubbard Brook streamflow with the reciprocal of discharge (note that D stands for total discharge $= Q_t$). Reprinted with permission from Johnson et al., 1969. Copyright by the American Geophysical Union.

The decreasing concentrations with discharge can be explained as dilution of high concentrations in deeper soil water which are a result of weathering of silicate rock. The increasing Al-concentrations were related to high concentrations of this element in the acid topsoil. The low concentrations of NO_3^- and K^+ during summer were attributed to a biological uptake which balances the decomposition of organic matter. The increase of these elements with discharge in winter was related to the relatively large availability of these ions from decaying organic matter in the topsoil.

This study is of interest, because it shows that flow components can be linked to specific chemical and biological properties of a watershed (Al^{3+} from the acid topsoil; also NO_3^- and K^+ when not used by vegetation). The K^+ increase with discharge has been found in many other studies, and PO_4^{3-} often displays a similar behaviour. Both elements are heavily cycled in the biomass, and can be almost completely and selectively removed from root zone water by vegetation. They are washed from vegetation or litter by rain and flushed into surface runoff. Other studies have also related specific elements to particular flowpaths. Blavoux and Mudry (1983) used Ca^{2+} and Mg^{2+} concentrations to discern 'matrix-flow' (similar to diffuse flow) and 'karst-river' (similar to conduit-flow, see Chapter 4). Haubert (1976) used Sr^{2+} in a similar study. Increasing Ca^{2+} and Mg^{2+} concentrations with discharge from springs in the Dolomites could be linked with more rapid contributions from forest slopes (Chapter 4).

However, problems arise when the chemical changes are interpreted in terms of quantitative hydrological contributions (Hall, 1970). The runoff separation only works in very simple and completely homogeneous areas. In most watersheds, different vegetations lead to different amounts of direct runoff, and to different groundwater qualities. The rapid flow component undoubtedly will show changes in concentration during the runoff event. A further complicating factor is channel routing which can separate the hydrograph rise and the water body that originally caused the discharge peak, as will be discussed in the next section. Despite all these complications, it is often possible to relate chemistry and discharge at a single station with relatively simple equations that contain only few parameters. A number of relations which are useful for empirical fitting and extrapolation purposes are presented in Table 8.4.

Table 8.4. Concentration/discharge models (Hall, 1970; Haubert, 1976). Notation as before; subscripts t, b, s and g indicate total runoff, baseflow (before peak flow), surface runoff, groundwater, and a, b and n are fitting parameters.

(1)	$C_t = a\,Q_t^{-1/n}$
(2)	$C_t = a\,Q_t^{-1/n} + C_s$
(3a)	$C_t = b - a\log Q_t$
(3b)	$C_t = a\exp\left(-b\,Q_t + 1/n\right)$
(3c)	$C_t = b - a\,Q_t^{1/n}$
(4)	$C_t = C_g\,/\,(1 + a\,Q_t^{1/n})$
(5)	$C_t = (C_g - C_s)\,/\,(1 + a\,Q_t^{1/n}) + C_s$
(6)	$C_t = a\,(Q_t - Q_b)^{-1/n} + C_s$

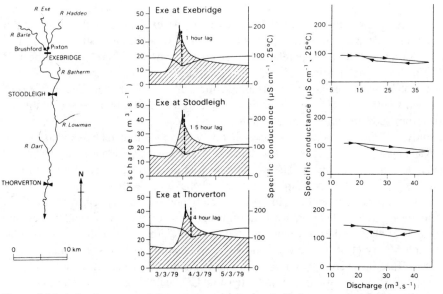

Figure 8.16. A lag between minimum concentration and discharge peak (resulting from channel routing) leads to hysteresis in the plot of concentration vs. discharge (from Walling & Webb, 1980).

8.3.3 *Lag effects*

A simple mixing model predicts that the largest change in concentration must be observed when discharge is at its maximum. In reality a time delay (lag) is often observed between the discharge maximum, and the concentration minimum in river water. Plots of concentration versus discharge, often show a clockwise loop (Figure 8.16), that can be attributed to channel routing: the rise of the hydrograph travels more rapidly than average water velocity in the stream channel. Hysteresis can also be attributed to a heterogeneous reaction of subareas in the watershed, or to a differentiated distribution of rainfall over the drainage basin.

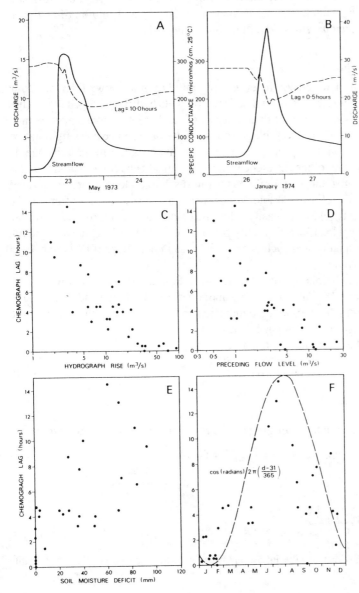

Figure 8.17. Lag effects in a drainage basin which are related to previous flow conditions (Walling and Foster, 1975).

Lag effects are more pronounced in larger rivers as a result of channel routing. A particular case in a small drainage basin has been described by Walling and Foster, 1975. These authors observed that the lag time between minimum concentration and maximum flow changed according to conditions as shown in Figure 8.17. The lag time shortened when the hydrograph rise was higher (Figure 8.17 C), when previous discharge was higher (Figure 8.17 D), or when soil moisture deficit was greater (Figure 8.17 E). These effects indicated that the lag was connected with the speed of translation of a rain peak to the discharge peak, and hence with soil moisture conditions. A wetter soil (evidenced by a higher previous discharge level) gives faster direct runoff, and so does a higher rainfall intensity. Rain on a relatively dry soil seemed to push out existing groundwater (soil moisture) first. Since the soil moisture conditions follow a seasonal cycle, being driest in the summer, wettest in the winter, and intermediate in the spring and autumn time, the lag could be described by an annual sinusoidal function (Figure 8.17 F).

8.3.4 *Water quality routing*

Our previous discussion of watershed heterogeneity and lag time showed that we could interpret the chemical response of a watershed if we knew its hydrological behavior. The a priori interpretation of chemical runoff data in terms of catchment response may be more difficult. The temporal variations of discharge and concentrations, noted at a single station, can be related accurately with empirical formulas, but this does not necessarily yield a physically acceptable picture of the differences in hydrological behavior of sub-areas. The actual measurement of area and temporal changes in discharge along a stream unravels these differences, and is therefore desirable. Water chemistry can sometimes offer a tool to obtain such measurements relatively easily. The procedure has been termed *EC*-routing (Appelo, et al., 1983, Appelo, 1988). The only prerequisite is that inflowing water quality differs from river water quality, and such is often the case in both large (O'Connor, 1976) and small (Webb, 1976) rivers.

Consider again the simple mixing model (Equations (8.5) and (8.6)), but now linked in series as shown in Figure 8.18. Capital letters Q and A indicate discharge and water quality in the main stream, and small letters q and a indicate similarly discharge and water quality of contributing streams, springs or seepage zones.

We start again with the continuity equation:

$$Q_{i+1} \cdot A_{i+1} = Q_i \cdot A_i + q_{i, i+1} \cdot a_{i, i+1} \tag{8.5a}$$

and

$$Q_{i+1} = Q_i + q_{i, i+1} \tag{8.6a}$$

When solving for discharge in the main stream, the equations are combined to give:

$$Q_{i+1} = Q_i \cdot [(A_i - a_{i, i+1})/(A_{i+1} - a_{i, i+1})] = Q_i \cdot R_{i+1} \tag{8.10}$$

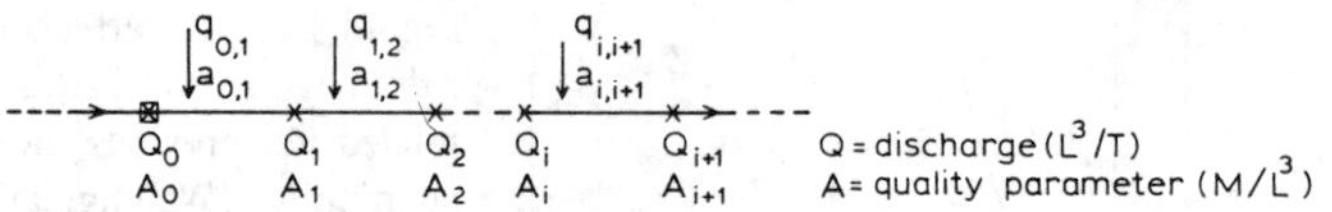

Figure 8.18. Stream reaches that receive inflow of different quality can be coupled in a routing procedure.

and for the side-inflow:

$$q_{i,i+1} = Q_i \cdot [(A_{i+1} - A_i)/(a_{i,i+1} - A_{i+1})] = Q_i \cdot S_{i,i+1} \tag{8.11}$$

When *EC* is routed along the stream, and discharge is measured at Q_0 only, a (fractional) variance is associated with the quantitative estimate of stream discharge:

$$\sigma^2_{Q_{i+1}} = \left(\frac{\Delta Q_0}{Q_0}\right)^2 + \sum_{j=1}^{i+1} \left(\frac{\Delta R_j}{R_j}\right)^2 \tag{8.12}$$

where ΔQ_i is the error of measurement of Q_i.

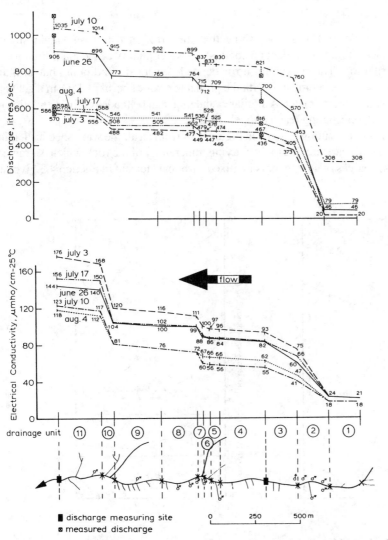

Figure 8.19. *EC* routings and calculated flow profiles in the Grossklausenbach (Ahrntal, N.Italy) (Appelo et al., 1983).

The variance $(\Delta R_j / R_j)^2$ can be obtained by differentiation of R_j to individual components in the equation and multiplication with the squared error of measurement. (It is assumed that errors in the *EC*-measurement are without correlation). Thus the variance of R_{i+1} is the sum of:

$$\left(\frac{\partial R_{i+1}}{\partial A_i}\right)^2 \cdot \frac{(\Delta A_i)^2}{R_{i+1}^2} = \frac{(\Delta A_i)^2}{(A_i - a_{i,\,i+1})^2} \tag{8.13a}$$

$$\left(\frac{\partial R_{i+1}}{\partial a_{i,\,i+1}}\right)^2 \cdot \frac{(\Delta a_{i,\,i+1})^2}{R_{i+1}^2} = \frac{(\Delta a_{i,\,i+1})^2}{(A_{i+1} - a_{i,\,i+1})^2} \cdot \frac{(A_i - A_{i+1})^2}{(A_i - a_{i,\,i+1})^2} \tag{8.13b}$$

$$\left(\frac{\partial R_{i+1}}{\partial A_{i+1}}\right)^2 \cdot \frac{(\Delta A_{i+1})^2}{R_{i+1}^2} = \frac{(\Delta A_{i+1})^2}{(A_{i+1} - a_{i,\,i+1})^2} \tag{8.13c}$$

The combined equations (8.13a+b+c) show that the error in estimating main stream discharge is almost entirely dependent on contrasts in water quality of the main stream and the lateral contribution. The maximum error depends on local conditions, but is often surprisingly low, also when compared with the probable error of perhaps 15% in discharge measurements with a current meter. A similar calculation can be made for the estimation of discharge from the contributing reach (Appelo et al., 1983).

An example of an *EC*-routing is given in Figure 8.19 from an alpine watershed in N. Italy. Contrasts in *EC* in this case are due to the varying mineralogy of the rock which is drained. (The example on sources of constituents in mixed carbonate terrain in Section 4.3 gives an

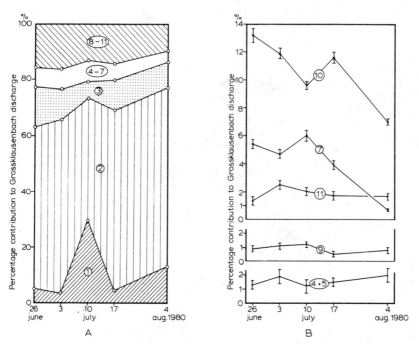

Figure 8.20. Partial area contributions to Grossklausenbach discharge in response to different hydrological conditions (snowmelt on June 26, July 3, Aug. 4; rainfall on July 10, 17) (Appelo et al., 1983).

interpretation of the observed water qualities). From the EC profiles and from the EC's of the contributing springs and seepage zones, the partial contributions to the stream were calculated. The resulting flow profiles are also given in Figure 8.19. It is of interest that even small increases of less than 1% in total flow are detected so that a very detailed picture of the buildup of stream discharge can be obtained.

It is also possible to account for sub-area response to changing hydrological conditions. Figure 8.20 shows the percentage contribution from different drainage units of the stream. The change in response results from varying reactions to snowmelt and to a very intense rainfall on July 9. It is obvious that unit 1, which is the cirque area in the uppermost reach of the stream, showed the most pronounced reaction, with an increase in contribution from 5% before to 30% of total runoff after the storm. This surprisingly strong reaction countered earlier ideas of a sluggish reaction of the numerous scree and slope deposits in this sub-area. Error bars in Figure 8.20B indicate the error in estimating the partial contribution. The precision varies, and is low for areas which contribute only 1-2% of total flow via diffuse sources.

EXAMPLE 8.6: *Calculate discharge along a stream from EC-measurements*
Discharge in a stream, measured with a current meter, is 100 l/s, $EC = 200\ \mu S/cm$. Lateral inflow from a spring has $EC = 500\ \mu S/cm$, and the stream downflow from the spring has $EC = 280\ \mu S/cm$. Estimate discharge of the stream after confluence, and give the error of the estimate. EC is measured accurately to 10 $\mu S/cm$.

ANSWER
From Equation (8.10) we find $Q_{i+1} = 100 \cdot (100\text{-}500)/(280\text{-}500) = 136$ l/s. The fractional variance is the sum of the three parts given in Equation (8.13):

(8.13a): $10^2/(200\text{-}500)^2 = 0.0011$,
(8.13b): $10^2/(280\text{-}500)^2 \cdot (200\text{-}280)^2/(200\text{-}500)^2 = 0.00015$, and
(8.13c): $10^2/(280\text{-}500)^2 = 0.002$

This gives a variance $\sigma^2 = 0.00325$, and hence an error of $\sqrt{0.00325} = 0.057$, which is 5.7%.

PROBLEMS

8.1. In a spring from a dolomite reef, water from two flowlines is mixed. Water from the first flowline flows through soil with $P_{CO_2} = 0.01$, dissolving dolomite. Water from the second originates from snowmelt and has $P_{CO_2} = 10^{-3.5}$, then infiltrates in the reefstock where it dissolves dolomite.
  a. Calculate Ca^{2+}, Mg^{2+}, HCO_3^-, CO_3^{2-} concentrations, P_{CO_2} and pH of water from the two flowlines, and the SI with respect to dolomite.
  b. The mixing-ratio of the flowlines at the spring is 1:1, calculate the composition. Is the water saturated with respect to dolomite and calcite?
  c. At another spring, water with composition calculated under **b** is mixed with water from a third flowline. This water flows through layers of gypsum, is saturated with respect to gypsum and has $P_{CO_2} = 0.01$, calculate the composition of water from the third flowline.
  d. The contribution of the third flowline is 10% to spring discharge. Calculate the composition of the spring water. Is the water saturated with respect to dolomite, calcite and gypsum?

8.2. Derive the relation between Ca^{2+} and $H_2CO_3^*$ concentrations in water in which calcite dissolves. Draw a plot of Ca^{2+} vs. $H_2CO_3^*$ concentrations, and explain the effect of mixing on the saturation index.

8.3. Derive the relation between Na^+ and $H_2CO_3^*$ concentrations in water when albite weathers to kaolinite (cf. Reaction 6.2). Explain the effect of mixing on the saturation index for this reaction.

8.4. Draw a plot of Ca^{2+} vs. SO_4^{2-} concentrations in water saturated with gypsum. Explain the effect of mixing on the saturation index.

8.5. A groundwater sample in the Nile delta has as composition (cf. Problem 5.6):

Na^+	Mg^{2+}	Ca^{2+}	HCO_3^-	Cl^-	SO_4^{2-}	pH	
8.0	0.2	0.7	2.7	4.5	?	8.1	mmol/l

a. Estimate the SO_4^{2-} concentration.

b. Calculate the CO_2 pressure of the water, and the saturation index with respect to calcite.

c. The water is used as irrigation water, and irrigation return flow is 10% (i.e. 90% of the water evaporates). Calculate the Ca^{2+} and HCO_3^- concentration, when calcite precipitates; use the same CO_2 pressure and *SI* for calcite as in the original groundwater.

d. Calculate *SAR*, and the exchangeable Na-X, in percentage, before and after evaporative concentration. Assume $K_{Mg\backslash Ca} = 1$, $K_{Na\backslash Ca} = 0.4$, Gaines and Thomas convention.

REFERENCES

Appelo, C.A.J., 1988, Influences on water quality in the Hierdensche Beek catchment (in Dutch), Inst. Earth Sciences, Free University, 100 pp.

Appelo, C.A.J., Becht, R., Spierings, T.C. and Van de Griend, A.A., 1983, Buildup of discharge along the course of a mountain stream. *J. Hydrol.* 66, 305-318.

Blavoux, B. and Mudry, J., 1983, Separation des composantes de l'écoulement d'un exutoire karstique à l'aide des méthodes physico-chimiques. *Hydrogéologie* 1983, 269-278.

Bögli, A., 1978, *Karsthydrographie und physische Speläologie.* Springer, Berlin, 292 pp.

Brons, H.J., Hagen, W.R. and Zehnder, A.J.B., 1991, Ferrous iron dependent nitric oxide production in nitrate reducing cultures of *Escherichia coli. Arch Microbiol.* 155, 341-347.

Christopherson, N., Neal, C., Hooper, R.P., Vogt, R.D. and Andersen S., 1990, Modeling streamwater chemistry as a mixture of soilwater end-members – a step towards second-generation acidification models. *J. Hydrol.* 116, 307-3.

Collins, A.G., 1975, *Geochemistry of oilfield waters.* Elsevier, Amsterdam, 496 pp.

Clayton, R.N., Friedman, I., Graf, D.L., Mayeda, T.K., Meents, W.F. and Shimp, N.F., 1966, The origin of saline formation waters. 1. Isotopic composition. *J. Geophys. Res.* 71, 3869-3882.

Drever, J.I., 1988, *The geochemistry of natural waters.* 2nd ed. Prentice-Hall, New Jersey, 437 pp.

Frape, S.K. and Fritz, P., 1987, Geochemical trends for groundwaters from the Canadian Shield. In P. Fritz and S.K. Frape (eds), *Saline waters and gases in crystalline rocks.* Geol. Ass. Canada Spec. Paper 33, 19-38.

Gilman, K. and Newson, M.D., 1980, Soil pipes and pipeflow: a hydrological study in upland Wales. British Geomorphol. Res. Group. Monogr. 1. Geobooks, Norwich, 110 pp.

Graf, D.L., 1982, Chemical osmosis, reverse chemical osmosis and the origin of subsurface brines. *Geochim. Cosmochim. Acta* 46, 1431-1448.

Gregory, K.J. and Walling, D.E., 1973, *Drainage basin form and process – a geomorphological approach.* Edward Arnold, London, 456 pp.

Hall, F.R., 1970, Dissolved solids – discharge relationships. I – Mixing models. *Water Resour. Res.* 6, 845-850.

Harder, H., 1976, Nontronite synthesis at low temperatures. *Chem. Geol.* 18, 169-180.

Hardie, L.A., 1991, On the significance of evaporites. *Ann. Rev. Earth Planet. Sci.* 19, 131-168.

Hardie, L.A. and Eugster, H.P., 1970, The evolution of closed-basin brines. *Miner. Soc. Am. Spec. Publ.* 3, 273-290.

Hanshaw, B.B. and Back, W., 1985, Deciphering hydrological systems by means of geochemical processes. *Hydrol. Sci. J.* 30, 257-271.

Harvie, C.E. and Weare, J.H., 1980, The prediction of mineral solubilities in natural waters: the Na-K-Mg-Ca-Cl-SO$_4$-H$_2$O system from zero to high concentration at 25° C. *Geochim. Cosmochim. Acta* 44, 981-997.

Haubert, M., 1976, Bilan hydrochimique d'un bassin versant de moyenne montagne: La Drause de Bellevaux (Brévon), Haute-Savoie. *Hydrogéologie* 1976, 5-30.

Herrmann, A. and Stichler, W., 1980, Groundwater-runoff relationships. *Catena* 7, 251-263.

Hooper, R.P., Christopherson, N. and Peters, N.E., 1990, Modeling streamwater chemistry as a mixture of soilwater end-members – an application to the Panola mountain catchment, Georgia, USA. *J. Hydrol.* 116, 321-343.

Horton, R.E., 1945, Erosional development of streams and their drainage basins: hydrophysical approach to quantitative morphology. *Geol. Soc. Am. Bull.* 56, 275-370.

Jankowski, J. and Jacobson, G., 1989, Hydrochemical evolution of regional groundwaters to playa brines in central Australia. *J. Hydrol.* 108, 123-173.

Johnson, N.M., Likens, G.E., Bormann, F.H. and Pierce, R.S., 1969, A working model for the variation in stream water chemistry at the Hubbard Brook Experimental Forest, New Hampshire. *Water Resour. Res.* 5, 1353-1363.

Jones, J.A.A., 1987, The effects of soil piping on contributing areas and erosion patterns. *Earth Surf. Proc. Landf.* 12, 229-248.

Kharaka, Y.K. & Berry, F.A.F., 1973, Simultaneous flow of water and solutes through geological membranes. *Geochim. Cosmochim. Acta* 37, 2577-2603.

Kharaka, Y.K. & Carothers, W.W., 1986, Oxygen and hydrogen isotope geochemistry of deep basin brines. In P. Fritz and J.Ch. Fontes (eds); *Handbook of environmental isotope geochemistry*. Elsevier, Amsterdam, 305-360.

Kunkle, G.R., 1965, Computation of groundwater discharge to streams during floods, or to individual reaches during base flow, by use of specific conductance. U.S. Geol. Survey Prof. Paper, 525-D, 207-210.

Mazor, E., 1983, Groundwater mixing – a key to understand water-rock interactions. *Proc. 4th Water-Rock Interaction Symp., Misasa, Japan* 333-336.

Mazor, E., 1992. Interpretation of water-rock interactions in cases of mixing and undersaturation. In Y.K. Kharaka and A.S. Maest (eds), *Water-Rock Interaction*, Proc 7th Water-Rock Interaction Symp. Balkema, Rotterdam, 233-236.

Nakamura, J., 1971, Runoff analysis by electrical conductance. *J. Hydrol.* 14, 197-212.

O'Connor, D.J., 1976, The concentration of dissolved solids and river flow. *Water Resour. Res.* 12, 279-294.

Parkhurst, D.L., Plummer, L.N. and Thorstenson, D.C., 1982, BALANCE – A computer program for calculating mass transfer for geochemical reactions in groundwater. U.S. Geol. Surv. Water Resour. Inv. 82-14, 92 pp.

Parkhurst, D.L., Thorstenson, D.C. and Plummer, L.N., 1980, PHREEQE – A computer program for geochemical calculations. U.S. Geol. Surv. Water Resour. Inv. 80-96, 210 pp.

Pilgrim, D.H., Huff, D.D. and Steele, T.D., 1978, A field evaluation of subsurface and surface runoff. II Runoff processes. *J. Hydrol.* 38, 319-341.

Pinder, G.F. and Jones, J.F., 1969, Determination of the groundwater component of peak discharge from the chemistry of total runoff. *Water Resour. Res.* 5, 438-445.

Pitzer, K.S., 1987, A thermodynamic model for aqueous solutions of liquid-like density. In I.S.E. Carmichael and H.P. Eugster (eds), Reviews in Mineralogy 17, Mineral. Soc. Am., 97-142.

Plummer, L.N., 1975, Mixing of sea water with calcium carbonate groundwater. *Geol. Soc. Am. Mem.* 42, 219-236.

Plummer, L.N. and Parkhurst, D.L., 1990, Application of the Pitzer equations to the PHREEQE geochemical model. In D.C. Melchior and R.L. Basset (eds); *Chemical modeling of aqueous systems II*. ACS Symp. Ser. 416, Am. Chem. Soc., 128-137.

Plummer, L.N., Prestemon, E.C. and Parkhurst, D.L., 1991, An interactive code (NETPATH) for modeling net geochemical reactions along a flow path. U.S. Geol. Survey Water Resour. Inv. 91-4078.

Postma, D., 1990, Kinetics of nitrate reduction by detrital Fe(II)-silicates. *Geochim. Cosmochim. Acta* 54, 903-908.

Riha, M., 1980, Subterranean groundwater treatment. *Proc. Hydrol. & Water Resour. Symp., Adelaide, Australia*, 49.

Sanford, W.E. and Konikow, L.F., 1989, Simulation of calcite dissolution and porosity changes in saltwater mixing zones in coastal aquifers. *Water Resour. Res.* 25, 655-667.

Sklash, M.G. and Farvolden, R.N., 1979, The role of groundwater in storm runoff. *J. Hydrol.* 43, 45-65.

Toler, L.G., 1965, Relation between chemical quality and water discharge in Spring Creek, Southwestern Georgia. U.S. Geol. Survey Prof. Paper, 525-C, 209-213.

Van Beek, C.G.E.M., 1980a, A model for the induced removal of iron and manganese from groundwater in the aquifer. *Proc. 3rd Water-Rock Interaction Symp., Edmonton, Canada*, 29-31.

Van Beek, C.G.E.M., 1980b, Removal of iron and manganese in the aquifer (in Dutch), H_2O 13, 635-638.

Van Beek, C.G.E.M. and Brandes, M.C., 1977, Regeneration of wells. KIWA-VWN coll. aug. 1977, KIWA (in Dutch).

Van de Griend, A.A. and Arwert, J.A., 1983, The mechanism of runoff generation from an alpine glacier during a storm traced by oxygen $^{18}O/^{16}O$. *J. Hydrol.* 62, 263-278.

Vanek, V., 1990, In situ treatment of iron-rich groundwater by the addition of nitrate. Dept. Ecology, University Lund, 33 pp.

Van Gaans, P.F.M., 1990, *The Pitzer model applied to aqueous GaCl3 solutions with evaluation of regression methods*. Ph.D. Thesis, Utrecht, 155 pp.

Walling, D.E. and Foster, I.D.L., 1975, Variations in the natural chemical concentration of river water during flood flows, and the lag effect: some further comments. *J. Hydrol.* 26, 237-244.

Walling, D.E. and Webb, B.W., 1980, The spatial dimension in the interpretation of stream solute behaviour. *J. Hydrol.* 47, 129-149.

Webb, B.W., 1976, Solute concentrations in the baseflow of some Devon streams. *Rep. Trans. Devon Ass. Adv. Sci.* 108, 127-145.

CHAPTER 9

Solute transport in aquifers

To predict the fate of chemicals during their transport in groundwater has become an ever more important job for hydrogeologists and geochemists. The problem is to define the flowlines of groundwater in the aquifer, the travel times of water along the flowlines, and to predict the chemical reactions which alter concentrations during transport. There are various numerical models available which can be used to describe flowlines, and some models include an option to model dispersion and adsorption or desorption. In the last decade a number of chemical computer models have become available as well; when transport is coupled to these chemical models a fairly complete picture of 'aquifer-reality' can be given. We will show in this chapter how flowlines and residence times can be calculated manually, and how retardation and dispersion of chemicals can be included in the calculations. The methods allow us to obtain solute travel times for simple systems, but for more complex situations or chemical relations, we must have recourse to numerical models. We will show how chemical reactions of any kind can be included in numerical transport codes.

9.1 GROUNDWATER FLOW

Groundwater flow at a given location depends on the permeability of the subsoil, and the *potential* or *hydraulic* gradient. The specific discharge is given by Darcy's law as:

$$v_D = -k \, dh/dx \qquad (9.1)$$

where v_D is specific discharge or *Darcy velocity* (m/day), k is permeability or hydraulic conductivity (m/day), and dh/dx is the (1D) hydraulic gradient.

The piezometric level of groundwater (as measured in a well) gives, after corrections for density differences etc., the potential (h) for groundwater flow. The discharge (or flux) is obtained by multiplying specific discharge with the surface area perpendicular to flow:

$$Q = v_D \cdot A = -kA \cdot dh/dx \qquad (9.2)$$

where Q is discharge (m^3/day), and A is *total* surface area, pore space and grain-skeleton together.

Aquifer permeability is usually calculated from permeameter measurements of the parameters used in Equation (9.2). The Darcy velocity is defined as specific discharge per unit area A of the aquifer. However, only the pore space contributes to water flow, so that the

Table 9.1. Permeability (k) and porosity (ε) of different sediments.

	k, m/day	ε (fraction)
Gravel	200 –2000	0.15–0.25
Sand	10 – 300	0.20–0.35
Loam	0.01– 10	0.30–0.45
Clay	10^{-5} – 1	0.30–0.65
Peat	10^{-5} – 1	0.60–0.90

actual velocity of water movement through the pores must be larger. The actual velocity is known as *pore water flow velocity*. When the porosity is a fraction ε, the pore water velocity is:

$$v = \frac{v_D}{\varepsilon} = -\frac{k}{\varepsilon} \cdot \frac{dh}{dx} \tag{9.3}$$

The pore water flow velocity can also be calculated from discharge and the cross section $A \cdot \varepsilon$:

$$v = Q/(A \cdot \varepsilon) \tag{9.4}$$

Some information on actual groundwater flow velocities can be obtained using permeabilities and porosities of sediments as given in Table 9.1. For example, the actual distance covered by groundwater in a sandy aquifer with porosity $\varepsilon = 0.3$, and permeability $k = 50$ m/day, amounts to 60 m/yr when the potential gradient is 0.001 (1 m per 1 km). Permeabilities vary strongly for different sediments, but porosities vary only from values of 0.2 for coarse, unsorted sands to 0.65 for clay. Hydraulic gradients follow primarily the local relief, and are further influenced by recharge and discharge rates, and groundwater withdrawal (Verruyt, 1969; Davis and De Wiest, 1967; Freeze and Cherry, 1979; Domenico and Schwartz, 1990).

Layers of clay, peat and loam, which are considered impermeable, are termed *aquicludes* or, when almost impermeable, *aquitards*. Such layers can separate more permeable (sandy) *aquifers*. If the uppermost aquifer extends to the earth's surface it is called a *phreatic* or unconfined aquifer. Aquifers which, on the other hand, are confined by aquicludes or aquitards are termed confined aquifers. Groundwater levels from a number of wells can be combined in a map of *isohypses*, which are the lines connecting points with equal groundwater elevations or potentials. The groundwater flow direction is at right angles to the isohypses (cf. Figure 9.1) if the aquifer is *isotropic*, which means that properties are equal in all directions.

The groundwater flow velocity can, in principle, be calculated from Equation (9.1), when the hydraulic gradient and permeabilities are known. However, aquifer sediments often show such variation in permeability that a water balance, based on Equation (9.4) may provide a better idea of average, or *effective* parameters. Effective parameters may differ from the ones which can be measured directly. An effective porosity is the porosity which contains water that partakes in flow, in contrast to the total porosity which includes the pores filled with stagnant water. An effective permeability is similarly obtained from the quotient of discharge over hydraulic gradient.

Water balances are also used in combination with measured groundwater potentials in numerical models, to optimize the *transmissivity* (which is the product of permeability and

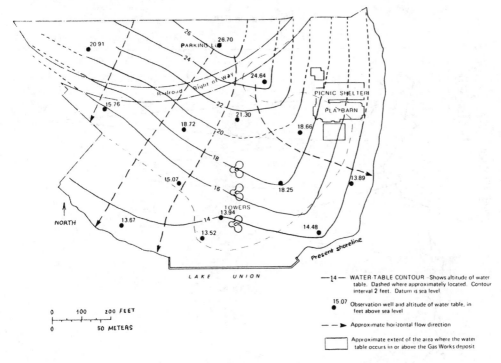

Figure 9.1. Water table contours (isohypses) and groundwater flow directions. From Turney and Goerlitz, 1990.

aquifer thickness). The transmissivity is a sufficient parameter for classical hydrological studies where only the quantity of (horizontal) water transport is of importance. Increased pollution has aroused interest for deciphering the actual flowlines in vertical profiles of the aquifer. Due to large variations in permeability, pollution may spread over much larger distances than predicted by average transport. We will show in the following sections how a water balance can be used to obtain groundwater flowlines and velocities.

9.1.1 *Flow velocity in the unsaturated zone*

Water in the unsaturated zone percolates vertically downward along the maximal gradient of soil moisture potential, when relief is moderate. A simple mass balance can give the rate of percolation at steady state:

$$v = P / \varepsilon_w \qquad (9.5)$$

where P is the precipitation surplus (m/yr), and ε_w the water filled porosity. This relationship has been verified in the field by Andersen and Sevel (1974) using (environmental) tritium profiles in a 20 m thick unsaturated zone in sandy sediments. A water flow velocity that is simply determined by mass balance means that infiltrating water pushes the old water ahead, a type of flow which is adequately termed *piston flow*. Figure 9.2 shows a piston type percolation of tritium in the unsaturated zone in chalk at three locations in England. The actual flow mechanism in that situation involves more than just simple piston

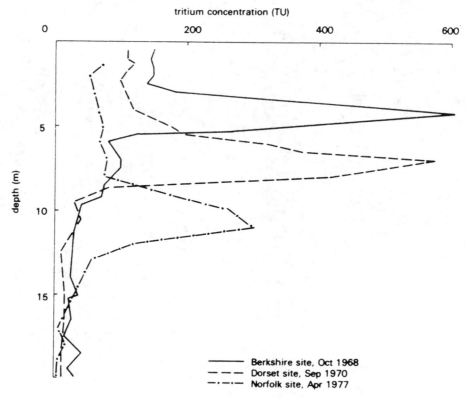

Figure 9.2. Piston flow displacement of environmental tritium in the English Chalk. From Foster and Smith-Carrington, 1980.

flow, however. The observed displacement of ca. 1 m/yr was too low compared with local precipitation and the porosity of the chalk, and rapid bypass flow through macro-fissures was suggested to contribute significantly to downward percolation (Foster and Smith-Carrington, 1980).

In arid or semi-arid areas with very small recharge rates, more complex situations may be encountered. The complexity is, amongst others, apparent in variations in Cl$^-$ concentrations with depth in the unsaturated zone. A simple concentration by evapotranspiration would give a uniform Cl$^-$ concentration below the depth where roots extract water (and reject part of Cl$^-$ contained in the soil moisture). However, it has been observed that Cl$^-$ in some cases may decrease again at larger depth, as shown in Figure 9.3. This decrease has been attributed to bypass flow of water along preferential flow paths down to levels beneath the root zone (Sharma and Hughes, 1985), and also to changes in recharge with time (Allison et al., 1985). At shallow depth, where temperature fluctuations can be significant, water vapor transport may be important as well. A recent observation of depth profiles of ^{36}Cl (mainly from 1950's atomic bomb testing) and of tritium (or ^{3}H, mainly from 1960's hydrogen bomb testing) in the New Mexico desert showed that ^{3}H is now found at greater depth than the earlier infiltrated ^{36}Cl, which can only be attributed to water vapor transport (Phillips et al., 1988).

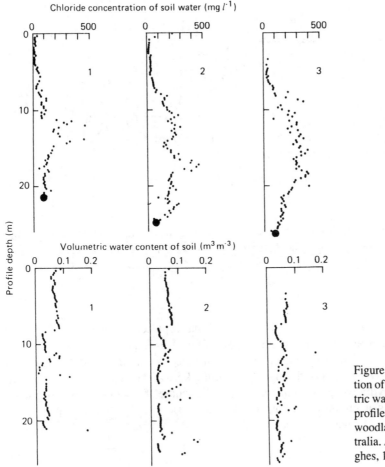

Figure 9.3. Depth distribution of chloride and volumetric water content for three profiles beneath Baksia woodland in Western Australia. After Sharma and Hughes, 1985.

9.1.2 *Flowlines and flow velocities in aquifers*

The water balance is the easiest way to visualize how groundwater flows in aquifers (Ernst, 1973; Gelhar and Wilson, 1974; Hoeks, 1981; Dillon, 1989). Imagine a phreatic aquifer shown in Figure 9.4, with a homogeneous sediment so that porosity and permeability are equal everywhere. The rectangular aquifer shown in Figure 9.4 has (very nearly) equal flow velocity at all depths along a vertical line. (Why does an aquifer not show the velocity profile of a river?). The point along the upper reach where water infiltrates, and depth in the aquifer are then related proportionally:

$$\frac{x}{x_0} = \frac{D}{D-d} \tag{9.6}$$

where D is the thickness of the aquifer. Water infiltrated at a point x_0, upstream of x, is at a given time found at depth d in the aquifer. Above d flows water that infiltrated between point x_0 and x, below d flows water that infiltrated upstream of x_0.

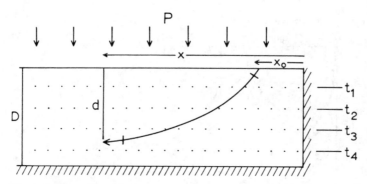

Figure 9.4. A profile of the homogeneous, phreatic aquifer.

Now, let a precipitation surplus of P m/yr enter the aquifer along its upper reach, so that $Q = P \cdot x$ m²/yr flows through the aquifer at point x. Using Equation (9.4), the flow velocity at point x is:

$$v = \frac{dx}{dt} = \frac{Px}{D\varepsilon} \qquad (9.7)$$

where ε is effective porosity. Note that thickness D has replaced surface area A, since we consider flow in a profile (per unit width of the aquifer). It is important to understand that the groundwater flow velocity increases with distance from the divide, since more precipitation must be discharged through the same thickness. The increase of flow velocity requires that the hydraulic gradient increases with the square of distance, as is explained in standard textbooks on groundwater flow (cf. Problem 9.4).

Integration of (9.7) gives, when water infiltrates at $t = 0$ at $x = x_0$:

$$\ln \frac{x}{x_0} = \frac{Pt}{D\varepsilon} \qquad (9.8)$$

or the distance x reached in time t:

$$x = x_0 \exp \left(\frac{Pt}{D\varepsilon} \right) \qquad (9.9)$$

We may also substitute the proportionality relationship (9.6) into (9.8), which yields:

$$\ln \left(\frac{D}{D - d} \right) = \frac{Pt}{D\varepsilon}$$

or:

$$d = D \left(1 - \exp \left[\frac{-Pt}{D\varepsilon} \right] \right) \qquad (9.10)$$

This is surely a remarkable result, since it demonstrates that a homogeneous aquifer with uniform supply from above, must contain water of equal age at each depth, independent of location. In other words, the *isochrones* are horizontal, and water resides in the aquifer in planes of equal age. Figure 9.5 illustrates how age and depth are related in a specific aquifer. It also shows how a uniformly infiltrated tracer like tritium from nuclear explosions in the

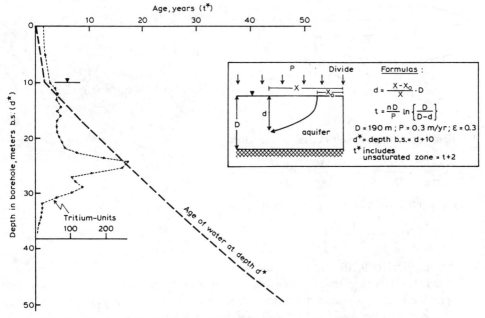

Figure 9.5. Age/depth relationship in an aquifer with a tritium profile for 1981.

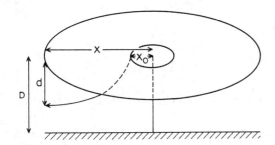

Figure 9.6. Radial flow in the uniform, phreatic aquifer.

atmosphere in the early sixties, is distributed with depth at a given time.

The treatment of proportionality can be further extended, for example to a case with radial divergent flow from a hill, as shown in Figure 9.6. At a point x (distance x from the hill), water flows through the aquifer with velocity:

$$v = \frac{dx}{dt} = \frac{P\pi x^2}{2\pi x D\varepsilon}$$

$$= \frac{Px}{2D\varepsilon} \tag{9.11}$$

which gives for the distance x, reached in time t:

$$x = x_0 \exp\left(\frac{Pt}{2D\varepsilon}\right) \tag{9.12}$$

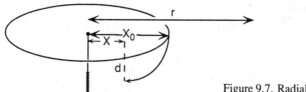

Figure 9.7. Radial flow towards a well.

The depth of infiltration shows a proportionality:

$$\frac{\pi x^2}{\pi x_0^2} = \frac{D}{D-d} \qquad \frac{x}{x_0} = \sqrt{\frac{D}{D-d}}$$

and the depth/time relationship becomes:

$$d = D\left(1 - \exp\left[\frac{-Pt}{D\varepsilon}\right]\right) \tag{9.13}$$

which is identical to (9.10).

For radial converging flow, i.e. towards a well, similar equations can be derived. For example, the ultimate distance at the surface from where water may enter the well is r, while x and x_0 are measured from the well as indicated in Figure 9.7.

Then, the flow velocity is:

$$v = -\frac{dx}{dt} = \frac{P\,(r^2 - x^2)}{2\,x\,D\varepsilon}$$

which can be integrated to give:

$$\ln\left(\frac{x^2 - r^2}{x_0^2 - r^2}\right) = \frac{Pt}{D\varepsilon} \tag{9.14}$$

With this equation the travel time to the well (at $x = 0$) can be calculated, for a pollutant infiltrated at distance x_0 from the well.

The equations which we have derived here are valid for an aquifer of infinite length, or in practice, not too close to a discharge point of the aquifer (in formula: $0 < x < (L - D)$, where L is distance from divide to discharge). Ernst (1973) and Gelhar and Wilson (1974) have also given formulas which can be used near the discharge point, where upward flow into a spring or a seepage zone occurs. Flow velocities are high at such a discharge point, and the 'planes of equal age' merge into a single mixed water composition. Calculations of age distribution and depth are then no longer feasible, but concentrations should rather be averaged (see Section 9.1.4).

EXAMPLE 9.1: *Calculation of infiltration depths and flow velocities in an aquifer*
An aquifer has thickness $D = 50$ m; $P = 0.3$ m/yr; $\varepsilon = 0.3$. At $900 - 1100$ m from the divide infiltrates polluted water in the aquifer, and a private well at 2000 m from the divide may be affected. Calculate the thickness of the tongue of polluted water, its mean depth, and arrival time at the well point.

ANSWER
The infiltration reach of 200 m is distributed proportionally over depth, and obtains a thickness of $200/2000 \cdot 50 = 5$ m. Mean depth is $1000/2000 \cdot 50 = 25$ m; travel time is, from Equation (9.8), for water infiltrated at 1000 m: $t = 34.7$ yr, and for water infiltrated at 900 m: $t = 40$ yr.

9.1.3 *Effects of non-homogeneity*

Spatial variations in recharge in an area may be caused by different drainage into surface water, or differences in precipitation surplus. For example, it has been observed that forests have a substantially smaller precipitation surplus than a grass cover, mainly as result of interception of rain (Calder, 1990). The depth of flowlines can be corrected for such variations as indicated in Figure 9.8. Two different precipitation surpluses P_1 and P_2 will give a difference in infiltration depth $\Delta(d)$:

$$\Delta(d) = (P_2 - P_1) A D / Q_{tot} \tag{9.15}$$

Most aquifers are not homogeneous, and large differences in permeability are usually found. The distribution of flowlines can be calculated manually by proportioning the volume flow according to the transmissivity difference. The correction is performed in the vertical plane as shown in Figure 9.9. If two layers exist of thickness d_1 and d_2, porosities ε_1 and ε_2, and with permeabilities k_1 and k_2 (m/day), the discharge through the vertical section can be divided in two portions Q_1 and Q_2. The ratio of the two is (assuming equal hydraulic gradient in both layers):

$$\frac{Q_1}{Q_2} = \frac{k_1 d_1}{k_2 d_2} \tag{9.16a}$$

and the velocities in the layers are:

$$\frac{v_1}{v_2} = \frac{k_1}{k_2} \frac{\varepsilon_2}{\varepsilon_1} \tag{9.16b}$$

Another case is an impermeable clay layer that divides the aquifer in a phreatic part and a confined part. The flow pattern in the confined part remains intact when the confining layer

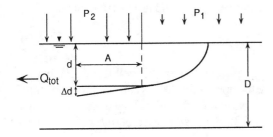

Figure 9.8. Effect of a different precipitation surplus on infiltration depth ($P_2 > P_1$).

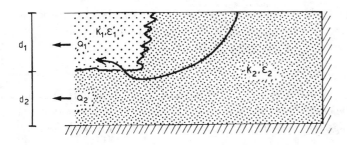

Figure 9.9. Effect of permeability-changes on flowlines.

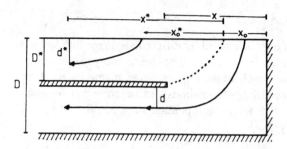

Figure 9.10. A clay layer separates the aquifer in two parts.

is strictly impermeable (Figure 9.10). For the upper, phreatic aquifer the formulas (9.6) to (9.10) apply again. For the confined part of the aquifer the age distribution with depth is conserved. The flowtime at point x^* in the confined part is (see Figure 9.10 for explanation of symbols):

$$t = \frac{D\varepsilon}{P} \ln\left(\frac{D}{D - d - D^*}\right) + \frac{(x^* + x_0 - x)(D - D^*)\varepsilon}{Px_0} \tag{9.17}$$

9.1.4 *The aquifer, an ideal mixed reservoir*

Concentrations of dissolved substances have been considered up to now as depth related entities, which can be measured using depth specific sampling devices (Chapter 1). Depth integrated sampling on the other hand, yields the average of concentrations over depth. Assume, for example in Figure 9.4, that the concentration of NO_3^- in infiltrating water increases from an initial C_i at time $t < 0$, to C_0 at time $t = 0$. When the concentration front has arrived at depth d, the average concentration is:

$$C = \frac{C_0 d + C_i(D - d)}{D} \tag{9.18}$$

Now, assume that initial concentration in the aquifer $C_i = 0$, and use (9.10) to obtain:

$$C = C_0 \frac{d}{D} = C_0\left(1 - \exp\left[\frac{-Pt}{D\varepsilon}\right]\right) \tag{9.19}$$

which shows that the concentration in a depth averaged sampling well (for example used for drinking water production) increases with time to the final value by an exponential function. It is of interest that the same function is valid for the 'ideal mixed reservoir' which is used by chemical engineers (Levenspiel, 1972). For such a reservoir a *residence time* is defined as:

$$t_R = V/Q \tag{9.20}$$

where t_R is the residence time (s), V the volume of the reservoir (m³), and Q the steady state inflow (= outflow) (m³/s). When a step input C_0 is introduced at time $t = 0$ in the input stream, the concentration in the exit stream increases as:

$$C = C_0(1 - e^{-t/t_R}) + C_i e^{-t/t_R} \tag{9.21}$$

With $C_i = 0$, Equation (9.21) is identical with (9.19) when $t_R = D\varepsilon/P$. This is true since $V = D\,\varepsilon\,x$, and $Q = Px$. The comparison fits, and is logical since the outflow at point x is considered to give an average concentration over depth; with uniform flow (no velocity variations with depth) similar averaging is obtained as in the mixed reservoir. Equation (9.19) has been used to derive characteristics of the groundwater reservoir from tritium measurements in spring water (Yurtsever, 1983; Maloszewski and Zuber, 1985). When the residence time is known, either from tracer measurements or from hydrogeological conditions, it is easy to obtain a first estimate of changes in concentration in time with Equation (9.21), for example of NO_3^- in production wells when application of excess fertilizer at the surface has been banned.

EXAMPLE 9.2: *Flushing of NO_3^- from an aquifer*
Estimate the time required to reduce NO_3^- concentration from 150 mg/l to 50 mg/l (the drinking water limit) in an aquifer after fertilizer input has been stopped. $D = 10$ m, $\varepsilon = 0.3$, $P = 0.6$ m/yr.

ANSWER
The residence time in the aquifer is $t_R = D\varepsilon/P = 5$ yr. With $C_0 = 0$, we find from (9.21) that $50/150 = e^{-t/t_R}$, or $t = -t_R \cdot \ln(\frac{1}{3}) = 5.5$ years.

9.1.5 *Interpretation of tracer transport: tritium profiles*

An aquifer is never ideally homogeneous, and shows never the ideal characteristics as considered so far. Figure 9.5 showed an ideal tritium profile for a borehole in a homogeneous aquifer, but observed tritium profiles deviate as result of inhomogeneities. The actual measured tritium contents in that profile are presented in Figure 9.11. It can be seen that the depth of the peak varies from borehole to borehole, or may even be absent. The deviations can be used to deduce flow conditions in the neighborhood of the borehole.

In an aquifer profile the following mass balance must hold for a borehole at distance x_i from the divide, at time t_i:

$$[\text{tritium input}]_0^{t_i} \cdot x_i = [\text{tritium in aquifer}]_{t_i,0}^{x_i} + [\text{tritium discharge}]_{x_i,0}^{t_i}$$

or:

$$\int_0^{t_i} TU_P(t)\, P dt = \int_0^{x_i}\int_0^{D} \frac{TU(z)\varepsilon}{x_i}\, dz dx + \int_0^{t_i}\int_0^{D} \frac{TU(z)v(z)\varepsilon}{x_i}\, dz dt \qquad (9.22)$$

where $TU_P(t)$ is tritium content of precipitation, $v(z)$ is groundwater flow velocity at depth z, and $TU(z)$ is tritium content at depth z.

(Note that both sides of Equation (9.22) were divided by x_i). The first term on the right-hand side of (9.22) is the mean tritium content of boreholes upstream of x. It becomes a constant when a minimum distance from the divide has averaged heterogeneities. Hence the second term must also become independent of x. Or at time t_i:

$$\int_0^{D} \frac{TU(z)v(z)\varepsilon}{x_i}\, dz = \text{constant} \qquad (9.23)$$

The tritium content in a borehole is often equated to the depth integrated tritium:

$$\int_0^{D} TU(z)\varepsilon\, dz \qquad (9.24)$$

Actually, Equation (9.23) would be the right choice for a comparison of groundwater tritium and input by rain. By using (9.24) instead, we have neglected variations in groundwater flow velocity with depth, and assumed that the flow is uniform in the profile. The point now is, that deviations in the mass balance can be interpreted as due to groundwater flow variations. For example, the tritium content of the boreholes shown in Figure 9.11 is given in Table 9.2, and amounts to an average 1306 TU.m. Multiplying this

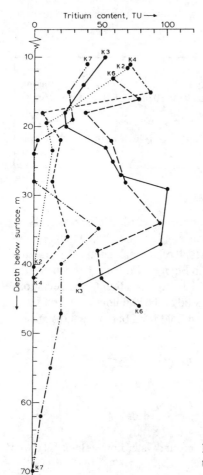

Figure 9.11. Tritium profiles measured in a sandy aquifer in the Netherlands (Appelo and Van Ree, 1983).

Table 9.2. Tritium contents of boreholes in the Netherlands' Veluwe area (1981 measurements; TU.m equals the integrated input with rain, and the depth integrated tritium in groundwater). From Appelo and Van Ree, 1983.

Borehole	TU.m
K2	459
K3	2055
K4	847
K6	2146
K7	1023
average	1306

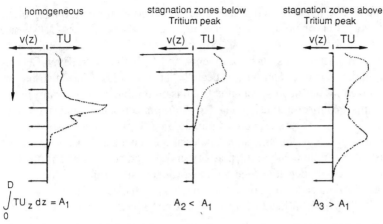

Figure 9.12. Deformation of tritium profiles by varying flow velocities with depth. Left: ideal homogeneous profile; center: stagnant zones below tritium peak; right: stagnant zones above tritium peak.

average with a porosity of 0.33 gives a mass of 430 TU.m in the groundwater, which is much lower than 620 TU.m input with rain over the period from 1953 to 1981. However, any agreement would be fortuitous given the large spread amongst individual boreholes. This spread should be attributed to inhomogeneities in the flow profile; it can be deduced that a borehole with less than 620 TU.m has stagnant zones below the expected peak depth, whereas boreholes with more than 620 TU.m must have stagnant zones above this depth (cf. Figure 9.12). The observed deviations from the ideal tritium profiles can also be used to model flow conditions in the complete profile (Herweijer et al., 1985). Lateral heterogeneities, however, may well be even more important than vertical changes in permeability, but require a two- or three- dimensional approach for adequate modeling. We return to this problem in Section 9.5.

9.2 RETARDATION OF CHEMICALS

Transport of chemicals with groundwater is normally affected by sorption to the solid aquifer material. We recall here that sorption is simply the change of mass of a chemical component in the solid aquifer sediment (Chapter 5). Changes in concentration of a chemical in groundwater at location x (= $\partial C / \partial t$) can be caused by displacement of a concentration gradient (= $\partial C / \partial x$), after correction for amounts which are adsorbed or desorbed:

$$(\partial C/\partial t)_x = -v\,(\partial C/\partial x)_t - S \qquad (9.25)$$

where S is a source (desorption) or sink (adsorption) of the chemical. (You may find it worthwhile to draw a plot of C vs. t for various C vs. x, to appreciate the meaning of the − signs in this equation). Adsorbed concentrations can be related to concentrations in water with Henry's law:

$$s/C = K'_d \qquad (9.26)$$

K'_d is the *distribution coefficient* which relates the adsorbed concentration (s, often expressed in mg/kg dry soil, or ppm) to the concentration in water (C in mg/l). When adsorption is absent, e.g. for anions like Cl^-, $K'_d = 0$. In practice, almost any chemical can be adsorbed on the solid particles of soils or sediments, and K'_d is then > 0. (Even Cl^- can become adsorbed under some conditions, especially at low pH). The distribution coefficient relates concentrations in different units, but it is preferable to use the same units, since adsorbed amounts are more obviously related to solute concentrations and vice versa. An easy way to do this is to express all concentrations on a total volume basis (i.e. the volume which includes pore water and solid material). This means that pore water concentrations must be multiplied with ε, the porosity, and sorbed concentrations must be multiplied with the ρ_b, the bulk density of the sediment (the density with air-filled pores). The net result is that adsorbed concentrations can also be expressed as mg/l porewater, by multiplying with ρ_b/ε. The distribution coefficient K'_d becomes dimensionless (is 'scaled') with the formula:

$$K_d = \frac{\rho_b s}{\varepsilon C} = \frac{q}{C}$$

(9.27)

where q is the adsorbed concentration expressed as mg/l porewater. When s is given in mg/kg, and C in mg/l, the conversion factor is for average conditions, cf. Chapter 5:

$$\rho_b/\varepsilon \approx 6 \, \text{kg/l}$$

(9.28)

The source or sink term S can now be expressed with the distribution coefficient in terms of changes in the solute concentration:

$$S = (\partial q/\partial t)_x = K_d (\partial C/\partial t)_x$$

which is substituted in (9.25) to give:

$$(1 + K_d)(\partial C/\partial t)_x = -v(\partial C/\partial x)_t$$

(9.29)

For $C(x,t) = $ constant we have by definition:

$$dC(x,t) = 0 = \frac{\partial C}{\partial x} dx + \frac{\partial C}{\partial t} dt$$

or

$$\left(\frac{\partial C}{\partial t}\right)_x = -\left(\frac{\partial C}{\partial x}\right)_t \left(\frac{\partial x}{\partial t}\right)_C$$

Substitute in Equation (9.29), divide both sides of the equation by $(\partial C/\partial x)$, and find:

$$(1 + K_d)\left(\frac{\partial x}{\partial t}\right)_C = v$$

The expression $(\partial x/\partial t)_C$ represents the velocity of a constant concentration of a chemical i:

$$\left(\frac{\partial x}{\partial t}\right)_C = v_{C_i}$$

(9.30)

(You may, again, find it worthwhile to draw plots of C vs. t for given C vs. x, and also an x vs. t plot of v_{C_i}, and compare the slopes of the lines).

The velocity of i with respect to water velocity is therefore:

$$v_{C_i} = \frac{v}{1 + K_d} = \frac{v}{R}$$ (9.31)

where $1 + K_d = R$ is the *retardation factor*. (When the distribution coefficient is expressed in units l/kg, you find: $R = 1 + \rho_b/\varepsilon \, K_d'$).

The velocity of chemicals which are adsorbed to sediment ($K_d > 0$) is clearly lower than the porewater flow velocity. This can be envisaged to be a result of adsorption sites which must first be 'filled' by the chemical to conform to the required distribution, before further transport is possible. When sorption is stronger (higher K_d), more water with pollutant is needed before the sorption sites are fully occupied, and retardation increases. The use of a retardation factor (or distribution coefficient) is presently a common option to model transport of pollutants. Pollutant chemicals which are sorbed by sediments (and hence show retardation) are heavy metals and many organic chemicals. A distribution coefficient can be estimated for heavy metals and organic pollutants as shown in the next sections.

9.2.1 *Distribution coefficients for organic micropollutants*

Many organic pollutants are *hydrophobic*, which indicates that these substances have a low affinity for solution in water (a polar liquid), and prefer solution in apolar liquids. These pollutants are readily taken up in organic matter of sediments. It has been shown that the tendency to become sorbed (i.e. the distribution coefficient for these organic chemicals) is related to the distribution coefficient of the chemical between water and an apolar liquid, like octanol. The latter is also termed the *partition constant* or *extraction coefficient* by analytical chemists; the soil sorption process, by analogy, is also envisaged as a partition process, where the hydrophobic pollutant partitions itself between water and the soil organic matter (Chiou et al., 1979; Chiou, 1989; Karickhoff, 1984). Such dissolution of the hydrophobic compound into organic matter is appropriately termed *ab*sorption, rather than *ad*sorption.

The distribution coefficient between water and octanol is obtained in a separatory funnel as shown in Figure 9.13. The organic chemical is introduced into a funnel containing water and octanol; the funnel is shaken and the two phases are separately collected. Analysis of the concentrations in the water and octanol phases gives C_w and C_o respectively, from which the distribution coefficient $K_{ow} = C_o/C_w$ is readily obtained. This distribution

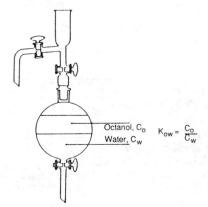

Figure 9.13. Separatory funnel which can be used to obtain octanol/water distribution coefficients.

Table 9.3. Partition coefficients for octanol-water (K_{ow}) and organic carbon-water (K_{oc} (Karickhoff, 1981).

Compound	$\log K_{ow}$	$\log K_{oc}$	Compound	$\log K_{ow}$	$\log K_{oc}$
Hydrocarbons and chlorinated hydrocarbons			*Carbamates*		
3-methyl cholanthrene	6.42	6.09	Carbaryl	2.81	2.36
Dibenz[a,h]anthracene	6.50	6.22	Carboturan	2.07	1.46
7, 12-dimethylbenz[a]anthracene	5.98	5.35	Chlorpropham	3.06	2.77
Tetracene	5.90	5.81			
9-methylanthracene	5.07	4.71	*Organophosphates*		
Pyrene	5.18	4.83	Malathion	2.89	3.25
Phenanthrene	4.57	4.08	Parathion	3.81	3.68
Anthracene	4.54	4.20	Methylparathion	3.32	3.71
Naphthalene	3.36	2.94	Chlorpyrifos	3.31	4.13
Benzene	2.11	1.78			
1, 2-dichloroethane	1.45	1.51	*Phenyl ureas*		
1, 1, 2, 2-tetrachloroethane	2.39	1.90	Diuron	1.97	2.60
1, 1, 1-trichloroethane	2.47	2.25	Fenuron	1.00	1.43
Tetrachloroethylene	2.53	2.56	Linuron	2.19	2.91
γ HCH (lindane)	3.72	3.30	Monolinuron	1.60	2.30
α HCH	3.81	3.30	Monuron	1.46	2.00
β HCH	3.80	3.30	Fluometuron	1.34	2.24
1, 2-dichlorobenzene	3.39	2.54			
pp'DDT	6.19	5.38	*Miscellaneous compounds*		
Methoxychlor	5.08	4.90	13Hdibenzo[a,i]carbazole	6.40	6.02
22', 44', 66'PCB	6.34	6.08	2, 2'biquinoline	4.31	4.02
22', 44', 55'PCB	6.72	5.62	Dibenzothiophene	4.38	4.05
			Acetophenone	1.59	1.54
Chloro-s-triazines			Terbacil	1.89	1.71
Atrazine	2.33	2.33	Bromacil	2.02	1.86
Propazine	2.94	2.56			
Simazine	2.16	2.13			
Trietazine	3.35	2.74			
Ipazine	3.94	3.22			
Cyanazine	2.24	2.26			

Table 9.4. Estimation of K_{oc} from K_{ow} by the expression $\log K_{ow} = a \cdot \log K_{oc} + b$ (Schwartzenbach and Westall, 1985).

Regression coefficient		Correlation coefficient	Number of compounds	Type of chemical
a	*b*			
0.544	1.337	0.74	45	Agricultural chemicals
1.00	−0.21	1.00	10	Polycyclic aromatic hydrocarbons
0.937	−0.006	0.95	19	Triazines, nitroanilines
1.029	−0.18	0.91	13	Herbicides, insecticides
1.00	−0.317	0.98	13	Heterocyclic aromatic compounds
0.72	0.49	0.95	13	Chlorinated hydrocarbons alkylbenzenes
0.52	0.64	0.84	30	Substituted phenyl ureas and alkyl-N-phenyl carbamates

coefficient is highly correlated with the distribution coefficient between organic carbon and water, K_{oc} (Karickhoff, 1981; Schwartzenbach and Westall, 1985).

Karickhoff (1981) suggests

$$\log K_{oc} = \log K_{ow} - 0.35$$

for the chemicals listed in Table 9.3. For groups of these chemicals, Schwartzenbach and Westall (1985) propose a linear regression of the form:

$$\log K_{oc} = a \log K_{ow} + b$$

Values of a and b are given in Table 9.4. Since the nature of the organic matter in soils and aquifer sediments may vary and thus have an effect upon the value of K_{oc} (e.g. Garbarini and Lion, 1986; Chiou et al., 1987), it may be appropriate to use simply the octanol-water coefficient as a first guess of K_{oc} when experimental data are not available for the system which must be described. More values of octanol/water distribution coefficients and other related data can be found in Hansch and Leo (1979), Tewari et al. (1982) and Verschueren (1983), and especially Lyman et al. (1990) is recommended for guidelines in estimating properties of organic pollutants.

The $\log K_{oc}$ refers to partitioning between water and a 100% organic carbon phase; the actual distribution coefficient for the soil or sediment is then obtained as

$$K_d' = K_{oc} \cdot f_{oc} \tag{9.32}$$

where f_{oc} is the fraction of organic carbon. This relationship holds when $f_{oc} > 0.001$, since otherwise sorption on non-organic solids can become relatively important (although giving only low K_d's, Karickhoff, 1984; Curtis et al., 1986). The solubility of the (solid) organic compound in water is another parameter which has been suggested in order to estimate the organic carbon partitioning coefficient. It is clearly related to the octanol/water distribution coefficients as shown in Figure 9.14.

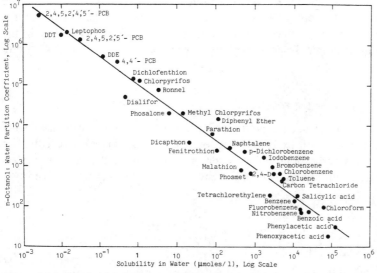

Figure 9.14. Relationship between solubility of a number of organic compounds and their octanol/water distribution (or partition) coefficient. Reprinted with permission from Chiou et al., 1977. Copyright American Chemical Society.

EXAMPLE 9.3: *Retardation of Lindane and PCB*
Calculate the retardation of lindane and of PCB with respect to groundwater flow in a sediment with 0.3% organic matter.

ANSWER
From Table 9.3, find the distribution coefficient K_{oc} for lindane, which is $10^{3.3}$, and obtain with (9.34):

$$\log K'_d = 3.3 + \log 0.003 = 0.8$$

$$K'_d = 6.3 \text{ ml/g, and } K_d = 6.3 \cdot \rho_b/\varepsilon \approx 38$$

Similarly for PCB:

$$\log K'_d = 5.6 + \log 0.003 = 3.1$$

Hence

$$K'_d \approx 1200 \text{ ml/g, and } K_d = 1200 \cdot \rho_b/\varepsilon \approx 7200$$

Lindane (γ-HCH) would have a velocity which is about 40 times less than the water velocity, whereas PCB would move imperceptibly slowly at 7200 times less than water velocity.

When sediments contain more than 0.1% organic carbon, the sorption of nonionic organic chemicals is wholly attributed to organic carbon. Peat layers, buried soils, and in general clayey sediments with a high organic carbon content should thus be particularly effective sorbers which diminish the spreading of organic pollutants. These layers also have relatively low permeabilities, so that water flow through them is restricted if they form part of an aquifer in which more permeable beds are also present. This aspect is important, especially in aquifer clean-up schemes which focus on flushing of the more permeable layers (we return to this subject in Chapter 11). One might expect that the deposition environment has an influence on the organic matter content, e.g. organic matter will be low in sediments deposited in glacial times, and higher during interglacial periods. The scarce data available on sandy sediments in The Netherlands show no definite trend in this respect (Figure 9.15). This figure shows that organic matter contents are in fact so low in the common aquifer sediments, that the approximations mentioned above may not be strictly valid.

9.2.2 *Distribution coefficients for trace metals*

A distribution coefficient for trace quantities of heavy metals can be estimated when adsorption is considered an ion exchange reaction (cf. Chapter 5). We recall that ion exchange is the process whereby ions replace other ions in organic matter, or on the surface of minerals. For example the replacement of Ca^{2+} by a heavy metal M^{2+} may proceed as:

$$M^{2+} + Ca\text{-}X_2 \leftrightarrow M\text{-}X_2 + Ca^{2+}$$

with

$$K_{M\backslash Ca} = \frac{[M\text{-}X_2][Ca^{2+}]}{[Ca\text{-}X_2][M^{2+}]} \tag{9.33}$$

The activity on the exchanger is expressed as equivalent fraction of *CEC*, cf. Chapter 5.

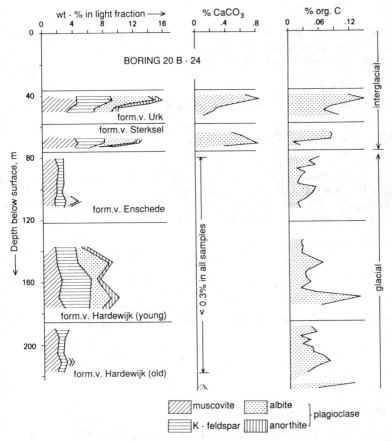

Figure 9.15. Reactivity of Pleistocene sediments in The Netherlands estimated from their contents of $CaCO_3$, weatherable silicates, and organic matter. Parameters plotted against depth.

Hence we find the adsorbed concentration of the trace metal as $meq_{M-X_2} = [M-X_2] \cdot CEC$ in meq/100 g.

When the metal is present in trace quantities only, it does not affect $[Ca-X_2]$, which thus becomes constant. When also $[Ca^{2+}]$ is constant, and solution activity equals molality, Equation (9.33) can be written as:

$$K'_d = \frac{[M-X_2] \cdot CEC}{2\,m_{M^{2+}}} = K_{M\backslash Ca}\frac{[Ca-X_2] \cdot CEC}{2 \cdot m_{Ca^{2+}}} \tag{9.34}$$

where K'_d has units 1/100 g. The distribution coefficient may thus be estimated from exchange coefficients (as given in Table 5.5), and exchangeable and solute concentrations of the major element in water.

EXAMPLE 9.4: *Retardation of Zn^{2+}*
Calculate the retardation of Zn^{2+} with respect to fresh water flow in a sediment which has a cation exchange capacity (CEC) = 0.5 meq/100 g. Zn^{2+} concentration is 100 ppb (or $100/65 \approx 1.5$ μmol/kg H_2O). Calculate the retardation for a Ca^{2+} concentration of 2 and 10 meq/l.

ANSWER

In fresh water Ca^{2+} is the dominant cation, and Ca^{2+} is also the dominant adsorbed ion. A distribution coefficient for Zn^{2+} can be estimated from the exchange of Ca^{2+} against Zn^{2+}. The exchange coefficient of Ca^{2+} for Zn^{2+} on soil or sediment is found from:

$$Zn^{2+} + Ca\text{-}X_2 \leftrightarrow Zn\text{-}X_2 + Ca^{2+}$$

with an exchange coefficient $K_{Zn\backslash Ca}$, recalculated from the values with respect to Na^+ from Table 5.5:

$$K_{Zn\backslash Ca} = \frac{[Ca^{2+}][Zn\text{-}X_2]}{[Zn^{2+}][Ca\text{-}X_2]} = \left(\frac{0.4}{0.4}\right)^2 = 1.0$$

We assume that the exchange complex of the sediment is wholly saturated with Ca^{2+}, so that $meq_{Ca\text{-}X_2} = 1 \cdot CEC = 0.5$ meq/100 g. The Ca^{2+} concentration in fresh water is 2 meq/l, or molality ≈ 0.001 mol/kg H_2O. The amount of $meq_{Zn\text{-}X_2}$ can be calculated, when adsorption of trace amounts of Zn^{2+} does not change the amount of $meq_{Ca\text{-}X_2}$ on the sediment:

$$meq_{Zn\text{-}X_2} = K \cdot CEC \cdot \frac{m_{Zn^{2+}}}{m_{Ca^{2+}}} = 1 \cdot 0.5 \cdot \frac{1.5 \cdot 10^{-6}}{1 \cdot 10^{-3}}$$

or:

$$meq_{Zn\text{-}X_2} = 7.5 \cdot 10^{-4} \text{ meq/100 g, or } 0.38 \text{ } \mu\text{mol/100 g.}$$

The distribution coefficient can be obtained directly from Equation (9.34), or from the ratio $meq_{Zn\text{-}X_2}/2\,m_{Zn^{2+}}$:

$$K'_d = \frac{meq_{Zn\text{-}X_2}}{2\,m_{Zn^{2+}}} = \frac{0.75}{2 \times 1.5} = 0.25 \text{ kg } H_2O/100 \text{ g} \approx 0.25 \text{ l/100 g.}$$

which is scaled:

$$K_d = K'_d \cdot 10\rho_b/\varepsilon = 0.25 \text{ (l/100 g)} \cdot 60 \text{ (100 g/l)} = 15.$$

The retardation factor is $R = 1 + K_d = 16$, or v_{Zn} is about 16 times less than v_{H_2O}.

Similarly for a Ca^{2+} concentration of 10 meq/l, $K_d = 3$, and the retardation factor $R = 4$.

We have made quite a number of simplifications in order to estimate the distribution coefficient from ion exchange equilibrium only. Particularly, we have neglected any pH effect, which influences exchange selectivity and capacity for minor elements especially. Changes in major ion concentrations also have an important effect (Reardon, 1981), as will be clear from the way in which we obtained a distribution coefficient. Such changes occur in fresh/salt water transition zones, for example in an estuary where a river discharges suspended solids into the sea, or in an aquifer where the fresh/salt water interface moves in response to pumping of fresh water, but also when leachate from a waste site enters an aquifer. The effects can be calculated with multicomponent ion exchange formulae as shown in Chapter 5.

A popular technique has been to obtain a distribution coefficient for desorbing heavy metals, by equilibrating a sample of aquifer sediment with a solution containing major ions and the trace element of interest. Centrifuging the sediment slurry, and replacing the decanted solution with distilled water induces the desorption. However, refilling with distilled water lowers concentrations of the major cations, and, to maintain equilibrium according to Equation (9.33), the concentration of the trace ion as well. The lower trace element concentration should not be attributed to 'difficult desorption' of the trace element without proper calculations, and K_d values obtained in this way should be considered suspect. When the replacing solution contains the major ions in the same concentrations as used for spiking the sediment, and is free of the trace element only, this problem is circumvented.

9.2.3 *Sorption isotherms and their effect on retardation*

The foregoing calculations of distribution coefficients for heavy metal adsorption already indicated that the distribution coefficient need not be a constant, nor to be independent of the concentration in water. Distribution coefficients for many organic compounds are smaller for high than for low concentrations in solution. (Compare the concentration dependency of the Sr^{2+} distribution coefficient in Chapter 5). The simple linear relation as used in the retardation Equation (9.31) cannot be expected to work correctly in all cases, and a curved plot of adsorbed versus solute concentrations as shown in Figure 9.16 will often be more appropriate. Such a relationship between solute and sorbed concentration is termed '*sorption isotherm*', since it applies at a single temperature.

The linear adsorption isotherm stems from Henry's law. Other isotherms which are often used, are the *Freundlich* and the *Langmuir* isotherm (Weber and Miller, 1989):

Freundlich isotherm: $s = K_F C^n$ (9.35)

Langmuir isotherm: $s = s_m K_L C / (1 + K_L C)$ (9.36)

The linear isotherm, and the Freundlich isotherm show no maximum in adsorption sites, but the Langmuir formula has a maximum s_m. Langmuir derived this isotherm to describe adsorption on solids as follows. Assume that the solid surface contains a fixed number of adsorption sites, s_m. All sites have identical sorption characteristics, and each site acts independently on the sorbate. At equilibrium with a given concentration of the sorbate in solution, a fraction s/s_m of the sites is occupied, and a fraction $1 - s/s_m$ is not occupied. Now, consider adsorption and desorption as two kinetic processes which must cancel at equilibrium. The flux from the surface, Q_d, is proportional to the fraction of sites covered

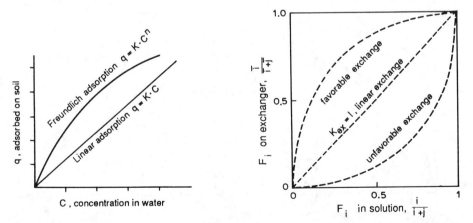

Figure 9.16. Sorption isotherms. (left) linear isotherm and convex (Freundlich) isotherm; (right) exchange isotherms.

by the sorbate, with a rate constant k_d:

$$Q_d = k_d \cdot s/s_m$$

The flux to the surface, Q_a, is likewise proportional to the fraction of surface sites that is empty, $1 - s/s_m$, and to the concentration in solution, C, with a rate constant k_a:

$$Q_a = k_a \cdot C \cdot (1 - s/s_m)$$

At equilibrium, $Q_d = Q_a$, and setting $k_a/k_d = K_L$, Equation (9.36) is obtained.

The Freundlich isotherm is more general than the Langmuir isotherm because it allows for surface heterogeneity. It can be derived by dividing the surface into groups of identical sites and expressing the adsorbate's surface coverage for each group of identical sites by a Langmuir isotherm (Hayward and Trapnell, 1964). The non-linear equations often describe experimental data better than the linear sorption equation, although much of this is largely a result of the presence of two coefficients which can be used to fit the equation to experimental data. Experimental data are often fitted to some linearized form of these equations, but it has been shown that such a procedure gives biased results, and that a non-linear least squares fit (as can be obtained with standard PC packages such as Statgraphics) should be preferred (Kinniburgh, 1986).

The shape of adsorption isotherms is normally *convex* (in the Freundlich isotherm: $0 < n < 1$), indicating that adsorption becomes less at higher concentrations in solution. Ion exchange isotherms can be *convex* or *concave*. Figure 9.16 shows the isotherm for binary exchange, where the axes of the plot express fractions in solution and on the exchanging surface. Linear adsorption or exchange is represented by the straight line. Favorable exchange means that the fraction adsorbed is larger than the fraction in solution, and unfavorable exchange indicates the reverse. A distribution coefficient can be calculated from an adsorption isotherm for each concentration C, irrespective of type of adsorption. The distribution coefficient is:

$$K_d = (q / C) = \tan \beta \tag{9.37}$$

where β is the angle of the sorption isotherm with the solute axis as shown in Figure 9.17. The distribution coefficient decreases with larger concentrations in a convex isotherm

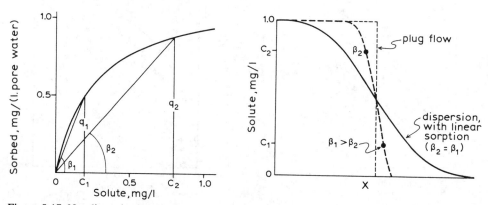

Figure 9.17. Non-linear isotherms give variable distribution coefficients for different concentrations (left); these affect front dispersion (right).

(Figure 9.17). Retardation for low concentrations is more than for high concentrations (with lower distribution coefficient). In other words, low concentrations run slower than high concentrations. Normally the front of a chemical moving through an aquifer shows spreading as result of dispersion. When low concentrations are retarded more than high concentrations, however, there is a *self sharpening* tendency as the front moves on (Figure 9.17). A concave isotherm has the reverse effect. In that case retardation decreases for low concentrations, low concentrations run faster, and the front is smeared out.

Whether exchange for heavy metals is favorable or unfavorable can be estimated from the exchange coefficient against the major cation in solution. Favorable exchange of the heavy metal takes place if, at equal concentrations of the heavy metal and the dominant cation, the heavy metal occupies a larger fraction of the exchange complex. For fresh water the dominant cation is obviously Ca^{2+}, for salt water it will be Na^+. The exchange coefficient for almost all heavy metals is > 1 for exchange with Ca^{2+} (an exception may be Cd^{2+}, when complexed to Cl^-), so that adsorption of the heavy metal is favoured. Since Na^+ is less strongly adsorbed than Ca^{2+} as long as total salt concentration remains below 2 mol/l, it can be safely assumed that heavy metals are also favorably adsorbed in salt water. For organic compounds, where a Freundlich or Langmuir isotherm relates solute and sorbed concentrations, the shape of the isotherm determines the sharpness of the front. For many compounds the value of the coefficient n of the Freundlich equation is smaller than one, indicating that the front will be self sharpening when the pollutant enters the aquifer.

Self sharpening upon initial contact means that the front becomes smeared out when the pollutant is flushed. This has not been appreciated properly in schemes which have been proposed for aquifer restoration by flushing with water. Although there is at present not enough knowledge to determine the optimum composition of the flushing solution which could do the job most quickly, it is of interest to evaluate influences of the isotherm on flushing possibilities, and we will take up this subject again in Chapter 11.

EXAMPLE 9.5: *Retardation calculated from Freundlich isotherm data*
Sorption of diuron (a herbicide) on sand with 0.95% organic carbon was found to follow a Freundlich isotherm, $s = 6.0\,C^{0.67}$ (Nkedi-Kizza, et al., 1983), with s in µg/g soil, C in µg/ml. Calculate the retardation of diuron when $C = 10$ µg/ml, resp. 1 µg/ml; assume $\rho_b/\varepsilon = 6$ g/ml. Also calculate the K_{oc}.

ANSWER
For $C = 10$ µg/ml we obtain $s = 28$ µg/g; $q = 168$ µg/ml porewater, and $K_d = q/C = 17$, so that $R = 18$.
  Similarly for $C = 1$ µg/ml. $K_d = 36$, and $R = 37$, i.e. twice higher than when $C = 10$ µg/ml.
Recalculating the sorption to the organic carbon fraction yields: $K_{oc} = K_F / f_{oc} = 6.0/0.0095 = 630$. This value may be compared to the whole soil $K_{oc} = 730$ (Nkedi-Kizza et al., 1983), where the whole soil contained 3.34% organic carbon, and K_{oc} is similarly calculated from a separately obtained K_F for the whole soil. It shows that sorption depends completely on organic carbon content.

9.3 DISPERSION AND DIFFUSION

Dispersion indicates the scattering or spreading of particles. A front with different concentrations at both sides becomes diffuse because the solute particles step over the front

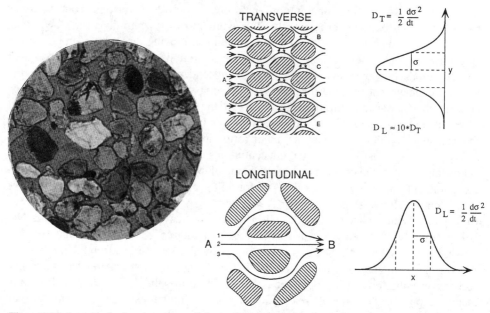

Figure 9.18. Longitudinal and transversal dispersion viewed at the microscopic scale.

by Brownian motion. The term *diffusion* is used for spreading of particles in stagnant water, whereas *dispersion* is used for flowing water. Dispersion in an aquifer develops because water is continuously forced to flow around sediment grains as illustrated in Figure 9.18. Here, the thin section of a sandy sediment indicates that contact points of individual sand grains in a sediment are few. Water can, in fact, move freely around the grains, but two types of dispersion are thereby induced which can be traced in the movement of chemicals. *Longitudinal* or *mechanical* dispersion (symbolized by D_L) is a result of differences in travel time along flowlines which split at grain boundaries (or larger obstacles), whereas *transversal* dispersion (D_T) is caused by stepover onto adjacent flowlines by diffusion. Also indicated in Figure 9.18 is the statistical nature of dispersion; the choice just before a grain to go left or right leads automatically to a probabilistic interpretation of the process.

Basic to the calculation of diffusion or dispersion are Fick's laws. The first relates the flux of a chemical to a concentration gradient:

$$F = -D \frac{\partial C}{\partial x} \tag{9.38}$$

where F is flux (mol/s.m^2); D is diffusion coefficient (m^2/s), and C is concentration, in units of e.g. (mol/m^3).

Note that activities, instead of concentrations, should be used to calculate the potential gradient, since differences in the chemical potential $\mu = \mu^0 + RT \ln[i]$ are the real driving force for diffusion just as for chemical reactions (cf. Appendix A). For example, imagine two compartments separated by a membrane permeable to Ca^{2+} ions, but impermeable to Ca^{2+} complexes with humic acids. Each compartment can have different total concentrations, but must have identical activities when diffusion has equalized the cation activity

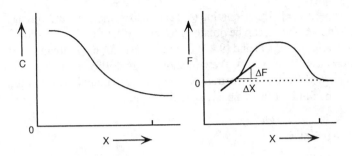

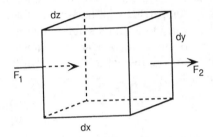

Figure 9.19. Diffusion develops as a result of a concentration gradient.

Figure 9.20. Concentration changes in a cube as result of diffusion.

gradient. In the following we will, for simplicity, assume that concentration can be considered equal to activity.

Figure 9.19 shows how a concentration gradient can induce a flux according to Fick's law. When the flux varies with x, the concentration may change in time. The change in concentration can be found from flux differences over the sides of a cube, as shown in Figure 9.20. The areal flux F_1 through the left side (at right angles to the gradient) is:

$$F_1 = -D\,\mathrm{d}z\mathrm{d}y\,\frac{\partial C}{\partial x} \tag{9.39}$$

and through the right side:

$$F_2 = -D\,\mathrm{d}z\mathrm{d}y\left(\frac{\partial C}{\partial x} + \frac{\partial}{\partial x}\frac{\partial C}{\partial x}\,\mathrm{d}x\right) \tag{9.40}$$

and therefore the change in number of moles n in the cube is:

$$\frac{\partial n}{\partial t} = F_1 - F_2 = -D\,\mathrm{d}z\mathrm{d}y\,\frac{\partial C}{\partial x} + D\,\mathrm{d}z\mathrm{d}y\,\frac{\partial C}{\partial x} + D\,\mathrm{d}z\mathrm{d}y\,\frac{\partial^2 C}{\partial x^2}\,\mathrm{d}x = D\,\mathrm{d}z\mathrm{d}y\,\frac{\partial^2 C}{\partial x^2}\,\mathrm{d}x$$

which gives a change in concentration: $\delta C = \delta n/(\mathrm{d}z\,\mathrm{d}y\,\mathrm{d}x)$. We therefore obtain:

$$\frac{\partial C}{\partial t} = D\,\frac{\partial^2 C}{\partial x^2} \tag{9.41}$$

which is known as Fick's second law. When advective flow also moves the concentration gradient, the change in concentration becomes:

$$\frac{\partial C}{\partial t} = D_L\,\frac{\partial^2 C}{\partial x^2} - v\,\frac{\partial C}{\partial x} \tag{9.42}$$

The longitudinal dispersion coefficient D_L in Equation (9.42) is also termed the hydrodynamic dispersion coefficient. The velocity v can be obtained for an aquifer flowline as shown in Section 9.1. The Equations (9.38), (9.41) and (9.42) form the basis for the mathematical treatment of the transport process (Bear, 1972). We will show in the following pages how the contribution of diffusion and dispersion to transport can be calculated with simple formulas which may already be familiar from a basic statistics course.

9.3.1 *Diffusion as a random process*

Fick's second law, Equation (9.41), is a partial differential equation, which gives the change of C in space (x) and time (t). We are normally more interested in the actual concentration at a given time and location, since it allows for a direct comparison with measured values. This means that we must integrate the equation. Let us try to obtain a solution that will contain integration constants which are determined by the initial or boundary conditions.

For example, we can readily verify by differentiation with respect to t and x, that the integration of Equation (9.41) yields a solution such as:

$$C = A \cdot t^{-\frac{1}{2}} \exp\left(\frac{-x^2}{4Dt}\right) \tag{9.43}$$

where A is a constant that must be found from the initial and boundary conditions. Let us have the initial condition that no chemical is present at time $t < 0$, and that at $t = 0$, N moles are injected at the origin, $x = 0$. This is known as a *single shot input* or *Dirac delta function*. It gives, as $t \to 0$, everywhere $C = 0$ except at the origin where $C \to \infty$. A boundary condition is that mass is conserved, so that for any time t,

$$N = A \cdot t^{-\frac{1}{2}} \int_{-\infty}^{\infty} \exp\left(\frac{-x^2}{4Dt}\right) dx$$

We substitute $x^2/4Dt = s^2$, which gives $dx = \sqrt{4Dt}\, ds$, and obtain:

$$N = A \cdot \sqrt{4D} \int_{-\infty}^{\infty} e^{-s^2} ds$$

The integral now contains a familiar form, closely akin to the error function erf(z):

$$\text{erf}(z) = \frac{2}{\sqrt{\pi}} \int_0^z e^{-s^2} ds \tag{9.44}$$

The error function is tabulated, and numerical approximations are known, as given in Examples 9.7 and 9.13, and Problem 9.12. The error function is 0 for $z = 0$, and 1 for $z = \infty$. Furthermore the error function is symmetric around $z = 0$, so that $\text{erf}(-z) = -\text{erf}(z)$. We can thus evaluate the integral as:

$$\int_{-\infty}^{\infty} e^{-s^2} ds = 2\frac{\sqrt{\pi}}{2} \text{erf}(\infty) = \sqrt{\pi}$$

which gives for the constant A

$$A = N/(4\pi D)^{\frac{1}{2}}$$

and the solution of Equation (9.41) for the conditions stated here is:

$$C(x, t) = \frac{N}{\sqrt{4\pi Dt}} \exp\left(\frac{-x^2}{4Dt}\right) \qquad (9.45)$$

where N is the input mass (moles) at time $t = 0$ at $x = 0$. Note that $C(x, t)$ is expressed as mol/m in this one-dimensional concept.

It is of interest that Equation (9.45) is fundamentally equal to the normal density function (the Gauss-curve):

$$n(x) = \frac{N}{\sqrt{2\pi\sigma^2}} \exp\left(\frac{-(x - x_0)^2}{2\sigma^2}\right) \qquad (9.46)$$

where x_0 is the average location (in Equation (9.45), $x_0 = 0$), and σ^2 is the variance of the distribution. Diffusion or dispersion can thus be seen basically as a statistic process. This is illustrated in Figure 9.21, where the spreading of particles is shown as they do (computed) random steps. One thousand particles ($N = 1000$) start at $x = 0$, and make a number of times a horizontal step of which the magnitude varies randomly from -2.5 to 2.5. The curve above the swarm sums the particles for each x. For a set of random numbers ranging

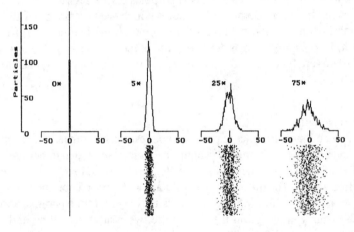

Figure 9.21. Thousand particles do steps of a size which varies randomly from -2.5 to 2.5.

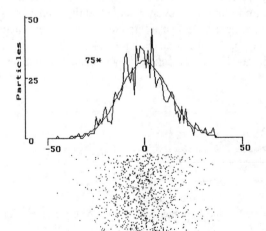

Figure 9.22. After 75 rounds of random steps only poor agreement remains with the ideal Gauss curve.

Table 9.5. Spread of a point source through diffusion ($D_f = 10^{-5}$ cm^2/s), compared with advective transport ($v = 10$ m/year).

t(s)	Diffusion, σ (cm)	Advection, x_0 (cm)
60 (= 1 min)	0.03	0.002
3600 (= 1 hr)	0.27	0.11
86400 (= 1 day)	1.3	2.7
$31.5 \cdot 10^6$ (= 1 year)	25	1000

from 0 to 1 the variance should be $\sigma^2 = 1/12$ (Miller, 1981). This value, multiplied with the square of the step size (= 5^2), gives the increase of the variance for each round (i) of random steps in Figure 9.21:

$$\sigma^2 = i \cdot 25/12$$

The thousand particles portrayed in Figure 9.21 only roughly approximate the smooth Gauss curve of (9.46), as the enlargement after 75 steps in Figure 9.22 indicates.

Computed random motions require many thousands of particles before a reasonable fit is obtained within a few percent deviation. Computer programs which use the 'random walk' based on particle tracking in combination with a random step (e.g. Prickett et al., 1981; Uffink, 1988; Kinzelbach, 1992), are known as devourers of computer time.

Equations (9.45) and (9.46) are fundamentally the same, and the variance σ^2 is related to the diffusion coefficient by:

$$\sigma^2 = 2Dt \tag{9.47}$$

where σ has the dimension length.

This simple formula is used time and again to provide rapid estimates of mean diffusion and dispersion lengths, and not only of chemicals in aquifers but in gases and in minerals as well (Moore, 1972). For example, for chemicals in water, $D_f \approx 10^{-5}$ cm^2/s (D_f is the diffusion coefficient in free water). The diffusion of a pollutant with time can be calculated using this value in Equation (9.47). Table 9.5 shows results for different time periods, and for comparison also the distance travelled in the same time by an advective flow of 10 m/year. It is of interest to see that diffusion over small time periods can keep up with advective transport, but that over larger periods (and hence also over larger distances) advective flow becomes more and more important. However, when advective transport is very small, as one hopes to achieve for example by constructing clay liners under waste sites, there may still be an appreciable transport by diffusion (Section 9.4).

The dispersion of a point source which travels with advective transport can be calculated

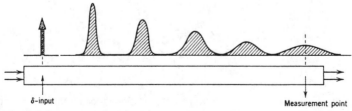

Figure 9.23. Propagation of a Gauss curve along a flowline. Reprinted with permission from Levenspiel, 1972. Copyright John Wiley & Sons.

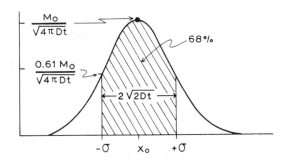

Figure 9.24. A Gaussian (normal) density function.

in quite the same way, the only difference being that x_0 now changes with time as the distance covered by the moving fluid, i.e. $x_0 = vt$. Figure 9.23 shows the propagation of a series of Gauss curves which corresponds to this principle.

Different possibilities for determining the value of σ, and hence of D_f or D_L, from measured data, are illustrated in the Gaussian density function of Figure 9.24. A value of 2σ gives the distance where 68% of the original point source is still present. When a variance σ^2 can be calculated from a series of observation points, either on time or on distance basis, the value of the diffusion or dispersion coefficient can be calculated according to (9.47). A time-based variance can be recalculated to become dimensionless, or to a distance basis, and vice versa as shown in Example 9.6.

EXAMPLE 9.6: *Dispersion coefficient from a single shot input*
Calculate the dispersion coefficient from a single shot input and measured output concentration at the end of a column. Output concentration is measured as a function of t. This example is from Levenspiel (1972).

Time t, min	Tracer Output Concentration, g/l
0	0
5	3
10	5
15	5
20	4
25	2
30	1
35	0

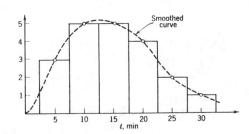

ANSWER
The variance of the measured output distribution can be calculated as the variance of a number of measurements, i.e.

$$\frac{(\text{Individual measurements})^2}{(\text{No. of observations})} - (\text{Average})^2 = \sigma^2$$

Hence:

$$\sigma_t^2 = \frac{\Sigma t_i^2 C_i}{\Sigma C_i} - (\bar{t})^2 = \frac{\Sigma t_i^2 C_i}{\Sigma C_i} - \left[\frac{\Sigma t_i C_i}{\Sigma C_i}\right]^2$$

$$\Sigma C_i = 3 + 5 + 5 + 4 + 2 + 1 = 20$$

$$\Sigma t_i C_i = (5 \times 3) + (10 \times 5) + \dots + (30 \times 1) = 300 \text{ min}$$

$$\Sigma t_i^2 C_i = (25 \times 3) + (100 \times 5) + \dots + (900 \times 1) = 5450 \text{ min}^2$$

Therefore

$$\sigma_t^2 = \frac{5450}{20} - \left(\frac{300}{20}\right)^2 = 47.5 \text{ min}^2$$

and

$$\sigma^2 = \frac{\sigma_t^2}{(\bar{t})^2} = \frac{47.5}{(15)^2} = 0.211$$

On a distance basis: $x = v \cdot t$, e.g. $v = 1 \text{ cm/min}$, $x = 1 \cdot 15 = 15 \text{ cm}$, and:

$$\sigma_x^2 = \sigma_t^2/(\bar{t})^2 \cdot x^2 = 0.211 \cdot 225 = 47.5 \text{ cm}^2.$$

since

$$\sigma_x^2 = 2 D_L t, \quad D_L = \sigma_x^2/(2t) = 47.5/(2 \cdot 15) = 1.58 \text{ cm}^2/\text{min}.$$

This example assumes that the breakthrough curve is symmetric around midpoint breakthrough. This is only true when a front is measured over its length at one time, by (plastically termed) 'through the wall measurements'. Such measurements are possible when radioactive or color tracers are used, or by dissecting the column when the front has penetrated a certain length (Bond and Phillips, 1990). However, it is more common to collect outflow in 'mixing cups' (as in this example) at the column outlet and analyze the contents separately. In that case the transformation can only be made when dispersion coefficients are small (and observations obey, anyhow, the normal distribution).

9.3.2 *Column breakthrough-curves*

When a tracer is supplied continuously at concentration $C = C_0$ in a column which was initially tracer-free, the front will move with the average water flow velocity through the column. At the same time dispersion occurs in the front, of which the essentials are shown in Figure 9.25. The concentrations are scaled in this figure from 0 to 1, by dividing all concentrations by C_0. The conditions of such a step-input are expressed mathematically as:

$$\left(\frac{\partial C}{\partial t}\right)_x = D_L \left(\frac{\partial^2 C}{\partial x^2}\right)_t - v \left(\frac{\partial C}{\partial x}\right)_t \tag{9.48}$$

subject to:

when $t = 0$: $C = C_0$ for $x < 0$; $C = C_i$ for $x > 0$;
$C = (C_0 - C_i)/2$ for $x = 0$
when $t > 0$; $C = C_0$ for $x = -\infty$; $C = C_i$ for $x = \infty$

The solution for these boundary conditions can be obtained similarly to Equation (9.45), and is:

$$C(x, t) = C_i + \tfrac{1}{2}(C_0 - C_i) \, \text{erfc}\left(\frac{x - vt}{\sqrt{4D_L t}}\right) \tag{9.49}$$

where $\text{erfc}(z)$ is the complementary error function (i.e. $1 - \text{erf}(z)$). Note that $\text{erfc}(-z) = 1 + \text{erf}(z)$, so that Equation (9.49) is symmetric around $x_{0.5} = vt$. The concentrations from C_i to $1/2 \cdot (C_0 - C_i)$ run ahead of the average front position at $x_{0.5} = vt$, and the argument of the error function complement is positive. The concentrations from $1/2 \cdot (C_0 - C_i)$ to C_0 on the other hand lag behind $x_{0.5}$, and the argument is negative.

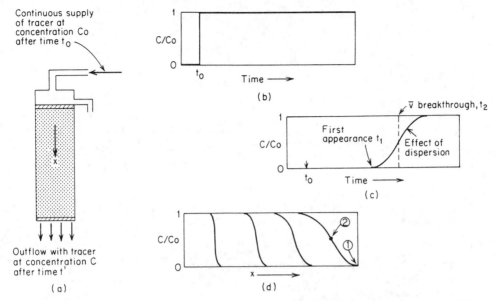

Figure 9.25. Longitudinal dispersion of a tracer passing through a column of porous medium. (a) Column with steady flow and continuous supply of tracer after time t_0; (b) step-function tracer input relation; (c) relative tracer concentration in outflow from column (dashed line indicates plug flow condition and solid line illustrates effect of mechanical dispersion and molecular diffusion); (d) concentration profile in the column at various times. Reprinted with permission from Freeze and Cherry, 1979. Copyright Prentice Hall, Englewood Cliffs, New Jersey.

EXAMPLE 9.7: *Front dispersion in a column*
Calculate the Cl^- concentration front in a column which is injected with 1 mmol NaCl/l. Flow velocity is $2.8 \cdot 10^{-4}$ cm/s, $D_L = 10^{-4}$ cm²/s; the initial Cl^- concentration in the column $C_i = 0$ mmol/l. Calculate the concentrations after 5 hr injection time at the advective front, and at $x = vt \pm \sqrt{2D_L t}$.

ANSWER
The advective front has arrived at $v \cdot t = 2.8 \cdot 10^{-4} \cdot (5 \times 60 \times 60) = 5.04$ cm. The concentration at this point, where $x = vt$, is $1/2$ erfc(0) = 0.5 mmol Cl^-/l (clearly just half of initial and input concentrations).
The concentration at $x - vt = \sqrt{2D_L t}$ is:

$$0.5 \, \text{erfc} \left[\frac{\sqrt{2D_L t}}{\sqrt{4D_L t}} \right] = 0.5 \, (1 - \text{erf} \, [\sqrt{\tfrac{1}{2}} \,]) = 0.5(1 - 0.68) = 0.16$$

where the value of erf $[\sqrt{\tfrac{1}{2}} \,]$, as found from tables, is 0.68.
The concentration at $x - vt = -\sqrt{2D_L t}$ is similarly:

$$0.5 \, \text{erfc} \left[\frac{-\sqrt{2D_L t}}{\sqrt{4D_L t}} \right] = 0.5 \, (1 + \text{erf} \, [\sqrt{\tfrac{1}{2}} \,]) = 0.5 + 0.5 \cdot 0.68 = 0.84$$

It will be noted that the spread between $C = 0.16$ and $C = 0.84$ is obtained over a length just equal to 2σ, and in fact encompasses 68% of the concentrations (0.84 − 0.16 = 0.68) around midpoint breakthrough, as illustrated in the figure. The distances from midpoint breakthrough are resp. $\pm \sqrt{2D_L t} = \pm 1.9$ cm.

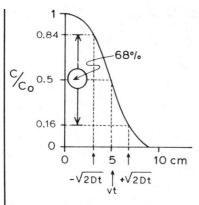

An alternative for tables of the error function is to use a numerical approximation, e.g. as given by Abramowitz and Stegun (1964). For example no. 7.1.26 reads:

$$erf(x) = 1 - (a_1 b + a_2 b^2 + a_3 b^3 + a_4 b^4 + a_5 b^5) \exp(-x^2)$$

where $b = 1/(1 + a_6 x)$, and $a_{1, ..., 6}$ are constants, given as:

$a_1 = 0.25482\ 9592$
$a_2 = -0.28449\ 6736$
$a_3 = 1.42141\ 3741$
$a_4 = -1.45315\ 2027$
$a_5 = 1.06140\ 5429$
$a_6 = 0.32759\ 11$

which can be easily programmed on a calculator (cf. Example 9.13), and is accurate to $1.5 \cdot 10^{-7}$. Note that the approximation calculates the error function for the absolute value of x, and that $erf(x)$ gets the sign of x.

The solution for a dispersion front, as given in Equation (9.49), holds for an infinite column, or, in practice, not too close to the column outlet. Outflow of water at the column outlet stops the back-diffusion at that point, so that other boundary conditions apply. For description of breakthrough curves at the end of laboratory columns a *flux-type* is considered most appropriate (Van Genuchten and Parker, 1984), which yields:

$$C(L, t) = C_i + \frac{C_0 - C_i}{2} \left(erfc \left[\frac{L - vt}{\sqrt{4D_L t}} \right] + \exp \left[\frac{Lv}{D_L} \right] erfc \left[\frac{L + vt}{\sqrt{4D_L t}} \right] \right) \qquad (9.50)$$

where L is column length. This equation is applied in Example 9.13, for checking the accuracy of a numerical calculation of a retarded front. The equation can also be used to obtain values for the dispersion coefficient from breakthrough experiments. An excellent program, CXTFIT by Parker and Van Genuchten (1984b), can do the job with a least squares fit, but it may sometimes be advantageous to have a visual fit as well, since some data points from the outlet stream may be less reliable than others. A visual fit is easily obtained by programming (9.50) in a spreadsheet program, and comparing the output values with observed concentrations for varying D_L. We mention here that solutions for more complex situations are obtained via Laplace transforms (cf. Boas, 1983, for an introduction on this mathematical technique, and continue with Crank, 1975, Carslaw and Jaeger, 1963, and Van Genuchten, 1981).

9.3.3 *Values for dispersion coefficients; dispersivity*

Values of dispersion coefficients are obtained in experimental setups, in columns (as noted above) or in field experiments. In Figure 9.26 values from column experiments are plotted against Peclet number, *Pe* (Bear, 1972). The Peclet number is a dimensionless number, defined as:

$$Pe = v\,d\,/\,D_f \tag{9.51}$$

where *d* is diameter of the particles in the column (m), D_f is diffusion coefficient in pure water (m^2/s), and *v* is pore water flow velocity (m/s).

The particle diameter can be compared with the median grain size of an aquifer sediment, but has strict application only in columns which are packed with spheres. In aquifers it should be compared with the size of inhomogeneities. Figure 9.26 shows on the vertical axis the ratio of measured (hydrodynamic, longitudinal) dispersion coefficient D_L over the value of the diffusion coefficient D_f in pure water. Two regions can be separated where the dispersion coefficient D_L has different values (Bear, 1972; Plumb and Whitaker, 1988). At low Peclet numbers (= low flow velocity), the ratio of D_L/D_f is constant. At higher Peclet numbers from 1 to 10^5, the value of D_L/D_f increases linearly with the Peclet number.

The interpretation of these regions is as follows. At low Peclet numbers, and thus at low flow velocities, the diffusion in water is moving chemicals faster than advective flow moves the concentration gradients (which exist at the front). The dispersion is said to be diffusion controlled. The dispersion coefficient obtains the value $D_L = D_f/\theta^2$, where θ is the *tortuosity* of the porous medium. The tortuosity is defined as the ratio of the actual path length taken by a solute in a porous medium, over the straight line distance. The quadratic dependence of diffusion on tortuosity results from the following reasoning (the Kozeny theory). Imagine a column which has its ends in two different solutions, so that a steady

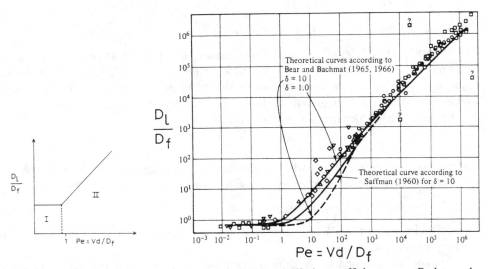

Figure 9.26. A plot of the ratio of dispersion coefficient/diffusion coefficient versus Peclet number. (From Bear, 1972).

state flux of some ion passes through the column. The sediment in the column has pores which are connected to tortuous capillaries of length L_a. Now mould the column in such a way that the capillaries are straightened out to length L, without affecting the volume of the pore space, the pore geometry, or the flux of the ion through the column. The concentration gradient in the moulded column will be smaller by (L_a/L). The volume of the column has not changed, and therefore the cross-sectional area will also be smaller by (L_a/L). Since the total flux through the column did not change either, the apparent diffusion coefficient in the straightened capillaries must be larger by $(L_a/L)^2 = \theta^2$.

A tortuous path, at an average angle of 45° with the straight line, gives obviously a tortuosity $\theta = L_a/L = 2/\sqrt{2} = 1.4$, and hence $D_L = D_f/2$. This is similar to the value found in the column experiments at low Peclet numbers displayed in Figure 9.26. If the value of the dispersion coefficient is compared with the variance σ^2 of the spread of a front in the column under these flow conditions then

$$D_L = D_f/\theta^2 \tag{9.52}$$

or $\sigma^2 = 2 D_L t = \text{constant} \times t,$

or $\sigma^2 \equiv \text{constant} / v$ (over a given travel length).

The variance hence increases with time, or the spread of a front at the end of a specific column of length L increases when flow velocity decreases. The upper boundary for diffusion controlled dispersion is found at a Peclet number of 0.5, as indicated by the experimental values in Figure 9.26. When d varies from 0.4 to 2 mm (values for medium to coarse sands), then a flow velocity lower than 37 to 8 m/yr yields diffusion controlled dispersion.

At higher Peclet numbers (equivalent to higher flow velocities, as we noted before), the ratio D_L/D_f increases in a ratio of 1:1 with the Peclet number, or in other words:

$$D_L = D_f \cdot Pe = d\,v \tag{9.53}$$

or $\sigma^2 = 2 D_L t = 2 d \cdot v t = 2 d x,$

or $\sigma^2 \equiv \text{constant}$ (over a given travel length).

If a specific column of length L is used, the variance or the spread of a front at the end of the column will be independent of the flowtime in the column. This holds with Peclet numbers > 1, or with flow velocities above around 50 m/yr. The spread of a front now depends on the traveled length only, and dispersion can be seen as a characteristic property of the medium for which the term *dispersivity* is used:

$$\alpha_L = D_L /v \tag{9.54}$$

where α is the dispersivity (m), which is also defined in a longitudinal (α_L) and transversal (α_T) sense. The variance can be calculated with this parameter to be:

$$\sigma^2 = 2D_L t = 2\alpha_L \cdot vt = 2\alpha_L \cdot x \tag{9.55}$$

i.e. dependent on x only. This important formula may be very efficient for calculating the spread of pollutant fronts.

If we consider additionally the diffusion dominated region (low Peclet number) and the transition zone, the value for the hydrodynamic, longitudinal dispersion coefficient becomes:

$$D_L = D_f/\theta^2 + \alpha_L v \tag{9.56}$$

The dispersivity in columns is equal to the representative grain diameter used in packing the column. When aquifer sediments are used in the column it would be represented by the diameter d_{10} below which 10% of particles fall (Perkins & Johnston, 1963). The latter authors suggest the relationship:

$$\alpha_L = 3.5 \, d_{10} \tag{9.57}$$

where 3.5 is a shape factor, which increases somewhat with smaller grain size. The dispersivity in aquifers is generally much higher, which means that the diffusion contribution in Equation (9.56) becomes negligible when flow velocity is more than 1 m/yr.

At this point a remark must be made about the *column Peclet number* that is used to characterize the dispersive behavior of a specific column of length L. The definition differs from (9.51) and is:

$$Pe^* = vL/D_L$$

On the basis of our previous discussion, we can translate this definition to $Pe^* = L/\alpha$ when the flow velocity in the column is sufficiently high to cause a dispersion dominated spreading. In the latter definition, L is used for flow domains in a general sense, for example for the grid size in a numerical model.

9.3.4 *Macrodispersivity*

The larger dispersivity in aquifers is due to the heterogeneous structure of natural sediments and aquifer rocks, with alternating sands of different permeability, and intercalated clays, silts or gravels. Here, it is no longer the individual grain size but rather the varying flow time along low-permeability structures which determines the dispersivity. This effect is known as *macrodispersivity*.

Until now, we have treated dispersion as a statistical process, which means that flow is redirected around obstacles so many times, that a normal (or Gaussian) distribution of flowpaths is obtained. When a simulation with a number of obstacles is performed, it is obvious that a few thousand of such obstacles (or flow-redirections) are necessary before a normal distribution results. This is easily reached in a column with homogeneously packed grains, but less so in an aquifer. Figure 9.27 gives an impression of the effect of geological structures on the Dirac delta function ('the single shot input'), and also shows that the sampling method can be an important contributor to field dispersivities. When large filters are used of more than 1 m length, a certain mixing of water layers is inevitable, leading to increased dispersion coefficients. Thus, field dispersivities are generally lowest when small filters or depth specific sampling techniques are used (Schröter, 1984). It is, in fact, possible to obtain dispersivity values in an aquifer which are of the same magnitude as found in column experiments, when a point sampling method is employed and a single flowline is monitored. Pickens and Grisak (1981), for example, found dispersivities which are only 7 mm, and remained constant over distances from 0.36 to 4 m. Even more stringent constancy of 1.6 cm over distances up to 40 m was observed by Taylor and Howard (1987), who used gamma-ray counting of a radio tracer in dry access tubes.

A depth integrated sampling includes the effects of the geological heterogeneities, and is thus more apt to give an overall picture of transport properties of an aquifer. Figure 9.28

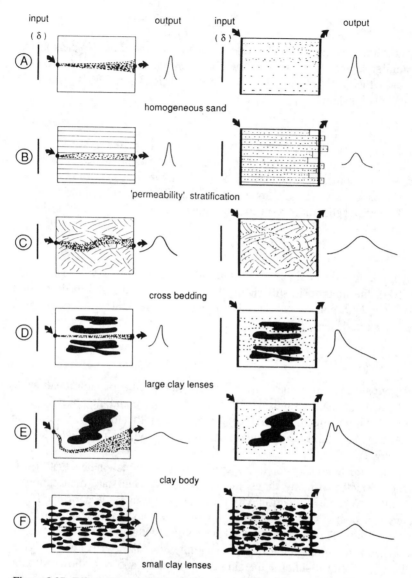

Figure 9.27. Effect of sedimentological structures on dispersion of a shot input. At left: point specific sampling, at right: depth integrated sampling in the same aquifer.

shows that such depth integrated dispersivities in aquifers may be orders of magnitude larger than values determined in columns. Even more important is that dispersivity appears to increase with distance. It is, in other words, a *scale dependent* property, in contrast to dispersivities obtained from single point or column observations. The general trend suggested by Figure 9.28 is that longitudinal dispersivity is about 10% of the traveled distance:

$$\alpha_L = 0.1\, x \tag{9.58}$$

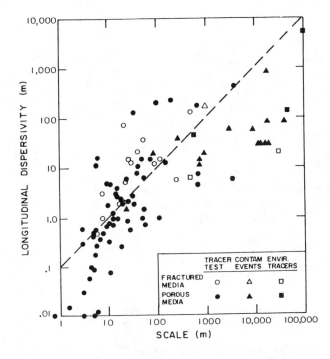

Figure 9.28. Values for dispersivities obtained in field experiments (from Gelhar et al., 1985).

If we introduce this trend in Equation (9.55), we obtain:

$$\sigma^2 = 2\alpha_L \cdot x = 2 \cdot 0.1 \, x \cdot x = 0.2 \, x^2 \tag{9.59}$$

for the variance of a pollutant front in aquifers (cf. Example 9.8).

It seems logical that the larger field dispersivity is caused by variation of arrival times at the observation well, which means that the permeability distribution in the aquifer lies at the heart of the process. Let us consider in detail the concept of a permeability stratified aquifer, developed by Mercado (1967; 1984). An aquifer, made of uniform layers with different permeabilities, is depicted in Figure 9.29. Assume that flow is confined to individual layers, and cannot cross a boundary between layers (transversal dispersion is absent). Tracer from the (fully penetrating) injection well arrives at the observation well in a stepwise fashion according to the permeability of the layer. In Figure 9.29 layers 1, 4, 8, 9 and 10 are contributing tracer to the observation well, and layer 5 has just started to contribute. We may reshuffle the layers, and order them from top to bottom according to increasing permeability (Figure 9.29, right). This orders at the same time the distance traveled by the chemical according to the permeability distribution. An average transmissivity is obtained from:

$$k_{0.5}D = \Sigma k_i d_i \tag{9.60}$$

where individual layers have thickness d_i and permeability k_i, and D is total thickness of the aquifer, with average permeability $k_{0.5}$. It is assumed that porosity ε is equal for all layers (cf. Equation 9.16).

Now suppose that permeability variations amongst the layers follow a normal distribution, with standard deviation:

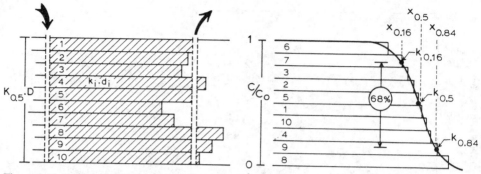

Figure 9.29. A permeability stratified aquifer in which the layers contribute successively to tracer breakthrough (left); ordering the layers according to permeability shows more clearly that travel distance follows the permeability distribution (right).

$$\sigma_k = k_{0.84} - k_{0.5}$$

or

$$\sigma_k / k_{0.5} = k_{0.84} / k_{0.5} - 1 \qquad (9.61)$$

where $k_{0.84}$ is the permeability which is larger than 84% of all values.

The tracer front is spread according to distances covered in individual layers, and the spread is related to dispersivity:

$$\sigma_x = x_{0.84} - x_{0.5} = \sqrt{2\alpha_L x_{0.5}} \qquad (9.62)$$

The distance covered by tracer in just half of the layers is

$$x_{0.5} = k_{0.5} \left(-\frac{dh}{dx} \frac{\Delta t}{\varepsilon} \right)$$

while the distance for 84% of the layers is less than

$$x_{0.84} = k_{0.84} \left(-\frac{dh}{dx} \frac{\Delta t}{\varepsilon} \right)$$

If we assume that the potential gradient $(- dh/dx)$ is equal for all layers, then

$$x_{0.84} = \frac{k_{0.84}}{k_{0.5}} x_{0.5} \qquad (9.63)$$

Combining Equations (9.61), (9.62) and (9.63) allows to derive an expression for the longitudinal dispersivity:

$$\alpha_L = \frac{1}{2} (\sigma_k / k_{0.5})^2 x_{0.5} \qquad (9.64)$$

Recall that $x_{0.5}$ is the distance covered by tracer in half of the layers, which is thus equal to $v \cdot t$. Equation (9.64) therefore suggests that dispersivity increases linearly with distance, as in fact indicated by Figure 9.28. If we assume that the permeability is log-normally distributed we obtain in a similar way:

$$\alpha_L = \tfrac{1}{2} (\exp[\sigma_{\ln(k)}] - 1)^2 x_{0.5} \tag{9.65}$$

where $\sigma_{\ln(k)}$ is the standard deviation of the log-normally distributed permeabilities.

It is of interest that the value for $\tfrac{1}{2}(\sigma_k / k_{0.5})^2$ as coefficient in Equation (9.64) can indeed be around 0.1 (Pickens and Grisak, 1981; Molz et al., 1983; random sample of Smith, 1981), required to conform with Equation (9.58). For log-normally distributed permeabilities, the value of $\sigma_{\ln(k)}$ is 0.4 for the Borden aquifer (Sudicky, 1986; however, cf. Example 9.9), which also leads to $\alpha_L = 0.1x$ by Equation (9.65).

A remarkable point of the permeability stratified aquifer is that dispersion diminishes again when flow direction is inverted (when a tracer is introduced in a slug in an injection well and after some time of further injection, water is pumped from the well, the tracer would return in the same step form as introduced). This is different from the usual diffusion concept, in which spreading increases with the square root of time irrespective of flow inversion. Of course, there are assumptions in the Mercado concept that require attention. Most important is perhaps the neglect of transversal dispersion. Transversal dispersion causes mixing amongst the layers, and diminishes the differences in travel distances, with the result that the increase in dispersivity becomes smaller. More intricate models in which horizontal space variability is allowed, as well as transversal dispersion, have been developed (Gelhar and Axness, 1983; Dagan, 1989), and the reader should consult the work by Dagan (1989), and especially the application by Sudicky (1986) for further details on this matter.

A word is needed here on transversal dispersivity; it is studied less, although it is known to be smaller than the longitudinal dispersivity. A rule of thumb states that transversal dispersivity is about 10% of the longitudinal dispersivity.

EXAMPLE 9.8: *Pollutant spreading during transport in an aquifer*
An aquifer has thickness $D = 50$ m; $P = 0.3$ m/yr; porosity $\varepsilon = 0.3$. Calculate the area which contains 68% of a point source pollutant, infiltrated 1 km from the divide, after 1 km flow (compare Example 9.1). Also estimate for this point (at 2 km from the divide) the time during which 68% of the pollutant mass with highest concentrations passes.

ANSWER
Assume that dispersivity is 10% of travel length, and obtain $\sigma^2 = 0.2\ x^2$ (Equation 9.59); with $x = 1000$ m, is $\sigma = 447$ m. Hence we estimate 68% of our pollutant to be within 2447 and 1553 m (from the divide). We next want the time period during which the highest concentrations pass, covering 68% of the pollutant mass. This is the period between the arrival time of $(\sigma + x)$ at 2000 m, and the time when $(x - \sigma)$ passes the point. Now, our estimated $\sigma = x\sqrt{0.2}$, and we find that $x = 2000/(1 + \sqrt{0.2}) = 1382$ m. This distance is attained in $50 \ln(1382/1000) = 16.2$ years. Similarly, we find from $x - \sigma = 2000$, that $x = 3618$ m, and the time necessary is 64.3 years. Hence the time period for 68% of pollutant mass, with highest concentrations, to pass the 2 km point is $64.3 - 16.2 = 48.1$ years.

EXAMPLE 9.9: *Longitudinal dispersivity in the Borden aquifer*
A point injection of a number of chemicals in an 8 m thick aquifer was followed over two years with a dense network of depth specific samplers (Mackay et al., 1986). Figure 9.30 shows the *depth averaged* Cl^- cloud after 462 days, with a midpoint at 41 m from the injection point. Estimate the longitudinal dispersivity.

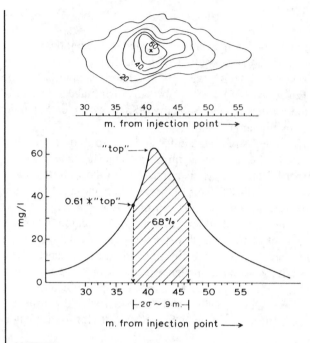

Figure 9.30. Vertically averaged chloride concentrations in the Borden experiment, 462 days after injection, and cross section over the plume. (Plume delineation after Sudicky, 1986).

ANSWER

The cross section suggests approximate Gaussian behavior, and we may apply estimation techniques from Figure 9.24, in addition to a specific calculation of variance as in Example 9.5. If we take the value of (0.61 · top) as border of the 68% area which encompasses 2σ, we obtain $2\sigma = 9$ m, and $\alpha_L = 0.25$ m. Calculating the variance as in Example 9.6 gives:

$$\sigma^2 = \frac{\sum x_i^2 C_i}{\sum C_i} - \left(\frac{\sum x_i C_i}{\sum C_i}\right)^2 = 25 \text{ m}^2 = 2D_L t$$

where x_i is the distance (m) covered by concentration C_i.

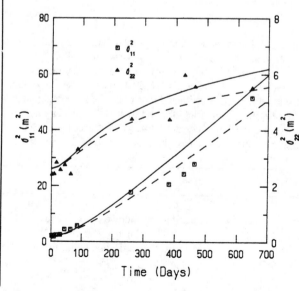

Figure 9.31. Development of longitudinal and transversal variance of the spread of Cl^-, injected in the Borden aquifer (from Sudicky, 1986, Copyright by the American Geophysical Union).

With $\sigma^2 = 2\alpha_L \cdot vt = 2\alpha_L \cdot 41$, we obtain $\alpha_L = 0.3$ m. This value should be compared with 0.36 m, given by Freyberg (1986). It is of interest to show here the development of the longitudinal spread σ_{11}^2 and the transversal spread σ_{22}^2 as observed when the plume was transported with groundwater flow. The dashed curves are calibration fits by Freyberg (1986) (note that the longitudinal variance increases linearly with time, and hence with travel distance), and the solid lines indicate model calculations based on permeability variations in the aquifer by Sudicky (1986). The dispersivity calculated with Equation (9.65) is 4.9 m, or more than 10 times too high.

9.3.5 *Diffusion influences on transport*

We noted that the diffusion coefficient in 'free' water divided by a tortuosity factor is adequate for describing dispersion in column experiments when the flow velocity is very low (Equation 9.52, or 9.56). This behavior can be extrapolated to field situations with very low advective flow velocities (i.e. in clays, or in peat layers). For example, profiles of Cl^- concentration in the Champlain Sea clay could be adequately simulated with diffusion of uniformly distributed Cl^- in the originally brackish clay (1/3 of seawater concentration) into an upper, weathered zone, where the salt is flushed (Desaulniers and Cherry, 1988). Diffusion in the clay occurred over the 10 000 years since regression of the sea, with an effective diffusion coefficient D_e of $2 \cdot 10^{-6}$ cm^2/s. This effective diffusion coefficient in a porous medium can be compared with a 'free' water diffusion coefficient of ca. $2 \cdot 10^{-5}$ cm^2/s at 10°C. Similar effective diffusion coefficients have been used by Volker (1961), Sjöberg et al. (1983) and Beekman (1991), in clays in which geological changes induced a Cl^- diffusion over about 300 to 500 years (Figure 9.32). In all these studies Fick's second law is used:

$$\frac{\partial C}{\partial t} = D_e \frac{\partial^2 C}{\partial x^2} \tag{9.41}$$

which can be solved for the boundary conditions:

$$C(x, t) = C_i \qquad \text{for } x > 0; t = 0;$$
$$C(x, t) = C_0 \qquad \text{for } x = 0; t > 0;$$
$$C(x, t) = C_i \qquad \text{for } x = \infty; t > 0$$

to give:

$$C(x, t) = C_i + (C_0 - C_i)\, \text{erfc}\left(\frac{x}{\sqrt{4D_e t}}\right) \tag{9.66}$$

Since a geological time period is interpreted, the product $D_e t$ is used to model the data (Sjöberg et al., 1983), rather than a singled out time and an a priori known effective diffusion coefficient. It therefore requires some idea of the timescale, before a value can be obtained for the field diffusion coefficient. However, values as quoted above for Cl^-, have also been found valid under well defined field conditions over a short timescale of about 5 years (Johnson et al., 1989), and in laboratory experiments operating with diffusion only (Li and Gregory, 1974). There is considerable controversy on how the effective diffusion coefficient can be calculated from the free water value, and related to sediment properties (Lerman, 1979). In the Kozeny theory, it is related to the tortuosity (Equation 9.52), but it can also be obtained from the measurement of the *formation factor F*, which is the ratio of

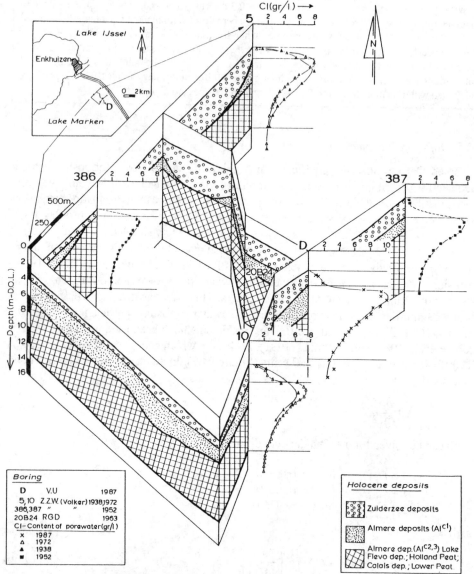

Figure 9.32. Diffusion profiles of Cl⁻ from the brackish Zuiderzee into fresh water clay and peat, and back diffusion of salt into the fresh water lake which developed after construction of a closure dam. Modeled curve is for 360 years of salt diffusion into the sediment, and 55 years out into the fresh water lake (Beekman, 1991).

the specific electrical resistance of a sediment with a solution, and the specific resistance of the free solution: $F = R_{sed} / R_w$. The formation factor is thus defined for a unit area of porous medium, i.e. pore volume and grain skeleton. When the sediment grains are perfect insulators, the tortuosity is related to the formation factor as (Childs, 1969):

$$\theta^2 = F \cdot \varepsilon$$

$$(9.67)$$

but a linear relationship $\theta = F \cdot \varepsilon$ has been proposed as well (cf. Bear, 1972), which neglects the surface area dependency expressed in the Kozeny theory. It is appropriate to stress that the formation factor is a parameter that can be measured, and then gives rigorously $D_e = D_f/(F \cdot \varepsilon)$. The formation factor itself is also a function of porosity. Empirical relationships between formation factor and porosity take the form of Archie's law:

$$F = 1/\varepsilon^n$$

where the exponent n varies from 1.4 to 2.0 (McNeil, 1980). If we take a high value $n = 2.0$, then $F = \varepsilon^{-2}$, $\theta^2 = \varepsilon^{-1}$, and the effective diffusion coefficient would follow from:

$$D_e = D_f/\theta^2 = D_f \cdot \varepsilon \tag{9.68}$$

where D_f is the diffusion coefficient in free water. With $\varepsilon = 0.5$, the relation gives $D_e = 0.5 \, D_f$, which is in the range of the values found in the column experiments compiled in Figure 9.26. It should be noted that when the diffusive flux is calculated according to Fick's first law, D_e must be counted for the surface over which diffusion occurs, i.e. must be multiplied by ε (Example 9.11).

Diffusion also has important consequences for transport in consolidated (rock) aquifers, in which most flow takes place along faults or fissures (conduits in Chapter 4), but in which the bulk rock is porous, and contains stagnant water that is in continuous exchange via diffusion with mobile water in the fissures. We encountered the phenomenon in this chapter in relatively slow tritium transport in the unsaturated zone in English Chalk (Figure 9.2). A porous formation with these characteristics is termed a *dual porosity* medium, or a rock with mobile-immobile zones. Aggregated soils in which transport occurs both by advective flow along shrinkage fissures, and by diffusion into the peds, have similar transport characteristics.

Transport in such a medium is characterized by a rapid component along the fissures, and a slow component into the bulk rock or soil aggregates. A single fissure exhibits a flow behavior with transport being attenuated by diffusion into the rock matrix; such sloweddown flow can be beneficial if radioactive waste migrates along fissures, since additional time for decay becomes available (Tang et al., 1981), and fluctuations in input concentrations are smoothened (Lowell, 1989). Breakthrough curves in laboratory experiments with soils sometimes show effects which can be adequately modeled on the basis of the dual porosity concept (Van Genuchten and Cleary, 1982; Rao et al., 1982; Brusseau and Rao, 1989). It is typical to observe a tailing when a pulse injection has passed the observation point, due to back-diffusion, again out of the matrix into the channels which contain the mobile fluid. Although the physical basis of the process is diffusion, it is generally modeled as a kinetic exchange amongst two distinct zones in the solid, and it can be shown that the result for breakthrough curves, with certain restrictions, is similar (Bolt, 1982).

9.4 COMBINING THE PROCESSES

Let us combine now the processes in one equation which describes the change in concentration due to transport, dispersion, and sorption (or reactions in general):

$$\left(\frac{\partial C}{\partial t}\right)_x = -v\left(\frac{\partial C}{\partial x}\right)_t + D_L\left(\frac{\partial^2 C}{\partial x^2}\right)_t - \left(\frac{\partial q}{\partial t}\right)_x \tag{9.69}$$

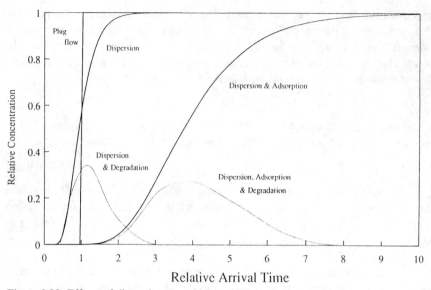

Figure 9.33. Effects of dispersion, retardation and degradation on column breakthrough of chemicals (after McCarty et al., 1981).

where C is solute concentration (mol/l), v is pore water flow velocity (m/s), D_L is hydrodynamic dispersion coefficient (m^2/s), q is concentration on solid (mol/l pore water).

In words, Equation (9.69) states that 'change in concentration at location x' is due to 'transport of change in concentration' + 'transport by dispersion' + 'change in solid concentration at x (expressed in terms of pore water concentration)'.

Figure 9.33 illustrates the effects which the different components of this equation have on breakthrough of chemicals during a column experiment. Degradation of the chemical may be of any kind, viz. radioactive decay, or biodegradation. Such effects may be thought of as being included in the change $(\partial q / \partial t)$, and will not be treated here, since transport of radionuclides and degradation in general involves complex chemistry which is better tackled with the models discussed in Section 9.5 and Chapter 10.

Analytical solutions for Equation (9.69) are simple when q can be expressed as a linear function of C, as derived in Section 9.2. Analytical solutions can also be obtained for complex relations between q and C, even for multicomponent influences, but only if the hydrodynamic dispersion coefficient $D_L = 0$; this possibility will be considered in Chapter 11. Let us now simplify Equation (9.69) by replacing $(\partial q/\partial t)$ by $K_d(\partial C/\partial t)$ so that we obtain (compare (9.29)):

$$(1 + K_d)\left(\frac{\partial C}{\partial t}\right)_x = -v\left(\frac{\partial C}{\partial x}\right)_t + D_L\left(\frac{\partial^2 C}{\partial x^2}\right)_t \tag{9.70}$$

or:

$$\left(\frac{\partial C}{\partial t}\right)_x = -\frac{v}{R}\left(\frac{\partial C}{\partial x}\right)_t + \frac{D_L}{R}\left(\frac{\partial^2 C}{\partial x^2}\right)_t \tag{9.71}$$

where $(1 + K_d)$ is replaced by the retardation factor R, and used to divide left and right side. If the retardation factor is constant (i.e. distribution coefficient K_d is independent of concentration), then Equation (9.71) is simply identical with the earlier advection-dispersion equation (9.48). All solutions given for that equation are valid when only the coefficients D_L/R and v/R are used instead of D_L and v. When the contribution of diffusion to the dispersion coefficient is negligible as is the case mostly in aquifers, the solutions are in fact exactly equal for each location; retarded chemicals show exactly the same amount of spreading at all points as a conservative chemical does, only arrive later depending on the retardation factor. We may illustrate this with the formula which describes front spreading in an infinite column (9.49). With retardation it would become:

$$C(x, t) = C_i + \tfrac{1}{2}(C_0 - C_i)\,\mathrm{erfc}\left(\frac{x - vt/R}{\sqrt{4D_L t/R}}\right) \tag{9.72}$$

and we apply the usual simplifications $D_L = \alpha v$, and $vt = x_{0.5}$ to get:

$$C(x, t) = C_i + \tfrac{1}{2}(C_0 - C_i)\,\mathrm{erfc}\left(\frac{x - x_{0.5}/R}{\sqrt{4\alpha\, x_{0.5}/R}}\right) \tag{9.73}$$

Transport in an aquifer of a chemical which shows linear retardation can thus be calculated very simple, and completely analogous to a conservative compound. Let us compare this possibility with results from a numerical model in Example 9.10.

EXAMPLE 9.10: *Calculation of aquifer pollution by waste site leachate*
Figure 9.34 shows results of numerical calculation of pollutants leached from a waste site in an aquifer profile (Pickens and Lennox, 1976). Effects of dispersivity and different retardations are illustrated by iso-concentration lines (contours) of 0.1, 0.3, 0.5, 0.7 and 0.9 (fractions of final concentration). Compare the spread of the front with an analytical guess. Data: $vt = 415$ m; $\alpha_L = 10$ m; $\rho_b = 1.8$ g/cm^3; $\varepsilon = 0.3$; $K'_d = 0, 0.1$ and 1 ml/g. Initial concentration $C_i = 0$ mg/l, the waste site is a continuous source with $C_0 = 1$ mg/l in the leachate.

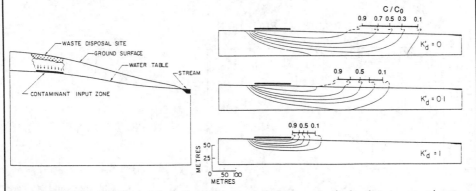

Figure 9.34. Effects of dispersivity and varying retardations on waste site leachate transport in an aquifer. Numerical calculations from Pickens and Lennox, 1976; approximate longitudinal spreading of 0.1/0.9, and 0.3/0.7 contour lines as calculated here, is indicated by the bar.

ANSWER
Assume that a flowline originating at the downstream end of the waste site is horizontal, and that transversal dispersion is absent. Distance travelled of the $C/C_0 = 0.5$ concentration is $x_{0.5} = x_0/R =$

$(v/R) \cdot t$. The concentration $C/C_0 = 0.1$ is at $x_{0.1}$, a distance further ahead that can be obtained from Equation (9.73): $C(x_{0.1},t) = 0.1 = 1/2 \operatorname{erfc}[(x_{0.1} - x_{0.5})/ \sqrt{4\alpha_L x_{0.5}}] = 1/2 \operatorname{erfc}[f(x)]$.

If we know the inverse complementary error function of 0.2 (from Abramowitz and Stegun, 1964), we obtain $f(x)$, and can calculate $x_{0.1}$. Thus, $f(x) = \operatorname{inverfc}[0.2] = 0.906 = (x_{0.1} - x_{0.5}) / \sqrt{4\alpha_L x_{0.5}}$, which gives $x_{0.1}$ as function of α, vt and R. The concentration $C/C_0 = 0.9$, at $x_{0.9}$, is found similarly from $0.906 = -(x_{0.9} - x_{0.5})/ \sqrt{4\alpha_L x_{0.5}}$. For $C/C_0 = 0.3$ and 0.7 the procedure is identical, with the value of $\operatorname{inverfc}[0.6] = 0.371$. The results are summarized as:

	R	vt/R	$(x_{C/C_0} \pm vt/R)$ for $C/C_0 =$	
			0.1 and 0.9	0.3 and 0.7
$K'_d = 0$	1	415	117	47.8
$K'_d = 0.1$	1.6	259	92	37.8
$K'_d = 1.0$	7	59	44	18.1

Advective flow in clays or peats (aquicludes and aquitards) is often negligibly small, so that diffusion is the only transport mechanism of water and chemicals. Equation (9.71) reduces to Fick's law, for which again the usual solutions are obtained (for example Equation 9.66), with the coefficient D_e being replaced by D_e/R. A specific example, showing diffusion of Cl^- and toluene in thick clays underlying a waste deposit, is given in Figure 9.35. The diffusion coefficients used for simulating the profile are indicated in the figure; the best fit for Cl^- and toluene is obtained with respectively $5 \cdot 10^{-6}$ and $5 \cdot 10^{-8}$ cm^2/s. Johnson et al. (1989) calculated diffusion coefficients for Cl^- and a number of organic chemicals from free water values, and found reasonable agreement after correcting for tortuosity and retardation.

Figure 9.35 shows that adsorption can slow down diffusive fluxes of pollutants which initially contact the clay layer. When sorption sites have been filled, retardation is no longer effective, and diffusive fluxes of conservative elements and adsorbed elements are just the same. Thus, when for example a 1 m thick clay liner is used to act as a barrier against

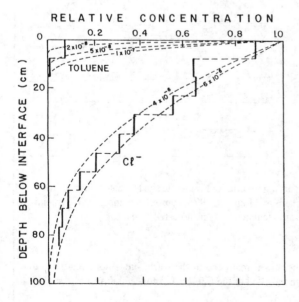

Figure 9.35. Chloride and toluene concentrations in interstitial solutions obtained from a clay core below a waste site. Simulation data for 1800 days and indicated diffusion coefficients. Reprinted with permission from Johnson et al., 1989. Copyright American Chemical Society.

downward percolation of leachates from a waste site, the steady state flux may be considerable, and cause extensive pollution of groundwater in aquifers below the site (Example 9.11).

EXAMPLE 9.11: *Diffusive flux through a clay barrier (after Johnson et al., 1989)*
Estimate steady state flux of benzene through a 1 m thick clay liner below a 10^4 m^2 waste site. Concentration of benzene in the waste site is 1 g/l; groundwater flow below the liner keeps the concentration at 0.01 g/l; The clay has porosity $\varepsilon = 0.5$; free water diffusion coefficient of benzene $D_f = 7 \cdot 10^{-6}$ cm^2/s. Also estimate maximum time to arrive at steady state conditions, when $K_{oc} = 65$ ml/g; $f_{oc} = 0.01$; $\rho_b = 1.6$ g/cm^3.

ANSWER
An effective diffusive coefficient is calculated from Equation (9.68): $D_e = D_f \cdot \varepsilon = 7 \cdot 10^{-6} \times 0.5 = 3.6 \cdot 10^{-6}$ cm^2/s. From Fick's first law obtain a flux

$$F = -D_e \cdot \varepsilon \cdot (\partial C / \partial x) = -3.6 \cdot 10^{-6} \cdot 0.5 \cdot (1 - 0.01) / -100 = 1.8 \cdot 10^{-8} \text{ mg/s.cm}^2$$

This amounts to 56 kg benzene/yr, a quantity which has the potential to contaminate $11.2 \cdot 10^9$ liters of water at the EPA drinking water limit of 0.005 mg/l. Note that we multiplied the flux with porosity to obtain the flux through the open, pore space.
　　The maximum time necessary to reach steady state can be calculated from the amount of adsorption that must occur. The distribution coefficient $K_d = K_{oc} \cdot f_{oc} \cdot (\rho_b / \varepsilon) = 65 \times 0.01 \times (1.6/0.5) = 2.08$. Then, adsorbed with 1 g/l and 0.01 g/l in solution are resp. 2.08 and 0.02 g/l porewater, or the average for the whole clay liner: $1.05 \cdot 1 \cdot 10^4 \cdot 0.5 = 5252$ kg benzene. Add the amount in solution, which is 2525 kg, to find a total quantity of 7777 kg benzene in the clay liner (one may wonder whether this is available in the waste). This amount is carried into and through the clay in approximately 140 years by the steady state flux. The initial diffusion profile is concave, and consequently exhibits a larger gradient, and a larger flux; the calculated time space to reach steady state is therefore a maximal one. The time for steady state can also be calculated with an analytical solution (as done by Johnson et al., 1989), but the student may already look forward to using more flexible numerical techniques which will be treated in Section 9.5.

9.5 NUMERICAL MODELING OF TRANSPORT

Simple distribution functions may not be realistic when geochemical problems must be tackled. If for example, precipitation of calcite is modeled along a flowline, then q_{Ca} is clearly a function of pH, $[HCO_3^-]$, and also of complexes and activity coefficients (Chapter 3 and 4). It is, in principle, possible to write out the transport equation for each component, include equations for reactions among components, and solve the whole set with a differencing scheme. This approach has been used by Miller and Benson (1983), Lichtner (1985), and Selim et al. (1987). Another way to get a numerical solution is to use the 'mixing cells in series' concept. It can be visualized as a number of batches in which reactions are calculated during a timestep, and the contents of each batch are then poured into the next (Figure 9.36). The transport parts of Equation (9.69) are calculated separately from the reaction part $(\partial q / \partial t)$ with this method (Cederberg et al., 1985; Schulz and Reardon, 1983; Van Ommen, 1985; Appelo and Willemsen, 1987). Since the method is easy to program on small computers, it can be an advantage to use this numerical technique to model transport situations with complex initial conditions.

After 1 timestep :

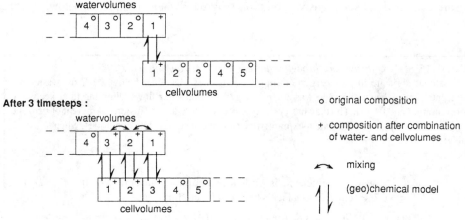

Figure 9.36. The mixing cell in series concept for simulating transport and chemistry.

9.5.1 *Only diffusion*

Let us split up the problem again in three parts, corresponding to the three terms of the right-hand side of the advection-dispersion-reaction equation (9.69). If we separate the dispersion term from the transport part, it is more appropriately termed diffusion (Section 9.3). Hence, the change in concentration as result of diffusion is:

$$\left(\frac{\partial C}{\partial t}\right)_x = D_f \left(\frac{\partial^2 C}{\partial x^2}\right)_t \tag{9.74}$$

where we want to evaluate the development of C at some location x in time. It requires calculation of the second derivative of C with x. An approximation is obtained by combining two Taylor expansions of a function $f(x)$. First in a forward sense:

$$f(x_n + \Delta x) = f(x_n) + f'(x_n)\Delta x + \frac{f''(x_n)}{2}(\Delta x)^2 + \frac{f'''(x_n)}{6}(\Delta x)^3 + \dots \tag{9.75}$$

where f' is the first derivative, f'' is the second derivative, etc., Δx is the stepsize.
And a similar Taylor expansion in a backward sense gives:

$$f(x_n - \Delta x) = f(x_n) - f'(x_n)\Delta x + \frac{f''(x_n)}{2}(\Delta x)^2 - \frac{f'''(x_n)}{6}(\Delta x)^3 + \dots \tag{9.76}$$

Adding both Equation (9.75) and (9.76) gives the finite difference approximation of the second derivative:

$$\frac{f(x_n + \Delta x) - 2f(x_n) + f(x_n - \Delta x)}{(\Delta x)^2} = f''(x_n) + O(\Delta x)^2 \tag{9.77}$$

Our approximation of the derivative of $f(x)$ at x_n is obtained by combining both a backward and a forward step, and it is therefore termed a *central difference* equation. The error term $O(\Delta x)^2$ can be calculated with formulas derived in standard books on numerical techniques; here it is of the order of the square of the stepsize Δx (Gerald and Wheatley, 1989).

The $f(x)$ as used in the Taylor formula is nothing but the concentration $(C)_x^t$, i.e. the concentration at x, at time t. Thus, the second derivative in Equation (9.74) is approximated as:

$$\left(\frac{\partial^2 C}{\partial x^2}\right)_{t1} = f''(x)_{t1} = \frac{C_{x-1}^{t1} - 2C_x^{t1} + C_{x+1}^{t1}}{(\Delta x)^2} \tag{9.78}$$

We can also difference the first derivative with respect to time:

$$\left(\frac{\partial C}{\partial t}\right)_x = \left(\frac{\Delta C}{\Delta t}\right)_x = \frac{C_x^{t2} - C_x^{t1}}{\Delta t} \tag{9.79}$$

which amounts to taking steps forward in time. The solution for the diffusion equation (9.74) is then obtained with a scheme which is central in space, and forward in time by combining Equations (9.78), (9.79) and (9.74):

$$
\begin{aligned}
C_x^{t2} &= C_x^{t1} + \frac{D_f \Delta t}{(\Delta x)^2}(C_{x-1}^{t1} - 2C_x^{t1} + C_{x+1}^{t1}) \\
&= mixf \cdot C_{x-1}^{t1} + mixf \cdot C_{x+1}^{t1} + (1 - 2\,mixf) \cdot C_x^{t1}
\end{aligned} \tag{9.80}
$$

where $mixf$ is $D_f \Delta t / (\Delta x)^2$.

The parameter $mixf$ is called a *mixing factor*, since its size determines the amount of mixing of adjacent cells. When $mixf = 1/2$, Equation (9.80) simplifies to:

$$C_x^{t2} = \frac{C_{x-1}^{t1} + C_{x+1}^{t1}}{2} \tag{9.81}$$

but with $mixf > 1/2$, a problem would be that more is mixed from the central cell (x) than is available at $t = t1$. In such case *numerical instability* sets in, which can easily lead to *oscillations* in the results. Sometimes instabilities occur at values of $mixf$ between $1/3$ and $1/2$, especially when sharp concentration changes exist between adjacent cells.

Figure 9.37 shows such a case, with the central cell having concentration $C_x^{t1} = 1$, surrounded by cells with $C_{x\pm1}^{t1} = 0$. Diffusion in time, of course, smears out the sharp transition and gives a smooth, Gaussian distribution. However, concentrations oscillate when calculated at $t = t2$ and $t3$ with $mixf = 1/2$. When the mixing factor remains smaller than $1/3$, such oscillations are prevented, since never more is mixed out of a cell, than

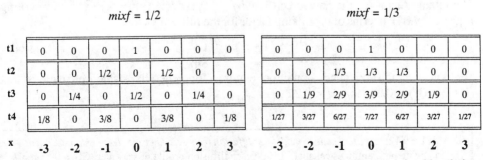

Figure 9.37. Stability in a central difference scheme depends on the mixing factor $mixf$; stable solutions are obtained with $mixf < \frac{1}{3}$.

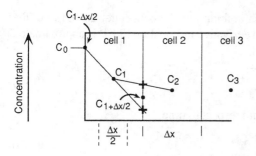

Figure 9.38. When concentrations are fixed at one boundary, the difference formulas at that point must be adapted.

remains behind. This stability criterion thus requires that:

$$mixf = \frac{D_f \, \Delta t}{(\Delta x)^2} < 1/3$$

or

$$\Delta t < \frac{(\Delta x)^2}{3D_f} \tag{9.82}$$

Boundary conditions may demand slightly different formulas for the mixing factor. When concentration is a fixed quantity C_0 at one boundary (for example when seawater diffuses into a fresh water sediment), a better approximation is obtained if we fix the concentration at that point as such. This can be arranged as illustrated in Figure 9.38. We halve the first cell, and calculate the concentration C_1 from central differences with $C_{1 - \frac{1}{2}\Delta x}$ and $C_{1 + \frac{1}{2}\Delta x}$. Clearly, $C_{1 - \frac{1}{2}\Delta x} = C_0$ (Figure 9.38). The concentration $C_{1 + \frac{1}{2}\Delta x}$ can be approximated as an average of two extremes. The first extreme is the midpoint concentration between the first two cells (which is $\frac{1}{2}(C_1 + C_2)$), and the second one is the expected concentration when the line $C_0 - C_1$ is extrapolated (which gives $C_1 - (C_0 - C_1)$). The average of the two is:

$$C_{1 + \frac{1}{2}\Delta x} = C_2/4 - C_0/2 + 5/4 C_1 \tag{9.83}$$

Once more we use (9.80) to calculate the concentration C_1 at the next timestep from C_1, $C_{1 - \frac{1}{2}\Delta x}$ and $C_{1 + \frac{1}{2}\Delta x}$, and obtain:

$$C_1^{t2} = \frac{1}{2} \, mixf^* \, C_0^{t1} + \frac{1}{4} \, mixf^* \, C_2^{t1} + (1 - \frac{3}{4} \, mixf^*) \, C_1^{t1} \tag{9.84}$$

The mixing factor for the halved cell is $mixf^* = D_f \, \Delta t \, /(\frac{1}{2}\Delta x)^2 = 4 \, mixf$, and rewriting Equation (9.84) in terms of the mixing factor for the full cell yields:

$$C_1^{t2} = 2 \, mixf \, C_0^{t1} + mixf \, C_2^{t1} + (1 - 3 \, mixf) \, C_1^{t1} \tag{9.85}$$

We note that this boundary condition requires for generally stable calculations that $(1 - 3 \, mixf) > 0.33$, or $mixf < 0.22$.

EXAMPLE 9.12: *A Pascal code to model Cl⁻ diffusion from seawater into fresh water sediment*
The following program lines calculate a 100 years diffusion profile of seawater ($Cl^- = 0.56$ mol/l) into a 10 m thick fresh water sediment ($Cl^- = 0.0$ mol/l). The lower end boundary is a solid stop (a mirror for Cl^- ions, cf. Problem 9.17). The program variables are self-explanatory.

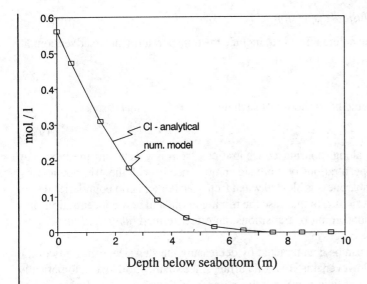

Figure 9.39. A 100 years of seawater diffusion in fresh water sediment; comparison of central difference and analytical solution.

```
Program diffuse;
const  Ncel = 10;
var  i, j, k, Nmix                           : integer;
     Totx, Tott, Delx, Delt, De, mixf, x     : real;
     Ct1, Ct2                                : array[0..Ncel] of real;
{MAIN}
begin De: = 1e-5 {cm2/s};
   Totx: = 1000.0 {cm}; Tott: = 100*365*24*60*60.0 {s}; Delx: = Totx/Ncel;
   for i: = 1 to Ncel do Ct1[i]: = 0.0; Ct1[0]: = 0.56;
   mixf: = De*Tott/(Delx*Delx); Nmix: = 1+trunc(mixf/0.22);
   if Nmix<10 then Nmix: = 10;                          {min. no. of mixes = 10}
   Delt: = Tott/Nmix; mixf: = mixf/Nmix;                {mixf is now < 0.22}
   for i: = 1 to Nmix do
      begin                                             { mix into Ct2-cells ...}
        Ct2[1]: = 2*mixf*Ct1[0]+mixf*Ct1[2]+(1-3*mixf)*Ct1[1];    { .. first cell}
        for j: = 2 to Ncel-1 do
          Ct2[j]: = mixf*(Ct1[j-1]+Ct1[j+1]) + (1-2*mixf)*Ct1[j]; { .. inner cells }
        Ct2[Ncel]: = mixf*Ct1[Ncel-1] + (1-mixf)*Ct1[Ncel];       { .. end cell }
        for j: = 1 to Ncel do Ct1[j]: = Ct2[j];         { .. copy back to Ct1-cells }
      end;
   writeln(' Depth below seabottom (m); Cl (mol.dm-3) after',Tott/(31536e3):8:2,' years');
   writeln('       0     ;',Ct1[0]:8:4); x: = Delx/2;
  for i: = 1 to Ncel do
     begin
      writeln('          ',x/100:9:3,Ct1[i]:8:4); x: = x+Delx;
     end;
end.{Diffuse.}
```

Figure 9.39 shows a comparison of the results of this program with the analytical solution:

$$C_{x,t} = 0.56 \, \mathrm{erfc}\left(\frac{x}{\sqrt{4D_e t}}\right) \tag{9.86}$$

The agreement is excellent, also in the first cell near the seawater boundary.

9.5.2 *Advection and diffusion/dispersion*

The second step in our modeling effort is to include the transport term due to advection:

$$\left(\frac{\partial C}{\partial t}\right)_x = -v\left(\frac{\partial C}{\partial x}\right)_t \tag{9.87}$$

The simplest solution would be to choose Δt such that $v\Delta t = \Delta x$. In that case is:

$$C_x^{t2} = C_x^{t1} - (C_x^{t1} - C_{x-1}^{t1}) = C_{x-1}^{t1} \tag{9.88}$$

Thus, we simply move along, pouring concentrations from one cell into the next one. Fronts with different concentrations on both sides move neatly with the grid boundaries, and remain sharp. Such sharpness is blurred when front transfer and grid boundaries do not correspond, i.e. when $v\Delta t \neq \Delta x$. In this case the mixing of old and new concentrations in a cell leads to gradual smoothening of transitions, an effect termed *numerical dispersion*. Figure 9.40 illustrates the process.

Numerical dispersion can be counteracted by decreasing the step size, especially of the grid (i.e. decrease Δx). However, this is a costly affair, since smaller grid size automatically decreases the timestep, and thus *increases* the number of timesteps as well. Numerical dispersion can also be used to mimic physical (field) dispersion by choosing the discretization steps Δt in relation to Δx in such a way that numerical dispersion is just equal to the physical dispersion which must be modeled (Van Ommen, 1985). However, front retardation as result of reactions must also be taken into account (Willemsen, 1992). It can be shown (Herzer and Kinzelbach, 1989, Notodarmojo et al., 1991, Willemsen, 1992), that the numerical dispersivity amounts to:

$$\alpha_{num} = \frac{\Delta x}{2} - \frac{v\Delta t}{2R} \tag{9.89}$$

where α_{num} is numerical dispersivity (m), and R is the retardation factor.

The retardation factor can here be calculated as the relative change of concentrations through reactions (sorption, precipitation, dissolution, etc.) within a cell:

$$R = \frac{C_{x,\,new} - C_{x,\,old}}{C_{x,\,new + react} - C_{x,\,old}} \tag{9.90}$$

where $C_{x,\,old}$ is the old concentration in a cell, $C_{x,\,new}$ is concentration in the fluid which enters the cell, and $C_{x,\,new + react}$ is the end concentration after reactions.

	$v\Delta t = \Delta x$						$v\Delta t = \tfrac{1}{2}\Delta x$			
t1	1	0	0	0	0		1	0	0	0	0
t2	1	1	0	0	0		1	1/2	0	0	0
t3	1	1	1	0	0		1	3/4	1/4	0	0
t4	1	1	1	1	0		1	.875	.5	.125	0
x	1	2	3	4	5		1	2	3	4	5

Figure 9.40. Numerical dispersion as result of discretization.

We recall here that (field) dispersivity forms part of the hydrodynamic dispersion coefficient (Section 9.3):

$$D_L = D_e + \alpha_L v \qquad (9.56)$$

and must be included in the calculation of the mixing factor when advective flow, in addition to diffusion, takes a part ($v \neq 0$). The actual value of α_L, comprised in D_L, can be corrected by subtracting the numerical dispersivity as result of transport and reaction:

$$\alpha_{mixf} = \alpha_L - \left(\frac{\Delta x}{2} - \frac{v \Delta t}{2R} \right) \qquad (9.91)$$

and the mixing factor is obtained from:

$$mixf = \frac{D_e \, \Delta t}{(\Delta x)^2} + \frac{\alpha_{mixf} v \Delta t}{(\Delta x)^2} \qquad (9.92)$$

Noting that *mixf* should be larger than zero, it is of interest that Equations (9.91) and (9.92) provide an estimate of maximum Δx, and hence of the minimum number of cells which is necessary for a calculation that is free of numerical dispersion. Assume $D_e = 0$, and $v \Delta t = \Delta x$ (the latter premise is anyhow part of the sensible strategy to maintain $v \Delta t$ as large as possible, for efficient calculations and also to keep α_{mixf} positive in Equation (9.91)), which means that in Equation (9.92), $\alpha_{mixf}/\Delta x > 0$. We combine with Equation (9.91), and obtain:

$$mixf > 0,$$

if

$$\Delta x < \frac{\alpha_L 2R}{R - 1} \qquad (9.93a)$$

or with *Ncell* = *Totx*/Δx:

$$mixf > 0$$

if

$$Ncell > \frac{Totx}{\alpha_L} \left(\frac{1 - 1/R}{2} \right) \qquad (9.93b)$$

where *Totx* is the length of the (laboratory) column or flowline.

Hence, we have gained a measure for the maximal timestep (from $v \Delta t < \Delta x$) and the minimal grid spacing (Equation 9.93), both from the point of view of numerical stability and arithmetical accuracy. In this scheme we have assumed that reactions can be considered linear on the scale of the cell. However, corrections for non-linearity are in a sense *automatically* included by small grid size and timestep dictated by the explicit scheme. This high-resolution discretization is at the same time a major disadvantage with respect to implicit schemes (which use next timestep solutions implicitly as part of the arithmetics, and which are unconditionally stable when retardation is linear, and therefore preferable when such is the case). For presentation of e.g. Crank-Nicolson schemes which can be used with simple reactions, the reader is referred to clear discussions in Gerald and Wheatley (1989).

EXAMPLE 9.13: *Numerical model of linear retardation of γ-HCH in a laboratory column*
Model the linear retardation of γ-HCH in a 5 cm long, and 5 cm ∅ laboratory column with a sandy
aquifer sample. The sand is low in organic carbon (0.05%), has porosity of 30%, and dispersivity is
7 mm. Injected into the column are 450 ml water containing 20 µg γ-HCH/l, at a rate of 10 ml/hr.
Assume that diffusion is negligible (cf. Problem 9.18).

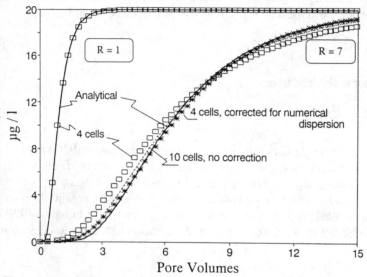

Figure 9.41. Breakthrough curves for conservative element, and retarded γ-HCH; comparison of
finite difference and analytical solutions.

Preliminary calculations provide the following data. With $K_{oc} = 2 \cdot 10^3$ (Table 9.3) for γ-HCH, the
distribution coefficient $K'_d = 5 \cdot 10^{-4} \cdot 2 \cdot 10^3 = 1$ ml/g, and the normalized distribution coefficient
$K_d = 1.8/.3$ (from $\rho_b/\varepsilon) \cdot 1 = 6$. Hence the retardation factor $R = 7$. Pore volume (Porv) in the
column is $\pi r^2 L \cdot \varepsilon = 29.5$ ml; porewater flow velocity (veloc) is 1.7 cm/hr.
    From Equation (9.93) follows that *Ncell* (=Ncel) $> 5/0.7 \cdot (7-1)/14 = 3.1$, hence Ncel gets the
integer value = 4. Then Δx (= Delx) = Totx/Ncel = 1.25.
    The maximum value of Δt (= Delt) is when $v\Delta t = \Delta x$, hence $\Delta t < 1.25/1.7 = 0.735$ hr. Total
injection time of 45 hr is subdivided in 45/0.735 = (integer value) = 62 timesteps (= Nshift). Thus
$\Delta t = 45/62 = 0.726$ hr.
    We may subdivide our program into a number of subroutines. Thus:

dspvty   : calculates initial mixing factors mixf.
dspcor   : corrects the mixing factors for mixing and retardation.
dffdsp   : mixes the cells with corrected mixing factors.
flush   : performs the advective transport.
distri   : calculates distribution of chemical over solid (q) and water (Ct).
analyt   : calculates the analytical solution.
{MAIN}   : sets the variables, and calls the subroutines in right order.

```
Program Columnperc;
Const  Nel = 1;    Ncelmax = 15;
Var    i, j, k                              : integer;
       Ncel, Nshift, Ish, Nmix, Mi         : integer;
```

```
        Porv, Veloc, Tott, Delt, Totx, Delx        : real;
        Mixf, D, Disp, R, Dum, Dum1, Dum2          : real;
        Cin                                        : array[0..Nel] of real;
        Ct1, Ct2, Q, Mix                           : array[0..Nel, 1..Ncelmax] of real;
        Pv                                         : array[1..200] of real;
        Cout                                       : array[0..Nel+2, 1..200] of real;

procedure dspvty;                                         { calculate mixfactors ... }
begin
  Mixf: = (D + Disp*Veloc)*Delt/sqr(Delx)*(1+2/Ncel);
                            { ... (1+2/Ncel) is correction factor for not mixing of end cells }
  Nmix: = 1 + trunc(3.0*Mixf); Mixf: = Mixf/Nmix;         { .. make sure that   Mixf<1/3 }
  for j: = 1 to Ncel do   for i: = 0 to Nel do   Mix[i,j]: = Mixf;
end;


procedure dspcor;                                         { correct mixfactors ... }
begin
  for j: = 1 to Ncel do
    begin   Mix[0,j]: = Mixf-(1-Veloc*Delt/Delx)/2/Nmix * (1+2/Ncel);
                                               { ... conservative element }
    for i: = 1 to Nel do                       { and also others ... }
      begin   R: = 7;   Mix[i,j]: = Mixf – (1-Veloc*Delt/R/Delx)/2/Nmix *(1+2/Ncel);
        if Mix[i,j]<0 then begin Mix[i,j]: = 0; writeln('Num. dispersion'); end;
      end;
    end;
end;


procedure dffdsp;                                         { perform actual mixing ... }
Var Rest : real;
begin
  for j: = 2 to Ncel-1 do   for i: = 0 to Nel do          { mix inner cells ... }
    begin   Dum1: = Mix[i,j];   Rest: = 1.0-2*Dum1;
      Ct2[i,j]: = Rest*Ct1[i,j] + Dum1*Ct1[i,j-1] + Dum1*Ct1[i,j+1];
    end;
  for i: = 0 to Nel do                                    { mix end cells ... }
    begin   Dum1: = Mix[i,1];   Ct2[i,1]: = (1-Dum1)*Ct1[i,1] + Dum1*Ct1[i,2];
      Ct2[i,Ncel]: = (1-Dum1)*Ct1[i,Ncel] + Dum1*Ct1[i,Ncel-1];
    end;
  for j: = 1 to Ncel do   for i: = 0 to Nel do   Ct1[i,j]: = Ct2[i,j];
end;


procedure flush;                                          { advective transport ... }
begin
  Dum1: = Veloc*Delt/Delx;
  for i: = 0 to Nel do
    begin   for j: = Ncel downto 2 do
              Ct1[i,j]: = Dum1*Ct1[i,j-1] + (1-Dum1)*Ct1[i,j];
      Ct1[i,1]: = Dum1*Cin[i] + (1-Dum1)*Ct1[i,1];
    end;
end;


procedure distri;                                  { distribute chemical over water and solid ... }
begin
```

```
    for j: = 1 to Ncel do    for i: = 1 to Nel do
      begin Dum1: = Q[i,j]+Ct1[i,j];
      Ct1[i,j]: = Dum1/R; Q[i,j]: = Dum1-Ct1[i,j];
      end;
  end;

  procedure analyt;                                        { analytical solution ... }
  var P, a1, a2, a3, a4, a5, e, er, er1, er2 : real;
                                                           { error function ... }
  function erfc(Dum1:real) : real;
  begin erfc: = (e*(a1+e*(a2+e*(a3+e*(a4+e*a5)))))*exp(-Dum1*Dum1) end;

  begin
    P: = 0.3275911; a1: = 0.254829592; a2: = -0.284496736;
    a3: = 1.421413741; a4: = -1.453152027; a5: = 1.061405429;
    for i: = 1 to Nshift do
      begin
        Dum: = (i+0.5)/Ncel*Totx;                          { conservative element ... }
        Dum1: = (Totx-Dum)/2/sqrt(Disp*Dum);    e: = 1/(1+P*abs(Dum1));
        if Dum1>0 then er: = erfc(Dum1) else er: = 2-erfc(Dum1);    er1: = er;
        Dum1: = (Totx+Dum)/2/sqrt(Disp*Dum);    e: = 1/(1+P*abs(Dum1));
        er: = erfc(Dum1); er2:=exp(Totx/Disp)*er;
        Cout[Nel+1,i]: = (er1+er2)*Cin[0]/2;              { .. conc. of conservative elem. }
                                                          { now the retarded elements ... }
        Dum1: = (R*Totx-Dum)/2/sqrt(Disp*R*Dum); e: = 1/(1+P*abs(Dum1));
        if Dum1>0 then er: = erfc(Dum1) else er: = 2-erfc(Dum1); er1: = er;
        Dum1: = (R*Totx+Dum)/2/sqrt(Disp*R*Dum); e: = 1/(1+P*abs(Dum1));
        er: = erfc(Dum1); er2: = exp(Totx/Disp)*er;
        Cout[Nel+2,i]: = (er1+er2)*Cin[0]/2;             { .. conc. of retarded elem. }
      end;
  end;

  {MAIN}
  begin D: = 0.0 {cm2/s}; Disp: = 0.7 {cm}; R: = 7; Totx: = 5.0 { cm };
    Ncel: = 1 + trunc(Totx/Disp*(1-1/R)/2);
                      { include this statement when minimum Ncel = 10 ... if Ncel < 10 then Ncel: = 10;}
    Delx: = Totx/Ncel;   Porv: = pi*sqr(2.5)*Totx*0.3;   Veloc: = Totx*10/Porv/3600 {cm/s};
    Nshift: = 1+trunc(45*3600*Veloc/Delx);   Delt: = 45*3600/Nshift {s};
    for i: = 0 to Nel do   for j: = 1 to Ncel do   begin Ct1[i,j]: = 0; Q[i,j]: = 0; end;
    Cin[0]: = 20; Cin[1]: = 20 {ug/1};
    dspvty; dspcor;                                      { call dspvty and dspcor }
    for Ish: = 1 to Nshift do
      begin
        flush; distri; for Mi: = 1 to Nmix do begin dffdsp; distri; end;
        Pv[Ish]: = (Ish+0.5)*Delt*Veloc/Delx/Ncel;
        for i: = 0 to Nel do Cout[i,Ish]: = Ct1[i,Ncel];
      end;   analyt;                                      { calculate analytical Cout ...}
    write('Pore_Vol El: 0,'); for i: = 1 to Nel do write('    ',i,',    '); writeln;
    for j: = 1 to Nshift do
      begin write(Pv[j]:6:3,' ');
        for i: = 0 to Nel+2 do write(Cout[i,j]:8:3); writeln;
      end;
  end. { Columnperc}
```

It will be noted that the subroutine 'dspcor' is called only once, since in this example the correction of the mixing factor is constant (since retardation is constant for a linear distribution). When retardation is variable, the subroutine must be called each time after 'flush' and 'distri', before subroutine 'dffdsp' is executed. The cellsize Δx also determines resolution of the outflow profile, and it may be useful to include a statement that Ncel must have a lower value (for example 10 as in the program). Output of the example is shown in Figure 9.41 for Ncel of 4 and 10, and the figure includes the analytical solution:

$$C_{L,t} = \frac{C_0}{2} \left[\mathrm{erfc} \left(\frac{RL - vt}{\sqrt{4D_L Rt}} \right) + \exp \left(\frac{vL}{D_L} \right) \mathrm{erfc} \left(\frac{RL + vt}{\sqrt{4D_L Rt}} \right) \right] \tag{9.94}$$

Element 0 is a conservative substance which is subjected to transport with dispersion only (e.g. Cl^-) and numerical calculation with only 4 cells shows already excellent agreement with the analytical solution. When HCH-elution is not corrected for numerical dispersion, an outflow profile is calculated which is too diffuse compared with the analytical solution. Decrease of grid size by increasing Ncel to 10, gives a steeper curve and a better approximation, but even better results are obtained when the calculations with Ncel = 4 are corrected for numerical dispersion. The correction is, of course, highly profitable in terms of execution time when compared with the other somewhat blunt (but easy) possibility of decreasing the cell size.

The boundary conditions in this scheme are of the flux-type, and a point should be noted. Concentrations are calculated midway of each cell, also in the two end cells. A 'flush' moves concentrations in and out of the column, without considering dispersion over the distance $\frac{1}{2}\Delta x$ to the column boundary. This neglect gives substantial deviations when few cells are used, but can be compensated for by multiplying the model mixing factor with (1 + 2/Ncel).

9.5.3 *Advection, diffusion/dispersion and non-linear reactions*

The model structure which has been introduced here is easily extended to more complicated chemical systems. Whenever a model is available for calculating the distribution of the chemicals over solution and solid in a batch system, it can be introduced in the procedure 'distri' and then applied to transport along a flowline. Models for batch systems have been made for non-linear sorption (an example will be given next), for kinetic sorption of chemicals by the solid material (Brusseau and Rao, 1989, have provided a recent review of such models), and for diffusion in solid aquifer material (Rao et al., 1982; Weber and Miller, 1988; Wu and Gschwend, 1986). Also, complete geochemical models have been programmed which allow the calculation of complicated geochemical reactions (Sposito and Mattigod, 1979; Parkhurst et al., 1980; Wolery, 1983), of which we will show applications in the next chapter. Let us consider here an example of non-linear sorption, as described by the Freundlich equation:

$$q = K_F C^n \tag{9.35}$$

where the exponent n determines the non-linearity of the sorption equilibrium. When, in a batch of water and solid, the total quantity of a chemical (T) and the sorption equation for that chemical is known, the amount in water and in sediment can be calculated using Newton's iterative method. Newton's method uses derivatives at a given estimate to obtain successive better approximations of the variables which determine the function. In our case we seek:

$$f(C) = C + q - T = C + K_F C^n - T = 0 \tag{9.95}$$

At a certain initial estimate C_1, there is a derivative of $f(C)$:

$$\frac{f(C_2) - f(C_1)}{C_2 - C_1} = f'(C_1) \qquad (9.96)$$

Now, the value of C_2 should be such that $f(C_2) = 0$; hence from Equation (9.96):

$$C_2 = C_1 - f(C_1)/f'(C_1) \qquad (9.97)$$

This simple formula shows generally rapid convergence for smooth, continuous functions. The derivative $f'(C)$ can be obtained analytically, or numerically as in Example 9.14.

EXAMPLE 9.14: *Effect of the Freundlich exponent on breakthrough curves from a column.* Determine the effect for a range of n, varying from $n = 1$ to 0.2, all with identical $K_F = 1$.

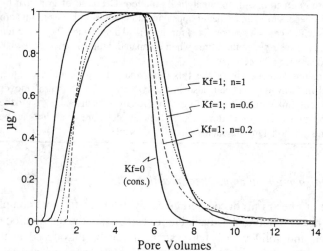

Figure 9.42. Effect of the exponent n in the Freundlich equation on column breakthrough curves.

The program in Example 9.13 needs only small extension to solve this problem. The procedure 'distri' now calculates the distribution of the chemical over solid and solution iteratively, and stops when the mass balance of Equation (9.95) is solved to be better than 10^{-6}. A function 'frndl' calculates the sorbed concentration q at each solute concentration C. Since the value of the retardation which is used to correct numerical dispersion is calculated in 'distri', a preliminary run with 3 cells is made to determine its maximal value. This maximal value is then used to estimate the number of cells for a dispersion-free calculation. The parameter 'numdisp' is set to 'true' whenever numerical dispersion occurs, i.e. when averaging of two mixing factors would give a negative result. We found that it was still necessary to increase this first estimate of Ncel by 1 (which is probably a result of very small concentrations that have a very high retardation factor when $n = 0.2$, and run farther ahead than is calculated in the 3-cell model).

```
Program Columnperc2;
Const Nel = 1;   Ncelmax = 40;
Var    Numdisp                                    : boolean;
       i, j, k                                    : integer;
       Ncel, Nshift, Ish, Nmix, Mi               : integer;
       Porv, Veloc, Tott, Delt, Totx, Delx, Cort  : real;
       Mixf, D, Disp, R, Rmax, Dum, Dum1, Dum2, Time : real;
```

```
    Cin, Kf, nf                                    : array[0..Nel] of real;
    Ct1, Ct2, Ctold, Q, Mix                        : array[0..Nel, 1..Ncelmax] of real;
    Pv                                             : array[1..200] of real;
    Cout                                           : array[0..Nel, 1..200] of real;

procedure dspvty;                                              { initial mixfactors ... }
begin
  Cort: = 1+2/Ncel;                               { correction term for not mixing end-cells }
  Mixf: = (D + Disp*Veloc)*Delt/sqr(Delx) * Cort;
  Nmix: = 1 + trunc(3.0*Mixf);    Mixf: = Mixf/Nmix;
  for j: = 1 to Ncel do    for i: = 0 to Nel do    Mix[i,j]: = Mixf;
end;

procedure dspcor;                                               { correct mixfactors ... }
begin                                             { Retardation is calculated as R ... }
  for j: = 1 to Ncel do                           { correct conservative element ... }
    begin Mix[0,j]: = Mixf-(1-Veloc*Delt/Delx)/2/Nmix*Cort;
    for i: = 1 to Nel do
    begin                                                       { correct others ... }
      R: = 1.0;
      Dum2: = Ct2[i,j]-Ctold[i,j]; Dum 1:=Ct[i,j]-Ctold[i,j];
      if abs(Dum1)>1e-9 then R: = Dum2/Dum1;
                                 { .. correct only if, after distri, Ct2 < Ct1 < Ctold }
      if R<1 then R:=1.0;
      if Rmax<R then Rmax: = R;
      Mix[i,j]: = Mixf-(1-Veloc*Delt /R/Delx)/2/Nmix*Cort;
    end;
    end;
end;

procedure dffdsp;                                         { perform actual mixing ... }
var   Rest  : real;
  begin                                                        { mix inner cells ... }
    for j: = 2 to Ncel-1 do    for i: = 0 to Nel do       { average mixing factors of neigboring cells .. }
    begin Dum1: = (Mix[i,j-1]+Mix[i,j])/2;    Dum2: = (Mix[i,j]+Mix[i,j+1])/2;
      if Dum1<0 then begin Dum1: = 0; Numdisp: = true; end;
      if Dum2<0 then Dum2: = 0;
      Rest: = 1.0-Dum1-Dum2;
      Ct2[i,j]: = Rest*Ct1[i,j]+Dum1*Ct1[i,j-1]+Dum2*Ct1[i,j+1];
    end;                                                        { mix end cells ... }
    for i: = 0 to Nel do
      begin Dum1: = (Mix[i,1]+Mix[i,2])/2; if Dum1<0 then Dum1: = 0;
      Ct2[i,1]: = (1-Dum1)*Ct1[i,1]+Dum1*Ct1[i,2];
      Dum2: = (Mix[i,Ncel-1]+Mix[i,Ncel])/2; if Dum2<0 then Dum2: = 0;
      Ct2[i,Ncel]: = (1-Dum2)*Ct1[i,Ncel]+Dum2*Ct1[i,Ncel-1];
      end;
    for j: = 1 to Ncel do    for i: = 0 to Nel do    Ct1[i,j]: = Ct2[i,j];
end;

procedure flush;                                          { advective transport ... }
begin
  Dum1: = Veloc*Delt/Delx;
  for i: = 0 to Nel do
```

```
   begin for j: = Ncel downto 2 do
     begin                                              { .. keep old conc's for dspcor .. }
       Ctold[i,j]: = Ct1[i,j];                 { .. now flush, and mix with old cell-contents ... }
       Ct1[i,j]: = Dum1*Ct1[i,j-1] + (1-Dum1)*Ct1[i,j];   Ct2[i,j]: = Ct1[i,j];
     end;
       Ctold[i,1]: = Ct1[i,1];
       Ct1[i,1]: = Dum1*Cin[i] + (1-Dum1)*Ct1[i,1]; Ct2[i,1]: = Ct1[i,1];
     end;
 end;

 function frndl (Kf, nf, C : real) : real;                         { Freundlich eqn ... }
 begin if C > 1e-10 then frndl: = Kf*exp(nf*ln(C)) else frndl: = 0; end;

 procedure distri;
         { distribution of q and new C according to Freundlich-isotherm; Ct and q in ug/(Pore V) .. }
 Var M1, T, M : real;    Ic, Ie, it : integer;
 begin
   for Ic: = 1 to Ncel do    for Ie: = 1 to Nel do
     begin                        { T(otal) is redistributed over Ct1 and Q with Newton's method ...}
       T: = Ct1[Ie,Ic] + Q[Ie,Ic];
       M: = T/(Kf[Ie]+1); It: = 0;
       repeat Dum2: = M; M1: = M; Dum: = frndl(Kf[Ie], nf[Ie], M1)+M1-T;
         M1: = M+1e-9; Dum1: = frndl(Kf[Ie], nf[Ie], M1) + M1-T;
         if abs(Dum1-Dum)>0 then M: = M-Dum/((Dum1-Dum)/1e-9);
         if M<1e-9 then M: = 0.0;   if M>T then M: = T;
         It: = It+1;
         if It>20 then begin writeln('FREUNDLICH equation, iteration > 20'); halt; end;
       until (abs(Dum) < = 1e-6) or ((Dum2+M) = 0);
       Ct1[Ie,Ic]: = M; Q[Ie,Ic]: = (T-M);
     end;
 end;

 procedure fill;                                              { set initial values ... }
 begin
   Delx: = Totx/Ncel;
   for i: = 0 to Nel do    for j: = 1 to Ncel do
     begin Ct1[i,j]: = 0; Q[i,j]: = 0; end;                { .. or, if Ct1>0, then Q = frndl(.,.,.) }
   Cin[0]: = 1; Cin[1]: = 1 {ug/l};
 end;

 {MAIN}
 begin D: = 0.0 {cm2/s}; Disp: = 0.5 {cm}; Totx: = 5.0 { cm };
   Porv: = pi * sqr(2.5) * 5 * 0.3; Veloc: = 5.0*10/Porv/3600 {cm/s};
   Time: = Totx/Veloc*5;                                  { Time length of injection, s};
   Kf[1]: = 1; nf[1]: = 0.5;                   { initial run for estimating minimum no. of cells ...}
   Ncel: = 3; Delt: = Totx/Veloc/3 {s}; fill; Rmax: = 1; dspvty; Numdisp: = false;
   for Ish: = 1 to 3 do begin flush; distri; dspcor; dffdsp; distri; end;
   if Numdisp = true then Ncel: = 2 + trunc(Totx/Disp*(1-1/Rmax)/2);
                                                           { Now start realistic simulation ...}
   fill; Rmax: = 1; Numdisp: = false;
   Nshift: = 1+trunc(Time*Veloc/Delx); Delt: = Time/Nshift {s};
   dspvty;
   for Ish: = 1 to Nshift do
     begin
```

```
flush;                              { .. new conc's in Ct1 and Ct2; old conc's in Ctold }
distri;                                      { .. Ct1 has new conc's, after distri }
dspcor;                        { .. difference amongst Ct1, Ct2, Ctold gives num. dispersion }
                                         { mix, to model remaining dispersion ... }
  for Mi: = 1 to Nmix do    begin dffdsp; distri; end;
  Pv[Ish]: = (Ish+0.5)*Delt*Veloc/Delx/Ncel;
  for i: = 0 to Nel do Cout[i,Ish]: = Ct1[i,Ncel];
 end;
write('Pore_Vol El: 0,'); for i: = 1 to Nel do write('   ',i,',   ');
writeln;
for j: = 1 to Nshift do
 begin write(Pv[j]:6:3,' ');
   for i: = 0 to Nel do write(Cout[i,j]:8:3); writeln;
 end;
end. { Columnperc2 }
```

Output of this program is shown in Figure 9.42. The results have been calculated for 5 pore volumes entering the column with the chemical at a concentration Cin = 1, and 10 pore volumes flushing the column. It is of interest to note that fronts of the incoming chemical are sharper when the exponent n is smaller; a smaller n gives intially rapid elution, but leads to long tails. We will consider this subject again in Chapter 11, in the context of flushing of pollutants from aquifers.

9.5.4 *Permeability variations in flow models*

One dimensional situations are applicable for modeling a flowline, a laboratory column, or a diffusion profile. However, we concluded at the end of Section 9.1 that transport in aquifers requires a two- or three-dimensional approach, because inhomogeneities may have a large influence on flow patterns and flow velocities. Bear and Verruyt (1987) and Kinzelbach (1992) provide an excellent and practical introduction to numerical modeling of groundwater flow, and we will not discuss all of that subject here. An important point remains as to how variations in aquifer properties should be introduced in the numerical model. With the help of geophysical measurements and geological arguments, an estimate of subsoil properties can often be obtained in relative terms of 'permeable' and 'impermeable' layers. The range in permeabilities in even the 'permeable' aquifer layers is very large, and it is obvious that these variations should be introduced into the flow model to obtain a better estimate of their possible effects. One way is to consider more explicitly the stochastic variability of sediment permeabilities on flow patterns (Smith and Schwartz, 1981). The permeabilities will show gradual transitions in space within a single geological unit, and possibly more abrupt changes at the boundary with another formation. In particular the gradual transitions are difficult to detect by classical geological and geophysical methods.

The properties of a sedimentary unit, a formation, or on a larger scale of a set of formations, can be characterized by the *covariance* between two points which are separated by Δx:

$$\text{COV}\,[\ln k_{x+\Delta x}, \ln k_x] = \frac{1}{n}\sum_{x=1}^{n}(\ln k_{x+\Delta x} - \overline{\ln k})(\ln k_x - \overline{\ln k}) \qquad (9.98)$$

where $\overline{\ln k}$ is the mean of a log-transformed set of permeability measurements. With the covariance known, a correlation coefficient Γ_c can be found if the distribution is log-normal:

$$\Gamma_c = \frac{COV}{\sigma^2_{\ln k}} \tag{9.99}$$

where $\sigma^2_{\ln k}$ is the variance of the permeability distribution.

The correlation coefficient has been found to show an exponential behavior:

$$\Gamma_c \sim \exp(-|\Delta x|/L) \tag{9.100}$$

with L known as the *correlation length* or *integral scale*.

The correlation length can be obtained from spatially distributed data (e.g. De Marsily, 1986). An example of a set of data which follow this relation is shown in Figure 9.43.

The correlation length can be introduced into the flow model with statistical techniques such as the turning bands method (Mantoglou and Wilson, 1982). A very easy and generally applicable technique is illustrated in Figure 9.44 (Gehrels and Appelo, 1991). The permeability grid of a model is filled cell by cell by taking randomly a value from the set of permeabilities, and checking the correlation with respect to an earlier obtained value for the neigboring cell. A correlation criterion is obtained from $\Delta x/L$, as shown in Figure 9.44. When the correlation criterion is not met, a new random value is drawn from the list; if the criterion is met, the next cell is addressed. Correlation lengths can be independently stated for vertical and horizontal permeabilities, and also for porosities or any other parameter which influences flow. Anisotropy is therefore easily introduced.

A few examples of flowlines in a 50×10 grid model are shown in Figure 9.45, and illustrate the influence of the correlation length on the flow pattern. The correlation length is obtained from the final distribution of permeabilities in the model. The figure shows that there is a development of geological structure when the horizontal correlation length increases. For a geologist, a picture is obtained which has an intuitive appeal, and which gives a feeling for probable flowlines and their variations.

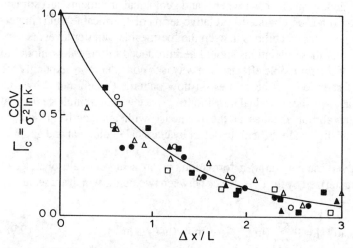

Figure 9.43. Correlation function for a set of (logarithmic) permeabilities from 8 boreholes in the Mt. Simon aquifer. (from Bakr, 1976, cited by Gelhar, 1984).

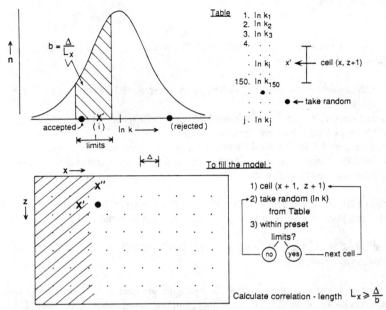

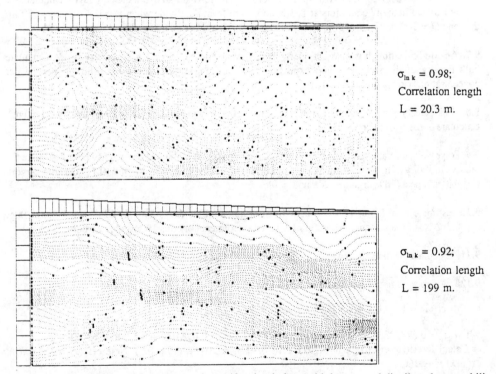

Figure 9.44. Filling the grid of a flow model with a set of permeabilities which have some correlation length.

$\sigma_{\ln k} = 0.98$;
Correlation length
$L = 20.3$ m.

$\sigma_{\ln k} = 0.92$;
Correlation length
$L = 199$ m.

Figure 9.45. Examples of flow patterns in aquifer simulations with log-normal distributed permeabilities and variable correlation lengths (Gehrels and Appelo, 1991).

PROBLEMS

9.1. Calculate the downward velocity of pore water in chalk, when $P = 0.3$ m/yr, and $\varepsilon_w = 0.1$. Compare the results with Figure 9.2.

9.2. Draw a sketch of the hydrological situation described in Example 9.1, where an aquifer has $D = 50$ m; $P = 0.3$ m/yr; $\varepsilon = 0.3$.
   a. Calculate porewater velocity and Darcy velocity in the aquifer at 1000 and 2000 m;
   b. Calculate the travel time of water between these two points;
   c. What is the age of water at $d = 10, 20, 30, 40$ m below the phreatic surface?
   d. Compare results with the formula $t = d \cdot \varepsilon_w/P$ which is valid for an unsaturated zone (but gives approximate results for the saturated zone as well).

9.3. A municipal well in a 100 m thick phreatic aquifer, $\varepsilon = 0.3$, $P = 0.3$ m/yr, draws water from up to $r = 5$ km. Calculate the travel time for water from a factory situated 1 km from the well.

9.4. Derive the parabolic relation of h versus x for a uniform phreatic aquifer (Hint: replace v in Equation (9.7) by the Darcy Equation).

9.5. Redraw Figure 9.10 on transparent paper; draw the hydraulic gradient in the upper and lower aquifer; indicate the direction of water flow through the confining layer, when it is leaky. Draw isochrones (lines connecting water with equal age) in the confined aquifer.

9.6. The clay layer in the aquifer of Figure 9.10 starts at 1 km from the divide, and is 1 km long. $D^* = 2/3 \cdot D$; what is x_0? The two aquifers merge again downstream from the clay layer; calculate the hydraulic head in the two aquifers at 1, 1.5, and 2 km from the divide when $P = 0.3$ m/yr, $k = 50$ m/day, and $D = 100$ m. Compare travel times for water just above and below the clay layer.

9.7. An aquifer sediment is made of quartz grains, with density 2.65 g/cm³; porosity is 0.2. Calculate bulk density of the sediment, and express sorbed concentrations (in mg/kg solid) in pore water concentrations (in mg/l pore water).

9.8. Bulk density of a limestone is 2.0 g/cm³; calcite, the only mineral in the rock has density 2.7 g/cm³. Calculate porosity of the rock.

9.9. Polycyclic aromatic hydrocarbons (PAH) have been found in the soil below gasworks. Estimate the relative mobility of naphthalene and tetracene with respect to water flow in sand (0.1% organic carbon; $\varepsilon = 0.3$) and peat (70% organic carbon; $\varepsilon = 0.6$).

9.10. Estimate retardation of Cd^{2+} in a fresh water aquifer; $K_{Cd\backslash Ca} = 1.2$; $CEC = 1.0$ meq/100g; $\varepsilon = 0.3$. Also of the insecticide parathion, when the sediment contains 1% organic carbon.

9.11. Calculate the distance covered by diffusion in 10 years in clay, for $D = 10^{-6}$ and 10^{-5} cm²/s.

9.12a. An approximation for the error function is

$$\text{erf}(x) = 1 - (1 + a_1 x + a_2 x^2 + a_3 x^3 + a_4 x^4)^{-4}$$

where $a_1 = 0.278393$, $a_2 = 0.230389$, $a_3 = 0.000972$, $a_4 = 0.078108$, and accurate to 0.0005.
   Calculate erf(0); erfc(0); erfc(0.707); erf(−0.707); erfc(0.5); erfc(−0.5). Note erfc(x) = 1 − erf(x); erf($-x$) = −erf(x).

9.12b. A single shot of 10 g NaCl is injected in a clay, there is no water flow. Calculate the mass of Cl^- between $x = x_0$, and $x = x_0 + 1$ m after 10 years; $D = 10^{-5}$ cm^2/s. Hint: Integrate Equation (9.45), and substitute the error function.

9.13. Derive the formula for the dispersivity in a permeability stratified aquifer with log-normal distributed permeabilities (Equation 9.65).

Hint: $\sigma_{\ln(k)} = \ln(k_{0.84}) - \ln(k_{0.5})$, use $x_{0.5} = \exp[\ln(k_{0.5})] \cdot (-dh/dx \cdot \Delta t)$ and similar for $x_{0.84}$.

9.14. Calculate the value of the dispersion coefficient for a column with dispersivity $\alpha_L = 5$ mm, for $v_{H_2O} = 10^{-4}, 10^{-5}, 10^{-6}$ m/s while $D_f = 10^{-9}$ m^2/s. Porosity is 0.3.

9.15. Estimate macro dispersivity for an aquifer, 50 m thick, $k_{0.5} = 50$ m/day, $\sigma_k = 10$ m/day. Also for an aquifer with log-normal distributed k, $\sigma_{\ln k}^2 = 0.3$.

9.16. Repeat the calculations of Example 9.11 for trichloroethane, with the same dimensions of the waste site, and concentrations in water, but find K_{oc} from Table 9.3, and use smaller porosity of compacted clay, $\varepsilon = 0.2$.

9.17. Derive the mixing factors in a diffusion model, when the end-point is a solid-stop.

9.18. Calculate the contribution of diffusion on the mixing factor in Example 9.13. Take $D_f = 10^{-5}$ cm^2/s. (Hint: see program of Example 9.12).

9.19. A waste-dump is situated as illustrated in the figure; Cd^{2+} and TCE are leached from the waste with precipitation excess in concentrations of resp. 1.12 and 7.1 mg/l, and enter the aquifer. Fronts and depths of leachates must be calculated, as well as a remedial pumping scheme, given the following hydrological parameters and aquifer properties: $P = 0.3$ m/yr; porosity = 0.36; bulk density = 1.8 g/cm^3; dispersivity = 10% of flowlength. The aquifer is homogeneous.

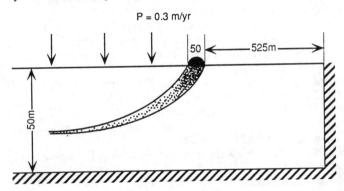

a. Calculate distance and depth of the front for a conservative solute (i.e. without decay or adsorption) after 25 and 100 years; calculate the thickness of the plume at these fronts. Compare the distances, covered after 25 and 100 years.

b. TCE has an adsorption isotherm $q = 6 + 6\ln(c)$, where q is adsorbed TCE, mg/l groundwater, and c is solute TCE, mg/l groundwater.
Estimate a linear distribution coefficient for Cd^{2+} from exchange equilibrium with Ca^{2+}, when $CEC = 1$ meq/100g, $Ca^{2+} = 10$ mmol/l, and the exchange coefficient $K_{Cd\backslash Ca} = 1$.

c. Calculate the location of the fronts for Cd^{2+} and TCE after 25 and 100 years (distance and depth).

d. Calculate the distance over which 68% of the Cd^{2+} front has spread after 100 years as a result of dispersion, and give the concentration profile in 3 points. Will the TCE-front be sharper or more spread out?

REFERENCES

Abramowitz, M. and Stegun, I.A., 1964, *Handbook of mathematical functions.* Dover Pub., New York 7th print, 1064 pp.

Allison, G.B., Stone, W.J. and Hughes, M.W., 1985, Recharge in karst and dune elements of a semi-arid landscape as indicated by natural isotopes and chloride. *J. Hydrol.* 76, 1-25.

Andersen, L.J. and Sevel, T., 1974, Six years' environmental tritium profiles in the unsaturated and saturated zones, Grønhøj, Denmark. *Isotope techniques in groundwater hydrology.* IAEA, Vienna, 1, 3-18.

Appelo, C.A.J. and Van Ree, C.C.D.F., 1983, Groundwater flowpaths in an ice-pushed ridge traced with tritium (in Dutch). H_2O 16, 35-39.

Appelo, C.A.J. and Willemsen, A., 1987, Geochemical calculations and observations on salt water intrusions, I. A combined geochemical/mixing cell model. *J. Hydrol.* 94, 313-330.

Bear, J., 1972, *Dynamics of fluids in porous media.* Elsevier, Amsterdam, 764 pp.

Bear, J. and Verruyt, A., 1987, *Modeling groundwater flow and pollution.* Reidel, Dordrecht, 414 pp.

Beekman, H.E., 1991, *Ion chromatography of fresh and salt water intrusion.* Ph.D. Thesis, Vrije Universiteit, Amsterdam, 198 pp.

Boas, M.L., 1983, *Mathematical methods in the physical sciences,* 2nd ed. Wiley & Sons, New York, 793 pp.

Bolt, G.H. (ed.), 1982, *Soil chemistry, B. Physico-chemical models.* Elsevier, Amsterdam, 527 pp.

Bond, W.J. and Phillips, I.R., 1990, Cation exchange isotherms obtained with batch and miscible displacement techniques. *Soil Sci. Soc. Am. J.* 54, 722-728.

Brusseau, M.L. and Rao, P.S.C., 1989, Sorption nonideality during organic contaminant transport in porous media. *Crit. Rev. Env. Control* 19, 33-99.

Calder, I.R., 1990, *Evaporation in the uplands.* Wiley & Sons, New York, 148 pp.

Carslaw, H.S. and Jaeger, J.C., 1963, *Operational methods in applied mathematics.* Dover Publ. New York, 359 pp.

Cederberg, G.A., Street, R.L. and Leckie, J.O., 1985, A groundwater mass transport and equilibrium chemistry model for multicomponent systems. *Water Resour. Res.* 21, 1095-1104.

Childs, E.C., 1969, *The physical basis of soil water phenomena.* Wiley & Sons, New York, 493 pp.

Chiou, C.T., 1989, Theoretical considerations of the partition uptake of nonionic organic compounds by soil organic matter. In B.L. Sawhney and K. Brown (eds), *Reactions and movement of organic chemicals in soils.* Soil Sci. Soc. Am. Spec. Publ. 22, 1-29.

Chiou, C.T., Freed, V.H., Schmedding, D.W. and Kohnert, R.L., 1977, Partition coefficient and bioaccumulation of selected organic chemicals. *Environ. Sci. Technol.* 11, 475-478.

Chiou, C.T., Kile, D.E., Brinton, T.I., Malcolm, R.L., Leenheer, J.A. and MacCarthy, P., 1987, A comparison of water solubility enhancements of organic solutes by aquatic humic materials and commercial humic acids. *Environ. Sci. Technol.* 21, 1231-1234.

Chiou, C.T., Peters, L.J. and Freed, V.H., 1979, A physical concept of soil-water equilibria for non-ionic organic compounds. *Science* 206, 831-832.

Crank, J., 1975, *The mathematics of diffusion.* 2nd ed. Oxford Press, 414 pp.

Curtis, G.P., Reinhard, M. and Roberts, P.V., 1986, Sorption of hydrophobic compounds by sediments. In J.A. Davis and K.F. Hayes (eds), *Geochemical processes at mineral surfaces.* Adv. Chem. Series, 323, Am. Chem. Soc., 191-216.

Dagan, G., 1989, *Flow and transport in porous formations.* Springer Verlag, Berlin, 465 pp.

Davis, S.N. and De Wiest, R.J.M., 1967, *Hydrogeology.* Wiley & Sons, New York, 463 pp.

De Marsily, G., 1986, *Quantitative Hydrogeology.* Academic Press, London, 440 pp.

Desaulniers, D.E. and Cherry, J.A., 1988, Origin and movement of groundwater and major ions in a thick deposit of Champlain Sea clay near Montreal. *Can. Geotechn. J.* 26, 80-89.

Dillon, P.J., 1989, An analytical model of contaminant transport from diffuse sources in saturated porous media. *Water Resour. Res.* 25, 1208-1218.

Domenico, P.A. and Schwartz, F.W., 1990, *Physical and chemical hydrogeology.* Wiley & Sons, New York, 824 pp.

Ernst, L.F., 1973, Determining the traveltime of groundwater (in Dutch). ICW-nota 755, 42 pp., Wageningen.

Foster, S.S.D. and Smith-Carrington, A., 1980, The interpretation of tritium in the chalk unsaturated zone. *J. Hydrol.* 46, 343-364.

Freeze, R.A. and Cherry, J.A., 1979, *Groundwater*. Prentice-Hall, Englewood Cliffs, New Jersey, 604 pp.

Freyberg, D.L., 1986, A natural gradient experiment on solute transport in a sand aquifer. 2, Spatial moments and the advection and dispersion of nonreactive tracers. *Water Resour. Res.* 22, 2031-2046.

Garbarini, D.R. and Lion, L.W., 1986, Influence of the nature of soil organics on the sorption of toluene and trichloroethylene. *Environ. Sci. Technol.* 20, 1263-1269.

Gehrels, J.C. and Appelo, C.A.J., 1991, Heterogeneity in permeability and its effect on flow pattern and travel time (in Dutch). H_2O, 24, 750-755.

Gelhar, L.W., 1984, Stochastic analysis of flow in heterogeneous media. *NATO Adv. Sci. Inst. Ser. E.* 82, 673-717.

Gelhar, L.W. and Axness, C.L., 1983, Three-dimensional stochastic analysis of macrodispersion in aquifers. *Water Resour. Res.* 19, 161-170.

Gelhar, L.W. and Wilson, J.L., 1974, Ground-water quality modeling. *Ground Water* 12, 399-408.

Gelhar, L.W., Mantoglou, A., Welty, C. and Rehfeldt, K.R., 1985, A review of field-scale physical solute transport processes in saturated and unsaturated porous media. EPRI, Palo Alto, CA 94303.

Gerald, C.F. and Wheatley, P.O., 1989, *Applied numerical analysis*, 4th ed. Addison-Wesley, 679 pp.

Hansch, C. and Leo, A.J., 1979, *Substituent constants for correlation analysis in chemistry and biology*. Wiley & Sons, New York.

Hayward, D.O. and Trapnell, B.M.W., 1964, *Chemisorption*. 2nd ed. Butterworths, 323 pp.

Herweijer, J.C., Van Luin, G.A. and Appelo, C.A.J., 1985, Calibration of a mass transport model using environmental tritium. *J. Hydrol.* 78, 1-17.

Herzer, J. and Kinzelbach, W., 1989, Coupling of transport and chemical processes in numerical models. *Geoderma* 44, 115-127.

Hoeks, J. 1981, Analytical solutions for transport of conservative and nonconservative contaminants in groundwater systems. *Water Air Soil Poll.* 16, 339-350.

Johnson, R.L., Cherry, J.A. and Pankow, J.F., 1989, Diffusive contaminant transport in natural clay: a field example and implications for clay-lined waste disposal sites. *Environ. Sci. Technol.* 23, 340-349.

Karickhoff, S.W., 1981, Semi-empirical estimation of sorption of hydrophobic pollutants on natural sediments and soils. *Chemosphere* 10, 833-846.

Karickhoff, S.W., 1984, Organic pollutant sorption in aquatic systems. *J. Hydraul. Eng.* 110, 707-735.

Kinniburgh, D.G., 1986, General purpose adsorption isotherms. *Environ. Sci. Technol.* 20, 895-904.

Kinzelbach, W., 1986, *Groundwater modelling*. Elsevier, Amsterdam, 333 pp.

Kinzelbach, W., 1992, *Numerische Methoden zur Modellierung des Transports von Schadstoffen im Grundwasser*, 2 Auflage. Oldenbourg, München, 343 pp.

Lerman, A., 1979, *Geochemical processes*. Wiley & Sons, New York, 481 pp.

Levenspiel, O., 1972, *Chemical reaction engineering*. Wiley & Sons, New York, 857 pp.

Li, Y.-H. and Gregory, S., 1974, Diffusion of ions in seawater and in deep-sea sediments. *Geochim. Cosmochim. Acta* 38, 703-714.

Lichtner, P.C., 1985, Continuum model for simultaneous chemical reactions and mass transport in hydrothermal systems. *Geochim. Cosmochim. Acta* 49, 779-800.

Lowell, R.P., 1989, Contaminant transport in a single fracture: periodic boundary and flow conditions. *Water Resour. Res.* 25, 774-780.

Lyman, W.J., Reehl, W.F. and Rosenblatt, D.H., 1990, *Handbook of chemical property estimation methods*. Am. Chem. Soc., Washington.

Mackay, D.M., Freyberg, D.L., Roberts, P.V. and Cherry, J.A., 1986, A natural gradient experiment on solute transport in a sand aquifer. 1, Approach and overview of plume movement. *Water Resour. Res.* 22, 2017-2029.

Maloszewski, P. and Zuber, A., 1985, On the theory of tracer experiments in fissured rocks with a porous matrix. *J. Hydrol.* 79, 333-358.

Mantoglou, A. and Wilson, J.L., 1982, The turning band method for simulation of random fields using line generation by a spectral method. *Water Resour. Res.* 18, 1379-1394.

McCarty, P.L., Reinhard, M. and Rittmann, B.E., 1981, Trace organics in groundwater. *Environ. Sci. Technol.* 15, 40-51.

McNeil, J.D., 1980, Electrical conductivity of soils and rocks. Geonics Ltd, Missisauga, 22 pp.

Mercado, A., 1967, The spreading pattern of injected water in a permeability stratified aquifer. *IASH Pub.* 72, 23-36.

Mercado, A., 1984, A note on micro and macrodispersion. *Ground Water* 22, 790-791.

Miller, A.R., 1981, *BASIC programs for scientists and engineers*. Sybex, Berkeley, 318 pp.

Miller, C.W. and Benson, L.V., 1983, Simulation of solute transport in a chemically reactive heterogeneous system: model development and application. *Water Resour. Res.* 19, 381-391.

Molz, F.J., Güven, O. and Melville, J.G., 1983, An examination of scale-dependent dispersion coefficients. *Ground Water* 21, 715-725.

Moore, W.J., 1972, *Physical chemistry*. 5th ed. Longman, London, 977 pp.

Nkedi-Kizza, P.S.C., Rao, P.S.C. and Johnson, J.W., 1983, Adsorption of diuron and 2,4,5-T on soil particle-size separates. *J. Env. Qual.* 12, 195-197.

Notodarmojo, S., Ho, G.E., Scott, W.D. and Davis, G.B., 1991, Modelling phosphorous transport in soils and groundwater with two-consecutive reactions. *Wat. Res.* 25, 1205-1216.

Parker, J.C. and Van Genuchten, M.Th., 1984a, Flux-averaged and volume-averaged concentrations in continuum approaches to solute transport. *Water Resour. Res.* 20, 866-872.

Parker, J.C. and Van Genuchten, M.Th., 1984b, Determining transport parameters from laboratory and field tracer experiments. Virginia Agric. Experim. Station 84-3, 96 pp.

Parkhurst, D.L., Thorstenson, D.C. and Plummer, L.N., 1980, PHREEQE – a computer program for geochemical calculations. US Geol. Surv. Water Resour. Inv. 80-96, 210 pp.

Perkins, T.K. and Johnston, O.C., 1963, A review of diffusion and dispersion in porous media. *Soc. Petrol. Eng. J.*, March, 70-84.

Phillips, F.M., Mattick, J.L., Duval, T.A., Elmore, D. and Kubik, P.W., 1988, Chlorine 36 and tritium from nuclear weapons fallout as tracers for long-term liquid and vapor movement in desert soils. *Water Resour. Res.* 24, 1877-1891.

Pickens, J.F. and Grisak, G.E., 1981, Scale-dependent dispersion in a stratified granular aquifer. *Water Resour. Res.* 17, 1191-1211.

Pickens, J.F. and Lennox, W.C., 1976, Numerical simulation of waste movement in steady groundwater flow systems. *Water Resour. Res.* 12, 171-180.

Plumb, O.A. and Whitaker, S., 1988, Dispersion in heterogeneous porous media. I. Local volume averaging and large-scale averaging. *Water Resour. Res.* 24, 913-926.

Prickett, T.A., Naymik, T.G. and Lonnquist, C.G., 1981, A random-walk solute transport model for selected groundwater quality evaluations. Bull. 65, Illinois State Water Survey, Champaign.

Rao, P.S.C., Jessup, R.E., Addiscott, T.M., 1982, Experimental and theoretical aspects of solute diffusion in spherical and nonspherical aggregates. *Soil Sci.* 133, 342-349.

Reardon, E.J., 1981, K_d's – Can they be used to describe reversible ion sorption reactions in contaminant transport? *Ground Water* 19, 279-286.

Schröter, J., 1984, Micro- and macrodispersivities in porous aquifers (in German). *Meyniana* 36, 1-34.

Schulz, H.D. and Reardon, E.J., 1983, A combined mixing cell/analytical model to describe two-dimensional reactive solute transport for unidirectional groundwater flow. *Water Resour. Res.* 19, 493-502.

Schwartzenbach, R.P. and Westall, J.C., 1985, Sorption of hydrophobic trace organic compounds in groundwater systems. *Water Sci. Technol.* 17, 39-55.

Selim, H.M., Schulin, R. and Flühler, H., 1987, Transport and ion exchange of calcium and magnesium in an aggregated soil. *Soil Sci. Soc. Am. J.* 51, 876-884.

Sharma, M.L. and Hughes, M.W., 1985, Groundwater recharge estimation using chloride, deuterium and oxygen-18 profiles in the deep coastal sands of Western Australia. *J. Hydrol.* 81, 93-109.

Sjöberg, E.L., Georgala, D. and Rickard, D.T., 1983, Origin of interstitial water compositions in postglacial black clays (Northeastern Sweden). *Chem. Geol.* 42, 147-158.

Smith, L. 1981, Spatial variability of flow parameters in a stratified sand. *Mathem. Geol.* 13, 1-21.

Smith, L. and Schwartz, F.W., 1981, Mass transport. I. A stochastic analysis of macroscopic dispersion. *Water Resour. Res.* 16, 303-313.

Sposito, G. and Mattigod, S.V., 1979, GEOCHEM: a computer program for calculating chemical equilibria in soil solutions and other natural water systems. Univ. California, 110 pp.

Sudicky, E.A., 1986, A natural gradient experiment on solute transport in a sand aquifer: spatial variability of hydraulic conductivity and its role in the dispersion process. *Water Resour. Res.* 22, 2069-2082.

Tang, D.H., Frind, E.O. and Sudicky, E.A., 1981, Contaminant transport in fractured porous media: analytical solution for a single fracture. *Water Resour. Res.* 17, 555-564.

Taylor, S.R. and Howard, K.W.F., 1987, A field study of scale-dependent dispersion in a sandy aquifer. *J. Hydrol.* 90, 11-17.

Tewari, Y.B., Miller, M.M., Wasik, S.P. and Martire, D.E., 1982, Aqueous solubility and octanol/water partition coefficient of organic compounds at 25°C. *J. Chem. Eng. Data* 27, 451-454.

Turney, G.L. and Goerlitz, D.F., 1990, Organic contamination of ground water at Gas Works Park, Seattle, Washington. *Ground Water Monitoring Rev.* 10, 187-198.

Uffink, G., 1988, Modeling of solute transport with the random walk method. In E. Custodio, A. Gurgui, J.P. Lobo Ferreira (eds), *Groundwater flow and quality modelling*. D. Reidel, Dordrecht, 247-265.

Van Genuchten, M.Th., 1981, Analytical solutions for chemical transport with simultaneous adsorption, zero-order production and first-order decay. *J. Hydrol.* 49, 213-233.

Van Genuchten, M.Th. and Cleary, R.W., 1982, Movement of solutes in soil: computer-simulated and laboratory results. In G.H. Bolt (ed.), *Soil chemistry*, Part B, Elsevier, Amsterdam, 349-386.

Van Genuchten, M.Th. and Parker, J.C., 1984, Boundary conditions for displacement experiments through short laboratory soil columns. *Soil Sci. Soc. Am. J.* 48, 703-708.

Van Ommen, H.C., 1985, The mixing cell concept applied to transport of non-reactive and reactive components in soils and groundwater. *J. Hydrol.* 78, 201-213.

Verruyt, A., 1969, *Theory of groundwater flow*. Macmillan, London, 190 pp.

Verruyt, A., 1972, Transport phenomena during disposal in the subsoil (in Dutch). Techn. Univ., Delft.

Verschueren, K., 1983, *Handbook of environmental data on organic chemicals*. 2nd ed. Van Nostrand Reinhold Cy., 1336 pp.

Volker, A., 1961, Source of brackish ground water in Pleistocene formations beneath the Dutch polderland. *Econ. Geol.* 56, 1045-1057.

Weber, J.B. and Miller, C.T., 1988, Modeling the sorption of hydrophobic contaminants by aquifer materials. I Rates and equilibria. *Water Res.* 22, 457-464.

Weber, J.B. and Miller, C.T., 1989, Organic chemical movement over and through soil. In B.L. Sawhney and K. Brown (eds), *Reactions and movement of organic chemicals in soils*. Soil Sci. Soc. Am. Spec. Publ. 22, 305-334.

Willemsen, A., 1992, Coupled geochemical transport model. Int. Energy Agency, Annex 7, Subtask A, in press.

Wolery, T.J., 1982, EQ3NR: a computer program for geochemical aqueous speciation-solubility calculations. Lawrence Livermore Lab., Cal., 191 pp.

Wu, S.-C. and Gschwend, P.M., 1986, Sorption kinetics of hydrophobic compounds to natural sediments and soils. *Environ. Sci. Technol.* 20, 717-725.

Yurtsever, Y., 1983, Models for tracer data analysis. *Guidebook on nuclear techniques in hydrology*, IAEA, Vienna, 381-402.

CHAPTER 10

Hydrogeochemical transport modeling

If we look closely in the pores of a topsoil (Figure 10.1), we see that many reactions are concerting while groundwater is generated from rainwater. The most important reactions are dissolution and precipitation of solids and minerals, redox reactions with organic matter, and ion exchange and sorption on clay minerals and organic matter. Previous chapters in this book treated these reactions separately, and showed how to calculate a water composition which is the typical result of one, single reaction. If you have gone through the exercises, you will have come to realize that doing these calculations is rewarding, though time consuming. When several reactions must be combined, and activity corrections have to be included because the answer should be accurate, the calculations become excessively laborious. Perhaps you lost the hope of finding a solution for the reaction scheme which you consider responsible for the water qualities you are studying.

This chapter introduces recently developed computer models that can do many of the geochemical calculations which we have discussed in earlier chapters, and more. A user guide is provided for PHREEQE, a geochemical model developed by Parkhurst, Thorstenson and Plummer (1980), together with many examples of application. Geochemical models calculate equilibrium compositions in a batch of water containing reactants, e.g. the water composition in a beaker with water and calcite, and with CO_2 gas bubbled through. They go a step beyond the calculation of activities and saturation states for a given water analysis, as is done by WATEQP (Chapter 3). A geochemical model can also calculate how water composition changes in response to reactions (dissolution of minerals, gases, etc.) or a change in temperature. A further step is to combine the geochemical model with a transport code. It was shown in Chapter 9 how an equilibrium model for a single component (γ-HCH) can be incorporated in a transport model. A multicomponent model does not pose more difficulties, and we give examples in this chapter of modeling transport with complex geochemical reactions.

It will become clear from this chapter that a hydrogeochemical model is more than just an easy tool to do calculations. It also provides the ability to test a concept or a suite of reactions for a geochemical or environmental problem. The comparison of field or laboratory data with geochemical model calculations allows us to *validate* the numerical procedures as well as the reaction scheme. It prevents gross errors or the violation of basic chemical laws, and shows our level of understanding. The possibilities are immense, and it may be difficult to suppress the desire to obtain a perfect fit by including more and more parameters. There are allowable 'knobs' however, for example the rate coefficients in kinetic processes, the varying solubilities of ill-defined oxides or hydroxides, or the

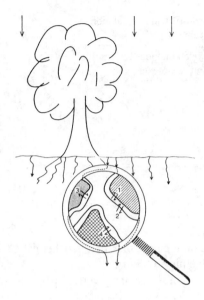

1. Mineral reactions
2. Gas exchange
3. Organic matter reactions
4. Cation exchange

Figure 10.1. A close-up view of reactions which are important for natural water quality.

complexation constants with organic acids. The thermodynamic and kinetic data for these substances are not available in a form that warrants direct inclusion in a general computer program. This means that the user is forced to select his own data. A fortunate aspect is here that the careful and critical thinking associated with that selection normally leads to a more complete understanding of the hydrogeochemical system that is studied.

10.1 GEOCHEMICAL MODELS, WHAT CAN THEY DO?

The propulsion of space rockets appears to be the first problem for which a comprehensive chemical model was developed. Product and reactant gases in the rocket fuel can be calculated as a function of temperature and pressure, if one knows the mass action coefficients for all possible reactions at equilibrium (Smith and Missen, 1982). The mass action and mass balance relations give a set of non-linear equations for which a solution can be found with Newton-Raphson iteration (as explained for one component in Chapter 9). Reactions in natural water are about as complex, and even more components may be involved, so that the use of such models can also be fruitful in our discipline. Equilibrium models have served to show the evolution of water quality influenced by silicate weathering (Helgeson et al., 1970; Fritz, 1975; Lichtner, 1985); groundwater quality as influenced by carbonate reactions (Plummer et al., 1983); effects of acidification and buffering reactions (Appelo, 1985; Cosby et al., 1985a; Eary et al., 1989); ore deposition in groundwater systems, and also leaching of tailings from the ore mining industry (Garven and Freeze, 1984; Walsh et al., 1984; Liu and Narasimhan, 1989); cation exchange with salt/fresh water displacements in aquifers and in column experiments (Appelo and Willemsen, 1987; Appelo et al., 1990); complexation of heavy metals in water and adsorption on solids (Westall and Hohl, 1980; Felmy et al., 1984); water treatment for aquifer thermal energy storage (Willemsen, 1990); denitrification in an aquifer (Postma et

Table 10.1. Calcite dissolution at P_{CO_2} = 0.01 atm; comparison of hand calculation with a geochemical model. Concentrations in mmol/l, and μS/cm.

	Hand $Ca^{2+} = 10^{-2.19} \cdot P_{CO_2}^{1/3}$	Numerical model, includes activity corrections and complexes
HCO_3^-	2.78	3.27*
Ca^{2+}	1.39	1.64*
$CaCO_3^0$	–	0.005$
$CaHCO_3^+$	–	0.049$
EC	278	327

*Total concentration, includes complexes.
$Actual complex concentration.

al., 1991). In fact, one of the models should be present on the hard disk of a geochemist's computer, and used on a routine basis for helping to solve his problems.

Consider, as an example, dissolution of calcite in a glass of water, at a given temperature and pressure of CO_2 gas. Calcite dissolves according to:

$$CO_{2(g)} + CaCO_{3(cc)} + H_2O \rightarrow Ca^{2+} + 2HCO_3^- \qquad (10.1)$$

and the mass action equation is:

$$K_{(10.1)} = \frac{[Ca^{2+}][HCO_3^-]^2}{P_{CO_2}} \qquad (10.2)$$

Other ions and complexes which exist in the solution will be H^+, OH^-, $CaHCO_3^+$, CO_3^{2-}, H_2CO_3. However, we may neglect their existence and try to calculate the composition by hand. Electroneutrality in solution requires:

$$2m_{Ca^{2+}} = m_{HCO_3^-} \qquad (10.3)$$

Combination of Equations (10.2) and (10.3), and setting activity $[i] = m_i$, allows to calculate the Ca^{2+} concentration as a function of CO_2 pressure. The outcome can be compared with the calculation of a geochemical model in which the same $K_{(10.1)}$ is used, and in which moreover activity coefficient corrections are included, as well as all other complexes and ions in solution. Table 10.1 gives the results of the two calculations, for a CO_2 pressure of 0.01 atm, at 25° C. The constants stem from the database given in Appendix B, and differ slightly from the ones used in Chapter 4 where the effects of complexes and activity coefficients were coarsely compensated by increasing the solubility product of calcite from $10^{-8.48}$ to $10^{-8.3}$.

The comparison shows that the inclusion of complexes and activity corrections gives 18% higher concentrations than are predicted by simple hand calculation. This illustrates sufficiently that using a geochemical model is a worthwhile step to a more accurate description of natural chemical reactions.

10.1.1 *Structure of the models and governing equations*

The example with calcite shows that we must calculate complexes, activity coefficients, equilibria with minerals and gas pressures, all as a function of temperature and for all kinds of hydrogeochemical conditions. It can be deduced that the geochemical model should minimize residuals in a number of equations (i.e. the solution is accepted when these residuals approach zero, or are smaller than some criterion). The residuals comprise:

Mass balance for components i:

$$\sum_i (R_1)_i = Tot_i - m_i - \sum_j cplx_j \cdot c_{j,i} \qquad (10.4)$$

Electroneutrality:

$$R_2 = \sum cplx_j \cdot z_j + \sum m_i \cdot z_i \qquad (10.5)$$

Electron balance:

$$R_3 = Tot\, e - \sum m_i \cdot e_i - \sum cplx_j \cdot e_j \qquad (10.6)$$

Mineral equilibria for minerals k:

$$(R_4)_k = \log (IAP)_k - \log K_k \qquad (10.7)$$

with

$$(IAP)_k = \prod_i (m_i \cdot \gamma_i)^{c_{k,i}} \qquad (10.8)$$

where Tot_i is the total (molal) concentration of component i, $cplx_j$ is the (molal) concentration of complex j, and is obtained from the mass action law:

$$cplx_j = \frac{K_j}{\gamma_j} \prod_i (m_i \cdot \gamma_i)^{c_{j,i}} \qquad (10.9)$$

K_j is the association constant for complex j, or the solubility product for mineral k, m_i is the molal concentration of free, uncomplexed i, γ_i is the activity coefficient, $c_{j,i}$ is the stoichiometric coefficient of i in the jth complex or mineral, z_i is the charge of i, and e_i is the redox status of i.

The problem, to reduce the residuals $R_1 .. R_4$ within some preset limit, involves solution of the non-linear equations, for example with Newton-Raphson iteration. The non-linearity stems from the mass action relations with complexes, and with mineral equilibria in combination with the mass balance equations. The concentrations of components and of complexes can easily vary over 30 orders of magnitude, and a log-transformation is therefore normally performed. Other tricks which can speed up convergency have been disclosed by Holm (1989). Redox problems are the most problematic since the concentrations can vary over the largest intervals in response to pe variations. Achievement of equilibrium with a mineral k may require dissolution or precipitation of an amount m_k of the mineral, and hence the total concentration of solute i is augmented with $m_k \cdot c_{k,i}$. Likewise, a reaction simply adds an amount to the total concentration. The activity coefficients γ are relatively small perturbations that can be calculated separately from the Newton-Raphson iterations. The charge and electron balance can be used to calculate concentrations of two components, for which usually pH, and pe are chosen.

The definition of a *component* requires some discussion; components in the thermodynamic sense are the independent identities from which the chemical system under consideration can be constructed. In the geochemical model PHREEQE the term 'master species' is used which is identical to the more common thermodynamic term. For thermodynamic purposes, the choice of components of a system is often arbitrary and depends on the range of conditions for the problem under consideration (Pitzer and Brewer, 1961). The elements of the periodic system with their usual valence, and additionally the oxidation potential (or pe), would be sufficient as components for a general hydrogeochemical system. In PHREEQE, the component O_2 (oxygen) has been replaced by the component H_2O, and O_2 is calculated as a complex from the reaction:

$$2H_2O \leftrightarrow O_{2(g)} + 4H^+ + 4e^- \tag{10.10}$$

or:

$$P_{O_2} = K_{(10.10)} \cdot \frac{[H_2O]^2}{[H^+]^4[e^-]^4} \tag{10.11}$$

Note that this requires H^+ and e^- to be 'components' as well, and these are obtained in PHREEQE from electroneutrality and the electron balance.

The component N_2 has similarly been replaced by the component NO_3^-, which is advantageous since NO_3^- is normally the analyzed form of nitrogen in water. The case of nitrate may illustrate the difference between total concentration of the component, and the concentration of the complex or species which is actually analyzed. The *component* NO_3 can occur in the *species* NH_4^+, $N_{2(g)}$, NO_2^-, and NO_3^-. With pH and pe given, the model calculates the distribution of the total component concentration over these species, i.e. the actual concentration of each species. For example, at pH = 7 and pe = 0, the thermodynamically stable (i.e. highly predominant) species is NH_4^+. Under these conditions of pe/pH the input concentration of NO_3^- becomes perhaps surprisingly low in the output. When equilibrium among different species of one element is unrealistic, it is possible to separate them and define additional new elements. For example, the denitrification of NO_3^- can be rapid and convert NO_3^- to N_2 gas, but production of NH_4^+ may be sluggish. Two separate components may then be defined: a component NO_3 with species NO_3^-, NO_2^- and N_2, and a second component NH_4 which consists of NH_4^+ (and complexes of NH_4^+ such as $NH_4SO_4^-$).

The databases can be arranged in a matrix which contains the stoichiometric coefficients $c_{j,i}$ of the reaction, i.e. the *i*th component in the *j*th complex or mineral. For example, for the earlier stated calcite problem, the part of the database which is actually used during the calculation looks like the matrix given in Table 10.2.

The components are H^+, H_2O, Ca^{2+}, and CO_3^{2-}. All other species in this problem can be constructed from these components. For example, the complex $CaOH^+$ is constructed from the association reaction:

$$1H_2O + 1Ca^{2+} - 1H^+ \leftrightarrow CaOH^+ \qquad K = 10^{-12.78}$$

and the 'mineral' P-CO_2 dissociates into:

$$CO_{2(g)} \leftrightarrow 2H^+ - 1H_2O + 1CO_3^{2-} \qquad K = 10^{-18.149}$$

Carbon dioxide gas is used here as a 'mineral', which allows us to set the gas pressure of CO_2. The standard state of a gas is by definition equal to 1 atm. Equilibrium in solution with $CO_{2(g)}$ at 1 atm pressure means that the relation among species activities should be:

$$\frac{[H^+]^2[CO_3^{2-}]}{[H_2O]} = 10^{-18.149} \cdot P_{CO_2} = 10^{-18.149} \cdot 1$$

But the solution can be maintained at a different CO_2 pressure by specifying disequilibri-

Table 10.2. Matrix of stoichiometric coefficients used for reactions in the calcite/CO_2-system.

Species	H^+	H_2O	Ca^{2+}	CO_3^{2-}	$\log K_{ass}$ (25°)
H^+	1				
H_2O		1			
Ca^{2+}			1		
CO_3^{2-}				1	
OH^-	−1	1			−14.00
HCO_3^-	1			1	10.329
$H_2CO_3^*$	2			1	16.681
$CaOH^+$	−1	1	1		−12.78
$CaCO_3^0$			1	1	3.224
$CaHCO_3^+$	1		1	1	11.435
Minerals					$\log K_{diss}$ (25°)
$CaCO_3$			1	1	−8.48
P-CO_2	2	−1		1	−18.149

The first four columns are headed by Component: H^+, H_2O, Ca^{2+}, CO_3^{2-}.

um. For example, a CO_2 pressure of 10^{-2} is obtained if the solute species are maintained at:

$$\frac{[H^+]^2[CO_3^{2-}]}{[H_2O]} = 10^{-18.149} \cdot P_{CO_2} = 10^{-18.149} \cdot 10^{-2} = 10^{-20.149}$$

which means that the product of activities is 10^{-2} smaller. A CO_2 pressure of 10^{-2} atm is thus obtained when the solution is made a factor 10^{-2} undersaturated with respect to the 'pure mineral' CO_2: the saturation index, or $\log(IAP/K)$, should be set at −2.

10.1.2 *Available models*

A number of models have been developed which all have the ability to equilibrate a given aqueous solution with minerals, add reactants and calculate pH and pe as function of reactions, change temperature and gas pressures, and again calculate solution composition in response to such additions or changes. Reviews of the models appear regularly in the literature (Nordstrom et al., 1979; Grove and Stollenwerk, 1987; Engesgaard and Christensen, 1988; Mangold and Tsang, 1991). Table 10.3 lists some geochemical models which are available (e.g. through the International Groundwater Modeling Centre) and known to be operational. The availability is in itself a tribute to the development of science, and it is certainly justified to honor the developers. Selection of a model may be based on specific features which have been incorporated, or on the content of the database. This book uses PHREEQM, which has PHREEQE (Parkhurst et al., 1980) built into a mixing cell transport model, and also cation exchange according to the Gaines and Thomas convention as an additional feature.

   The model results depend completely on the values for the thermodynamic constants which are used in the database. Each complex and mineral must be given an equilibrium constant at 25°C, and for temperature dependence either the enthalpy of the reaction or a

Table 10.3. Available geochemical computer models.

Model	Features	Reference
PHREEQE	Fast convergence on carbonate problems; efficient and modular program	Parkhurst et al., 1980
EQ3	Fast convergence; mineral equilibrium fixes component concentration; EQ6 has path-finding ability; bulky program; large database	Wolery, 1983
GEOCHEM	New version available (1988); database with many species of interest for soil science	Sposito and Mattigod, 1979
MINTEQ	Large database, aimed at complexing of heavy metals; fancy models for adsorption on solid surfaces	Westall et al., 1976; Felmy et al., 1984
CHARON	Uses the RAND model discussed by Smith and Missen, 1982; has transport capacities; only executable version available	De Rooy, 1991

power series for log K is used. A comparison of speciation models (Nordstrom et al., 1979) gives some insight into deviations in the database. Although speciation models (such as WATEQP) only calculate the distribution for a given water analysis, the actual database can be identical with the one used in geochemical models and the comparison of Nordstrom et al. (1979) therefore remains valid for geochemical models as well. The Environmental Protection Agency of the USA has directed considerable effort into providing MINTEQ with an extensive and critical database, which includes data for many heavy metals and is updated regularly. The results of a recent comparison of the solubility products used in three of the models by Lichtner and Engi (1989) is shown in Figure 10.2. Abbreviated names, such as 'trem' for tremolite, or 'phlo' for phlogopite, are supplied with the minerals which show deviations larger than 2.5 in log K, but there is convergence of values for many of the

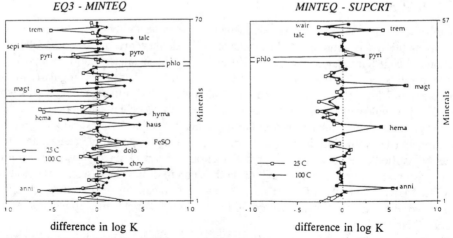

Figure 10.2. Comparison of solubility products used in the geochemical models MINTEQ, EQ3, and SUPCRT at two temperatures (the solubility products used in two models are subtracted). From Lichtner and Engi, 1989.

minerals. Most of these minerals are in a well crystallized form, with thermodynamic data extrapolated from high-temperature measurements to 25°C. The minerals which are encountered by hydrogeochemists in aquifers are generally less well defined, both with regard to composition and crystallinity. This fact and the frequently encountered slow kinetics of mineral reactions are the major causes of deviations among observed ground-water compositions and their modeled counterparts.

10.1.3 *Cation exchange in the geochemical models*

Cation exchange in the classical way, as treated in Chapter 5, has not been incorporated in a standard form into the models. A choice of refined electrostatic models is offered by MINTEQ (Westall and Hohl, 1980; Felmy et al., 1984) but the user must supply the constants which describe the electrostatic potential of the exchanger. A 'mineral' reaction which effectively fixes the ion ratios in water was suggested by Parkhurst et al. (1980) as a substitute for ion exchange. However, ion exchange has been included more formally when geochemical models were combined with a transport code. For example, Cederberg et al. (1985) dropped the potential term of the electrostatic model in MINTEQ, and modeled ion exchange with transport according to the Gaines and Thomas convention. This convention was also used by Jauzein et al., 1989. A Langmuir-type formulation has been used by Bryant et al. (1986), and Miller and Benson (1983) have included ion exchange according to the Vanselow convention in an explicitly written program which also embodies trans-port. It is not surprising that ion exchange has been included in geochemical transport codes, as the importance of ion exchange becomes manifest in *dynamic* situations, where flow of water of a different composition triggers the exchange process.

All the geochemical models use an ion-association (speciation) model to calculate complexes in solution. It is possible to introduce ion exchange as a combination of association reactions, without adaptation of the program's algorithm (Appelo and Willem-sen, 1987). For example, for Na^+-Ca^{2+} exchange an association reaction and a dissociation reaction can be added to give the cation exchange reaction:

$$2Na^+ + 2X^- \leftrightarrow 2Na\text{-}X \qquad\qquad (K_{NaX})^2$$

$$\underline{Ca\text{-}X_2 \leftrightarrow 2X^- + Ca^{2+}} + \underline{(K_{CaX_2})^{-1}}\Pi$$

$$2Na^+ + Ca\text{-}X_2 \leftrightarrow 2Na\text{-}X + Ca^{2+} \qquad (K_{Na\backslash Ca})^2 = (K_{NaX})^2/K_{CaX_2} \quad (10.12)$$

Here, X^- is the cation exchanger. The two 'half' reactions can be introduced directly into the geochemical model, but there are two problems. The first is that a free X^- site does not exist in a pure exchange model, since all X^- sites are covered by cations. The second is that the activity of the associated complex is calculated with respect to the usual standard state of 1 mol/kg H_2O for solute molecules, whereas it should be calculated as a fraction with respect to a standard state, with the exchanger fully covered by the cation.

The first problem can be countered by setting the association constants at high values. A high association constant implies that the constituent ion with the lower total concentration drops to a low 'free ion' concentration. For example, if we have

$$Na^+ + X^- \leftrightarrow NaX \qquad\qquad K_{NaX} = 10^{20}$$

and total X_t ($= NaX + X^-$) = 0.01 mol/kg H_2O, and total Na ($= NaX + Na^+$) = 0.02

mol/kg H_2O, then we find respectively NaX = 0.01; Na^+ = 0.01; X^- = 10^{-20} mol/kg H_2O. Thus, the 'free' X^- concentration is negligible as required. With all X^- tied up in neutral complexes, the ionic strength of the solution remains identical even though a high amount of X_t is present. The other calculations of the model (as influenced by activity coefficients) are therefore not affected. When the value of K_{NaX} is fixed at 10^{20}, the association constants for other cations are obtained from exchange equilibria with Na^+. For example, K_{CaX_2} is, from Reaction (10.12):

$$Ca^{2+} + 2X^- \rightarrow CaX_2$$

with

$$K_{CaX_2} = (K_{NaX})^2 / (K_{Na\backslash Ca})^2 = 10^{40}/(0.4)^2 = 10^{40.80}$$

(the value of $K_{Na\backslash Ca} = 0.4$ is taken from Table 5.5).

The second problem is to translate the activity of the associated complex from a reference state of 1 mol/kg H_2O to the exchanger reference state. However, this is easy. If the concentration of the complex is divided by the sum of all complexes of X^-, i.e. by m_{X_t}, and is multiplied with the stoichiometric coefficient of X^- in the complex, exactly the equivalent fraction β of the complex is obtained. For example for NaX:

$$\beta_{Na} = \frac{m_{NaX}}{m_{X_t}} = \frac{[NaX]}{\gamma_{NaX} \cdot m_{X_t}}$$

or for CaX_2:

$$\beta_{Ca} = \frac{2 \cdot m_{CaX_2}}{m_{X_t}} = \frac{2 \cdot [CaX_2]}{\gamma_{CaX_2} \cdot m_{X_t}}$$

where m_{NaX} and m_{CaX_2} are concentrations in mol/kg H_2O. The stoichiometric coefficient of X^- in the complex is a constant, and so is m_{X_t} for a given cation and soil or sediment with constant exchange capacity. It is therefore permissible to include these constants in the association constant of the complex. For example we *must* calculate for Na^+:

$$\frac{\beta_{Na}}{[Na^+][X^-]} = K_{NaX}$$

but we *can* calculate:

$$\frac{[NaX]}{[Na^+][X^-]} = K_{NaX} \cdot m_{X_t} \cdot \gamma_{NaX} = {}^c K_{NaX}$$

which obviously gives the same result.

Similarly for Ca^{2+}:

$$\frac{[CaX_2]}{[Ca^{2+}][X^-]^2} = K_{CaX_2} \cdot \frac{m_{X_t}}{2} \cdot \gamma_{CaX_2} = {}^c K_{CaX_2}$$

Here, $^c K$ is the *conditional association constant*, that is used in the actual calculations of the geochemical model. We assume, as in Chapter 5, that $\gamma_{NaX} = \gamma_{CaX_2} = ... = 1$. The option of using equivalent fractions of exchangeable cations is known as the Gaines and Thomas convention. Use of the Gapon convention is similar, and requires only that all association reactions be written with a single X^-. For example for Ca^{2+}:

$$0.5Ca^{2+} + X^- \leftrightarrow Ca_{.5}\text{-}X$$

The Vanselow convention is different since it uses molar fractions for the exchangeable cations. The total number of exchangeable cations is a variable in the case of heterovalent exchange, even if *CEC* is constant, which invalidates the above procedure for this convention. It can nevertheless, provide results over a limited concentration range (Shaviv and Mattigod, 1985), or the calculation to molar fractions can be performed as an activity correction in the geochemical model (Viani and Bruton, 1992).

EXAMPLE 10.1: *Calculation of Ca/Al cation exchange*
Exchange coefficients for Al^{3+} on clays are difficult to obtain, because they must be determined at a low pH where aluminum can be measured in solution (without all aluminum being precipitated as $Al(OH)_3$), and Al^{3+} is the dominant species. The pH may be so low that H^+ starts to compete for the exchanger sites, and the clay can dissolve and release additional Al^{3+}. There have moreover been reports in the literature that hydroxy complexes of Al^{3+}, such as $Al(OH)^{2+}$, may play a role in neutralizing charge on clay minerals and organic matter (Bloom et al., 1977; Tipping et al., 1988). A synthetic ion exchanger gives at least no problems with Al release from the clay, and may be used as a model substance. Anderson et al. (1991) added 0.094 gram DOWEX-50 in Ca^{2+} form to 30 ml solution containing 3.3 mmol/ l $CaCl_2$, 50 µmol/ l $AlCl_3$ and 0, 25, 50, and 100 µmol/ l NaF, and with pH adjusted to 3, 4 and 5. DOWEX-50 is a commonly used hard acid cation exchanger with 5 meq/g cation exchange capacity. Results of analysis of solution and exchanger composition are shown in Table 10.4a.

Table 10.4a. Aluminum and fluoride sorption to Ca^{2+} saturated DOWEX-50 at an initial Al concentration of 50 µmol/l Al^{3+}, and a background ionic strength equivalent to 10 mmol/l from $CaCl_2$ (from Anderson et al., 1991).

µmol/l	Aqueous, µmol/l		Sorbed, µmol/g	
$F_{initial}$	Al^{3+}	F^-	Al^{3+}	F^-
pH 3				
0	12.2	0	12.1	0
25	20.0	18.0	9.5	2.2
50	24.3	43.2	8.2	2.1
100	30.4	92.7	6.2	2.3
pH 4				
0	12.6	0	11.9	0
25	19.2	17.2	9.9	2.4
50	25.8	44.3	7.6	1.8
100	32.4	93.2	5.6	2.1
pH 5				
0	14.2	0	11.4	0
25	19.8	17.8	9.7	2.2
50	25.2	44.8	7.9	1.7
100	32.1	97.5	5.7	0.7

It is apparent from Table 10.4a that F^- is sorbed to the exchanger, and it seems logical to assume that a positive complex of F^- with Al^{3+} is sorbed on the negatively charged cation exchanger. Also, the activity of $[Al^{3+}]$ decreases from pH = 3 to 5, because of formation of aqueous hydroxy complexes. However, sorbed amounts of aluminum do not decrease at all in the fluoride-free solutions, suggesting that Al^{3+}-hydroxy complexes become sorbed at higher pH. The exchanger may therefore be covered by different Al^{3+} complexes. The fractions of these complexes can be summed:

$$\beta_{CaX_2} + \beta_{AlX_3} + \beta_{AlOHX_2} + \beta_{AlFX_2} + \beta_{AlF_2X} + \beta_{HX} = 1 \tag{10.13}$$

Using equivalent fractions, we have:

$$\left(\frac{\beta_{AlX_3}}{3} + \frac{\beta_{AlOHX_2}}{2} + \frac{\beta_{AlFX_2}}{2} + \beta_{AlF_2X} \right) \cdot X_t = Al_{sorb} \tag{10.14}$$

and also:

$$\left(\frac{\beta_{AlFX_2}}{2} + 2\,\beta_{AlF_2X} \right) \cdot X_t = F_{sorb} \tag{10.15}$$

where X_t is *CEC* in meq/g, and Al_{sorb} and F_{sorb} are sorbed concentrations of aluminum and fluoride, in mmol/g.

We can use the geochemical model to obtain solute activities for the analyzed aqueous concentrations, and then calculate exchange coefficients for the different pairs. The activities of the relevant solute species are given in Table 10.4b.

Table 10.4b. Activities of cations and complexes in the solutions from Table 10.4a.

pH = 3.00				
$F_{in.}$	0	25	50	100
Ca^{2+}	$2.17 \cdot 10^{-3}$	$2.17 \cdot 10^{-3}$	$2.17 \cdot 10^{-3}$	$2.17 \cdot 10^{-3}$
Al^{3+}	$4.58 \cdot 10^{-6}$	$1.94 \cdot 10^{-6}$	$2.72 \cdot 10^{-7}$	$2.45 \cdot 10^{-8}$
$AlOH^{2+}$	$4.55 \cdot 10^{-8}$	$1.92 \cdot 10^{-8}$	$2.70 \cdot 10^{-9}$	$2.43 \cdot 10^{-10}$
AlF^{2+}	–	$8.33 \cdot 10^{-6}$	$7.53 \cdot 10^{-6}$	$3.13 \cdot 10^{-6}$
AlF_2^+	–	$1.80 \cdot 10^{-6}$	$1.04 \cdot 10^{-5}$	$2.00 \cdot 10^{-5}$
pH = 4.00				
$F_{in.}$	0	25	50	100
Ca^{2+}	$2.21 \cdot 10^{-3}$	$2.21 \cdot 10^{-3}$	$2.21 \cdot 10^{-3}$	$2.21 \cdot 10^{-3}$
Al^{3+}	$4.66 \cdot 10^{-6}$	$1.75 \cdot 10^{-6}$	$2.15 \cdot 10^{-7}$	$1.03 \cdot 10^{-8}$
$AlOH^{2+}$	$4.63 \cdot 10^{-7}$	$1.74 \cdot 10^{-7}$	$2.13 \cdot 10^{-8}$	$1.02 \cdot 10^{-9}$
AlF^{2+}	–	$8.15 \cdot 10^{-6}$	$7.27 \cdot 10^{-6}$	$2.09 \cdot 10^{-6}$
AlF_2^+	–	$1.90 \cdot 10^{-6}$	$1.23 \cdot 10^{-5}$	$2.14 \cdot 10^{-5}$
pH = 5.00				
$F_{in.}$	0	25	50	100
Ca^{2+}	$2.21 \cdot 10^{-3}$	$2.21 \cdot 10^{-3}$	$2.21 \cdot 10^{-3}$	$2.21 \cdot 10^{-3}$
Al^{3+}	$2.89 \cdot 10^{-6}$	$1.14 \cdot 10^{-6}$	$1.53 \cdot 10^{-7}$	$6.40 \cdot 10^{-9}$
$AlOH^{2+}$	$2.87 \cdot 10^{-6}$	$1.13 \cdot 10^{-6}$	$1.52 \cdot 10^{-7}$	$6.36 \cdot 10^{-9}$
AlF^{2+}	–	$7.55 \cdot 10^{-6}$	$6.27 \cdot 10^{-6}$	$1.63 \cdot 10^{-6}$
AlF_2^+	–	$2.51 \cdot 10^{-6}$	$1.29 \cdot 10^{-5}$	$2.07 \cdot 10^{-5}$

Let us first calculate the exchange coefficients for Al^{3+} in the fluoride-free solutions. Since at pH = 3 and 4, the activity of Al^{3+} is 100 resp. 10 times greater than of $AlOH^{2+}$, it seems safe to assume that at least at these pH's Al^{3+} is the only Al-species that is competing for the exchanger sites. The activity of H^+ in the pH = 3 solution is 200 times larger than the Al^{3+} activity, and protons may also become adsorbed. For the reaction

$$\tfrac{1}{2}\,Ca^{2+} + H\text{-}X \leftrightarrow \tfrac{1}{2}\,Ca\text{-}X_2 + H^+ \tag{10.16}$$

The exchange coefficient is about $K_{Ca\backslash H} = 2.5$ for DOWEX-50, which at pH = 3, gives $\beta_{HX} = 0.007$, i.e. also a trace component. (For the sake of completeness, H-X has been included in the calculations, but the effect on Al/F-sorption of such a minor element is small). If only Al^{3+}, H^+ and Ca^{2+} are on the exchanger, we can calculate β_{AlX_3} from the analyzed sorbed Al-concentrations; at pH = 3 is $\beta_{AlX_3} = 3 \times 12.1/5000 = 0.00726$, and $\beta_{CaX_2} = 0.98569$. Calculating the exchange coefficient for the reaction:

$$\tfrac{1}{2}\, Ca^{2+} + \tfrac{1}{3}\, Al\text{-}X_3 \leftrightarrow \tfrac{1}{2}\, Ca\text{-}X_2 + \tfrac{1}{3}\, Al^{3+} \tag{10.17}$$

gives $K_{Ca\backslash Al} = 1.83$ at pH = 3 and 4 alike, but $K_{Ca\backslash Al} = 1.58$ for pH = 5. The value at pH = 5 decreases because Al_{sorb} remains equal, although the Al^{3+} activity has decreased. However, $AlOH^{2+}$ has at pH = 5 the same activity in solution as Al^{3+}, and we may assume that $AlOH^{2+}$ is also adsorbed. The exchange reaction for this complex is:

$$\tfrac{1}{2}\, Ca^{2+} + \tfrac{1}{2}\, AlOH\text{-}X_2 \leftrightarrow \tfrac{1}{2}\, Ca\text{-}X_2 + \tfrac{1}{2}\, AlOH^{2+} \tag{10.18}$$

with an exchange coefficient $K_{Ca\backslash AlOH}$.

Fixing the value of $K_{Ca\backslash Al} = 1.83$, allows us to obtain a value for $K_{Ca\backslash AlOH}$ from the data at pH = 5 using a combination of Equation (10.13), (10.14), and (10.16) to (10.18); it gives $K_{Ca\backslash AlOH} = 0.90$.

The next step is to obtain the exchange coefficients for the AlF-complexes. If AlF^{2+} is the only F complex on the exchanger, the coefficient $K_{Ca\backslash AlF}$ can be calculated from

$$\tfrac{1}{2}\, Ca^{2+} + \tfrac{1}{2}\, AlF\text{-}X_2 \leftrightarrow \tfrac{1}{2}\, Ca\text{-}X_2 + \tfrac{1}{2}\, AlF^{2+} \tag{10.19}$$

and F_{sorb} (Equation (10.15)). However, the value for $K_{Ca\backslash Al}$ is then also fixed since the remaining Al_{sorb} must consist of Al-X_3. Example results for the pH = 4.0 data are given in Table 10.5. They indicate that in this model $K_{Ca\backslash Al}$ is not a constant, but decreases by a factor of almost 5. In other words, the presence of F^- in solution lowers the activity of aqueous Al^{3+} appreciably, but relative to this lowered activity, more Al^{3+} is sorbed. Similar results are found for the other pH's, and the trends do not change when AlF_2^+ is included as exchangeable complex. Also note that the exchange coefficient $K_{Ca\backslash AlF}$ is lower in the 100 µmol/l F solution.

Table 10.5. Combined estimations of Ca/AlF and Ca/Al exchange coefficients on DOWEX-50 at pH = 4.0 (from data reported by Anderson et al., 1991).

µmol/l $F^-_{in.}$	$K_{Ca\backslash AlF}$	$K_{Ca\backslash Al}$
0	–	1.83
25	1.95	1.55
50	2.11	0.85
100	1.06	0.36

Although the experimental data do not allow the exact derivation of exchange complex constants for the $Ca^{2+}/Al^{3+}/AlF^{2+}/AlF_2^+$ model, we can use the coefficients for illustrative purpose. Table 10.6 gives exchange coefficients for Ca\$i$-pairs which have been incorporated in PHREEQE. The coefficients are conveniently recalculated to a monovalent standard ion, and then expressed as an association reaction from the components. Sodium has been made the standard monovalent ion for exchange in PHREEQE. For example, the $AlOH^{2+}$ exchange complex is derived from adding:

$$
\begin{array}{ll}
Ca\text{-}X_2 + AlOH^{2+} \leftrightarrow AlOH\text{-}X_2 + Ca^{2+} & (K_{AlOH\backslash Ca})^2 \\
Ca^{2+} + 2\, Na\text{-}X \leftrightarrow Ca\text{-}X_2 + 2\, Na^+ & (K_{Ca\backslash Na})^2 \\
2\, Na^+ + 2\, X^- \leftrightarrow 2\, Na\text{-}X & (K_{NaX})^2 \\
\underline{Al^{3+} + H_2O \leftrightarrow AlOH^{2+} + H^+} & \underline{(K_{AlOH})} \quad \Pi \\
Al^{3+} + H_2O + 2\, X^- \leftrightarrow AlOH\text{-}X_2 + H^+ & (K_{AlOHX_2})
\end{array}
$$

Note that the mass action constants for exchange reactions are based on a single X^-, and the constants involving $Ca-X_2$, etc. therefore appear as squared.

Table 10.6. Values for exchange coefficients of Ca^{2+} versus Al^{3+} and hydroxy/fluoride complexes with Al^{3+} on DOWEX-50.

$K_{Ca\backslash Al}$	$K_{Ca\backslash AlOH}$	$K_{Ca\backslash AlF}$	$K_{Ca\backslash AlF_2}$
1.83	0.90	2.03	3.00

With $K_{AlOH\backslash Ca} = 1.105$ (estimated here); $K_{Ca\backslash Na} = 2.5$; $K_{NaX} = 10^{20}$ (the reference association); $K_{AlOH} = 10^{-5}$ (Nordstrom et al., 1990), we obtain $K_{AlOHX_2} = 10^{35.88}$.

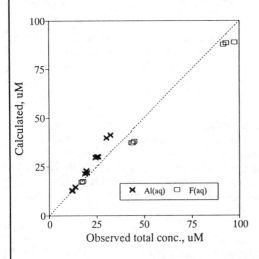

Figure 10.3. Observed and calculated aqueous (total) concentrations of Al^{3+} and F^-, for DOWEX-50 in Ca^{2+} form in a solution with (initially) 3.3 mmol/l $CaCl_2$, 50 μmol/l $AlCl_3$, and 0, 25, 50 and 100 μmol/l NaF.

Figure 10.3 shows the comparison of calculated (total) concentrations of Al^{3+} and F^- in solution with observed values. The differences are due to variations in exchange coefficients which we already observed when deriving the exchange coefficients for these data: with fixed Al^{3+} exchange coefficients, less Al^{3+} is adsorbed when F^- is added to the solution, and aqueous concentrations remain too high; because of these high Al-concentrations there is more F^- complexed to Al^{3+} in solution, and also more AlF-complex is bound on the exchanger.

10.2 GUIDE FOR THE MODELS PHREEQE/PHREEQM

Many hydrogeochemical problems can be calculated by hand, but the more complex ones are much easier handled with a geochemical model. The earlier obtained insight is, however, indispensable for choosing among the possibilities of such model. It seems reasonable to include a guide for the use of PHREEQE/PHREEQM here, as it may well become a standard tool for hydrogeochemical calculations. Leafing rather than reading through the following pages may familiarize you with the rich potential of the program, and the section will be of assistance when you make up the input file for your own problems. Many further details on PHREEQE can be found in the manual by Parkhurst et al. (1980)

Table 10.7. Examples of input variables used in a FORTRAN program.

A4	*Text*, 4 machine-readable alphanumeric characters, e.g. '*Qe#3*'
3 I2	3 *Integers*, each of 2 numbers, e.g. '99 87 ' is read as 99, 8, 70
2 F5.2	2 *Reals*, each with fixed field length of 5, and 2 decimals. Always use a decimal point with this format, and an exponent if necessary. These overwrite the indicated format if used elsewhere in the field, e.g. '*123.24.5e2*' is read as *123.2, 450*
5X	5 *spaces*

which can be ordered from the US Geological Survey (reference at the end of this chapter). PHREEQM can be ordered with the orderform enclosed elsewhere in the book (cf. p. 520).

It is important to note that PHREEQE is a FORTRAN program which demands input in a very strict format (e.g. McCracken, 1972). Problems are defined in an input file, with the variables located on lines 80 characters wide. Every variable starts at a fixed point on the line, and has a fixed field length. The type and field length of the variables is indicated by a code shown in Table 10.7.

A peculiarity is that blanks in reals and integers are read as zeros. Leading blanks give no problem, but tailing blanks translate in a multiplication by 10, as the example for 3 I2 in Table 10.7 illustrates. Blanks are no problem behind an exponent, e.g. '4.123e2' and '41.23e 1 ' are both equivalent to 412.3. Clearly, we must be very careful with the setup of the input lines, and errors which are simply due to wrong formats are frequent. It is good practice to use one of the test examples which resembles the problem at hand as a template in an editor program, in the insert mode, and to overwrite the numericals exactly in the field assigned to them. Recently PIP, the PHREEQM input procurer, has been developed by Smit and Appelo to facilitate the generation of input files for PHREEQE and PHREEQM. PIP is included on the disk with programs that can be obtained with the order form included in this book.

10.2.1 *The geochemical model PHREEQE*

A typical input file for PHREEQE is illustrated in Figure 10.4. The input file defines the chemical composition of one or two solutions, specifies chemical reactions, mineral equilibria, mixing, etc. The input file always starts with a TITLE line and a second line in which the OPTIONS are set. Then follow datablocks under specific headers (KEYWORDS) which define the solutions, the minerals, the reactions, etc. We will discuss how the input file must be constructed line by line.

```
Example - Calcite equilibrium at log PCO2 = -2.0.                    { title }
015001000  0 0              0.0              { options, nsteps, ncomps, V0 }
SOLUTION 1                                   { data blocks, headed by KEYWORDS }
Pure water                                             { a description }
 0  0 0           7.0       4.0     25.0      1.0
MINERALS                                            { block with mineral data }
FIX PCO2    2      4.0     -1.468   -4.776       0     -2.0    { 2 lines for CO₂ }
 35  1.0          3 -1.0
CALCITE     2    4.0      -8.482   -2.297       0      0.00    { and for calcite }
 15  1.0          4  1.0
END
```

Figure 10.4. An input file for PHREEQE. This problem equilibrates pure water with calcite at a CO_2 pressure of 0.01 atm.

TITLE, format (A80)

OPTIONS (IOPT(i), i = 1, .., 10), NSTEPS, NCOMPS, V0, format (10 I1, 2 I2, 6X, F10.5).

IOPT(1):   = 1 gives *printout of database*, = 0 suppresses this printout. Note that this option gives a simple line by line printout. Values which are actually read from the database and used by PHREEQE can be printed when the database is included in a problem.

IOPT(2):   = 0 *Electroneutrality is not adjusted* in the initial solution,
= 1 pH is used to obtain *electroneutrality*,
= 2 *Electroneutrality* is obtained *by adding a cation or anion*, as indicated under NEUTRAL (see below).

IOPT(3):   = 0 Calculates the *aqueous model only*, i.e. activities and saturation indexes. With IOPT(3) > 0 there may be an equilibration with minerals, and moreover if
= 1 SOLUTION 1 is *mixed* with SOLUTION 2 in a number of steps. Requires STEPS input, and a value for NSTEPS.
= 2 SOLUTION 1 is *titrated* with SOLUTION 2 in a number of titration steps. Requires input under STEPS, NSTEPS, and V0.
= 3 *Reagents are added* in specified reaction steps. Requires further input under REACTION, STEPS, NSTEPS, and NCOMPS where the stoichiometry and amounts are defined.
= 4 *Reagents are added in* NSTEPS equal increments. Also requires input under REACTION, STEPS, NSTEPS, and NCOMPS; the amount is entered under STEPS, and split in NSTEPS equal increments to be added to the solution.
= 5 SOLUTION 1 is *equilibrated with minerals*. Needs MINERALS input.
= 6 *Reagents are added until equilibrium* with the first mineral in the MINERALS data block. Requires input under REACTION, NCOMPS, and MINERALS.

IOPT(4):   = 0 The temperature is *constant*, or calculated linearly from the two end members when mixing or titrating.
= 1 The temperature is *different* from the initial solution; this temperature is read under TEMP.
= 2 The temperature is *varied* from T_0 to T_f during addition of reagents in NSTEPS equal increments. Requires T_0 and T_f (in °C) under TEMP, and furthermore input under NSTEPS, REACTION, STEPS, and NCOMPS (i.e. this option must be combined with IOPT(3) = 3 or 4). Effects of temperature on activity calculations and/or equilibrations with minerals can be efficiently calculated with this option; enter zero reaction amounts under STEPS, and the temperature effect on the unchanged initial solution is calculated NSTEPS times.
= 3 The temperature is *varied* during the reaction steps as specified by NSTEPS values (in °C) under keyword TEMP.

IOPT(5):   = 0 The pe is *constant*. Use this option whenever possible, it speeds up the calculations tremendously.

= 1 The pe is a *variable*, and changes from the initial solution may be due to reactions, temperature changes, or mineral equilibrations.

IOPT(6):   An *activity coefficient* is always calculated with the WATEQ Debije-Hückel formula if parameters are available (see below under SPECIES). If data are not available, then depending on IOPT(6):

= 0 The Debije-Hückel formula is used when the species has an ion-size parameter. Otherwise the Davies formula is used.

= 1 The Davies formula is used.
The different formulas are listed under keyword SPECIES.

IOPT(7):   = 0 Do not save the solution at the end of a simulation (but a printout of the solution is still made).

= 1 *Save* at the end in SOLUTION 1.

= 2 *Save* at the end in SOLUTION 2.
With the aqueous concentrations saved, a new simulation can be performed with that solution composition.

IOPT(8,9): = 0 No debugging printout.

= 1 Gives a long printout of the iteration process, used only in case of convergence problems.

IOPT(10):   = 0 Performs PHREEQE calculations as originally programmed by Parkhurst et al. (1980).

= 1 PHREEQM, geochemical *modeling with transport* in a mixing cell flow-tube. Gaines and Thomas convention for cation exchange. This option is further discussed in Section 10.2.2.

= 2 As 1, but with Gapon convention for cation exchange (needs a modification of the database as explained in the PHREEQM manual on disk).

NSTEPS: The number of reaction steps. A value is needed when IOPT(3) = 1 – 4, or when IOPT(4) = 2 or 3.

NCOMPS: The number of reagents added in a reaction.

VO: The volume of solution 1 that is titrated (IOPT(3) = 2).

The options in the line above define the problem that must be solved by PHREEQE. Hereafter follow datablocks with the problem-specific information. A datablock is headed by a KEYWORD. When similar data are repeated in a datablock, the last line of such a block is a blank line.

ELEMENTS format (A8)
This block normally resides in the database PHREEDA, but may need revision when new elements are modeled.

Table 10.8. The ELEMENTS database used in PHREEQM, and the corresponding master species with their charge (ZSP) and redox state (THSP).

ELEMENTS	No.	TGFW	Inp. Form.	Master Species	THSP
CA	4	40.08	Ca^{2-}	Ca^{2-}	0
MG	5	24.305	Mg^{2-}	Mg^{2-}	0
NA	6	22.9898	Na^-	Na^-	0
K	7	39.0983	K^-	K^-	0
FE	8	55.847	Fe^{2-}	Fe^{2-}	+2
MN	9	54.9380	Mn^{2-}	Mn^{2-}	+2
AL	10	26.9815	Al^{3-}	Al^{3-}	0
BA	11	137.33	Ba^{2-}	Ba^{2-}	0
SR	12	87.62	Sr^{2-}	Sr^{2-}	0
SI	13	60.0843	SiO_2	H_4SiO_4	0
CL	14	35.453	Cl^-	Cl^-	0
C	15	61.0171	HCO_3^-	CO_3^{2-}	+4
S	16	96.06	SO_4^{2-}	SO_4^{2-}	+6
N	17	62.0049	NO_3^-	NO_3^-	+5
B	18	10.81	B	H_3BO_3	0
P	19	94.9714	PO_4^{3-}	PO_4^{3-}	0
F	20	18.9984	F^-	F^-	0
LI	21	6.941	Li^-	Li^-	0
BR	22	79.904	Br^-	Br^-	0
NH4	23	18.0386	NH_4^-	NH_4^-	-3
X	30	1.0	X^-	X^-	0

{ The ELEMENTS block ends with a blank line }

TNAME, NELT, TGFW, format (A8, 2X, I2, 3X, F10.0).

These indicate the element name, the index number from 4 to 30, and the gram formula weight of the species entered in a solution. The index numbers for elements (NELT) are used in all examples in this chapter, and the ELEMENTS database is listed here in Table 10.8 for easy reference together with other basic information needed for definition of species.

SPECIES format (A8)

This block defines the components as well as the associated complexes, and resides normally in PHREEDA. The elements are repeated in this block under reserved numbers from 4 to 30 which correspond with the index numbers of the elements. In PHREEQE terminology they are called 'master species'. Numbers 1, 2 and 3 are the 'master species' H^+, e^- and H_2O respectively. A total of four lines is needed for each species, which indicate the charge, the association constant, etc. as follows:

a) i, format (I3). The index number of a species; 250 is the maximum.

b) SNAME, NSP, KFLAG, GFLAG, ZSP, THSP, DHA, ADHSP(1,2), ALKSP, format (A8, 2X, I3, 2 I1, 6 F10.3).

SNAME is the species name; NSP is the number of components (or 'master species' with index numbers from 1 to 30) in the association reaction; KFLAG = 0 uses Van't Hoff equation, KFLAG = 1 uses power series for temperature dependence of K_{ass}; GFLAG = 0 uses either Davies or simple Debije-Hückel for activity coefficients (as determined by setting of IOPT(6)), GFLAG = 1 uses WATEQ Debije-Hückel formula; ZSP is charge; THSP is redox status, i.e. the sum of THSP of the components times their stoichiometric coefficient; DHA is a_i^0 in the Debije-Hückel formula; ADHSP(1,2) are a_i and b_i in the WATEQ Debije-Hückel formula; ALKSP is the alkalinity of the species, i.e. equivalents of H^+ which the species can bind when it is titrated in a solution down to pH = 4.5.

The activity coefficient formulas are:

Debije-Hückel:

$$\log \gamma_i = \frac{-A\,z_i^2\,\sqrt{I}}{1 + Ba_i^0\,\sqrt{I}}$$

WATEQ Debije-Hückel:

$$\log \gamma_i = \frac{-A\,z_i^2\,\sqrt{I}}{1 + Ba_i\,\sqrt{I}} + b_i\,I$$

Davies:

$$\log \gamma_i = -A\,z_i^2 \left(\frac{\sqrt{I}}{1 + \sqrt{I}} - 0.3\,I \right)$$

where A and B are constants that depend on temperature, and I is ionic strength.

EXAMPLE 10.2: *The redox state (THSP) of PHREEQE species*

The redox concept of PHREEQE is based on bookkeeping, and similar to the concept of electron equivalents introduced in Section 7.5.3 (Table 7.3). A relative number of electrons, THSP, is assigned to the master species which participate in redox reactions. Normally, THSP of these master species is equal to the number of electrons which were lost when the ion was formed from the element, for example

$$Fe - 2e^- \rightarrow Fe^{2+}$$

gives $(0) - 1 \cdot (-2) = +2.0$ as THSP for Fe^{2+}.

However, H^+ and O^{2-} have THSP $= 0$ in the PHREEQE database, which means that

H_2 has THSP $= -2.0$ $(2H^+ + 2e^- \rightarrow H_2,$ $2 \cdot (0) + 2 \cdot (-1) = -2.0)$

and

O_2 has THSP $= 4.0$ $(2O^{2-} - 4e^- \rightarrow O_2,$ $2 \cdot (0) - 4 \cdot (-1) = 4.0)$

The choice of THSP $= 0$ for O^{2-} leads to THSP values of the oxy-anions, such as

CO_3^{2-} with THSP $= 4.0$ $(C + 3O^{2-} - 4e^- \rightarrow CO_3^{2-},$ $(0) + 3 \cdot (0) - 4 \cdot (-1) = 4.0)$

and

SO_4^{2-} with THSP $= 6.0$ $(S + 4O^{2-} - 6e^- \rightarrow SO_4^{2-},$ $(0) + 4 \cdot (0) - 6 \cdot (-1) = 6.0)$

A species must contain the THSP's of the master species (noted in Table 10.8) as well as the electrons in the required stoichiometry. For example,

Fe^{3+} has THSP $= 3.0$ $(Fe^{2+} - e^- \rightarrow Fe^{3+},$ $(2) - (-1) = 3.0)$

$FeSO_4^0$ has THSP $= 8.0$ $(Fe^{2+} + SO_4^{2-} \rightarrow FeSO_4^0,$ $(2) + (6) = 8.0)$

Lastly, we note that the ions which are not redox sensitive have simply THSP $= 0.0$, for example Na^+, K^+, and Ca^{2+} all have THSP $= 0.0$

---

c) LKTOSP, DHSP, (ASP(i), $i = 1, .., 5$), format (2 F10.3, 5 F12.5).

LKTOSP is the log K_{ass} at 25°C; DHSP is the enthalpy of the reaction in kcal/mol, used in the Van 't Hoff equation; alternatively a power series is used of the form:

$$\log K_{ass} = \text{ASP}(1) + \text{ASP}(2) \cdot T + \text{ASP}(3)/T + \text{ASP}(4) \cdot \log T + \text{ASP}(5)/T^2$$

where T is temperature in K.

d) $(LSP(i), CSP(i), i = 1, .., NSP)$, format 6 (I3, F7.3).

LSP(i) is the index number, CSP(i) is the stoichiometric coefficient for the component (master species) in the association reaction.

EXAMPLE 10.3: *Examples of SPECIES definition in PHREEQE*

```
SPECIES
15
CO3-2       101 -2.0      4.0         4.5         5.4       {  master species for Carbon }
0.0         0.0                                             0.0         2.0
15 1.0                                                  {  a master species has no K   , and
34                                                            has only its own component }
HCO3-       211 -1.0      4.0         4.5         5.4       0.0         1.0        { Species number }
10.329      -3.561    107.8871    0.03252849   -5151.79    -38.92561   563713.9
15 1.0       1 1.0                                              {  here the 2 master species }
36                                                          {  Species number }
CH4 AQ      400 0.0       -4.0        0.0                               0.0
41.071      -61.039                                             {  K    and enthalpy }
15 1.0       1 10.0     2 8.0      3 -3.0                      {  4 master species }
                      {  the SPECIES block ends with a blank line }
```

Note how the valence (THSP) of CH_4 is obtained from the master species in the reaction:

$$CO_3^{2-} + 10\,H^+ + 8\,e^- - 3\,H_2O \rightarrow CH_4$$

i.e.

$$1 \cdot (4) + 10 \cdot (0) + 8 \cdot (-1) - 3 \cdot (0) = -4.0$$

LOOK MIN (A8)

A block with mineral data, used for calculating saturation indexes of the solution. Up to 40 minerals can be included in PHREEQE, each mineral having from two up to four lines; the data reside normally in PHREEDA.

a) MNAME, NMINO, THMIN, LKTOM, DHMIN, MFLAG, SIMIN, format (A8, 2X, I2, 3X, 3 F10.2, 5X, I1, 9X, F10.3).

MNAME is the mineral name; NMINO is the number of species in the dissociation reaction; THMIN is the sum of the THSP's of the constituent species; LKTOM is the $\log K_{diss}$ at 25°C; DHMIN is the dissociation enthalpy in kcal/mol for use in the Van 't Hoff equation; MFLAG = 0 uses Van 't Hoff, MFLAG = 1 uses a power series for temperature dependence of the dissociation constant; SIMIN is the saturation index in the final solution when equilibrating.

b) $(LMIN(i), CMIN(i), i = 1, .., NMINO)$, format 5 (I4, F11.3).

LMIN(i) is the species number, and CMIN(i) is the stoichiometric coefficient in the dissociation reaction; note that complexes are possible. When more than 5 species make up the mineral, a second line can be used.

c) $(AMIN(i), i = 1, .., 5)$, format 5 F12.5.

Here the coefficients of the power series are given:

$$\log K_{diss} = AMIN(1) + AMIN(2) \cdot T + AMIN(3)/T + AMIN(4) \cdot \log T + AMIN(5)/T^2$$

where T is temperature in K.

EXAMPLE 10.4: *Examples of LOOK MIN mineral definitions*

```
LOOK MIN
Calcite    2    4.0         -8.480      -2.297        1          0.0
    15 1.0             4 1.0                    { 2 species CO₃ (= 15) and Ca (= 4) }
-171.9065   -.077993    2839.319     71.595                { power series for log Kdiss }
PCO2       2    4.0         -1.468      -4.776        1          0.0
    35   1.0           3 -1.0                  { diss. with H₂O (= 3) in H₂CO₃ (= 35) }
 108.3865    0.01985076  -6919.53    -40.45154     669365.0
                         { the LOOK MIN block ends with a blank line }
```

Note that for calcite two master species are used, i.e.

$$CaCO_3 \rightarrow Ca^{2+} + CO_3^{2-}$$

and for CO_2 gas the complex H_2CO_3 (with index number 35):

$$CO_{2(g)} \rightarrow H_2CO_3 - H_2O$$

SOLUTION i, format (A8, 1X, I1).

This block defines the starting solution. **i** can be 1 or 2, which are the two solutions used for mixing or titrating; note that mineral equilibration is always performed on SOLUTION 1. The block contains the following lines:

a) HEAD (A80), a title for the solution.

b) NTOTS, IALK, IUNITS, PH, PE, TEMP, SDENS, format (I2, I3, I2, 3X, 4F10.3).

NTOTS is the number of element concentrations to be read; may be zero for pure water; IALK = 0 indicates that total inorganic carbon (*TIC*) is entered, IALK = 15 (the index number for carbon) indicates that total alkalinity is input; note that IALK = 15 *requires* an alkalinity input > 0 with element 15; also note that electroneutrality balancing with pH (IOPT(2) = 1) necessitates *TIC* input; alkalinity input is not possible when IOPT(2) = 1, since alkalinity is adapted to obtain electroneutrality. Concentration units are indicated by IUNITS as follows:

IUNITS = 0   molality, mol/kg H_2O, alkalinity in eq/kg H_2O;

     = 1   mmol/l, alkalinity in meq/l;

     = 2   mg/l of the species for which TGFW has been entered under ELEMENTS; alkalinity in mg HCO_3^-/l (Check that ELEMENTS has HCO_3^- as carbon species);

     = 3   ppm, again of the species for which TGFW is resident in ELEMENTS. Alkalinity in ppm HCO_3^-;

     = 4   mmol/kg solution, and alkalinity in meq/kg solution.

Note that molality is the preferred unit (IUNITS = 0), as this prevents possible errors with weight conversions.

PH: the pH of the solution; may be approximate when pH is used to obtain charge balance (IOPT(2) = 1);

PE: enter a pe that is compatible with the dominant redox sensitive species. For reduced water at intermediate pH with dominant Fe^{2+}, use pe $\approx$ 2; aerobic water with dominant NO_3^-, use pe $\approx$ 12, but this may need some trial and error, and a good eye for NO_3/ NO_2/N_2 ratios. It is also possible to obtain pe by equilibrating with $O_{2(g)}$ at a given saturation index (Example 10.6) or by equilibrating with a H_2 pressure for reduced conditions. Note that $O_{2(aq)}$ and $H_{2(aq)}$ are calculated from pe, H_2O and pH, e.g.:

$$2\,H_2O - 4\,H^+ - 4\,e^- \rightarrow O_{2(aq)}$$

When pe has been fixed with IOPT(5) = 0, and pH-changes are large as a result of reactions, then $O_{2(aq)}$ or $H_{2(aq)}$ may increase to such high concentrations that they even influence the activity of H_2O. The lowest possible activity of H_2O in the model is 0.32, but model results are simply worthless when $[H_2O]$ has dropped as far as that. It is useful to vary input pe a bit in case of convergence problems, as this may sometimes bring the desired solution. TEMP is the temperature in °C; SDENS is solution density in kg/l.

c) (LT(*i*), DTOT(*i*), *i* = 1, .., NTOTS), format 5(I4, F11.3).

The concentrations are input on lines with 5 components each (or less for the last line). LT is the index number of the element, DTOT is the concentration in units specified by IUNITS. Note that the element index number must have been read beforehand under ELEMENTS and SPECIES.

EXAMPLE 10.5: *Estimate pH from TIC and electroneutrality*

```
Estimate pH from electroneutrality; pH_lab = 7.24; pH_field = 6.99.          { title }
010001000  0 0      0.0                                       { note IOPT(2) = 1 }
SOLUTION 1
Groundwater from CH4 producing zone, 14H34 Hoorn (Neth.); Alk = 14.0e-3
12  0 0    7.24        2.0       10.9        1.000            { pH_field = 6.99 }
    4 1.89e-3      5 3.24e-3     6 4.2e-3     7 0.206e-3     8 19.e-6
    9 11.e-6      13 0.772e-3   14 1.37e-3   15 16.6e-3     16 75.e-6
   19 132.e-6     23 0.615e-3
END
```

Output of PHREEQE gives a pH = 7.037 and *Alk* = 13.7 meq/kg H_2O. Measured (in the field) was pH = 6.99 and *Alk* = 14.0 meq/l.

MINERALS (A8)

This header precedes the minerals which the solution must be equilibrated with, as opted for by setting IOPT(3) > 0. A maximum of 10 minerals may be input for equilibration. Input formats are completely identical with the LOOK MIN formats discussed before, and it is advantageous to copy the minerals as far as possible from the LOOK MIN block in PHREEDA: only SIMIN must be specified to obtain the sought saturation index in the final solution. This block ends with a blank line.

EXAMPLE 10.6: *Calculation of the exact pe for water in equilibrium with atmospheric oxygen*

```
Obtain pe of acid rain from P_O2 equilibrium.                              { title }
0050110000 0 0      0.0                                        { note IOPT(3) = 5 }
SOLUTION 1
De Bilt, summer-1987 rain (KNMI/RIVM 1988).
10  0 0    4.57       16.5        18.0        1.0
    4 10.e-6       5 5.e-6       6 31.e-6     7 2.3e-6      14 37.e-6
   15 0.000       16 49.e-6     17 59.e-6    20 1.2e-6     23 102.e-6
MINERALS                                                    { equilibrate with }
P_O2         1        4.0       -2.96       -1.844     0       -0.7      { P_O2 = 10^-0.7 }
    32 1.0

END
```

The 'mineral' P-O_2 is defined with P_{O_2} = 0.2 atm, i.e. SIMIN is set at log (0.2) = -0.7.

Output provides pe = 16.64. Note that if initial pe is much different from the final pe, there may also occur some shift in pH as result of the redox reaction.

NEUTRAL (A8)

Defines the cation and anion that are used to obtain electroneutrality (IOPT(2) = 2). One line is input with:

LPOS, LNEG, format (2 I5). These indicate the index number of the cation (provides additional positive charge) and the anion (provides negative charge). 'Master species' must be used, i.e. species with index numbers between 4 to 30.

REACTION (A8)

Defines stoichiometry and redox status of reagents which are added to SOLUTION 1. NCOMPS (the number NCOMPS is given in the OPTIONS line) reagents are input as:

LREAC(i), CREAC(i), THMEAN(i), i = 1, .., NCOMPS), format 4 (I4, 2 F8.3).

LREAC is the 'master species' index number, i.e. from 4 to 30. The index number for $O_{2(aq)}$ (LREAC = 32, THMEAN = 4.0) or $H_{2(aq)}$ (LREAC = 33, THMEAN = − 2.0) may be used to change the redox status; CREAC is the coefficient of the component in the reaction, i.e. CREAC multiplies the number of moles read in STEPS; THMEAN is the redox status of the 'master species', free to choose, and either equal to THSP of the 'master species' or, if different, to be used to induce oxidation/reduction reactions. One line contains 4 reagents (at most); add as many lines as dictated by NCOMPS.

STEPS (A8)

Defines amounts in the reaction process to be modeled as depending on the option set with IOPT(3). This parameter and NSTEPS are entered in the options line.

IOPT(3) = 1 NSTEPS values are read of the fraction of SOLUTION 1 to be mixed with SOLUTION 2; these fractions are not additive;

IOPT(3) = 2 NSTEPS values are read of the volume of SOLUTION 2 to be titrated into SOLUTION 1.

IOPT(3) = 3 NSTEPS values are read of the number of moles to be added of the REACTION.

IOPT(3) = 4 One value is read of the total number of moles of REACTION. This amount is added in NSTEPS equal increments.

The values are read as:

XSTEP, format (8F10.3). A maximum of eight values can be on one line.

KNOBS (A8)

This input varies the limits which are used in the numerical procedure. It can be used in case of convergence problems, cf. Parkhurst et al. (1980) for details. The following parameters should be entered, in free format on a single line:

DMAX, DMIN, FUDGE, RMAX, RMIN, CNVRG1, CNVRG2, ITMAX, CHKMU. The default values are 10.0 0.7 1.0 20.0 0.9 0.1 10000.0 200 1000.0. The suggestion is that turning DMAX (from 0.4 − 20.0), DMIN (from 0.1 − 0.9), RMAX (from 0.4 − 20.0), RMIN (from 0.1 − 0.9), and the maximal number of iterations ITMAX may help when convergence is a problem. PHREEQM uses the default values given in Example 10.7.

EXAMPLE 10.7: NO_3^- *reduction in groundwater from agricultural land by organic matter*

```
Nitrate groundwater is reduced with CH2O.                              { title }
0030110000 4 1      0.0                                   { note IOPT(3) = 3 }
SOLUTION 1
Groundwater under corn (Appelo, 1985, H2O 18, 557).
 9  0 0   4.48        17.3587   11.0        1.0
   4 1.09e-3       5 0.35e-3       6 1.15e-3      7 0.89e-3     14 1.79e-3
  15 0.000        16 1.00e-3      17 2.35e-3     10 0.305e-3
MINERALS                                                     { equilibrate with }
Gibbsite    3         0.0     8.11    -22.8        0          0.00     { Gibbsite }
  10          1.0    3        3.0    1          -3.0


REACTION
  15    1.0     0.0                       { CH2O (15) is added with THMEAN = 0 }
STEPS
 2.546e-4 2.185e-3  3.191e-3  3.1922e-3              { in (NSTEPS = 4) steps }
KNOBS                                        { Newton-Raphson param's are modified }
17.0    0.4  1.0     17.0  0.8   0.1  10000.0 400    1000.0
END
```

The water has been given a pe which brings it in equilibrium with $P_{O_2} = 10^{-0.7}$ atm. CH_2O, which stands for organic matter, is added in 4 steps. The first step adds just sufficient CH_2O to reduce all $O_{2(aq)}$, the amount is calculated from:

$$CH_2O + O_2 \rightarrow H_2CO_3 \tag{10.20}$$

i.e. as much CH_2O is added as there is $O_{2(aq)}$. The second step adds more, to reduce additionally the NO_3^- content down to 0.806 mmol/l (= 50 mg/l, the drinking water limit). The amount is calculated from:

$$\tfrac{1}{5}H^+ + \tfrac{1}{5}NO_3^- + \tfrac{1}{4}CH_2O \rightarrow \tfrac{1}{10}N_2 + \tfrac{1}{10}H_2O + \tfrac{1}{4}H_2CO_3 \tag{10.21}$$

i.e. the amount of CH_2O is $\tfrac{5}{4}$ times the amount of NO_3^- that must be reduced. The final reduction is done in two steps, one leaving a tiny bit of NO_3^-, the other (the fourth and final step of this simulation) reducing all NO_3^- and more. The results are presented in Table 10.9.

Table 10.9. Changes in water composition as nitrate is reduced by organic matter (concentrations in mmol/l or pe/pH units).

Step	CH_2O	pH	pe	$O_{2(aq)}$	NO_3^-
0	0	4.48	17.36	0.255	2.35
1	0.255	4.31	16.81	$0.3 \cdot 10^{-6}$	2.35
2	2.185	6.00	14.32	$0.2 \cdot 10^{-9}$	0.806
3	3.191	6.31	13.33	$0.4 \cdot 10^{-12}$	0.001
4	3.192	6.31	-2.01	$<10^{-30}$	$<10^{-30}$

The pH-decline after the first step is a result of gibbsite precipitation, the initial solution is slightly supersaturated with respect to this mineral. Reduction of $O_{2(aq)}$ by CH_2O will decrease pH when the produced carbonic acid dissociates in H^+ and HCO_3^-. The reduction of NO_3^- by CH_2O, on the contrary, increases pH (at least if pH < 6.9); the normally quite acid leachate from agricultural land can obtain an apparently natural composition again after reduction of NO_3^-. The pH increases even further if equilibrium with gibbsite is omitted. Note how addition in the last step of just a tiny bit more C than necessary to reduce the available oxidants, leads to a great drop in pe. Geochemical models have some problems with such large changes in pe. The parameters for the Newton-Raphson method were changed (with KNOBS); the program converges more slowly with the default values, especially when gibbsite equilibrium is omitted in this problem.

SUMS (A8)

Defines the sum of any of the database species, by grouping together species of interest. Up to 10 different sums can be defined, each sum may contain up to 50 species.

a) SUNAME, NSUM, format (A8, 2X, I2).

SUNAME is the name of the sum, with NSUM species from the database. The NSUM species are defined on the next card:

b) (LSUM(i, j), $j = 1, .., $ NSUM), format (20I4).

LSUM is the species number; 20 species are read (at most) on one line. Add as many lines as indicated by NSUM.

The cards a) and b) are repeated for each sum; the input ends with a blank line.

EXAMPLE 10.8: *Examples of SUMS of species*
An example, showing how to obtain SO_4^{2-} and S^{2-} species separately, and also equivalents of Ca^{2+} and Mg^{2+} is:

```
SUMS
SO4t      20                                         { Total SO4²⁻ in M }
   16   40   52   78   88   96  100  108  109  126  127  128  128  141  154  155  155  168  173  179
S-2t       8                                         { Total S²⁻ in M }
   41   42   43  110  110  111  111  111      { Fe(HS)₃ (=111) is counted thrice }
Eq.Ca+Mg  10                                  { Ca + Mg in eq.(kg_H₂O)⁻¹ }
    4    4   75   77   83    5    5   85   87   92   { Ca²⁺ (=4), Mg²⁺ (=5) are counted twice }
                { blank line ends this data block }
```

TEMP (A8)

Specifies the temperature during the reaction steps. It is required input if IOPT(4) is greater than 0. The temperatures are input in °C as:

XTEMP, format (8F10.1). If IOPT(4) = 1, one value is read. If IOPT(4) = 2, two values are read, of T_0 and T_f. If IOPT(4) = 3, NSTEPS values are read, where NSTEPS is given in the options card.

END (A8)

This ends the data input for a simulation. A problem may follow in the same input file, using the result of the previous simulation, when the result is saved as SOLUTION 1 (with IOPT(7) = 1) or as SOLUTION 2 (with IOPT(7) = 2).

EXAMPLE 10.9: *Calcite precipitation from groundwater that is heated for use in Aquifer Thermal Energy Storage (ATES)*

```
Calcite precipitation from groundwater used for Aquifer Thermal Energy Storage { title }
0042010000 4 1      0.0                                { note IOPT(3, 4) }
SOLUTION 1
St.Paul ATES site groundwater (Holm, et al., 1987, Water Resour. Res. 6, 1005)
  8 15 0    7.46        4.0        12.0        1.0
     4 1.19e-3      5 0.87e-3      6 0.24e-3      7 0.19e-3      14 0.026e-3
    15 4.87e-3     16 0.10e-3     13 0.122e-3
MINERALS                                              { equilibrate with }
CALCITE     2    4.0       -8.480     -2.297        1        0.00       { calcite }
   15 1.0          4 1.0
-171.9065    -.077993    2839.319       71.595
```

```
REACTION                              { a dummy reaction to obtain different temperatures }
   4   0.0    0.0
TEMP
   12.0      100.0                     { T₀ and T_f; temp is changed in (T_f-T₀)/NSTEPS increm. }
STEPS
    0.0                                                        { just 0.0 is added }
END
```

Note that a dummy reaction was added to obtain equilibration with calcite at different temperatures. An alternative is to repeat the problem each time for a selected temperature with IOPT(4) = 1.

The results are given in Table 10.10.

Table 10.10. Theoretical amounts of calcite precipitated from heating St. Paul groundwater.

Temp. °C	mmol/l	mg/l	cm^3/m^3
34.0	0.088	8.8	23.9
56.0	0.252	25.2	68.4
78.0	0.462	46.2	125
100.00	0.686	68.6	186

Thus, almost 0.2 liter of precipitate might form from every cubic meter of groundwater heated to 100°C. The groundwater at St. Paul has been heated up to 110°C, but ATES operation faulted initially through clogging of the heat exchangers and the injection well by calcite scaling. Water treatment with ion exchange of Ca^{2+} for Na^+ was a succesful measure to prevent the precipitation (Perlinger et al., 1987).

10.2.2 *The hydrogeochemical transport model PHREEQM*

Models which calculate the distribution of a component over the aqueous phase and the solid phase can be inserted in a transport code as explained in the section on numerical models in Chapter 9. PHREEQE has been inserted as kind of an enlarged procedure '*distri*' in such a transport model: PHREEQE in a *M*ixing cell flowtube. The resulting code PHREEQM includes all the modeling capacities of PHREEQE, but now in a 1D transport code with dispersion and diffusion. The transport part allows for flexible modeling of 1D linear transport as well as radial flow. Transport by diffusion can also be modeled, and successive simulations on the same column are possible with change of input solution, reversal of flow direction, diffusion instead of advective transport, etc. The results of column simulations are written into files which can be read by spreadsheet programs. This allows for an easy graphical processing of the output results.

Some concepts used in the column simulation are illustrated in Figure 10.5. A column consists of at most 10 layers, each of which consists in turn of a number of cells. Solutions in cells react with minerals, an exchange complex, etc. Following initialization, the cells in one layer have similar characteristics, e.g. exchange capacity, composition, temperature, etc. The chemical characteristics can be defined for PHREEQM in a way that is similar to a batch operation. A cation exchange complex can be specified according to the Gaines and Thomas convention, or the Gapon convention. During a run, all cells in a layer maintain the same MINERAL equilibria and have the same REACTION specified. A mineral may become exhausted due to dissolution, or may on the other hand precipitate.

The transport part is understood most easily as a piston flow of solution through the cells which make up the flowtube, in combination with mixing of adjacent cells (cf. Chapter 9).

After 3 SHIFTS: PHREEQE

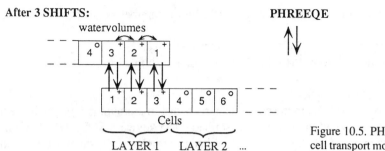

Figure 10.5. PHREEQE in a mixing cell transport model.

The solutions reside a time Δt in the cell. When $v \cdot \Delta t = \Delta x$, ($v$ is velocity in m/s, Δx is the cell length) the solutions *shift* to the next cell. Hence follows the notion that we may define transport in shifts, in combination with number of cells, and cell lengths. A shift means that all solutions in the column move down or up, exchange and react within each cell, and mix with their new neighbor cells (to simulate dispersion and diffusion). Note that 'downward' or 'upward' is entirely conceptual; the normally chosen shifts into higher numbered cells might refer to 'downward' flow. The number of shifts, divided by the number of cells, gives the number of pore volumes which flows through the column. Thus we define timestep and cell length, and the velocity of pore water follows from $v = \Delta x / \Delta t$. At one end of the column a flushing solution is defined, which keeps a constant composition. Each shift starts with downward or upward transport, followed by mixing of the aqueous concentrations in adjacent cells. Thereafter reference is made to the geochemical model (to introduce reactions, and calculate exchange, mineral equilibria, etc.). When diffusion is modeled the transport part of a shift is limited to mixing of adjacent cells only.

The mixing cell model is based on the central differences scheme that was discussed in Section 9.5. It is able to simulate dispersion of conservative ions excellently, and diffusion of conservative and retarded ions alike. Different numbers of cells for the same column length give identical breakthrough curves and diffusion profiles. For exchanging ions (retarded ions in general) there is a scale dependence with transport as a result of numerical dispersion which is not corrected for (i.e. the correction as described when modeling transport of γ-HCH in Example 9.13 has not been included). The numerical dispersion becomes negligible when a certain number of cells is used, and the general modeling procedure is to start simulations with few cells (to obtain rapid output of results which can be inspected to adapt the geochemical model as necessary) and to carry out a few final runs for a specific problem with larger number of cells.

In general, the input of PHREEQM is identical to that of PHREEQE (PHREEQM can do everything that PHREEQE can), but the number of solutions which can be read in is increased from 2 to 11. In this way, a maximum of 10 layers can be defined (actually, the solution used to initialize the exchange complex of cells in a layer is read, together with MINERAL and REACTION data). An additional solution is used as the flushing solution. Three more keywords with data blocks are introduced:

– LAYERSOL : similar to SOLUTION. Used for reading in a solution which is subsequently used for initializing a layer of the exchange column; includes specification of mineral equilibria and dimensioning of cells;

– TRANSPRT : used for defining transport (flushing) parameters as flow direction, dispersion, etc;
– MEDIUM : used for definition of diffusivity in water.
Set IOPT(10) to 1 (one) to inform PHREEQM of the column simulation.

LAYERSOL NLAY, format (A8, I2).

> NLAY = number of the layer in the column (starts with 1, maximum is 10). On the next lines a solution is input in the format described under keyword SOLUTION. In this solution one may include the amount of exchange complex (X^-) which is present in each cell of the layer. X^- is element nr. 30, and exchangeable cations which associate with X^- have reserved species numbers between 181 and 200 (see below under keyword SPECIES). PHREEQM divides the concentration of X^- by 10^{10} before the initial solution is calculated. Afterwards, when the cations complexed to this small amount of X^- have been calculated, the amounts of Na-X, Ca-X_2, etc. are multiplied again by 10^{10} in order to arrive at the specified quantity of X^-. The exchangeable cations are thus thought to be implicitly present when the solution composition and cation exchange capacity of the sediment (as X^-) have been stated. This procedure is permitted since ratios of the exchangeable cations are independent of amount of X^-. The value of X^- to be entered can be obtained in meq/l pore water from the cation exchange capacity of the aquifer sediment:

$$X^- \text{ (meq/l)} = 10 \cdot \frac{\rho_b}{\varepsilon} \cdot CEC$$

> where ρ_b is the bulk density (kg/l), ε is the water filled porosity (a fraction), and CEC is the cation exchange capacity (meq/100 g).
> In column operations X^- must be larger than 10^{-6} mol/kg H_2O, since PHREEQM only includes elements with concentrations above the machine precision of 10^{-16} mol/kg H_2O and the initial concentration of X^- is divided by 10^{10}.
> After the solution concentrations there follows a line with:

NCELL, IOPT(3), NMINEX, NCMPEX, EXSTEP, format (4 I3, F15.0).

> NCELL is the number of cells in the layer. The total number of cells of all layers together is termed NCOL. NCOL can not be more than 100;
> IOPT(3) for this layer. IOPT(3) may be 0, 3, 4 or 5, with the same meaning as in PHREEQE. (0 is just the aqueous model; 3 and 4 are equivalent since NSTEPS = 1, and indicate reaction input and/or mineral equilibrium; 5 means only mineral equilibrium);
> NMINEX is the number of minerals (max. 10) with which equilibrium must be maintained;
> NCMPEX is the number of reagents (max. 10) to be input in each cell of this layer, equal to NCOMPS in batch simulations;
> EXSTEP is the number of mol/kg H_2O of reaction added per cell, per shift, similar to XSTEP under STEPS in a batch operation.
> On the next line(s) the LENGTHs and DISPersivities of all cells in the layer are read:

(LENGTH(i), DISP(i), i = 1, NCELL). Both are expressed in meter. Free format can be used.

> Note that the mixing cell concept uses equal water volumes which are transported through the column. When linear flow is simulated, all cells should have the same length. With radial flow successive cells should have

LENGTH(n) = LENGTH(1) · ($\sqrt{n} - \sqrt{n-1}$),

where LENGTH(1) is length of the first cell. If total radius of the modeled injection is R, then LENGTH(1) = $R/\sqrt{\text{NCELL}}$.

In case of diffusion, there is no restriction to cell lengths, although more accurate results are obtained when all cells have the same length.

Mixing between adjacent cells is calculated from (cf. Section 9.5):

$$mixf = \frac{D_L \cdot \Delta t}{(\Delta x)^2} \qquad (10.22)$$

where D_L is the hydrodynamic dispersion coefficient (m^2/s), Δt is length of the timestep (= DELTAT, s), and Δx is cell length (= LENGTH(i), m).

We have also, cf. Equation (9.56):

$$D_L = D_e + \alpha_L \cdot v \qquad (10.23)$$

where D_e is the effective diffusion coefficient (DM, m^2/s), and α_L is the longitudinal dispersivity (DISP(i), m).

And lastly, in the model is:

$$v \cdot \Delta t = \Delta x \qquad (10.24)$$

On combining Equations (10.22), (10.23), and (10.24) we find that the mixing factor is obtained from:

$$mixf = \frac{\text{DISP}(i)}{\text{LENGTH}(i)} + \frac{\text{DM} \cdot \text{DELTAT}}{(\text{LENGTH}(i))^2} \qquad (10.25)$$

where DM is the diffusion coefficient (m^2/s) and DELTAT is the timestep (s). (These variables are further explained below). When adjacent cells have different LENGTH's or DISPersivities, the mixing factor is calculated for the average.

The restriction on mixing is, that *mixf* should be smaller than 0.33 to prevent numerical oscillations (i.e. each cell retains at least 1/3 of the original concentrations). When dispersivity DISP or diffusion coefficient DM times diffusion time DELTAT requires a larger *mixf*, more mixes are performed with a reduced *mixf'*. If for example, cell length = 0.01 m, and dispersivity = 0.009 m, then *mixf* = 0.9. In this case 3 mixes (NMIX = 3) are performed per shift, with *mixf'* = *mixf*/NMIX = 0.9/3 = 0.3. If, as another example, diffusion is modeled with DM = 10^{-9} m^2/s, cell length = 0.1 m, and DELTAT = 31.536·10^6 s (= 1 year), then *mixf* = 3.1536. Mixing is then done INTEGER(3.0 × 3.15) + 1 = 10 times for each shift, with a small mixing factor *mixf'* = 3.1536 / 10.

If NMINEX is not equal to 0, the next NMINEX lines should contain:
MNAME, SIMEX, AMTMIN, format (A8, 1X, 2 F10.3)
    MNAME is the name of the mineral as it appears under LOOK MIN in the data base PHREEDA;
SIMEX is the saturation index for this mineral;
AMTMIN is the initial amount of mineral in a cell (moles/kg H$_2$O). When 0, the initial amount of mineral is set to 10 moles/kg H$_2$O, i.e. near infinite. When AMTMIN is set

negative, the initial solution is not equilibrated with the mineral; precipitation may occur later. To obtain initial equilibrium, without substantial amount of mineral available for dissolution, set AMTMIN to a small positive number, e.g. 10^{-20}.

If NCMPEX was not equal to 0, the next line(s) should contain:

LREAC, CREAC and THREAC for each component (format (4 (I4, 2 F8.3)), identical to the data entered under REACTION in PHREEQE. Use as many lines as needed. When more mixes are required because of large dispersion, the program divides EXSTEP by the number of mixes per shift. In this way the reaction is split up, and added in small portions with each mix.

SOLUTION i, format (A8, I2)

Input here the composition of the solution which flows or diffuses into the column in the usual PHREEQE format. The number i should be equal to the number of layers LAYERSOL + 1.

TRANSPRT (format A8)

This data block is used for defining flow parameters. The next line after the one with TRANSPRT should contain:

NSHIFT, ISHIFT, IFRIX, IPREX, POR, DELTAT, SOLTOL, TMPTOL, format (4 I5, 4 F10.0).

NSHIFT is the *number of shifts* of the aqueous solution in the column. In case of *transport*, NSHIFT divided by NCOL (the total number of cells of the model) is equal to the number of column pore volumes that is injected. When only *diffusion* takes place, NSHIFT times DELTAT is the total diffusion time. Output of the equilibrium calculations is written after each shift.

ISHIFT is the *shifting direction*. 1 means that aqueous cell contents are transported to the next higher numbered cell ('to the right' or 'downward'), 0 means no flushing, only diffusion (stationary), −1 means transport to lower numbered cells ('to the left' or 'upward').

IFRIX determines *mixing of end cells*, the meaning depends on whether flow with dispersion is modeled, or diffusion only (i.e. the meaning is determined by the value of ISHIFT).

Diffusion, (ISHIFT = 0): With IFRIX = 0, the column is closed, i.e. there is no mixing with outer solutions. If IFRIX = 1, the first cell mixes with the upper solution, if −1, the bottom cell mixes with the end solution. Set this option to 1 when an upper solution diffuses into the column.

Dispersive flow, (ISHIFT = 1, −1): When IFRIX = 0, the mixing factor calculated from cell length and dispersivity/diffusion is multiplied with (1 + 2 / NCOL) to correct for not mixing the end-cells. (NCOL is the total number of cells in the column). With IFRIX ≠ 0, this correction is not performed. The correction term is necessary when modeling laboratory columns; mixing of the end-cells is not done to preserve mass balance in the column. When modeling transport in an 'infinite' medium (an aquifer), one may add a few additional cells in order to prevent end-cell problems. Set IFRIX to 1 in that case.

IPREX determines *printout details*. If IPREX = 1 an extended printout of species, etc. is given after each model calculation (a quick way to fill a hard disk). If IPREX = 0 only total molalities are output. Minimal information about the calculations only is given with IPREX = −1; with IPREX = −2, this is further confined to the last cell only (handy for modeling column experiments).

POR is *porosity* of a (any) cell, expressed as a fraction. All cells in the column have the same porosity;

DELTAT is *timestep* in seconds for each shift;

SOLTOL is a tolerance level used for skipping CP-time devouring calls to the geochemical model. If for any master species the summed, relative differences in concentrations are greater than this level (i.e. $|\Sigma \text{ dif/conc.}| > $ SOLTOL), the geochemical model is called. Differences in concentrations in a cell are summed over succesive shifts until the geochemical model is called for that cell. (Summed concentration differences imply that concentrations may vary around some average during several shifts before the tolerance is exceeded). Another check is performed to see whether concentrations have changed after a calculation. When the concentrations of all elements in a cell changed less than SOLTOL relative units, a message appears: 'COLUMN CELL NR. nr FLUSHED';

TMPTOL has the same meaning, but for the temperature. The present version of PHREEQM works for isothermal cases only.

Note: if during one shift all cells in the column were skipped or flushed, PHREEQM determines that the column is flushed with respect to the present tolerance levels. Consequently, it dumps the column to file EXCHDMP and continues with the next step, which may either be the reading of new instructions or a stop.

If the next line is blank, no spreadsheet file are written by PHREEQM.

If one does want a *spreadsheet* file to be output, enter the following two lines:

a) SSNAM, NSSDMP, SSNAM2, SSNAM3, format (A12, 1X, I2, 5X, A12, 3X, A12).

SSNAM is the name of the spreadsheet file which receives the *aqueous (total) concentrations* in each cell of the column;

SSNAM2 is the name of the spreadsheet file which receives the concentrations of *exchangeable cations* and the amounts of *minerals* precipitated/dissolved.

SSNAM3 is the name of spreadsheet file to write *summed species* into. Use keyword SUMS to indicate the species which need to be summed.

NSSDMP is the number of *master species* written on SSNAM and SSNAM2;

     Next line contains:

b) ISSDMP(i), i = 1, NSSDMP, format (9I4). Indexes of master species. Total concentrations of these master species will be written on the spreadsheet files.

If, during a specific run, one does not need any spreadsheet output, one can just enter the name of the spreadsheet file followed by a negative number for NSSDMP. The file is kept open in that case for succeeding runs. Note that blank lines are written on the files when output variables are not available (e.g. when the call to the geochemical model was skipped due to tolerance levels, so that summed species are not known).

MEDIUM (format (A8)

Defines the diffusion coefficient in the pore solution. The next line contains:

DM format (1F15.0), the effective molecular diffusion coeffcient of elements in the pore solution (equal for all elements) in m²/s (see Section 9.3.5 for a discussion on how to obtain a value for DM).

SPECIES

Exchangeable cations must be defined as species for columns in which cation exchange takes place. The exchangeable species is built into an association reaction between the adsorbed elements, and element X^- (= nr. 30). Species numbers 181 to 200 have been reserved for the exchangeable species. These species may have the activity coefficient of the cation, and are not counted in the sum of all species that PHREEQE uses to calculate the activity coefficient for neutral species. Adsorbed species may be built from several master species, including H^+, etc., as shown in Section 10.1.3. Note, however, that the resulting species must be electrically neutral. This is required to prevent an influence on ionic strength, and to comply with the restriction of PHREEQE to calculate pH from (predefined) electrical neutrality of the solution. Early versions of PHREEQM used the Gapon convention for calculating exchangeable cation activity. It is still possible to use this convention by setting IOPT(10) = 2, and changing the species to the ones given in Appendix V of the PHREEQM manual on disk. Example 10.10 shows the definition of exchangeable Na^+ and Ca^{2+}.

EXAMPLE 10.10: *A few exchangeable species defined for PHREEQM*

```
SPECIES
181        $ START OF BLOCK OF ADSORBED SPECIES. IBEGX=181; UP TO 200
NAX          201 0.0        0.0       4.0       4.0       0.075     0.0
20.00        0.0
  6 1.0      30 1.0
183
CAX2         201  0.0       0.0       6.0       5.0       0.165     0.0
40.80        0.0
  4 1.0      30 2.0
```

After each column simulation run, PHREEQM dumps all information concerning the column to a dumpfile called EXCHDMP. In a next run of PHREEQM, this dumpfile is automatically read back (if it is made available to the program, i.e. 'local' or in the same 'working directory'). Then the user can do the following:

– use a new flushing solution, by defining a new SOLUTION numbered (NLAY + 1). The keyword TRANSPRT with its parameters should be added;

– reverse the shifting direction or model diffusion only. In this case only TRANSPRT input is required. In case of diffusion only, mixing of the first and last cell with outer solutions is determined by the value of IFRIX. A new 'upper' solution (to be mixed with the first cell) can be defined with SOLUTION;

– change parameters of the layers, e.g. introduce new minerals, give other saturation indexes, define other reactions. Consult the appendix on the structure of EXCHDMP in the PHREEQM manual on disk for these manipulations.

Use of PHREEQM on IBM-PC's and compatibles

A first remark is that a mathematical coprocessor (8087, –287 or –387) is absolutely imperative. A second remark concerns the amount of memory PHREEQM needs: about 400 Kbyte of RAM. File buffers etc. increase this (see below). A third remark is that PHREEQM, like many other FORTRAN 77-programs, freely makes use of many simultaneously opened files. Therefore, your CONFIG.SYS file on the DOS-disk (boot-disk) should contain at least the following lines:

Files = 11
Buffers = 11
Break = on

The last statement makes it easier to halt PHREEQM in case of trouble. The number of file handles and buffers, 11, is an absolute but satisfactory minimum and may be set higher.

1. Make sure that the database, PHREEDA, is in the default directory. By means of setting global variables in the MS-DOS environment, other paths or file names can be used (see SET command in MS-DOS manual).

2. Be sure to save precious EXCHDMP files which should be kept for later runs, not the coming one, by renaming.

3. The spreadsheet files which appear under the keyword TRANSPRT overwrite any existing files with the same name and path.

4. Type: PHREEQM. After printing a banner on the screen, the program asks for the name of the file containing the input data and for the name of a file to receive the output. Typing 'con' for the latter sends output to the screen. Next, after reading PHREEDA and the input file, PHREEQM keeps you informed of what it is doing by printing short messages on the screen.

10.3 EXAMPLES OF GEOCHEMICAL TRANSPORT MODELING

This section contains examples of geochemical modeling using PHREEQE and PHREEQM. The examples are partly based on testcases of PHREEQE/PHREEQM, and have been selected to show the performance of the model. The examples also serve to illustrate certain aspects of geochemical transport modeling in more detail than in the manuals. The first two examples illustrate some features of batch/pathway modeling with PHREEQE, while the remainder is devoted to true transport modeling with PHREEQM.

EXAMPLE 10.11: *Evolution of water in a limestone/dolostone aquifer*
Parkhurst et al. (1980) have provided a number of test cases with PHREEQE for model testing. Test problem 4 is of interest as it shows the evolution of water composition in a dolomite/limestone aquifer that contains trace amounts of gypsum (or anhydrite), organic matter (CH_2O), pyrite and goethite. Dissolution of gypsum is simulated by adding Ca^{2+} and SO_4^{2-} to a water which has equilibrated with calcite and dolomite. Other reactions which are simulated are an increase of temperature, addition of organic matter and pyrite/goethite equilibrium, as summarized in Table 10.11. The input file which calculates part A (initial equilibration of water in the recharge area) and part E (the complete reaction scheme) is given in Table 10.12.

Table 10.11. Reactions simulated in PHREEQE test case 4.

Reaction characteristics	Part				
	A	B	C	D	E
Calcite equilibrium	X	X	X	X	X
Dolomite equilibrium	X	X	X	X	X
Dissolve 8 mmol $CaSO_4$		X	X	X	X
Increase temperature to 65°C			X	X	X
Add 4.0 mmol CH_2O				X	X
Goethite and pyrite equilibrium					X

Table 10.12. Input file for PHREEQE test case 4, part A and E. From Parkhurst et al., 1980.

```
TEST PROBLEM #4 PART A. CALCITE-DOLOMITE EQUILIBRIUM, LPCO2=-2.0, 25C.
015001100  0 0                0.0
SOLUTION 1
PURE WATER
 0   0 0           7.0        4.0        25.0        1.0
MINERALS
FIX PCO2    2       4.0       -1.468     -4.776         1                  -2.0
     35  1.0        3 -1.0
108.3865     0.01985076  -6919.53      -40.45154     669365.0
CALCITE     2       4.0       -8.482     -2.684         1
     15  1.0        4 1.0
504.987      0.115304    -15169.5       -200.841
DOLOMITE    3       8.0      -17.020     -8.29          0
      4  1.0        5 1.0           15 2.0

END
TEST PROBLEM #4 PART E - DE-DOLOMITIZATION,DELTA T,SULFATE REDUCTION,FE PRESENT
004211000   4 3                0.0
MINERALS
CALCITE     2       4.0       -8.482     -2.684         1
     15  1.0        4 1.0
504.987      0.115304    -15169.5       -200.841
DOLOMITE    3       8.0      -17.020     -8.29          0
      4  1.0        5 1.0           15 2.0
PYRITE      4       0.0      -18.48      11.3           0
      1 -2.0        2 -2.0        8 1.0         42 2.0
GOETHITE    3       3.0        0.5       -14.48         0
    115  1.0        3 2.0        1 -3.0

REACTION
     4       1.0        0.0  16     1.0       6.0  15      0.5       0.0
STEPS
       0.008
TEMP
      25.0       65.0
END
```

Part A starts with equilibration of pure water with calcite and dolomite at $P_{CO_2} = 10^{-2.0}$ atm. (note that OPTION(2) = 1, and OPTION(3) = 5). Part E adds 8 mmoles $CaSO_4$ and 4 mmoles CH_2O in 4 steps, while maintaining equilibrium with calcite, dolomite, pyrite and goethite; moreover, the temperature is increased stepwise to 65°C. The thermodynamic data for the minerals in this test case are from the 1980 version of PHREEQE, and there may be slight differences with saturation indexes calculated using updated versions of LOOK MIN. Output of the calculations is presented in Figure 10.6, and illustrates changes in water composition and the major mineral reactions.

The output illustrates the classical dedolomitization reaction (Back and Hanshaw, 1970; Plummer et al., 1990) in a carbonate aquifer (Chapter 4). Dissolution of gypsum brings Ca^{2+} in solution, making the water *super*saturated with respect to calcite, which precipitates. The loss of CO_3^{2-} brings the water to *sub*saturation with respect to dolomite, which dissolves. Equilibrium with both calcite and dolomite fixes the $[Mg^{2+}]/[Ca^{2+}]$ ratio in solution to about 0.8 (Chapter 4). The reactions lead to a decrease of pH, and an increase of CO_2 pressure (curves B in fig. 10.6). This seems paradoxical at first sight, since the reaction

$$1.8CaSO_4 + 0.8CaMg(CO_3)_2 \rightarrow 1.6CaCO_3 + Ca^{2+} + 0.8Mg^{2+} + 1.8SO_4^{2-} \qquad (10.26)$$

is independent of H^+. The pH decrease is an effect of increasing Ca^{2+} and Mg^{2+} concentrations in solution, slightly affected by differences in complex constants of Ca^{2+} and Mg^{2+} with HCO_3^- and SO_4^{2-}. We can demonstrate these effects on calcite/dolomite equilibration if we set the solubility of dolomite as twice that of calcite, i.e. $K_{dol} = (K_{calcite})^2 = 10^{-16.964}$. It means that for the reaction

$$Mg^{2+} + 2CaCO_3 \leftrightarrow CaMg(CO_3)_2 + Ca^{2+}$$

the equilibrium ratio is

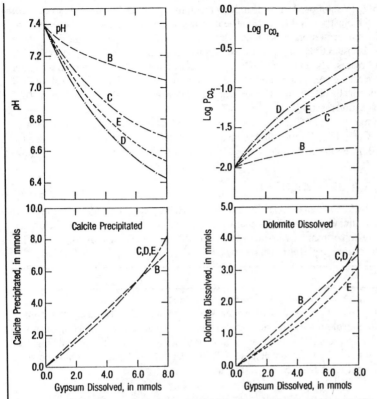

Figure 10.6. Evolution of water composition in a calcite/dolomite aquifer, with dissolution of gypsum. Curves B-E are from different modeling options (see Table 10.11).

$$K = [Ca^{2+}]/[Mg^{2+}] = 1$$

This implies that equilibration with dolomite only, would give exactly equilibrium with calcite as well – at least, if complexation effects are absent. In reality the $CaHCO_3^+$ complex is stronger than the $MgHCO_3^+$ complex, so that a tiny bit of calcite must dissolve to give $\Sigma Ca/\Sigma Mg > 1$. This changes when SO_4^{2-} is also present in solution, since the $MgSO_4^0$ complex is stronger than the $CaSO_4^0$ complex. In this case $\Sigma Ca/\Sigma Mg$ will become smaller than one. Our dolomite, dissolving in a SO_4^{2-} solution, will make the solution supersaturated with respect to calcite, and cause it to precipitate. Table 10.13 shows what is happening.

Table 10.13. Equilibrium of water with dolomite and calcite. Dolomite has a fictitious $K_{dol} = (K_{calc})^2 = 10^{-16.964}$. Eight mmol SO_4^{2-} is added as $SrSO_4$ in solution X, and as $CaSO_4$ in solution Y.

Solution	P_{CO_2}	pH	*TIC*	ΣSO_4^{2-}	Sr^{2+}	ΣMg^{2+}	ΣCa^{2+}	$\Sigma Ca/\Sigma Mg$
(A)	10^{-2}	7.40	4.554	0	0	1.05	1.06	1.009
(X)	$10^{-2.25}$	7.63	4.726	8	8	1.158	1.129	0.974
(Y)	$10^{-1.763}$	7.06	4.307	8	0	5.004	4.863	0.972

When $SrSO_4$ is added to solution A, which is closed with respect to CO_2, the complexation of Ca^{2+} and Mg^{2+} with SO_4^{2-} allows a further dissolution of dolomite (also aided by smaller activity coefficients in the more concentrated solution). More Mg^{2+} must come in solution than Ca^{2+}, to balance the larger quantity of $MgSO_4^0$ complex. *TIC* and pH also increase, and CO_2 is consumed.

Now, if we add $CaSO_4$ to solution A, it will induce precipitation of $CaCO_3$ and dissolution of dolomite (note the stoichiometry in Table 10.13). The increase of the Ca^{2+} and Mg^{2+} concentrations requires that the CO_3^{2-} concentration be smaller, while equilibrium with calcite and dolomite is maintained. More *TIC* is lost as CO_3^{2-} in calcite than enters the solution from dolomite, hence H_2CO_3 dissociates, and pH decreases. The addition of SO_4^{2-} again requires that the $\Sigma Ca/\Sigma Mg$ ratio is smaller than one, i.e. a bit more calcite precipitates, and a bit less dolomite dissolves than would be the case if both cations behave identically in solution. The pH drop is therefore enhanced by approximately 0.06 units due to the complexation reactions.

An increase in temperature to 65°C decreases the relative solubility of calcite more than of dolomite, so that the $[Ca^{2+}]/[Mg^{2+}]$ ratio must be lower than at 25°C; this enhances the pH decline (curve C in Figure 10.6). Partial reduction of SO_4^{2-} is a further H^+ producing process, as follows from:

$$SO_4^{2-} + 2CH_2O \rightarrow HS^- + 2HCO_3^- + H^+$$

and curve D in Figure 10.6 shows an even larger decrease of pH. However, high HS^- concentrations are not often found in natural water, since HS^- readily precipitates with Fe into a sulfide mineral such as pyrite. The iron may come from a hydroxide such as goethite (FeOOH), and dissolution of this mineral cancels out the change in pH, as illustrated by curve E.

EXAMPLE 10.12: *Evapotranspiration of acid rainwater*
Evapotranspiration increases the acidity of rainwater, since it concentrates H^+ just like the other elements. Concentrating a solution can be simulated by adding a reaction with all components in exactly the same ratio as present in the original solution. An alternative is to titrate with negative amounts of H_2O, as suggested by Parkhurst et al. (1980). An example of the last option is shown in Table 10.14. The rain is concentrated 3, 5, and 1000 times.

Table 10.14. Input file for concentrating acid rain by 'titrating'.

```
Concentrate acid rain by 'negative' titration.
0150112000 0 0      0.0
SOLUTION 1
Equilibrate pure water with atm. oxygen, store in 'titrating' SOLUTION 2.
 0  0 0    7.00      12.0      18.0        1.0
MINERALS
P_O2        1     4.0      -2.96      -1.844        0        -0.7
   32  1.0

END
Acid rain (Exam. 10.3); pe from P_O2 equil. Oxididize NH4; E.N. from pH.
0130111000 1 2      0.0
SOLUTION 1
De Bilt, summer-1987 rain (KNMI/RIVM 1988).
10  0 0    4.57      16.5      18.0        1.0
    4 10.e-6      5 5.e-6      6 31.e-6      7 2.3e-6      14 37.e-6
   15 0.000      16 49.e-6     17 59.e-6     20 1.2e-6     23 102.e-6
MINERALS                            { equilibrate with atm. oxygen }
P_O2        1     4.0      -2.96      -1.844        0        -0.7
   32  1.0

REACTION                                  { oxidize NH4 into NO3 }
  23  -1.0  -3.0     17  1.0    5.0
STEPS                             { add just the conc. of NH4 }
102.e-6
END
Concentrate solution 1 by titrating with negative amounts of solution 2.
0020110000 3 0      1.0
STEPS
-0.6666666-0.80      -0.999               { concentrate 3*, 5*, 1000* }
MINERALS                          { maintain equil. with atm. O2 }
P_O2        1     4.0      -2.96      -1.844        0        -0.7
   32  1.0

END
```

The 'titration' option that is used has a dangerous implication, since concentrations of H^+ and e^- are added in negative amounts as well. These concentrations (or rather, 'values' since it is difficult to speak of concentrations of electrons) should therefore be smaller in 'pure water' than in the titrated solution, or alternatively, be determined by electroneutrality and mineral equilibrium. Equilibrium with P_{O_2} has been included throughout all parts of the example here, and is sensible, since the change in pH can easily induce a redox reaction whereby NO_3^- is transformed into N_2. The problem at hand starts with the definition of 'pure water', which is saved in SOLUTION 2. Then the acid rain is defined, whereby NH_4^+ is oxidized into NO_3^- via a reaction input (these two nitrogen species have been uncoupled in the database of PHREEQM), and is saved in SOLUTION 1. The third and last part calculates the titration where addition of -0.666, -0.8 and -0.999 fraction of 'pure water' corresponds to concentration of the components in solution by 3, 5, and 1000 times. The pe of 'pure water' in equilibrium with P_{O_2} is lower than of the acid rain, and a negative concentration of e^- is calculated initially. However, the equilibrium with P_{O_2} prevents problems which can arise when the program calculates the logarithm of e^-.

Output of the program is shown in Table 10.15, and shows the expected decrease of pH.

Table 10.15. Concentrating acid rain by evapotranspi-ration.

	pH	NH_4^+	NO_3^-
Rain	4.53	0.102	0.059
Oxidize NH_4^+	3.64	0.0	0.161
Conc. 3 $\times$	3.17	0.0	0.483
Conc. 5 $\times$	2.96	0.0	0.805
Conc. 1000 $\times$	0.85	0.0	161.

EXAMPLE 10.13: *Flushing of Na^+ and K^+ from a column with $CaCl_2$ solution*
This example is from the PHREEQM test cases, and is useful to illustrate some aspects of transport calculations. An 8 cm long column has 1 mmol/l $NaNO_3$ and 0.2 mmol/l KNO_3 in the pore water. The sediment in the column has a CEC = 1.1 meq/l of pore water, occupied in equal proportion by Na^+ and K^+, and is flushed with $CaCl_2$ solution. The input file for this example is given in Table 10.16. The options line of the input file indicates that a column must be simulated (IOPT(10) = 1). Nitrate is used as an anion, and redox reactions which could affect the NO_3^- concentration are removed from the SPECIES block by writing an empty line after the species 48 (= NO_2^-) and 49 (= N_2). A block LAYERSOL defines the initial composition of the column pore fluid, and the sediment characteristics. In this case a solution is defined with Na^+, K^+ and NO_3^-, and X^-, the cation exchange capacity in the correct units. PHREEQM calculates the exchangeable cations which are in equilibrium with the solution, i.e. exchangeable cations need not further be defined. The column is split into 8 cells of 0.01 m, each with a dispersivity of 0.002 m. Then follows the eluting solution which contains $CaCl_2$, and a block with transport parameters. In TRANSPRT, the number of shifts (NSHIFT) is set to 25, i.e. all solutions are transported ('shifted') 25 times into the next cell. This means that 25 (= NSHIFT)/ 8 (= NCOL, the number of cells) = 3.125 pore volumes are injected (and eluted). The diffusion coefficient is set to 0 under keyword MEDIUM, and mixing between adjacent cells is determined by dispersivity only. The mixfactor *mixf* = (0.002/0.01) $\cdot$ (1 + 2/NCOL) = 0.25 is calculated by PHREEQM. Fractions of 0.25 of cell contents are mixed into neighboring cells with each shift, and we note that *mixf* remains below the criterion for numerical oscillations (Section 9.5). The timestep is set to 3600 s, which gives a velocity $v = \Delta x/\Delta t = 2.8 \cdot 10^{-6}$ m/s. However, note that the flow velocity is immaterial in this example, since dispersion is calculated from dispersivity only, and since it is assumed that the cation exchange reaction proceeds to equilibrium.

Output from the program is presented in Figure 10.7. It shows the chromatographic separation of Na^+ and K^+. Na^+, though present in equal proportion with K^+ on the exchanger, is weakly adsorbed, and is

Table 10.16. Input file for column flushing with $CaCl_2$ solution.

```
COLTES#1. COLUMN ELUTION OF NA+K BY CACL2.
0000010001 0 0       0.0
SPECIES                                 { Make NO3 a conservative ion }
 48                                                { remove NO2

 49                                     and N2 from database }

 181      $ START OF BLOCK OF ADSORBED SPECIES. IBEGX=181; UP TO 200
NAX           200  0.0      0.0      0.0                      0.0
20.00     0.0
  6 1.0     30 1.0
 182
KX            200  0.0      0.0      0.0                      0.0
20.70     0.0
  7 1.0     30 1.0
 183
CAX2          200  0.0      0.0      0.0                      0.0
40.80     0.0
  4 1.0     30 2.0

MEDIUM                                  { set diffusion to zero }
0.0
SUMS                                    { Summed species demonstrate
CA+NA+Keqs  4                              perfect mass balance
   4    4    6    7                      in the transport model }
CL+NO3eqs  2
  14   17
CEC         4
 181 182 183 183

LAYERSOL 1
FILLS COLUMN WITH INITIAL NA + K-NO3.
 4  0 1   7.00      8.000     25.0     1.000
   6 1.0           7 0.2        30 1.1       17 1.2
   8  0  0  0                { Column with 8 cells of 1 cm, disp = 0.2 cm }
 0.01  0.002  0.01  0.002  0.01  0.002  0.01  0.002  0.01  0.002
 0.01  0.002  0.01  0.002  0.01  0.002
SOLUTION 2                              { The injected solution }
CACL2 IN COLUMN.
  2  0 1   7.00      8.000     25.0     1.000
    4 0.6          14 1.2000
TRANSPRT                                { 25 shifts = 3.125 pore volume }
   25    1    0   -2      0.3     3600.    5.E-05      10.0
C:COLT1S     4     C:COLT1X     C:COLT1SU Sprdsheet files written in C: }
   4    6    7   14
END
```

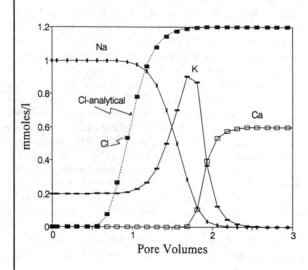

Figure 10.7. A column with Na^+ and K^+ on the exchange complex is eluted with a $CaCl_2$ solution. Symbols give the calculated concentration at the indicated pore volume. Dotted line for Cl^- gives the analytical solution for a conservative solute.

eluted from the column first. K^+ is more tenaciously held than Na^+, and appears retarded in the column effluent. The concentrations in the elution fluid are plotted against 'Pore Volume' of the column, which is obtained as follows. The numerical scheme uses cell-centered concentrations, which means that concentrations in the last cell (cell 8 of our example) belong to a point at 7.5 cm. This concentration arrives at the column outlet at 8 cm when half a cell volume has been shifted. Hence we can plot concentrations in the last cell against volume (m, in this 1D column) which has left (or entered) the column:

Volume (m) = (SHIFT + 0.5) · CELL LENGTH · POR

or in non-dimensional pore volumes of the column:

Pore Volumes = (SHIFT + 0.5) / NCOL

The eluate concentration of Cl^- is compared in Figure 10.7 with the analytical solution:

$$C_{L,t} = \frac{C_0}{2} \left[\mathrm{erfc} \left(\frac{L - vt}{\sqrt{4\alpha vt}} \right) + \exp \left(\frac{L}{\alpha} \right) \mathrm{erfc} \left(\frac{L + vt}{\sqrt{4\alpha vt}} \right) \right]$$

where L is column length, 0.08 m; $v \cdot t$ is front location, i.e. $0.08 \cdot$(SHIFT + 0.5)/NCOL; α is dispersivity, 0.002 m (note that the diffusion coefficient has been set to 0); C_0 is input Cl^- concentration, 1.2 mmol/l.

There is an excellent agreement of the numerical model and the analytical formula for Cl^-. The comparison with an analytical solution is not possible for exchanging cations, since the exchange isotherms are not linear. The isotherm of Ca^{2+} with respect to Na^+ and K^+ is convex (the tangent to the isotherm has always greater value than the isotherm itself), and the smaller concentrations of Ca^{2+} run slower than the higher concentrations. The effect is that Ca^{2+}, as the displacing ion, obtains a breakthrough curve that is sharper than can be calculated for a linearly retarded ion. Such a 'sharpening' front gives numerical problems, as can be seen when breakthrough curves for 8 cells (in the example) are compared with curves calculated with 40 cells (Figure 10.8).

Corrections for the numerical dispersion have not been programmed into PHREEQM (mainly because different total concentrations and reactions may disturb the correction procedure discussed in Section 9.5). However, when cell length Δx is smaller than dispersivity α, the numerical dispersion is reduced. The mixing factor $mixf = \alpha/\Delta x$ is then greater than 1, and at least 3 mixes per shift are performed in order to maintain the actual $mixf < 0.33$ (cf. Chapter 9). In other words, the timestep used for modeling diffusion and dispersion is at least three times smaller than the timestep used for advection. The crux now is that transport by diffusion extends over 6 neighboring cells, and

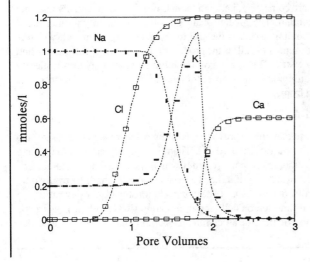

Figure 10.8. Comparison of an 8 and 40 cell model calculation. Results from the 8 cell model are indicated as single points (taken from Figure 10.7), the dotted lines are from the 40 cell model. Dotted line for Cl^- coincides for the analytical solution and the 40 cell model.

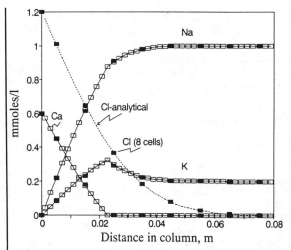

Figure 10.9. Diffusion of $CaCl_2$ solution in a column with Na^+ and K^+ in solution and on the exchanger. Filled rectangles are obtained with an 8 cell model, open rectangles with a 40 cell model. Chloride from the 40 cell model coincides exactly with the analytical solution.

predominates over advective transport. And such mixing, or transport by diffusion only, is *free of numerical dispersion*!

That the central difference scheme is free of numerical dispersion (when diffusion is modeled) follows from the discussion in Section 9.5, and we can illustrate it by a numerical example for our column. Figure 10.9 compares an 8 and a 40 cell model calculation, but now the $CaCl_2$ solution *diffuses* into the column. For the sake of completeness we compare the analytical solution for Cl diffusion,

$$C_{x,t} = C_0 \, \text{erfc} \left(\frac{x}{\sqrt{4Dt}} \right)$$

with the 8-cell calculation, and the agreement for this conservative anion is again excellent (Figure 10.9). However, in this case also the cations, which behave non-conservatively and participate in non-linear exchange reactions, show an almost perfect fit of 8 cell and 40 cell calculations.

The 40 cell grid provides a more detailed picture (note the small K^+ peak at 0.023 m), but the concentrations at the mid-cell points are coincident, i.e. both models provide about equally accurate results for the same points of the column. The greater detail goes hand in hand with increasing computing time. The most time-consuming aspect is the call to the geochemical model, which is made after each shift and/or mix. The number of calls is the product of the number of cells, shifts, and mixes per shift, i.e. NCOL · NSHIFT · NMIX. This is a maximum number which may lessen slightly when parts of the column become equilibrated, and concentrations do not change any further. The number of mixes, NMIX, is calculated from

$$\text{NMIX} = \text{INT} (3.0 \cdot mixf) + 1$$

where INT (..) calculates the integer value. In this expression *mixf* is inversely related to cell size, and directly related to the number of cells (via Equation 10.25). There is also a direct relation among number of cells and NSHIFT, via the amounts of pore volumes which must flush the column. The effect of increasing the number of cells by a factor of n (i.e. decreasing the cell length Δx by dividing by n), is therefore to multiply the number of calls by n^3. (The *cubic* relationship also follows from the *quadratic* dependence of the maximal timestep on cell length, Equation (9.82)). On a 16 Mhz AT computer it takes less than 6 minutes to calculate the 8-cell model, but more than 11 hours are needed for the 40-cell model.

EXAMPLE 10.14: *Flushing a cation exchange complex with SrCl$_2$ solution*
Exchangeable cations are routinely determined by removing the adsorbed ions with a cation which is not present in the soil solution, eg. with Ba^{2+} (Bascomb, 1964), NH$_4^+$ (Thomas, 1982), or Ag-thiourea (Maes and Cremer, 1982). A column experiment with Sr^{2+} as the displacing cation was modeled with PHREEQM, and the results provide an idea of the reliability of the results which can be obtained (Appelo et al., 1990). The sediment in the column was a clay from the river Rhine, with *CEC* = 10.0 meq/100 g. Figure 10.10 shows the analyzed eluted cations, as well as modeled concentrations.

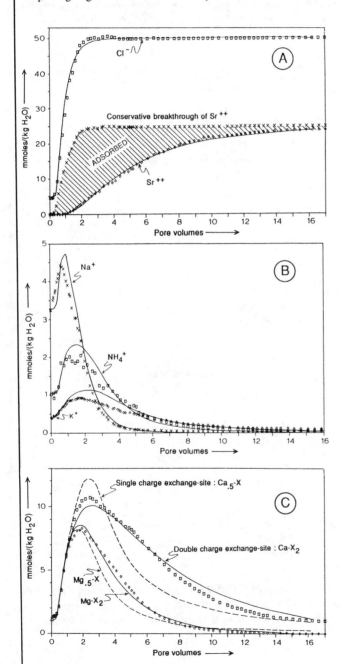

Figure 10.10. Elution of exchangeable cations adsorbed on a fresh water sediment with 50 meq/l SrCl$_2$ solution. Analyzed concentrations are shown as individual points, modeled concentrations are indicated by drawn lines. Single charge and double charge exchange site models correspond with Gapon, and Gaines-Thomas conventions resp. From Appelo et al., 1990.

The increase of Cl^- in Figure 10.10A indicates the conservative breakthrough of this ion from pore water concentration to 50 mM in the influent; it can be used to calculate a 'conservative' breakthrough of Sr^{2+}, also shown in Figure 10.10A. The measured concentrations of Sr^{2+} lag behind because the ion is taken up by the sediment, in exchange for the cations Na^+, K^+, NH_4^+, and Mg^{2+}, Ca^{2+}, shown in Figure 10.10B and C. Amounts of exchangeable cations are simply obtained by integrating the effluent curve, and subtracting one pore volume with the concentration of the cation in native pore water, which is analyzed in the initial effluent from the column. Activities of the aqueous cations are calculated for this native water, and exchange coefficients can be computed in the standard way; these are recalculated in association constants of the cation with X^-, and included in the input file as shown in Table 10.17.

Table 10.17. Input file for modeling elution of exchangeable cations with $SrCl_2$ solution.

```
Ketelmeer-sediment eluted with 50mN SrC12; J. Hydrol. 120, 225-250 (1990).
0000010001 0 0      0.0
SPECIES
181       $ START OF BLOCK OF ADSORBED SPECIES. IBEGX=181; UP TO 200
NAX          200  0.0      0.0      0.0                        0.0
20.00     0.0
  6 1.0    30 1.0
182                          { all exchange coeff. from experiment }
KX           200  0.0      0.0      0.0                        0.0
20.932    0.0
  7 1.0    30 1.0
183
CAX2         200  0.0      0.0      0.0                        0.0
41.234    0.0
  4 1.0    30 2.0
184
MGX2         200  0.0      0.0      0.0                        0.0
40.786    0.0
  5 1.0    30 2.0
189
NH4X         200  0.0     -3.0      0.0                        0.0
20.634    0.0
 23 1.0    30 1.0
200       $ LAST OF BLOCK OF ADSORBED SPECIES. ILASX=200.
SRX2         200  0.0      0.0      0.0                        0.0
41.00     0.0
 12 1.0    30 2.0

MEDIUM                                      { SETS DIFFUSION TO 0 }
0.0
LAYERSOL 1
Ketelmeer-water.
 9 15 0   6.30       4.000     10.0      1.000
    4 1.12e-3     5 1.24e-3      6 3.25e-3     7 0.39e-3     23 1.02e-3
   30 256.4e-3   14 4.5e-3      15 4.7e-3     16 0.07e-3
    4  0  0  0                    { COLUMN WITH 4 CELLS OF 2 CM, DISP = 1 CM.}
 0.0195  0.0098  0.0195  0.0098  0.0195  0.0098  0.0195  0.0098
SOLUTION 2
SrC12 in the column.
 2  0 0   7.00       4.000     10.0      1.000
 12 25.e-3      14 50.e-3
TRANSPRT                                    { 64 SHIFTS.}
   64   1    0   -2      0.3      3600.    5.E-05       10.0
c:kt1s.2        6        c:kt1x.2            { Sprdsheet files written in C: }
    4   5   6    7  23  14
END
```

The input file defines only 4 cells to model the 7.8 cm long column. The dispersivity of 0.98 cm then gives *mixf* = 0.98/1.95 = 0.503. This mixfactor is multiplied by $(1 + 2/NCOL) = 1.5$ as the correction factor for end-effects (IFRIX = 0 in the input file). All in all, it leads to 3 mixes per shift, and numerical dispersion is negligibly small as discussed before.

The exchange data of the model are completely derived from experiment, and this automatically gives an identical total mass for each element in the model and experiment, both in solution and as exchangeable cations. However, the modeled elution curves can still be different from those derived experimentally, and they provide a test of model assumptions. Comparing the elution curves of the monovalent cations shows close agreement for Na^+ (Figure 10.10B). Modeled concentrations of NH_4^+ and K^+ are too high for the initial elution peak and (necessary for mass balance) too low in the tail; the long tailing has been attributed to slow release of NH_4^+ and K^+ from closed interlayers of illite. Na^+ shows a similar but smaller discrepancy between analyzed and modeled concentrations, with a tail of 0.6 µmol/l Na^+. This ion should be lost quickly from the exchanger, as it is the most weakly held of all cations. It may well be that the tail of Na^+ originates from weathering of plagioclase feldspars; it remains stable at the 0.6 µmol/l level even when 2.5 l have been percolated through the column. The model predicts too high initial concentrations, and drops below the tail after 6 pore volumes, and a pure exchange model clearly gives a much more rapid and complete loss from the sediment.

Two model curves are shown for each of the divalent ions Ca^{2+} and Mg^{2+} (Figure 10.10C). One is for a single charge site as is used in the Gapon exchange equation, the other for a double charge exchange site, as in the Gaines and Thomas equation. The two formulations give remarkably different results, which can be explained as follows. The Gapon convention for Ca^{2+} and Mg^{2+} with respect to a single exchange site is:

$$\tfrac{1}{2}\,Mg^{2+} + Ca_{0.5}\text{-}X \leftrightarrow Mg_{0.5}\text{-}X + \tfrac{1}{2}\,Ca^{2+}$$

with an equilibrium constant:

$$K_{Mg\backslash Ca} = \frac{[Mg_{0.5}\text{-}X][Ca^{2+}]^{0.5}}{[Ca_{0.5}\text{-}X][Mg^{2+}]^{0.5}}$$

which relates a linear ratio on the exchanger with an unwanted square root of the ratio in solution. The result is that minor concentrations in solution are strongly favored by the exchanger. Figure 10.10C shows that this exchange model is not adequate for the divalent cations, and that the Gaines and Thomas convention is the preferable option. There is still a slight deviation, also with this convention, for Ca^{2+} when about 10 pore volumes have left the column. It may be that this is due to a variation of the exchange *coefficients*, but the good general fit allows us to be confident that use of exchange *constants* can give quite adequate results even in these multicomponent systems.

EXAMPLE 10.15: *Recovery of fresh water injected into a brackish water aquifer*
Radial problems can be translated into 1D terms by a transformation of coordinates, and this allows for much easier calculations (Boas, 1983). Radial problems can also be written in 1D finite difference terms (Gerald and Wheatley, 1989; Sauty, 1978), but the idea of shifting cell contents is applied in PHREEQM. A simple case of well injection and recovery is illustrated in COLTES3 of the PHREEQM test examples. Fresh $Ca(HCO_3)_2$ water is injected in a brackish water aquifer for storage purpose, remains there for 40 years, and is then recovered. The water is equilibrated with calcite and, moreover, a reaction is simulated which brings organic carbon into solution. The input file is shown in Table 10.18.

The input file feeds the problem in three parts to PHREEQM. The first part contains the initialization of the aquifer and the injection stage. The block LAYERSOL in this part defines a radial column, simply by adapting the cell-lengths in such a way that each cell has an equal volume. This can be arranged by giving successive cells the size:

$$\text{LENGTH}(n) = \text{LENGTH}(1) \cdot (\sqrt{n} - \sqrt{n-1})$$

where LENGTH(1) is the size of the first cell, adjacent to the well. The injection volume per shift is $\pi \cdot (\text{LENGTH}(1))^2 \cdot \text{POR} = 4.042\ m^2$. Four shifts inject $16.168\ m^2$. The block specifies furthermore the equilibrium with calcite (note that the thermodynamic data are read from the LOOK MIN data file), and the reaction which releases CH_2O. The second part is the period of storage; diffusion takes place

Table 10.18. Input file for radial well injection, storage, and recovery.

```
COLTES#3. INJECTION OF FRESH WATER IN BRACKISH DUNE AQUIFER.
0000110001 0 0      0.0
LAYERSOL 1
INITIATES AQUIFER WITH BRACKISH WATER. WELL INJECTION AND RADIAL FLOW.
 7 15 1    7.00       8.0       10.0       1.0
    4 2.00        5 5.5        6 50.0      7 1.00      14 62.0
   15 4.0       30 30.0
    5  3  1  1  1.0E-5        { 1 mineral equil.; reaction with 1 component }
 5.00  0.10   2.071 0.10   1.589 0.10   1.340 0.10   1.180 0.10
CALCITE  0.0             { mineral name should conform to LOOK MIN name }
   15  1.0     0.0                              { org. C is added }
SOLUTION 2
FRESH CAHCO3-WATER IS INJECTED.
 2 15 1    7.50       8.0       10.0       1.0
    4 1.500       15 3.000
TRANSPRT               $ 1 SHIFT = 8 HR.
     4     1     1    -1     0.3 28.800E3    5.E-005      10.0
c:COLT3S       6        c:COLT3X
    4    5    6    7   14   15
END
COLTES#3B. INJECTED WATER REMAINS 20 YR IN AQUIFER. DIFFUSES.
0000110001 0 0      0.0
TRANSPRT               $ 1 SHIFT = 20 YR.
     2     0     0    -1     0.3 6.3072E8    5.E-005      10.0
c:COLT3S       6        c:COLT3X
    4    5    6    7   14   15
END
COLTES#3C. INJECTED WATER IS WITHDRAWN.
0000110001 0 0      0.0
TRANSPRT               $ 1 SHIFT = 8 HR.
     3    -1     1    -1     0.3 28.800E3    5.E-005      10.0
c:COLT3S       6        c:COLT3X
    4    5    6    7   14   15
END
```

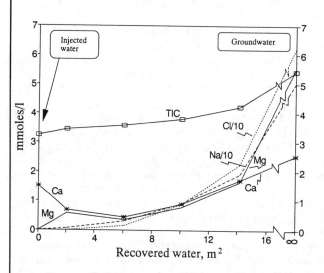

Figure 10.11. Modeled quality changes in fresh water, recovered after injection and storage in a brackish aquifer. A volume of 16.168 m^2 has been injected.

in a further stagnant water body, the release of CH_2O continues with 0.01 mmol/kg H_2O per shift, and equilibrium with calcite is maintained. The third and last part models the recovery; the water quality is plotted against recovered volume (in m^2) in Figure 10.11.

This example illustrates that injection and storage of water into an aquifer can affect the water quality. In this case there is mixing with brackish water, indicated by an increase of the Cl$^-$ concentration, which may well have been anticipated by most hydrologists. There is also cation exchange, indicated by a decrease of Ca^{2+}, and an increase of Mg^{2+} and thereafter of Na$^+$, which may perhaps be more surprising.

EXAMPLE 10.16: *Aquifer thermal energy storage (ATES) at St Paul, USA*
A high temperature ATES test site has been constructed on the campus of the university of Minnesota in St Paul. Water is injected into a quartzitic aquifer at 180 m depth. Tests with injection of hot water of up to 110°C have been performed and were described in the literature (Holm et al., 1987; Perlinger et al., 1987; McKinley et al., 1988). Heating of groundwater may induce precipitation of calcite, and the resulting scale can clog the heat exchanger and the injection well. Scaling of the injection well was initially circumvented at St Paul by installing a calcite precipitator after the heat exchanger (Holm et al., 1987, cf. Example 10.9). For a long term test during which about 92000 m³ was injected, the water was softened to a $NaHCO_3$-type water by ion exchange before the heat exchanger (Perlinger et al., 1987). Softening is one of the more efficient treatment techniques against scaling, because the intensity of the treatment will decrease when loading and unloading is repeated over the seasons. This is primarily a result of the increasing alkalinity which comes from renewed calcite dissolution in the cold part of the reservoir (Willemsen, 1990).

The exchange of Ca^{2+} for Na^+ enforces disequilibrium with the sediment's exchange complex, and cation exchange may be important. Recovered water in the long term test at St Paul showed an increase of Ca^{2+} and Mg^{2+} concentrations, and a decrease of Na^+. The changing composition has been described on the basis of carbonate and silicate reactions (Perlinger et al., 1987). However, the observed increase of Mg^{2+} could not be explained, and in fact a decrease is more normally observed when H_4SiO_4 concentrations increase through silicate weathering. The decrease is often interpreted as the result of precipitation of a Mg-silicate (Willemsen and Appelo, 1985; Holm et al., 1987). We can try and model the composition of recovered water as affected by cation exchange with the aquifer sediments.

Table 10.19. Parameters used for modeling chemistry of the long term test cycle at St Paul.

Basic data

Hydraulic parameters

Aquifer thickness:	30 m	Injected, cq. recovered:	92100 m³
Porosity:	0.25	Maximal injection radius, without dispersion:	62.5 m

Water quality, cq. exchangeable cations

	Na^+	K^+	Mg^{2+}	Ca^{2+}	
Groundwater	0.25	0.2	0.7	1.3	mmol/l
Exchangeable	0.08	0.31	7.40	22.1	meq/l
Injected	4.35	0.05	0.05	0.05	mmol/l
Exchangeable	5.59	0.32	9.30	14.8	meq/l

Fitted parameters

CEC	30 meq/l	Dispersivity	3.4 m

The cation exchange characteristics of the St Paul aquifer have not been determined, and neither are the hydraulic parameters (dispersivity, bulk density of the sediment) given in the literature. A few test runs with PHREEQM have been made with the standard selectivity constants, and a varying *CEC* and dispersivity, using the water quality of source groundwater given by Holm et al. (1987). The parameters are presented in Table 10.19. A *CEC* of 30 meq/l provided a reasonable fit with observed data (which translates to 0.41 meq/100g when bulk density is 1.9 g/cm³, and porosity is 25%). The input file is shown in Table 10.20.

The exchange coefficients have been assumed constant over the range of temperatures encountered when injected water is recovered (decreasing from 100 to 40°C). The observed and simulated concentrations of Na^+, Mg^{2+} and Ca^{2+} are shown in Figure 10.12, and the close agreement between

Table 10.20. Input file for ATES at St Paul.

```
ATES at St.Paul. Injection and recovery of softened water.
0000010001 0 0        0.0
MEDIUM
  0.0
SPECIES                                  { just standard exchange coeff. }
181      $ START OF BLOCK OF ADSORBED SPECIES. IBEGX=181; UP TO 200
NAX          200  0.0       0.0       0.0                        0.0
20.00      0.0
   6 1.0     30 1.0
182
KX           200  0.0       0.0       0.0                        0.0
20.70      0.0
   7 1.0     30 1.0
183
CAX2         200  0.0       0.0       0.0                        0.0
40.80      0.0
   4 1.0     30 2.0
184
MGX2         200  0.0       0.0       0.0                        0.0
40.60      0.0
   5 1.0     30 2.0

LAYERSOL 1
Initiates aquifer. Water from Holm et al., 1987, WRR 23, 1005.
  6 15 0   6.92        7.0         10.0         1.0
    4 1.3e-3       5 0.7e-3       6 2.5e-4       7 0.2e-3       15 4.13e-3
   30 30.e-3
   15  0  0  0  0.0
   19.77 3.4    8.189 3.4    6.284 3.4    5.297 3.4    4.667 3.4
   4.219 3.4    3.88  3.4    3.612 3.4    3.392 3.4    3.208 3.4
   3.051 3.4    2.916 3.4    2.796 3.4    2.691 3.4    2.596 3.4
SOLUTION 2
Softened water is injected. Average temp. = 60 deg. C.
  5 15 0   6.92        7.0         60.0         1.0
    4 0.05e-3      5 0.05e-3      6 4.35e-3      7 0.05e-3      15 4.13e-3
TRANSPRT                                        { 1 SHIFT = 2 days }
   10     1     1    -1      0.25 17.280E4    5.E-005     10.0
ST.PS           4        ST.PX
    4    5    6    7
END
Injected water is withdrawn.
0000010001 0 0        0.0
TRANSPRT                                        { 1 SHIFT = 2 days }
    9    -1     1    -1      0.25 17.280e4    5.E-005     10.0
ST.PS           4        ST.PX
    4    5    6    7
END
```

the two makes it probable that we have identified the dominant mechanisms to explain the observed water qualities.

Modeled K^+ concentrations are compared with observed concentrations in Figure 10.13. The observed values are significantly higher than modeled concentrations which may be caused by a combination of factors, i.e. dissolution of K-feldspar and K-containing clay minerals, and variation of the K/Ca exchange coefficient with temperature as found by Griffioen (1990). Part of the K^+ released from mineral dissolution will be active in the exchange process, but the amounts are small, and of not much influence on exchange of the other cations. The $H_4SiO_4^0$ concentrations were found to be limited by quartz saturation, so that actual amounts of dissolution of K-feldspar can not be calculated without additional assumptions regarding the amount of quartz precipitation.

It is of interest that about $1/3$ of Na^+ injected with the softened water was not recovered, although the amounts of injected and recovered water were identical. The discrepancy was explained as due to dispersion (Perlinger et al., 1987), but in that case one would expect a somewhat different trend of Na^+ concentrations versus cumulative flow than is observed. Cation exchange with a non-linear isotherm gives a broadening front for Na^+ when water is injected (Na^+ has difficulty in displacing Ca^{2+} from the exchanger sites), but when flow direction is reversed and water is pumped up again, a

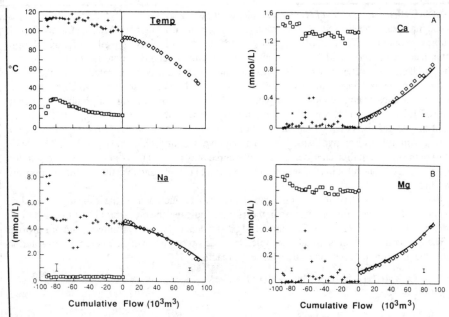

Figure 10.12. Observed concentrations of Na$^+$, Mg^{2+}, and Ca^{2+} during injection respectively recovery of softened and heated groundwater at St Paul. Squares indicate concentrations in source groundwater, crosses the same for this water after ion exchange and heating (negative flow); diamonds give the concentrations in recovered water (positive flow). Drawn lines are modeled concentrations in recovered water.

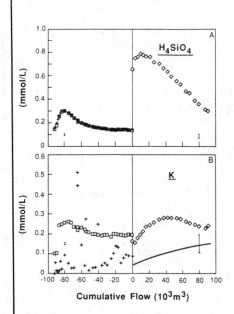

Figure 10.13. Observed K$^+$ and H$_4$SiO$_4^0$ concentrations in injected and recovered water during the long term test cycle at St Paul (Perlinger et al., 1987). Symbols as explained with Figure 10.12.

sharpening front for Na$^+$ develops when this ion is displaced again by Mg^{2+} and Ca^{2+}. The steepening aspect of all the concentration curves in the end of the pumping period is typical for such sharpening behavior, and it leads also to an incomplete recovery of injected Na$^+$.

EXAMPLE 10.17: *Reduction of MnO$_2$ by FeSO$_4$ solution*

This example calculates a redox push-front, where ferrous iron reduces MnO$_2$ in a column. It illustrates the option of PHREEQM to have restricted quantities of minerals available for dissolution. A column has been packed with MnO$_2$ (birnessite) – containing sand. The pore solution was initially flushed with distilled water, leaving low concentrations of Ca^{2+} and Mg^{2+} in solution, but these ions still dominate the exchange complex. Injected is a 5 mmol/l FeSO$_4$ solution, which reduces the birnessite, and subsequently elutes all manganese as Mn^{2+}. The input file for our problem is given in Table 10.21. Note that the column pore solution is equilibrated with MnO$_2$ at a P_{O_2} = 0.2 atm. The amount of 'O2 gas' is set very small at 10^{-20} mol/kg H$_2$O, so that only aqueous O$_2$ offers redox buffering. Amorphous iron hydroxide is set to a negative quantity, which means that 'Fe(OH)3a' is not part of the initial equilibrium assemblage, although it may precipitate as soon as the pore solution is supersaturated. Concentrations of selected elements and Fe^{2+}, obtained via the 'SUMS' option, are shown in Figure 10.14.

Sulfate is a conservative ion in this problem (it can be included in the exchange complex, however), and the concentration increase in the effluent indicates the breakthrough of injected fluid. There is a concomitant increase of Ca^{2+} and of Mg^{2+} (the latter element not shown), and of Fe$_t$ (total Fe). The iron concentration is initially completely made up of Fe^{3+}, which can remain in solution because the pH drops to a very low value due to the reaction:

$$MnO_2 + 4H^+ + 2Fe^{2+} \rightarrow Mn^{2+} + 2H_2O + 2Fe^{3+}$$

and:

$$2Fe^{3+} + 6H_2O \rightarrow 2Fe(OH)_{3,\,am} + 6H^+$$

Mn^{2+} and Fe^{3+} are retarded in the effluent, because these ions must first displace Ca^{2+} and Mg^{2+} from the exchange complex. As soon as birnessite is exhausted from the column, Fe^{2+} can remain in solution, the pH-drop terminates, and total Fe in solution consists almost completely of Fe^{2+}.

Table 10.21. Input file for the column elution of MnO$_2$ with Fe^{2+}.

```
COLTES#5. MNO2 IN COLUMN REDUCED BY FESO4.
0000110001 0 0      0.0
SUMS
Mn+3       1
 144
Fe+2       9                              { ferrous iron concentration }
   8 102 105 106 107 108 109 110 111


SPECIES
188
FEX3        300 0.0      3.0      0.0                      0.0
47.28    9.68
  8 1.0    30 3.0    2 -1.0


LAYERSOL 1
FILL COLUMN WITH CA + MGC12; MNO2-equilib.; P_O2 = 0.2 ATM., FE(OH)3 PRECIP'S.
  4  0 1   7.00        11.50      25.0     1.000
    4 0.1           5 0.02        30 50.           14 0.24
  3  5  3  0  0              $ COLUMN WITH 3 CELLS OF 4 CM, DISP = 1 CM.
 0.04  0.010  0.04  0.010   0.04  0.010  0.04  0.010  0.04  0.010
BIRNESSI 0.0        10.E-3                  { MnO2 present in 10 mM }
FE(OH)3a 0.0        -1.0        { Fe(OH)3 not initially present }
O2 gas   -0.7       1.0E-20                 { aq. O2 equil. only }
SOLUTION 2
FESO4-SOLUTION ENTERS COLUMN.
  2  0 1   7.00        1.500      25.0      1.000
    8 5.00           16 5.00
TRANSPRT          $ 30 SHIFTS.
   30    1   0    0     0.3    3600.    5.E-05       10.0
COLT5S          5      COLT5X     COLT5SU { also sprdsheet file for SUMS }
   4    5   8  16    9
END
```

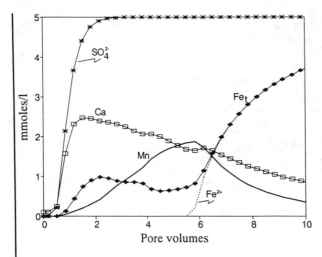

Figure 10.14. Modeled effluent concentrations from a column with MnO_2, flushed with $FeSO_4$ solution.

Manganese is still present in the effluent at this stage, even when all MnO_2 has been reduced, because the exchange complex acts as a source.

We must note here how the value for the exchange constant for $Fe-X_3$ was obtained. A 'half' exchange reaction:

$$Fe^{3+} + 3X^- \rightarrow FeX_3$$

has an estimated relative association $\log K = 60.3$, similar to the one for AlX_3. It must be written in terms of Fe^{2+}, the 'master species'. Hence the association reaction for species 115 (i.e. Fe^{3+}) is added to our exchange reaction:

$$Fe^{2+} \rightarrow Fe^{3+} + e^- \qquad \log K = -13.02 \quad \Delta H_r^0 = 9.68$$

to give:

$$Fe^{2+} + 3X^- \rightarrow FeX_3 + e^- \qquad \log K = 47.28 \quad \Delta H_r^0 = 9.68$$

The concentration pattern which we have found is typical for redox problems. The large solubility difference between oxidized and reduced forms of one element can easily give a concentration of redox sensitive elements in ore bodies, e.g. of iron-hydroxide. For example, uranium roll fronts are also a result of oxidation and reduction fronts along the same flowline (Walsh et al., 1984). The coupling of many reactions makes it very difficult to keep the overview on the important driving forces in such a chemical transport system, and geochemical modeling is the adequate tool here. However, the coupled reactions can also rather easily lead to numerical oscillations, in which 'waves' of chemicals appear to travel down the column. In the example here, such waves become apparent if the number of mixes is reduced to one (i.e. with dispersivity $\alpha < 0.008$ for the model cell length). The oscillations are countered by decreasing the cell length, i.e. increasing the number of mixes, as discussed in Example 10.13.

EXAMPLE 10.18: *Acidification of groundwater*
Acidification of soils, soil moisture and groundwater is one of those extremely complex interplays of hydro-bio-geochemical reactions where a geochemical model can be helpful (Appelo, 1985; Cosby et al., 1985a, b; Eary et al., 1989). The most important reactions are an overall concentration of the components of acid atmospheric deposition by evapotranspiration (cf. Example 10.12), displacement of basic cations from the exchange complex by H^+ and Al^{3+}, and buffering by dissolution of Al-hydroxide (Ulrich et al., 1979; Mulder, 1988). Let us model these reactions in a groundwater

aquifer. When flow is horizontal in the aquifer, the vertical infiltration can be calculated from effective precipitation and porosity as shown in Chapter 9. Thus, we may model just the vertical component of groundwater flow, affected by transversal dispersion only. We use typical data from a sandy aquifer in the Netherlands as the starting point. A column with 5 cells is initiated with rain cq. groundwater from the pre-acid era, based on data from Leeflang (1938). Water in the column is in equilibrium with gibbsite and a CO_2-pressure of 0.001 atm. Acid rain enters the column, concentrated three times by evapotranspiration. Table 10.22 shows the input file.

Table 10.22. Input file for the acidification problem.

```
COLTES#2. ACID RAIN IN VELUWE SEDIMENTS (CF. APPELO, (1985): H2O 18, P.557).
0000010001 0 0      0.0
MEDIUM                  $ SETS DIFFUSION TO 1E-9 M2/S
1.0E-009
SPECIES
  48                                      { Remove NO2- from database }

  49                                      { Remove N2 }

 181    $ START OF BLOCK OF ADSORBED SPECIES. IBEGX=181; UP TO 200
NAX          200  0.0        0.0       0.0                       0.0
20.00    0.0
   6 1.0    30 1.0
 182
KX           200  0.0        0.0       0.0                       0.0
20.70    0.0
   7 1.0    30 1.0
 183
CAX2         200  0.0        0.0       0.0                       0.0
40.80    0.0
   4 1.0    30 2.0
 185                          { Al data based on Anderson et al., 1991 }
ALX3         200  0.0        0.0       0.0                       0.0
60.41    0.0
  10 1.0    30 3.0
 189
ALOHX2       400  0.0        0.0       0.0                       0.0
35.89    0.0
  10 1.0    30 2.0     3 1.0     1 -1.0

LAYERSOL 1
INITIATES VERTICAL COLUMN WITH 3* CONCENTRATED, PRE-ACID RAIN.
  9 15 1    5.62        8.0        10.0         1.0
    4 0.12           5 0.09        6 0.21       7 0.01        14 0.27
   15 0.002         16 0.135      17 0.09      30 6.0
    5   5  2   0  0                   { horizontal flow, infiltrating 1 m/yr.}
   1.0  .06   1.0  .06   1.0  .06   1.0  .06   1.0  .06   { transv. disp.'s }
GIBBSITE 0.0                               { equil. with gibbsite
PCO2      -3.0                                 and CO2 }
SOLUTION 2
3* CONCENTRATED ACID RAIN ENTERS THE COLUMN.
  8  0 1   3.08        8.0        10.0         1.0
    4 0.042          5 0.027       6 0.21       7 0.01        14 0.255
   15 0.000         16 0.213      17 0.573
TRANSPRT                          { 1 shift = 1 yr, total 50 years }
   50     1    1    -1      0.3 31.536E06   5.E-005      10.0
COLT2S        6       COLT2X
   4    5    6    7   10   14
END
```

The transversal dispersivity has been set to 6 cm, in accordance with modeled tritium profiles (Herweyer et al., 1985). The cell length is 1m, which means a timestep of one year; note that exchange coefficients for Al^{3+} and $AlOH^{2+}$ have been obtained from the earlier Example 10.1. Results of the model, displayed as the change in concentration in time at 2.5 m depth (cell 3), are presented in Figure 10.15. These show that groundwater pH is initially buffered at a high value through cation exchange of Ca^{2+} by Al^{3+}. The Al^{3+} comes from gibbsite which dissolves in acid rain

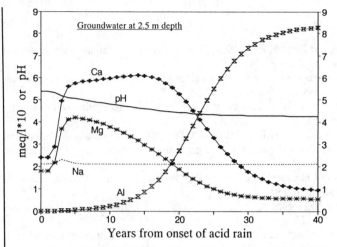

Figure 10.15. Modeled development of acid groundwater from acid rain; column cell 3, shifts are here translated into years of percolation.

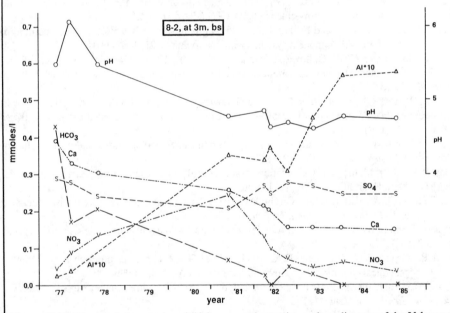

Figure 10.16. Observed changes in acidifying groundwater in sandy sediments of the Veluwe area, the Netherlands (Appelo, 1985).

groundwater. When the exchange buffer is exhausted, a uniform low pH of about 4.3 is attained.

It is of interest to compare the modeled results with analyses from periodic sampling of a borehole, shown in Figure 10.16. The downward trend of pH and concomitant upward trend of Al concentrations are remarkably similar. The acidification of groundwater lags a long time behind the onset of strong acidification in the years '50 and '60. It follows that cation exchange and dissolution of Al-hydroxide are able to retard the appearance of acidification effects a long time. Restoration, however, can be expected to take at least equally long time (Cosby et al., 1986).

REFERENCES

Anderson, M.A., Zelazny, L.W. and Bertsch, P.M., 1991, Fluoro-Aluminum complexes on model and soil exchangers. *Soil Sci. Soc. Am. J.* 55, 71-75.

Appelo, C.A.J., 1985, CAC, Computer aided chemistry, or calculating the composition of groundwater with a geochemical computer model (in Dutch). *H_2O* 18, 557-562.

Appelo, C.A.J. and Willemsen, A., 1987, Geochemical calculations and observations on salt water intrusions. I A combined geochemical/mixing cell model. *J. Hydrol.* 94, 313-330.

Appelo, C.A.J., Willemsen, A., Beekman, H.E. and Griffioen, J., 1990, Geochemical calculations and observations on salt water intrusions. II Validation of a geochemical transport model with column experiments. *J. Hydrol.* 120, 225-250.

Back, W. and Hanshaw, B.B., 1970, Comparison of chemical hydrogeology of the carbonate peninsula of Florida and Yucatan. *J. Hydrol.* 10, 330-368.

Bascomb, C.L., 1964, Rapid method for the determination of cation exchange capacity of calcareous soils. *J. Sci. Food Agric.* 15, 821-823.

Bloom, P.R., McBride, M.B. and Chadbourne, B., 1977, Adsorption of aluminum by a smectite. I. Surface hydrolysis during Ca^{2+}-Al^{3+} exchange. *Soil Sci. Soc. Am. J.* 41, 1068-1073.

Boas, M.L., 1983, *Mathematical methods in the physical sciences.* 2nd ed. Wiley & Sons, New York, 793 pp.

Bryant, S.L., Schechter, R.S. and Lake, L.W., 1986, Interactions of precipitation/dissolution waves and ion exchange in flow through permeable media. *AIChE J.* 32, 751-764.

Cederberg, G.A., Street, R.L. and Leckie, J.O., 1985, A groundwater mass transport and equilibrium chemistry model for multicomponent systems. *Water Resour. Res.* 21, 1095-1104.

Cosby, B.J., Hornberger, G.M., Galloway, J.N. and Wright, R.F., 1985a. Modeling the effects of acid deposition: assessment of a lumped-parameter model of soil water and streamwater chemistry. *Water Resour. Res.* 21, 51-63.

Cosby, B.J., Wright, R.F., Hornberger, G.M. and Galloway, J.N., 1985b. Modeling the effects of acid deposition: estimation of long-term water quality responses in a small forested catchment. *Water Resour. Res.* 21, 1591-1601.

Cosby, B.J., Hornberger, G.M., Galloway, J.N. and Wright, R.F., 1986, Times scales of catchment acidification. *Environ. Sci. Technol.* 19, 1144-1149.

De Rooy, N.M., 1991, Mathematical simulation of biochemical processes in natural waters by the model CHARON. Delft Hydraulics, R1310-10, 144 pp.

Eary, L.E., Jenne, E.A., Vail, L.W. and Girvin, D.C., 1989, Numerical models for predicting watershed acidification. *Arch. Environ. Contam. Toxicol.* 18, 29-35.

Engesgaard, P. and Christensen, T., 1988, A review of chemical solute transport models. *Nordic Hydrol.* 19, 183-216.

Felmy, A.R., Girvin, D.C. and Jenne, E.A., 1984, MINTEQ – A computer program for calculating aqueous geochemical equilibria. U.S. Environm. Protection Agency, Athens.

Fritz, B., 1975, Etude thermodynamique et simulation des réactions entre mineraux et solutions. *Sci. Geol. Univ. Strasbourg Mém.* 41, 153 pp.

Garven, G. and Freeze, R.A., 1984, Theoretical analysis of the role of groundwater flow in the genesis of stratabound ore deposits. *Am. J. Sci.* 284, 1085-1174.

Gerald, C.F. and Wheatley, P.O., 1989, *Applied numerical analysis.* 4th ed. Addison-Wesley, Reading, Mass., 679 pp.

Griffioen, J., 1990, Hydrogeochemical column experiments with sediment from a Dutch ATES test facility. Amsterdam, Institute of Earth Sciences, Free University, 103 pp., restricted.

Grove, D.B. and Stollenwerk, K.G., 1987, Chemical reactions simulated by groundwater quality models. *Water Resour. Bull.* 23, 601-615.

Helgeson, H.C., Brown, T.H., Nigrini, A. and Jones, T.A., 1970, Calculation of mass transfer in geochemical processes involving aqueous solutions. *Geochim. Cosmochim. Acta* 34, 569-592.

Herweyer, J.C., Van Luyn, G.A. and Appelo, C.A.J., 1985, Calibration of a mass transport model using environmental tritium. *J. Hydrol.* 78, 1-17.

Holm, T.R., 1989, Comment on 'Computing the equilibrium composition of aqueous systems: an iterative solution at each step in Newton-Raphson.' *Environ. Sci. Technol.* 23, 1531-1532.

Holm, T.R., Eisenreich, S.J., Rosenberg, H.L. and Holm, N.P., 1987, Groundwater geochemistry of short-term aquifer thermal energy storage test cycles. *Water Resour. Res.* 23, 1005-1019.

Jauzein, M, André, C., Margrita, C., Sardin, M. and Schweich, D., 1989, A flexible computer code for modeling transport in porous media: IMPACT. *Geoderma* 44, 95-113.

Leeflang, K.W.H., 1938, The chemical composition of precipitation in the Netherlands (in Dutch). *Chem. Wkbld* 35, 658-664.

Lichtner, P., 1985, Continuum model for simultaneous chemical reactions and mass transport in hydrothermal systems. *Geochim. Cosmochim. Acta* 49, 779-800.

Lichtner, P. and Engi, M., 1989, Data bases of thermodynamic properties used to model thermal energy storage in aquifers. IEA-Annex VI, restricted.

Liu, C.W. and Narasimhan, T.N., 1989, Redox-controlled multiple species chemical transport. *Water Resour. Res.* 869-910.

Maes, A. and Cremers, A., 1982, Cation exchange in clay minerals: some recent developments. In G.H. Bolt (ed.), *Soil Chemistry*, Elsevier, Amsterdam. Part B, 205-232.

Mangold, D.C. and Tsang, C.-F., 1991, A summary of subsurface hydrological and hydrochemical models. *Rev. Geophys.* 29, 51-79.

McCracken, D.D., 1972, *A guide to FORTRAN IV programming*. Wiley & Sons, New York, 288 pp.

McKinley, J.P., Jenne E.A. and Smith, R.W., 1988, Experimental investigation of interaction between heated groundwater and sandstone. *Proc. Jigastock 88, Versailles, France*, Vol. 2, 687-691.

Miller, C.W. and Benson, R.W., 1983, Simulation of solute transport in a chemically reactive heterogeneous system: model development and application. *Water Resour. Res.* 19, 381-391.

Mulder, J., 1988, *Impact of acid atmospheric deposition on soils: field monitoring and aluminum chemistry*. Ph.D. Thesis, Wageningen

Nordstrom, D.K., Plummer, L.N., Langmuir, D., Busenberg, E., May, H.M., Jones, B.F. and Parkhurst, D.L., 1990, Revised chemical equilibrium data for major water-mineral reactions and their limitations. In D.C. Melchior and R.L. Bassett (eds), *Chemical modeling of aqueous systems II*. ACS Symp. Ser. 416, 398-413.

Nordstrom, D.K., Plummer, L.N., Wigley, T.M.L., Wolery, T.J., Ball, J.W., Jenne, E.A., Bassett, R.L., Crerar, D.A., Florence, T.M., Fritz, B., Hoffman, M., Holdren Jr., G.R., Lafon, G.M., Mattigod, S.V., McDuff, R.E., Morel, F., Reddy, M.M., Sposito, G. and Thrailkill, J., 1979, A comparison of computerized chemical models for equilibrium calculations in aqueous systems. In E.A. Jenne (ed.), *Chemical modeling of aqueous systems*. ACS Symp. Ser. 93, 857-892.

Parkhurst, D.L., Thorstenson, D.C. and Plummer, L.N., 1980, PHREEQE – A computer program for geochemical calculations. U.S. Geol. Surv., Water Resour. Inv. 80-96, 210 pp (Copies can be purchased from: US Geol. Surv., Books and open-file reports Section, Box 25425, Federal Center, Denver, Colorado 80225-0425, USA).

Perlinger, J.A., Almendinger, J.E., Urban, N.R. and Eisenreich, S.J., 1987, Groundwater geochemistry of aquifer thermal energy storage: Long term test cycle. *Water Resour. Res.* 23, 2215-2226.

Pitzer, K.S. and Brewer, L., 1961, *Thermodynamics*. 2nd ed. (rev. of Lewis and Randall). McGraw-Hill, New York, 723 pp.

Plummer. L.N., Busby, J.F., Lee, R.W. and Hanshaw, B.B., 1990, Geochemical modeling in the Madison aquifer in parts of Montana, Wyoming and South Dakota. *Water Resour. Res.* 26, 1981-2014.

Plummer, L.N., Parkhurst, D.L. and Thorstenson, D.C., 1983, Development of reaction models for groundwater systems. *Geochim. Cosmochim. Acta* 47, 665-686.

Postma, D., Boesen, C., Kristiansen, H. and Larsen, F., 1991, Nitrate reduction in an unconfined sandy aquifer: water chemistry, reduction processes and geochemical modeling. *Water Resour. Res.* 27, 2027-2045.

Sauty, J.P., 1978, Identification des paramètres du transport hydrodispersif dans les aquifères par interprétation de traçages en écoulement cylindrique convergent ou divergent. *J. Hydrol.* 39, 69-103.

Shaviv, A. and Mattigod, S.V., 1985, Exchange equilibria in soils expressed as cation-ligand complex formation. *Soil Sci. Soc. Am. J.* 49, 569-573.

Smith, W.R. and Missen, R.W., 1982, *Chemical reaction equilibrium analysis.* Wiley & Sons, New York, 364 pp.

Sposito, G. and Mattigod, S.V., 1979, GEOCHEM: a computer program for calculating chemical equilibria in soil solutions and other natural water systems. Univ. California, 110 pp.

Thomas, G.W., 1982, Exchangeable cations. In A.L. Page et al. (eds), *Methods of soil analysis*, Am. Soc. Agronomy, part 2. 2nd ed. 159-166.

Tipping, E., Backes, C.A. and Hurley, M.A., 1988, The complexation of protons, aluminium and calcium by aquatic humic substances: a model incorporating binding-site heterogeneity and macroionic effects. *Water Resour. Res.* 22, 597-611.

Ulrich, B., Mayer, R. and Khanna, P.K., 1979, *Deposition von Luftverunreinigungen und ihre Auswirkungen in Waldökosystemen im Solling.* 2te Aufl. Sauerländer's Verlag, Frankfurt am Main, 291 pp.

Viani, B.E. and Bruton, C.J., 1992, Modeling ion exchange in clinoptilolite using the EQ3/6 geochemical modeling code. In Y.K. Kharaka and A.S. Maest (eds), *Water-Rock Interaction*, Proc. 7th Water-Rock Interaction Symp. Balkema, Rotterdam, 73-77.

Walsh, M.P., Bryant, S.L., Schechter, R.S. and Lake, L.W., 1984, Precipitation and dissolution of solids attending flow through porous media. *AIChE J.* 30, 317-328.

Westall, J.C. and Hohl, H., 1980, A comparison of electrostatic models for the oxide/solution interface. *Adv. Colloid Interface Sci.* 12, 265-294.

Westall, J.C., Zachary, J.L and Morel, F.M.M., 1976, MINEQL, a computer program for the calculation of chemical equilibrium composition of aqueous systems. Tech. Note 1, Dept Civil Eng., MIT, Cambridge.

Willemsen, A., 1990, Water treatment and environmental effects. In J.C. Hooghart and C.W.S. Posthumus (eds), *Hydrochemistry and energy storage in aquifers.* TNO-CHO Delft, 105-124.

Willemsen, A. and Appelo, C.A.J., 1985, Chemical reactions during heat storage in shallow aquifers in the Netherlands: laboratory experiments and geochemical modelling. Int. Ass. of Hydrogeologists, *Hydrology in the services of man*, part 4: energy sources and groundwater control, 68-78.

Wolery, T.J., 1983, EQ3NR: a computer program for geochemical aqueous speciation-solubility calculations. Lawrence Livermore Lab., California.

CHAPTER 11

Aspects of flushing and aquifer clean-up

The clean-up of polluted aquifers is a very costly affair and yet necessary to prevent present-day and future hazards. The hazards are the pollution of private or communal water wells or the seepage of polluted groundwater into surface water where it can lead to directly visible environmental effects. Aquifer remediation may take a long time, since the pore volume of the aquifer must be flushed a number of times to remove both solute and adsorbed pollutant. We consider in this chapter organic or inorganic pollutants in solution in water, and adsorbed to the solid aquifer material, e.g. heavy metals and hydrophobic organic chemicals. Oil, gasoline, dense or light fluids that are immiscible with water, are excluded, not because these are unimportant as pollutants, but because these require a hydraulic treatment (two-phase flow, or even three-phase under unsaturated conditions, cf. Domenico and Schwartz, 1990) or bioremediation, rather than the geochemical treatment we are investigating here.

There are a few aspects which are important in aquifer clean-up. The first is to try and enhance desorption as much as possible, since normally most of the chemical is bound to the solid, and must be made soluble for removal. If desorption is enhanced by a factor of 10 at the final concentration to be reached, flushing can be accelerated by a factor of 10. We will reconsider the retardation concept, and find that the slope of the adsorption or exchange isotherm at the final concentration determines the number of pore volumes that must flush the aquifer. This theory is an *equilibrium* theory, which means that solute and adsorbed concentrations are related according to an adsorption or desorption isotherm. The second aspect we must treat is deviation from equilibrium, either due to slow diffusion from stagnant zones (called *physical nonequilibrium*), or due to slow kinetics of the desorption reaction (called *chemical nonequilibrium*).

11.1 FLUSHING FACTORS FROM DESORPTION ISOTHERMS

The distribution coefficient was introduced in Chapter 9 to relate adsorbed and solute concentration. The pristine sediment adsorbs the ion (element, molecule) from the solution according to the relation

$$q = K_d \cdot C$$

Here, q is the sorbed concentration (mass/l pore water), K_d is the distribution coefficient (–), and C is the solute concentration (mass/l pore water). The distribution coefficient determines the retardation with respect to a conservative ion (Cl⁻, for example). The retardation factor $R = 1 + K_d$ indicates that the sediment needs to be percolated 'K_d' pore volumes more for breakthrough of the adsorbed ion than of the conservative ion (Section 9.2). The relation similarly determines the retardation during flushing: a conservative element is flushed after 1 pore volume, the adsorbed ion after $(1 + K_d)$ pore volumes. Acceleration of flushing is obtained when the distribution coefficient, or the isotherm slope, can be lowered, and we must therefore study the influence of the isotherm slope on flushing times.

The change of solute concentration at some point x has been found in Chapter 9 to be:

$$\left(\frac{\partial C}{\partial t}\right)_x = -v\left(\frac{\partial C}{\partial x}\right)_t - \left(\frac{\partial q}{\partial t}\right)_x \tag{9.25}$$

We can write the change in sorbed concentration (the sink term $\partial q/\partial t$) as a differential relation between (sorbed) q, and (solute) C:

$$\left(\frac{\partial q}{\partial t}\right)_x = \frac{dq}{dC}\left(\frac{\partial C}{\partial t}\right)_x \tag{11.1}$$

which gives in (9.25):

$$\left(1 + \frac{dq}{dC}\right)\left(\frac{\partial C}{\partial t}\right)_x = -v\left(\frac{\partial C}{\partial x}\right)_t \tag{9.29}$$

Implicit differentiation of $C_{x,t}$ = constant, gives, as in Chapter 9:

$$\left(\frac{\partial C}{\partial t}\right)_x = -\left(\frac{\partial C}{\partial x}\right)_t\left(\frac{\partial x}{\partial t}\right)_C$$

Insert this relation in Equation (9.29), divide both sides of the equation by $(\partial C/\partial x)$ and rearrange to obtain the transport velocity $v_{C_i} = (\partial x/\partial t)_C$ of a concentration C of i as:

$$v_{C_i} = \frac{v}{1 + \dfrac{dq}{dC}} \tag{11.2}$$

where v is the pore water flow velocity.

If we take a laboratory column or a flowtube in an aquifer of length L, we can relate the pore water flow velocity and the transport velocity of C_i to the number of pore volumes which must flush the column before this concentration arrives at the column outlet. After one pore volume has been injected, a step change in concentration of a conservative element in the injection fluid arrives at the outlet. A retarded element, being transported more slowly through the column, needs proportionally more pore volumes before arrival. Hence the relation is:

$$\frac{V_{C_i}}{V_0} = \frac{v}{v_{C_i}} \tag{11.3}$$

where V_0 is the pore volume (m³) of the column with length L, and V_{C_i} the volume (m³) which must flush the column for C_i to arrive at L. We can combine Equations (11.2) and (11.3) to obtain:

$$V^*_{C_i} = \frac{V_{C_i}}{V_0} - 1 = \frac{dq}{dC} \tag{11.4}$$

The variable V^* has been derived before, and is known as the Ψ-condition of Sillén (1951), or the throughput parameter T of chemical engineers (Vermeulen et al., 1984).

Equation (11.4) indicates that a simple relationship exists between the slope of the desorption isotherm at concentration C_i, and the number of pore volumes which must flush the column to arrive at that concentration: a change in sorbed amount δq is transported as δC in solution and mass balance requires that $\delta q = \delta C \cdot V^*$. The parameter V^* is thus appropriately termed a *flushing factor*. A desorption isotherm which is convex upwards (e.g. Freundlich equation $q = K_F C^n$, with an exponent $n < 1$) has a larger (dq/dC) at lower concentrations and lower concentrations need therefore progressively more pore volumes for elution. A broadening elution curve (non-sharpening front) is the result. Figure 11.1 compares the elution curves for a linear isotherm and for a curved isotherm, and demonstrates their construction.

The conservative element shown in Figure 11.1 exists in solution only, it is not supplied by the solid matter in the column. Such a substance is depleted after displacement of one pore volume of the column and its concentration falls then to zero. Substance A, which for all concentrations has a distribution coefficient of 0.6, has also a flushing factor of 0.6 for all concentrations. The *solute* mass of substance A is eluted after one pore volume, but the solid material furnishes an additional $6/10$ of the original solute mass. Hence 0.6 additional pore volumes are needed to flush the *sorbed* part of A from the column. The solute concentration remains at C_{A_1} during flushing of the sorbed mass, because the sorbed concentration is in equilibrium with that solute concentration. Substance B has a convex isotherm, and the slope of the isotherm at a concentration dictates when that concentration arrives at the outlet. This is illustrated for concentration C_{B_1} which has slope 0.5, and for concentration C_{B_2} with slope 1.2. It is important to note that C_{B_1} arrives earlier at the outlet

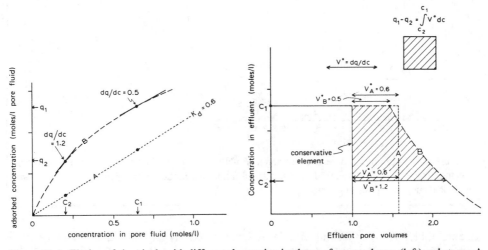

Figure 11.1. Elution of chemicals with different desorption isotherms from a column. (left): substance A with linear desorption isotherm, B with convex desorption isotherm. (right): Isotherm effects on breakthrough curves. Desorption isotherm can be obtained by graphical integration as shown for B.

than C_{A_1}, even though the distribution coefficient $K_d = q_1/C_1$ is larger. It must be realized that the distribution coefficient only determines the retardation in the case of linear distribution, i.e. when the distribution coefficient is equal to the slope of the isotherm.

EXAMPLE 11.1: *Analytical modeling of column elution*
A 4 cm long sediment column has been spiked with 37 µg/l β-HCH. The sorption isotherm is $q = 7 \cdot C^{0.7}$. Breakthrough of Cl⁻ gives a dispersivity of 3.4 mm. Estimate the elution of β-HCH with respect to conservative breakthrough.

ANSWER
Dispersion is neglected in the flushing factor theory. An approximate elution curve can nevertheless be estimated by relating V^* to a conservative breakthrough curve with the usual sigmoidal shape found in experiments, instead of to the ideal sharp front of plug flow. The flushing factor is:

$$V^* = \frac{V}{V_0} - V_{cons} = \frac{dq}{dC} = 4.9\,C^{-0.3}$$

where V_{cons} indicates the breakthrough pore volume of a conservative substance at the same relative concentration as the retarded chemical. The breakthrough of HCH is obtained from: $V_{HCH} = V^* + V_{cons}$. Figure 11.2 compares the calculated breakthrough curve with numerical model results obtained with the same sorption coefficients using the program of Example 9.14, and also gives measured β-HCH concentrations in a column experiment (Appelo et al., 1991). The analytical approximation is calculated for the points of the conservative curve given by Cl⁻ analyses. A model fit of conservative breakthrough with a dispersivity of 3.4 mm is also plotted. Note that dispersion initially induces a more rapid decrease of concentrations than is found with the analytical formula. Reasonable results are found for relative concentrations smaller than $C/C_0 \sim 0.4$

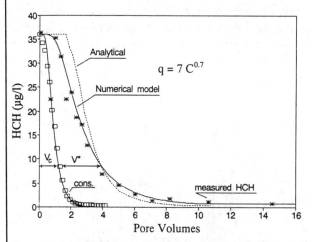

Figure 11.2. Elution of a sediment column spiked with 37 µg β-HCH/l. The conservative curve is obtained from Cl⁻ breakthrough. Experiment and numerical model from Appelo et al., 1991, the analytical curve from flushing factors.

Example 11.1 shows that our simple and very easy hyperbolic equation gives quite acceptable results when the relative concentrations are smaller than $C/C_0 \sim 0.4$. Higher relative concentrations can be found more accurately by taking the ratio of the retarded and the conservative breakthrough pore volumes, rather than the sum of the two (Griffioen et al., 1992). However, we would in general use a numerical model that includes diffusion and dispersion to calculate the breakthrough of such a laboratory column, and reserve the flushing factor theory to obtain a rapid insight, for example into what might occur in a polluted aquifer. The distribution of aquifer properties and of pollutant concentrations in different parts of the aquifer is normally not well-known, and we must have an understanding of the transport processes involved in clean-up to be able to assess the effects of heterogeneity. Our example indicates that the transport principles are understood sufficiently well to give quite reasonable 'first estimates'.

EXAMPLE 11.2: *Elution of TCE from an aquifer*
An aquifer has been polluted by TCE (trichloroethylene). Concentration in groundwater is 10 mg/l. How many pore volumes need to be flushed through the aquifer before the TCE concentration is below 1 mg/l?

ANSWER
A sorption isotherm is needed to relate TCE concentration in water and adsorbed on aquifer sediment. Field measurements of TCE in water and sorbed TCE at the same spot can be combined to a sorption isotherm. Mehran et al. (1987) have reported a number of such measurements, shown in Figure 11.3. The sorbed concentrations have been recalculated to mg/l pore water, by multiplying 'μg/g soil' with an estimated $\rho_b/\varepsilon = 6$. Note the large scatter in this plot, which seems unrelated to sediment properties.

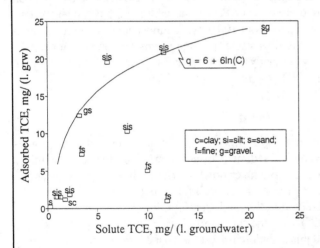

Figure 11.3. Adsorbed and solute TCE concentrations in an aquifer (data from Mehran et al., 1987)

An approximate sorption isotherm is plotted in Figure 11.3:

$$q = 6 + 6\ln(C)$$

which can be differentiated to give V^*:

$$V^* = \frac{dq}{dC} = \frac{6}{C_{tri}}$$

and $V^*_{(C = 1)} = 6$. Hence $1 + 6 = 7$ pore volumes need to be pumped from the area to lower TCE concentrations in pumped water to 1 mg/l. Use of a linear isotherm, with $K_d \approx 1.4$, would result in a much lower estimate of needed flushing volumes. In this case a pore volume is essentially the volume of polluted water with TCE > 1 mg/l, when clean water is injected, or withdrawn from the neighboring aquifer.

11.1.1 *Desorption isotherms from a single elution curve*

A single elution experiment is sufficient to obtain a desorption isotherm. The change of q from C_1 to C_2 can be obtained from Equation (11.4):

$$\int_{q_1}^{q_2} dq = q_2 - q_1 = \int_{C_1}^{C_2} V^* \, dC \tag{11.5}$$

which can provide q_2 (when q_1 is known) at C_2, by integrating the area of the breakthrough curve. With q known at each C, the full desorption isotherm is obtained easily. The integral is illustrated for substance B in Figure 11.1 (right). It should be noted that the integration is performed along the vertical axis, as this axis represents the solute concentration C. Also note that the integral is negative when performed from C_1 (high) to C_2 (low).

The application of this simple and attractive technique requires that equilibrium exists between solute and adsorbed concentration, and also that dispersion is small with respect to spreading caused by the increase in slope of the desorption isotherm (Grifioen et al., 1992; cf. Example 11.1). It may be noted that Equation (11.4) is based on conservation of mass at the outlet ($x = L$) only, and remains valid irrespective of possible non-equilibrium further upstream in the column. However, equilibrium is necessary to obtain the equilibrium desorption isotherm. Equilibrium is expected when flow velocities are low; too low velocities lead to diffusion in the breakthrough curve, and give a time dependent broadening of breakthrough of the desorbing substance. The lowest flow velocity where dispersion is still the major front broadening process is at a Peclet number of ≈ 0.5 (Chapter 9). This means that flow velocities from 40 up to about 100 m/yr are most appropriate for column elution experiments. Fortunately, this flow velocity range seems low enough for equilibrium to be reached for ion exchange in sandy (non aggregated) sediments (James & Rubin, 1979; Valocchi, 1985; cf. Section 11.3).

11.1.2 *C(x) and C(V) profiles and sharp fronts*

The use of the flushing factor equation also allows the calculation of concentration profiles within a column or along the flowpath in a sediment, by considering that the pore volume V_0 is a linear function of place x along the column or flowline. Hence, if we express distance x in terms of pore volume, we find $V/x = V^* = (dq/dC)$ for a concentration C_i, with V the injected volume. This relation holds for all x, and therefore, in a $V - x$ diagram we find simple linear relations $V = (dq/dC) \cdot x$ as the characteristic equation for the propagation of a concentration C_i. Sillén derived this equation in 1951, and used it in an enlightening way to solve ion exchange problems in the elution curves from columns. Figure 11.4, from Sillén's work, shows how a $C - x$ profile is obtained at given time t from the desorption isotherm and a $V - x$ diagram. The dotted line in Figure 11.4b indicates a constant time. The line has a slope in the V/x diagram of $-V_0$, the pore volume per unit length. Constant time is not simply given by a parallel to the x-axis (i.e. time is not presented directly by V) because we consider only flushing of the chemical's sorbed mass: one pore volume has passed already to flush the solute mass upstream from each point x. However, we could have increased the slope V^* with 1 (as was done for Figure 5.13), and in that case constant time would be indicated by constant V. The procedure is similar to the method of characteristics used by Charbeneau and coworkers (1982, 1988) for ion exchange in a V/V_0-diagram (x expressed in V_0, as in Equation 11.4).

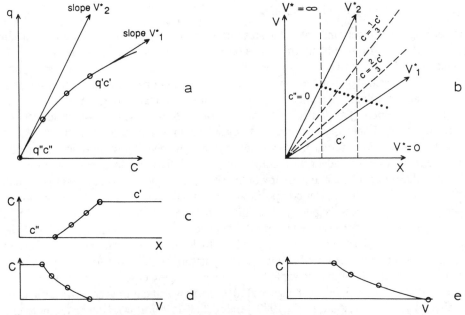

Figure 11.4. Desorption of a single solute. Initial state: q', C'. Final state: $q'' = C'' = 0$. V^* is the slope of the desorption isotherm which gives the characteristic velocity of a given concentration through the column. (a) Sorption isotherm $q(c)$; (b) $V - x$ diagram. Thick lines with $V/x = 0$, V_1^*, V_2^*, and ∞ are drawn. V_1^* and V_2^* are the slopes of the tangents in (a). On the broken line through the origin, $C = \frac{1}{3}C'$ and $\frac{2}{3}C'$; (c) $C(x)$ for a given time (dotted line in (b); this line has a slope of $V/x = -V_0$, the pore volume per unit length of x; (d) and (e) $C(V)$ for two different x (broken vertical lines in (b)). The circles are at $C = 0$, $\frac{1}{3}C'$, $\frac{2}{3}C'$ and C'. (Sillén, 1951).

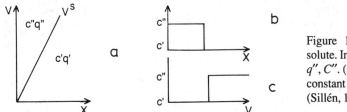

Figure 11.5. Sorption of a single solute. Initial state: q', C'. Final state: q'', C''. (a) V-x diagram; (b) $C(x)$ for constant t; (c) $C(V)$ for a given x. (Sillén, 1951).

It is also possible to use the diagrams for sharp fronts. Sharp fronts will develop during adsorption of a pollutant with a convex sorption isotherm. In that case dq/dC decreases as C_i increases, which would imply faster breakthrough (at the end of the column) of higher than of lower concentrations. This is impossible, of course, and only a sharp transition occurs with

$$V^s = \frac{\Delta q}{\Delta C} = \frac{(q'' - q')}{(C'' - C')} \tag{11.6}$$

as shown in Figure 11.5. It will be noted that this result is identical to use of the classical retardation factor.

11.1.3 *Acceleration of aquifer clean-up*

The flushing factor theory indicates that an aquifer remediation scheme will improve when the slope of the desorption isotherm can be decreased at the desired end-concentration in groundwater. Therefore we have to look for possibilities to manipulate the desorption isotherm. It has been shown in Section 9.2.1 that the distribution coefficient of hydrophobic chemicals can be calculated from a partitioning of the chemical among water and organic matter in the sediment. The partioning theory describes *ab*sorption as a dissolution process, where the hydrophobic organic chemical dissolves into soil organic matter. Such a dissolution process is different from *ad*sorption, since the penetration into organic matter may take more time than adsorption onto a surface and may thus lead to nonequilibrium behavior (cf. Section 11.3), and enthalpy of the absorption reaction is expected to be small or zero (Chiou, 1989). The water composition also has only a minor effect on absorption or desorption if the process is purely dissolution into organic matter (Karickhof, 1984). Desorption of cations or heavy metals can be enhanced by acids or salts, which replace the cations, but such replacement is of no use with a hydrophobic organic chemical. Current research in which optimization of aquifer clean-up is sought, therefore focuses on increasing the water solubility of the hydrophobic chemical by using cosolvents in water such as alcohols (Rao et al., 1985; Nkedi-Kizza et al., 1989), or by using humic acids or surfactants (Madhun et al., 1986; Enfield et al., 1989; Abdul et al, 1990). It is of interest to note that desorption through surfactants is enhanced especially when the added surfactant forms clusters (micelles), which indicates again that dissolution into organic matter occurs, but now into a form that can be transported as a colloid (Kile et al., 1990).

The stability of colloidal suspensions is well known to be a function of cation composition and salt concentration (Chapter 5). Monovalent cations such as Na^+ are effective in

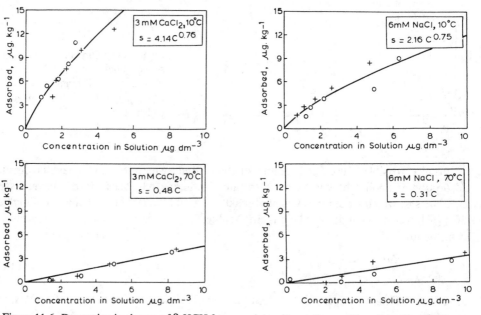

Figure 11.6. Desorption isotherms of β-HCH from sandy aquifer sediments (Appelo et al., 1990).

increasing the double layer repulsion, which is the major contributing factor to obtain stable peptized suspensions. The cation composition of the solution can indeed influence the measured distribution coefficient as shown in Figure 11.6 for HCH desorption from a sandy sediment (Appelo et al., 1990). The apparently lower distribution coefficient (i.e. the ratio of adsorbed over solute concentration in Figure 11.6) in NaCl-solutions can be attributed to peptization of organic matter. This process manifests itself in a green coloring of the NaCl solutions. Note also that an increase of temperature may enhance desorption of HCH's even more and injection of hot water can therefore be a profitable way to accelerate aquifer clean-up. Since injecting hot water may also induce calcite precipitation, there is an argument to combine heating and Na-treatment.

The Na-effect observed for HCH-desorption seems related to the so-called particle effect, i.e. the observation that K_d in laboratory batch studies declines when the ratio of solid material over water increases (Figure 11.7 from O'Connor and Connolly, 1980; Voice et al., 1983). The effect has been explained as resulting from increased concentrations of suspended solids which have adsorbed the hydrophobic chemical and are incompletely removed before analysis of the solution, e.g. by extraction in hexane (Chiou et al., 1984; Gschwend and Wu, 1985). The effect might be used to enhance clean-up schemes by adding humic acids as a carrier to water (Madhun et al., 1986; Abdul et al., 1990; Enfield et al., 1989). However, the conditions for formation and transport of suspended mineral and organic matter in natural aquifers should be evaluated carefully, so that the added humic acids remain stable in solution. It may well be that the Na$^+$ ion has an essential role here.

Use of a single K_{oc}, assumes that all natural organic matter can be lumped together. Absorption behavior of different types of organic matter is not well known, but differences of an order of magnitude have been observed among various humic and fulvic acids

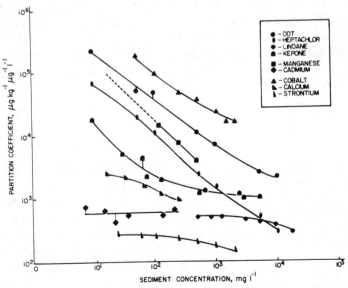

Figure 11.7. The particle effect, or as the ratio of solid material over water increases, lower distribution coefficients for chemicals are found. Reprinted with permission from O'Connor and Connolly, 1980, Copyright Pergamon Press PLC.

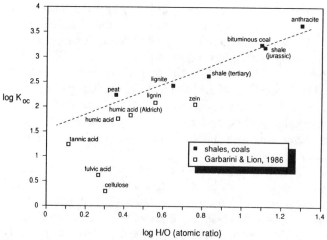

Figure 11.8. The distribution coefficient K_{oc} of trichloroethylene (TCE) increases with increasing H/O ratio. Reprinted with permission from Grathwohl, 1990, Copyright American Chemical Society.

(Garbarini and Lion, 1986). Grathwohl (1990) investigated organic matter from sediments with different ages, and a different degree of maturing or ripening, ranging from recent organic matter to precursors of oil. He found that a higher H/O-ratio in the organic matter gives an increase in the distribution coefficient for sorption of hydrophobic organic chemicals (Figure 11.8). This effect can be associated with lower polarity and higher hydrophobicity of organic matter as the H/O ratio increases in more mature organic matter. Recently it has been observed that an increase of pH lessened the sorption of naphthalene (Kan and Tomson, 1990), and this effect was linked to increases in the surface charge of humic acid at higher pH, and a resulting higher polarity. Such an effect can be used in a clean-up scheme by increasing the pH of the flushing water. Note however, that all manipulations induce chemical or physical changes in the aquifer which almost certainly implies that buffering reactions from water-rock interactions become operative. These are new techniques for which practical experience must still be gained.

11.2 ION EXCHANGE

For exchanging ions the same principles apply as for adsorption in transport calculations. If only two ions are involved the analogy is (almost) complete, because the concentration of one cation can be expressed as a function of the other when fractions are used, provided that the total solute concentration as well as the cation exchange capacity remain constant. For multicomponent exchange slightly more complicated algebra may be necessary, although the principles are identical. It is advantageous to scale the ion exchange equation by introducing fractions:

$$\alpha_i = z_i \cdot m_i / A_0 \quad \text{and} \quad \beta_i = q_i / CEC \tag{11.7}$$

where z_i is charge, m_i is mol/l, A_0 is total concentration of exchanging ions in solution (eq/l), and CEC is exchange capacity of the sorbent (which for simplicity is also expressed in eq/l pore water). The use of fractions has as consequence that:

$$\Sigma \alpha_i = 1 \quad \text{and} \quad \Sigma \beta_i = 1$$

Hence with binary exchange between i and j we may express the solute activity $[j] \approx m_j = (1 - \alpha_i) A_0 / z_j$. If we consider here monovalent ions ($z_i = z_j = 1$), the exchange reaction

$$i + j\text{-X} \leftrightarrow i\text{-X} + j \tag{11.8}$$

gives a mass action quotient:

$$K_{i \backslash j} = \frac{[j] \cdot [i\text{-X}]}{[i] \cdot [j\text{-X}]} = \frac{(1 - \alpha_i) \cdot A_0 \cdot \beta_i \cdot CEC}{\alpha_i \cdot A_0 \cdot (1 - \beta_i) \cdot CEC}$$

$$= \frac{(1 - \alpha_i) \cdot \beta_i}{\alpha_i \cdot (1 - \beta_i)} \tag{11.9}$$

Conservation of matter for ion i requires:

$$A_0 \left(\frac{\partial \alpha_i}{\partial t}\right)_x = - v \cdot A_0 \left(\frac{\partial \alpha_i}{\partial x}\right)_t - CEC \left(\frac{\partial \beta_i}{\partial t}\right)_x$$

We have

$$v_{\alpha_i} = \left(\frac{\partial x}{\partial t}\right)_{\alpha_i}$$

and obtain similar to Equation (11.2) the velocity of fraction α of i as:

$$v_{\alpha_i} = \frac{v}{1 + \dfrac{CEC}{A_0} \dfrac{d\beta_i}{d\alpha_i}} \tag{11.10}$$

The flow velocity of the fraction α_i is expressed relative to pore water flow velocity v, and hence in terms of pore volumes which must flush the column:

$$V^*_{\alpha_i} = \frac{V_{\alpha_i}}{V_0} - 1 = \frac{CEC}{A_0} \frac{d\beta_i}{d\alpha_i} \tag{11.11}$$

Figure 11.9 shows how the elution curve is constructed when an ion is displaced by a less tightly bound ion (unfavorable exchange). The number of pore volumes V^* needed to flush the column until fraction α_i is reached is found from the slope of the adsorption isotherm at α, multiplied by the ratio CEC/A_0.

Just as for desorption, it is possible to obtain a complete exchange isotherm from a single unfavorable exchange elution ('j' less tightly held than B, C or A). For example, the fraction of ion B, while eluted and removed from the exchanger is given by:

$$\beta_{B_2} - \beta_{B_1} = \frac{A_0}{CEC} \int_{\alpha_{B_1}}^{\alpha_{B_2}} V^* \, d\alpha \tag{11.12}$$

which allows the calculation of β_{B_2}. The exchange coefficient $K_{B \backslash j}$ for α_{B_2} is then obtained from (for homovalent exchange):

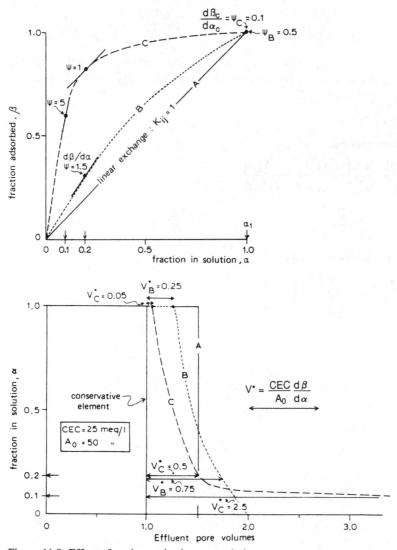

Figure 11.9. Effect of exchange isotherm on elution curve. (upper): Ions A, B and C have given exchange isotherm with ion *j*. Ψ indicates the slope dβ/dα. (lower): Elution of A, B and C can be constructed from the exchange isotherm (Or alternatively, the exchange isotherm can be obtained from the elution curve, see text).

$$K_{Bj} = \frac{\beta_{B_2} \cdot \alpha_{j_2}}{\beta_{j_2} \cdot \alpha_{B_2}} = \frac{\beta_{B_2} \cdot (1 - \alpha_{B_2})}{(1 - \beta_{B_2}) \cdot \alpha_{B_2}} \tag{11.13}$$

The exchange isotherms in Figure 11.9 show a stronger preference for C than for B, both with respect to the same eluting '*j*'. This has the result that initial flushing of C is faster than of B (the slope at high α is smaller for C than for B). Total surface under the curves B or C relative to a conservative element must remain equal, as this represents the ratio CEC/A_0. Small fractions of C therefore need much longer flushing.

The effect of an increase of A_0 (total concentration in flushing solution) and of *CEC* (total exchangeable ions) is immediately clear. A doubling of A_0, halves the volume which is needed to flush down to the desired concentration, etc. Another interesting feature illustrated in Figure 11.9, is that a low concentration of the exchanged ion (which needs to be flushed) can be reached with fewer pore volumes when the exchange is more 'unfavorable'. This principle is illustrated in Example 11.4, but first we will show the applicability of our formulas with a simple example (Example 11.3).

EXAMPLE 11.3: *Flushing of K^+ from a column*

Van Eijkeren and Loch (1984) packed a laboratory column with cation exchange resin beads. The column was equilibrated with 0.3 meq KCl/l, then flushed with 0.3 meq NaCl/l. Column length $L = 6.9$ cm; water flux = 56.7 cm/day; Dispersion coefficient $D_L = 11.3$ cm^2/day. Experimental points are shown in Figure 11.10. Let us model these data with our Equation (11.11).

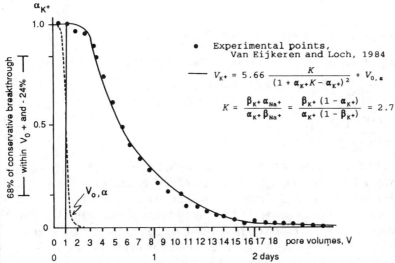

Figure 11.10. Potassium ions are exchanged from a laboratory column by sodium. Dots: relative potassium concentrations from experiment. Line: modeled results. (Experimental points from Van Eijkeren and Loch, 1984).

A small dispersion coefficient yields a variance $\sigma^2 = 2D_L t$ of concentrations around midpoint breakthrough. Flowtime for the column is $t = 6.9/56.7 = 0.12$ days. Hence $\sigma = \sqrt{2D_L t} = 1.67$ cm. Or, scaled to column length, $\sigma = 1.67 / L = 1.67/6.9 = 0.24$. For a conservative element, this σ indicates that + and − 34% of midpoint concentration ($C/C_0 = 0.5$), is eluted with the column pore volume V_0, + and − 24%. The (approximate) conservative breakthrough curve can thus be constructed analogous to the procedures outlined in Chapter 9, with $C/C_0 = 0.84$ at $V = 0.76\ V_0$; $C/C_0 = 0.5$ at $V = V_0$; $C/C_0 = 0.16$ at $V = 1.24\ V_0$; etc.; V = volume injected in the column. The conservative breakthrough is plotted in Figure 11.10 (line $V_{0,\alpha}$).

The next step is to obtain *CEC* of the column by integrating the area between conservative breakthrough and K^+ experimental points. This gives $CEC/A_0 = 5.66$. For homovalent exchange the slope of the exchange isotherm can be obtained from

$$\beta_i = \frac{K_{i \setminus j} \cdot \alpha_i}{1 - \alpha_i + K_{i \setminus j} \cdot \alpha_i}$$

and

$$\frac{d\beta_i}{d\alpha_i} = \frac{K_{i\backslash j}}{(1 - \alpha_i + K_{i\backslash j} \cdot \alpha_i)^2}$$

where it is assumed that $K_{i\backslash j}$ is constant. From Equation (11.11), $V^* = CEC/A_0 \cdot (d\beta/d\alpha)$, where V^* is the number of pore volumes that a given fraction of exchanged element arrives later than the same fraction of a conservative ion. V^* can be read for a few fractions from the plot of experimental points, and $K_{i\backslash j}$ is solved for these fractions. An average coefficient $K_{i\backslash j}$ can be calculated, which in this case is an almost constant $K_{K\backslash Na} = 2.7$. All variables have been found now, and the elution curve can be drawn as shown in Figure 11.10.

It must be noted that Van Eijkeren and Loch modeled the experimental data with a rather more intricate model, incorporating mobile and immobile fractions. They used a lower $K_{K\backslash Na} = 1.7$ obtained from batch experiments. It is interesting that the same discrepancy between K's from batch experiments and a flow experiment was found by Rainwater et al. (1987), who used essentially the same theory as given here. Rainwater et al. modeled a laboratory sand box with injection and withdrawal wells.

11.2.1 *Concentration effects in heterovalent exchange*

The ion exchange reaction can be written in generalized form for ions of any charge as:

$$1/j \cdot J^{j+} + 1/i \cdot I\text{-}X_i \leftrightarrow 1/j \cdot J\text{-}X_j + 1/i \cdot I^{i+} \qquad (11.14)$$

with

$$K_{J\backslash I} = \frac{[I^{i+}]^{1/i} [J\text{-}X_j]^{1/j}}{[J^{j+}]^{1/j} [I\text{-}X_i]^{1/i}} \qquad (11.15)$$

where $K_{J\backslash I}$ is the exchange constant when J^{j+} replaces I^{i+} on the exchanger, i is charge of ion I^{i+}, $[I^{i+}]$ is activity of I^{i+} in solution, and $[I\text{-}X_i]$ is equivalent fraction of I on the adsorber.

This equation uses the Gaines-Thomas-convention for the exchange reaction (cf. Chapter 5). A consequence of Equation (11.15) is that dilution of the solution will force the higher charged ions to enter the cation exchange complex (i.e. to become adsorbed on soil or sediment). The extent of favoring is equal to the amount of dilution raised to the inverse power of the charge ratio. For example for Na-Ca exchange:

$$Na^+ + \tfrac{1}{2}\,Ca\text{-}X_2 \leftrightarrow Na\text{-}X + \tfrac{1}{2}\,Ca^{2+}$$

$$K_{Na\backslash Ca} = \frac{[Na\text{-}X]\,[Ca^{2+}]^{\frac{1}{2}}}{[Ca\text{-}X_2]^{\frac{1}{2}}\,[Na^+]}$$

When Na^+ and Ca^{2+} activities in solution are both 1 (if we neglect activity corrections, concentrations are 1 mol/l), the ratio of exchangeable Na^+ over $\sqrt{Ca^{2+}}$ will be ca. 0.4. When both cations have an aqueous activity of 0.001 (concentrations are 1 mmol/l) the ratio of exchangeable Na^+ over $\sqrt{Ca^{2+}}$ will be $0.4\,\sqrt{0.001} = 0.013$. This is the *concentration effect of heterovalent exchange*. It can be used to advantage for desorption of heavy metals in an aquifer clean-up scheme (Example 11.4).

EXAMPLE 11.4: *Flushing of Cd^{2+} from a sediment with NaCl solution*
An aquifer has been polluted by a $CdCl_2$ spill. It is suggested that the adsorbed Cd^{2+} be removed with water to which road salt (NaCl) has been added. Find the number of pore volumes needed to lower the

Cd^{2+} concentrations in water to 2.5 mmol/l when total concentration of NaCl in the flushing solution is 1, 0.1 and 0.01 eq/l. Cation exchange capacity (*CEC*) = 10 meq/100g; bulk density = 2 g/cm³; porosity = 0.2. *CEC* is initially completely occupied by Cd^{2+}.

ANSWER
The flushing of all Cd^{2+} from the aquifer, both in water and adsorbed, clearly requires that the concentration of NaCl in the flushing solution is as high as possible (A_0 in Equation (11.11) must be high). However, when only a short term remedial action is desired, and the concentration of Cd^{2+} in water needs to be lowered rapidly, it can be advantageous to use a low concentration in the flushing solution for the following reasons:
– Cd^{2+} forms a fraction of total solutes, hence the absolute concentration of Cd^{2+} decreases with total concentration;
– a fixed desired end-concentration forms a larger fraction of the lower total solute concentration, which is always connected with smaller $d\beta/d\alpha$;
– because of uni-/bivalent exchange, the exchange isotherm is more convex when total solute content is lower: hence a lower Cd^{2+} concentration in solution can be in equilibrium with a higher Cd^{2+} concentration in the soil.
Exchange isotherms for Cd-Na exchange at different concentrations are obtained from the reaction:

$$Na^+ + \tfrac{1}{2} Cd\text{-}X_2 \leftrightarrow Na\text{-}X + \tfrac{1}{2} Cd^{2+}$$

with an exchange equilibrium:

$$K_{Na\backslash Cd} = \frac{[Na\text{-}X] \cdot [Cd^{2+}]^{\frac{1}{2}}}{[Cd\text{-}X_2]^{\frac{1}{2}} \cdot [Na^+]} = \frac{(1 - \beta_{Cd}) \cdot (\alpha_{Cd}/2 \cdot A_0)^{\frac{1}{2}}}{\beta_{Cd}^{\frac{1}{2}} \cdot (1 - \alpha_{Cd}) \cdot A_0} = 0.3$$

where α_{Cd} is equivalent-fraction of Cd^{2+} in solution, and A_0 is normality of the solution.

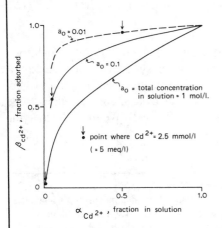

Figure 11.11. Isotherms for Na-Cd exchange at 3 different normalities.

Figure 11.11 gives the exhange isotherms for the 3 concentration levels. The concentration of Cd^{2+} = 2.5 mmol/l is indicated on the curves; the slope of the exchange isotherm gives $V^* = CEC/A_0 \cdot d\beta/d\alpha$, and can be obtained by differentiation:

A_0	$d\beta/d\alpha$	V^* [1]
1	3.0	3
0.1	3.74	37
0.01	0.11	11

[1] $V^* = d\beta/d\alpha \cdot CEC/A_0, CEC = 1$ eq/l

which shows that the solution with 0.1 mol/l NaCl needs far more pore volumes to arrive at Cd^{2+} = 2.5 mmol/l than either a 1 mol/l, or a 0.01 mol/l solution of NaCl. Actually, an aquifer clean-up for Cd^{2+} has been performed in The Netherlands by flushing with acid to a pH of 4 (Urlings et al. (1988). The use of acid has the advantage that adsorption to oxides is diminished, but can be applied only in a calcite-free sediment.

11.2.2 *Multicomponent exchange (sharp fronts)*

Multicomponent exchange requires that the flushing factors $V^* = CEC/A_0 \cdot d\beta/d\alpha$ are equal for all ions which participate in the exchange process. The solution of this problem, which involves integration along eigenvectors of the Jacobian matrix formed by $\{d\beta_i/d\alpha_j\}$, is possible by numerical integration (Charbeneau, 1988). Analytical solutions are available for homovalent multicomponent ion exchange, but they are outside the scope of the present treatment, and the interested reader should consult the work of Helfferich and Klein, 1970; Rhee et al., 1970; Pope et al., 1978; Hwang et al., 1988, and the recent book by Rhee et al., 1989.

A considerable simplification is possible for the case of favorable exchange, and may serve to illustrate one feature of the multicomponent interactions. Favorable exchange means that the incoming ions all form sharp fronts with

$$V^s = \frac{CEC}{A_0} \frac{\Delta\beta}{\Delta\alpha}$$

Note how this equation is again simply a mass balance: it relates a change in adsorbed amounts $CEC \cdot \Delta\beta$ to a change in solute amounts $A_0 \cdot \Delta\alpha$ by the flushing factor V^s:

$$CEC \cdot \Delta\beta = V^s \cdot A_0 \cdot \Delta\alpha$$

The number of generated sharp fronts is equal to the number of ions $n - 1$. Between the two fronts (transitions) a plateau with invariant concentrations is eluted, in which the equilibrium relations ((11.9), or similar) are valid. It is possible to obtain a set of equations which allows the complete sequence with sharp fronts as well as plateau concentrations to be calculated, irrespective of charge of the ions (Appelo et al., 1993). We may illustrate the procedure with an example.

EXAMPLE 11.5: *Retardation of a K^+, Rb^+, Cs^+ mixture*
The exchange strength increases within the series of alkaline and earth-alkaline elements when the elements become heavier (this is a result of smaller hydration energies). The heavier elements are often part of nuclear waste and may give environmental problems when released. Assume that a compacted clay is initially in equilibrium with NaCl-water, and that a mixture of K, Rb and Cs-Cl water is spilled. The exchange equations are:

$$K^+ + Na\text{-}X \leftrightarrow K\text{-}X + Na^+ \qquad K_{K\backslash Na} = 5$$

$$Rb^+ + Na\text{-}X \leftrightarrow Rb\text{-}X + Na^+ \qquad K_{Rb\backslash Na} = 10$$

$$Cs^+ + Na\text{-}X \leftrightarrow Cs\text{-}X + Na^+ \qquad K_{Cs\backslash Na} = 12.5$$

The equations are arranged in order of increasing exchange strength of the ion. The exchange capacity of the clay is 10 meq/100 g, bulk density = 2.0 g/cm³, porosity = 0.2. The concentrations in the spill are $K'' = 4.0$, $Rb'' = 5.5$, and $Cs'' = 0.5$ meq/l. Since these ions are more strongly bonded than Na^+, the spill drives off all Na^+. Within the spill the ions are also separated according to exchange strength, and three sharp fronts develop.

For the first front:

$$(V_K^s)_1 = \frac{CEC}{A_0} \frac{0 - \beta_K}{0 - \alpha_K} = \frac{CEC}{A_0} \frac{0 - 1}{0 - 1} = \frac{CEC}{A_0} \tag{11.16}$$

For the second front:

$$(V_{Rb}^s)_2 = \frac{CEC}{A_0} \frac{0 - \beta_{Rb\,(Rb,\,K)}}{0 - \alpha_{R\,(Rb,\,K)}} \tag{11.17a}$$

$$(V_K^s)_2 = \frac{CEC}{A_0} \frac{1 - \beta_{K\,(Rb,\,K)}}{1 - \alpha_{K\,(Rb,\,K)}} \tag{11.17b}$$

where $\beta_{Rb\,(Rb,\,K)}$ is fraction Rb^+ on exchanger with only Rb^+ and K^+ in solution, $\alpha_{Rb\,(Rb,\,K)}$ is fraction of Rb^+ in solution with only Rb^+, K^+.

And for the third and last front:

$$(V_{Cs}^s)_3 = \frac{CEC}{A_0} \frac{(0 - \beta_{Cs\,(Cs,\,Rb,\,K)})}{(0 - \alpha_{Cs\,(Cs,\,Rb,\,K)})} = \frac{CEC}{A_0} \frac{\beta''_{Cs}}{\alpha''_{Cs}} \tag{11.18a}$$

$$(V_{Rb}^s)_3 = \frac{CEC}{A_0} \frac{(\beta_{Rb\,(Rb,\,K)} - \beta_{Rb\,(Cs,\,Rb,\,K)})}{(\alpha_{Rb\,(Rb,\,K)} - \alpha_{Rb\,(Cs,\,Rb,\,K)})} = \frac{CEC}{A_0} \frac{(\beta_{Rb\,(Rb,\,K)} - \beta''_{Rb})}{(\alpha_{Rb\,(Rb,\,K)} - \alpha''_{Rb})} \tag{11.18b}$$

$$(V_K^s)_3 = \frac{CEC}{A_0} \frac{(\beta_{K\,(Rb,\,K)} - \beta_{K\,(Cs,\,Rb,\,K)})}{(\alpha_{K\,(Rb,\,K)} - \alpha_{K\,(Cs,\,Rb,\,K)})} = \frac{CEC}{A_0} \frac{(\beta_{K\,(Rb,\,K)} - \beta''_K)}{(\alpha_{K\,(Rb,\,K)} - \alpha''_K)} \tag{11.18c}$$

and the theory of coherence (Helfferich and Klein, 1970), maintains that $(V_K^s)_2 = (V_{Rb}^s)_2$, as well as $(V_K^s)_3 = (V_{Rb}^s)_3 = (V_{Cs}^s)_3$.

The different fractions on the exchange sites are a function of the solution composition, and can be obtained from combinations of Equation (11.15) (cf. Chapter 5):

$$\beta_I = \frac{[I^+]}{[I^+] + K_{JI} \cdot [J^+] + K_{KI} \cdot [K^+] + \dots}$$

The fraction of the exchangeable cations ($\beta''_{i,j,\dots}$) which are in equilibrium with solute ions ($\alpha''_{i,j,\dots}$) in the spill-mixture can be calculated with this equation. From β''_{Cs} and α''_{Cs} the value of $(V^s)_3$ is obtained. Further back-substitution of β''_{Rb} and β''_K gives the value for the second front. The terminology here is that we start 'upstream' with our calculation, and work downward to earlier fronts which have passed already the observation well.

For the third (and last) front we find:

$$\alpha''_{Rb} = 0.55 \qquad\qquad\qquad\qquad\qquad\qquad \alpha''_{Cs} = 0.05 \qquad \alpha''_K = 0.4.$$

$$\beta''_{Rb} = \frac{5.5 \cdot 10^{-3}}{5.5 \cdot 10^{-3} + 0.5 \cdot 4 \cdot 10^{-3} + 1.25 \cdot 0.5 \cdot 10^{-3}} = 0.677 \quad \beta''_{Cs} = 0.077 \qquad \beta''_K = 0.246.$$

and from (11.18a):

$$(V_{Cs}^s)_3 = 1.54 \cdot CEC/A_0.$$

The next step is to obtain $\beta_{Rb\,(Rb,\,K)}$ and $\alpha_{Rb\,(Rb,\,K)}$ from a combination of (11.18b), in which $(V_{Rb}^s)_3 = (V_{Cs}^s)_3$, and the exchange equilibrium among K^+ and Rb^+ in the solution which contains only these two ions (i.e. the plateau before the third front):

$$\beta_{Rb\,(Rb,\,K)} = \frac{\alpha_{Rb\,(Rb,\,K)}}{\alpha_{Rb\,(Rb,\,K)} + K_{K\backslash Rb}(1 - \alpha_{Rb\,(Rb,\,K)})} = \frac{\alpha}{\alpha + \frac{1}{2}(1 - \alpha)} = \frac{2\alpha}{\alpha + 1}$$

where in the last part the subscripts have been omitted. It gives in (11.18b):

$$1.54 = \frac{\frac{2\alpha}{\alpha + 1} - 0.677}{\alpha - 0.55}$$

and

$$\alpha_{Rb(Rb, K)} = 0.514, \qquad \beta_{Rb(Rb, K)} = 0.679.$$

With the values for $\beta_{Rb(Rb, K)}$ and $\alpha_{Rb(Rb, K)}$, and Equation (11.17a), the last unknown, $(V^s)_2 = 1.32 \cdot CEC/A_0$ is obtained. Recalculation in terms of relative pore volumes gives:

$$(V^s)_1 = 1 \cdot CEC/A_0 = 100, \qquad (V^s)_2 = 132, \qquad (V^s)_3 = 154$$

as the number of pore volumes after which resp. K^+, $K^+ + Rb^+$, and $K^+ + Rb^+ + Cs^+$ arrive at an observation well behind a conservative element.

It is also possible to construct an $A_0 \cdot V$ versus $CEC \cdot x$ diagram (Figure 11.12). From this diagram a transect over $CEC \cdot x$ at constant t can be obtained to show fractions adsorbed to the clay. A related transect over $A_0 \cdot V$ at constant $CEC \cdot x$ shows the development of concentrations at a distance x in the clay layer.

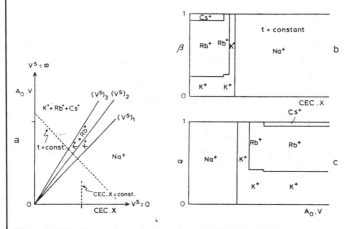

Figure 11.12. Separation of three ions, K^+, Rb^+, Cs^+ in a clay originally saturated with Na^+. (a) $A_0 \cdot V - CEC \cdot x$ diagram; (b) β ($CEC \cdot x$), transect through the clay showing adsorbed fractions at given time; (c) α ($A_0 \cdot V$), concentrations in time which pass point x in the clay.

Self-sharpening fronts occur during an adsorption process with a convex adsorption isotherm and with favorable exchange. This is a result of the fact that low concentrations have a larger K_d than the higher concentrations. Small concentrations which develop in the front due to spreading (dispersion) and run ahead, are therefore checked by adsorption. Of course, the front is never completely sharp, but the combined effect of dispersion and sharpening through cation exchange will tend to balance after some path length (an approximation for large K_d has been given by Bolt, 1982, p.315 a.f.).

11.3 PROCESSES THAT AFFECT SORPTION

There are a few processes which may prevent a straightforward application of simple isotherms under flow conditions. These processes have been summarized in an enlightening way by Van Genuchten and Cleary (Chapter 10 in Bolt, 1982), and a comprehensive summary of the recent literature has been prepared by Brusseau and Rao (1989). The description of sorption during flow is considerably simplified if the *local equilibrium assumption* (LEA) is valid. A high flow velocity sweeps solutes away too rapidly, before homogenization with respect to diffusion and reactions has been attained (James and Rubin, 1979; Valocchi, 1985; Brusseau and Rao, 1989). We have to analyze the reasons for *physical* or *chemical non-equilibrium* as related to flow velocity, and also of *hysteresis* as an additional term that causes deviations from ideality.

11.3.1 *Hysteresis*

The pollutant which shows hysteresis is more tenaciously held by the soil during desorption than during adsorption (or, K_d during desorption is larger than K_d during adsorption). In general, this behavior gives in the Freundlich isotherm relations such as:

$$q = k_a C^{n_a} \quad \text{differs from} \quad q = k_d C^{n_d}$$

where $k_a \neq k_d$, and $n_a \neq n_d$, and the subscripts a and d refer to adsorption and desorption respectively. Particularly organic chemicals are sometimes held with a greater tenacity during desorption, and an example is shown in Figure 11.13. The hysteresis leads to a tailing in the elution curve shown in Figure 11.14.

Hysteresis in soil moisture retention curves is easily explained from physical principles (e.g. the 'ink-bottle' effect explains why more water is held during moisture desorption than during adsorption at the same pF). For adsorption and desorption of chemicals the hysteresis is more difficult to explain, especially from the point of the partitioning theory. Most causes for an *apparent* hysteresis are linked to experimental problems, such as failure to reach equilibrium, losses due to transformations or accumulating errors in successive

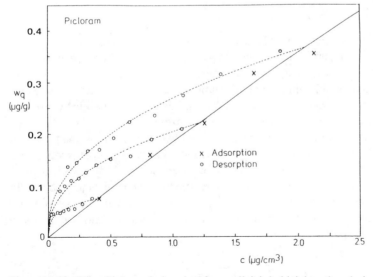

Figure 11.13. When Picloram is desorbed from soil, it is held tighter than during adsorption (hysteresis) (From Van Genuchten and Cleary, 1982).

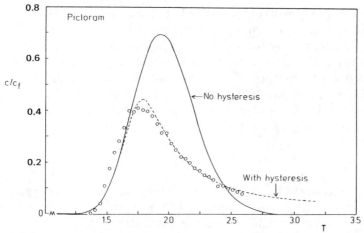

Figure 11.14. Calculated and observed effluent curves for Picloram movement through Norge loam. T is number of pore volumes. C/C_f indicates relative concentration with respect to input concentration. (From Van Genuchten and Cleary, 1982).

extractions (Brusseau and Rao, 1989). It has been found that the desorption of α-HCH from an approximately 20 year old waste, that contained calcite and gypsum, was extremely slow (Rijnaarts et al., 1990). This slow release could be related to an occlusion of the pollutant in precipitated calcite. Such occluded material is only released when the calcite dissolves, and it may be considered as *immobilized* or as *bound residue*, i.e. material which is only liberated when the most rigorous extraction technique is applied. When the waste has been reacting a long time, and precipitation reactions have occurred, such immobilization makes clean-up by simple pump-and-treat techniques impossible.

11.3.2 *Physical non-equilibrium*

Consider an adsorption experiment in which water with some chemical is injected into a sediment column. Different concentrations in a cross-section of the column arise when the flow sweeps the solutes rapidly past stagnant zones. Stagnant zones may be due to the cementation of grains, or the shielding of pores by large clods of organic matter and oxides. They are also present in aggregated soils in which the aggregates are accessible by diffusion only, and in the rock matrix of a consolidated rock. When diffusion into the stagnant zone is unable to equalize concentration gradients brought about by advective flow, the process is said to be in *physical non-equilibrium*. It may happen in column experiments when flow velocity is high and not all pores are penetrated with equal speed, and also in batch experiments when homogenization is inadequate and time for the chemical to reach all the pores has been too short. For conservative (e.g. Cl⁻) and sorbed substances alike, physical non-equilibrium is thus a result of immobile water that exists in dead-end pores or micropores in aggregated parts of the soil.

A *mobile fraction* Φ_m can be defined as

$$\Phi_m = \varepsilon_m / \varepsilon$$

where ε is total water filled pore volume, ε_m is the mobile part.

When $\Phi_m = 1$, there is no immobile water and all sorption sites are immediately accessible for the adsorption and desorption reaction. When a stagnant zone is present, the velocity of the liquid in the mobile region will increase because advective flow is confined to a smaller cross section and the breakthrough of conservative and retarded chemicals occurs earlier than is calculated from the flow velocity and the total water-filled porosity (the latter obtained as weight loss by drying). Diffusion from the mobile zone into the stagnant zone and vice versa determines the tail part of the effluent curve.

Figure 11.15 shows an example where the fraction of mobile water depends on aggregate size. The effect of stagnant zone is illustrated by tritium effluent curves. The asymmetrical breakthrough shown in the lower part of Figure 11.15 is attributed to a relatively high portion of stagnant water (the fraction of mobile water $\Phi_m = 0.53$). The tailing end results from diffusion out of stagnant water; the form of the tail is however similar to curves obtained with adsorption/desorption hysteresis (Figure 11.14).

Some idea of the pore size which is homogenized by diffusion can be obtained using the formula (Section 9.3.1):

$$\sigma^2 = 2Dt$$

where σ^2 is variance of the diffusion (Gaussian) curve, D is diffusion coefficient, t is time.

In free water $D \approx 10^{-5}$ cm^2/s. The distance covered by diffusion can be compared with

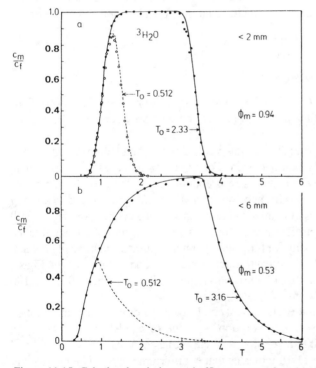

Figure 11.15. Calculated and observed effluent curves for tritium movement through a clay loam for small aggregates <2 mm (upper) and larger aggregates < 6 mm (lower). T = number of pore volumes; T_0 is injected volume of tritiated water, or change-over to unlabelled water. (Van Genuchten and Cleary, 1982).

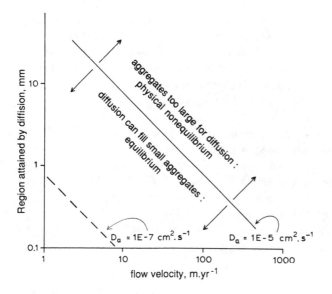

Figure 11.16. Comparison of diffusive transport and advective transport at different flow velocities. Lines for two values of the diffusion coefficient are shown.

the distance covered by advective transport as illustrated in Figure 11.16. This figure compares the distance traveled by diffusion as σ (which means 68% equalization in terms of the Gauss curve), and the distance traveled in equal time by advection. Lines are drawn for the relationship

$$\sigma = \sqrt{2Dt} \quad \text{(diffusive transport)}$$

$$= vt \quad \text{(advective transport)}$$

For example, with $D = 10^{-5}$ cm^2/s, is $\sigma = 0.1$ cm reached in $(0.1)^2 /(2 \cdot 10^{-5}) = 500$ s. This distance is covered in the same time period by advective transport when $v = 2 \cdot 10^{-4}$ cm/s, or 63 m/yr.

It will be clear that at normal groundwater flow velocities of around 20 m/yr, only small stagnant zones < 2 mm are fully homogenized. For a homogeneous sand the 'stagnant zone' would be the pore size, and this pore size is about equal to the particle size below which 10% of the particles exist (Perkins and Johnston, 1963). Stagnant zones in a sandy sediment can also exist as clusters of cemented grains, in which a central pore area is still accessible for cations via diffusion. The size of the stagnant zone is thus about equal to the dispersity, and an idea of flow velocity for full equilibrium in columns can be easily gained, when the dispersivity is obtained from breakthrough curves of a conservative tracer. Figure 11.16 shows that column dispersivities of a few mm require a flow velocity below ca. 100 m/yr.

In the finest pores where cation exchange takes place, or in clods of organic matter, diffusion velocity is further lessened by sorption. A value of ca. 10^{-7} cm^2/s has been deduced for partially collapsed interlayer spaces of clay illites or vermiculites (Keay and Wild, 1961; Freer, 1981). Figure 11.16 indicates that equilibrium may still be obtained at reasonable flow velocities in column experiments but only if these interlayers are smaller than a few tenths of microns. We can extrapolate the concept that stagnant zones and dispersivity are comparable entities to aquifers with much larger dispersivity, in the order of

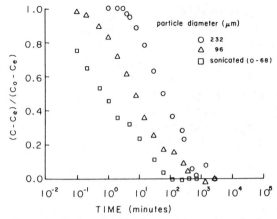

Figure 11.17. Sorption kinetics experimental results for tetrachlorobenzene on two size fractions, and sonicated whole sediment. The y-axis gives the relative deviation from equilibrium, see text. Reprinted with permission from Wu and Gschwend, 1986, Copyright American Chemical Society.

meters and more. Under such conditions no equilibrium in all parts of the aquifer can be attained, because diffusion in loam or peat layers of a few meters thick, is a slow process with respect to groundwater flow velocity.

The diffusion coefficient for sorbed substances is smaller than for non-sorbed, conservative chemicals by the retardation factor R (Section 9.4). For sorbed substances the conditions of non-equilibrium may be more pronounced because of slow diffusion into the sorbent matter. Such disequilibrium due to retarded diffusion is related to aggregate size as illustrated in Figure 11.17 from Wu and Gschwend (1986). These authors employed an elegant technique for determining ad- and desorption rates, by using a batch vessel with a closed-loop purging system and in-the-loop photo-ionization detection of the chemical. The time dependent adsorption of tetrachlorobenzene on three size fractions of a sediment is illustrated by plotting C, the actual concentration measured, relative to the equilibrium concentration C_e, and the initial concentration of the chemical C_0. Figure 11.17 illustrates the shift to longer equilibrium times when the size-fraction of the sediment increases. Small effective diffusion coefficients in large stagnant zones (clay, peat, loam layers) may retard total clean-up most drastically. It can be shown with the diffusion model of Example 9.12, that diffusion out of such layers takes much more time than the initial pollution into the layer, because the gradient for backward diffusion is smaller, and because additional diffusion occurs into the stagnant zone. It has been calculated that the loss by back-diffusion out of the layer is linearly (1:1) related to the effective diffusion coefficient (Van Leeuwen, 1991). If desorption is enhanced by e.g. temperature increase, the pollutant loss is more rapid, and clean-up is accelerated.

11.3.3 *Chemical disequilibrium*

Chemical disequilibrium results from slow kinetics of the sorption reaction compared to the flow of water which moves the chemical past the point where the reaction takes place. Reaction rates are easiest determined in batch experiments, since physical nonequilibrium and dispersion in columns produce results which are generally difficult to interpret in kinetic terms (Levenspiel, 1972). However, kinetics *can* play a role in column experiments, and thereby influence the breakthrough curves. For ion exchange, the kinetics of the exchange reaction have not been found to be rate limiting. Rather, diffusion through water films surrounding the adsorbing surface limits the reaction (Helfferich, 1962; Sparks, 1989). However, fixation of alkalis such as K^+ or Cs^+ can be a slow process, in which the clay structure itself may be subjected to physical changes. Such changes in clay structures are imperceptibly slow on the scale of laboratory experiments, and appear not to take place at all. They can be important in slowly moving groundwater systems, however.

Chemical nonequilibrium has been related to the nondimensional *Damkohler number* D_k:

$$D_k = k_{fb} L / v$$

where k_{fb} is the reaction rate (s^{-1}) for first order adsorption, L is distance (or column length, m), and v is flow velocity (m/s). Valocchi (1985) and Jennings (1987) have found that Damkohler numbers >100 are necessary for the local equilibrium assumption to be valid. It is of interest that a lower limit of the Damkohler number implies that, ultimately, equilibrium is always reached when the flowpath L is long enough. The reason lies in the fact that the concentration steps are smeared by non-equilibrium. When distance increases, the concentration steps may become so small that equilibrium can be reached.

Bahr and Rubin (1987) used an elaborate method in which all terms influenced by chemical kinetics are accounted for, so that different kinetic rate mechanisms can also be included. It is not easy to obtain kinetic data for more complex sorption reactions, but one described case where local equilibrium was attained had a flow velocity of 51 m/yr. This case, a column experiment carried out by Hornsby and Davidson (1973) showed a dispersion coefficient of $1.7 \cdot 10^{-5}$ cm^2/s, which is within the expected range for physical equilibrium as well.

11.3.4 *Modeling the processes*

Ideal breakthrough curves of sorbing chemicals have been modeled in Section 9.5. Whenever the experimental results cannot be explained with these simple models, explanations of the deviations are sought in one of the factors enumerated above. Physical nonequilibrium or intrasorbent diffusion are intuitively logical processes which can violate the local equilibrium assumption, and have been given much modeling attention. It has been demonstrated that chemical nonequilibrium, for reactions that follow simple linear rates, is mathematically equivalent to a simple model of physical nonequilibrium (Nkedi-Kizzi et al., 1984). Physical nonequilibrium and intrasorbent diffusion are both diffusion processes into stagnant zones, and have been modeled as such (Wu and Gschwend, 1986; Rao et al., 1982; Goltz and Roberts, 1988; Weber and Miller, 1988).

Alternatively the concept of a *two-box model* can be used. Here, the sorption process is visualized as taking place in two separate parts of the soil or sediment, which reach equilibrium with the pore water composition at different rates.

The models are visualized in Figure 11.18. The one-box-model has a mass-transfer rate:

$$\rho_b \cdot \partial s / \partial t = k_f \varepsilon \cdot C - k_b \rho_b \cdot s \tag{11.19}$$

where k_f and k_b are the sorption and desorption rates (s^{-1}). Since at equilibrium the change

Kinetics Models	Independent Kinetics Fitting Parameters (derived parameters)
One-box model 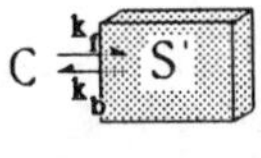	k_f $(k_b - k_f / K_p)$
Two-box model 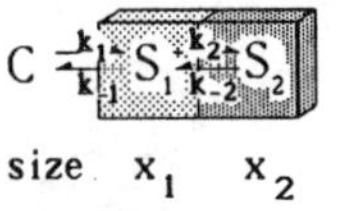 size x_1 x_2	k_1, k_2, x_1 $(k_{-1} - k_1/K_p)$ $(k_{-2} - k_2)$ $(x_2 - 1 - x_1)$
Diffusion model 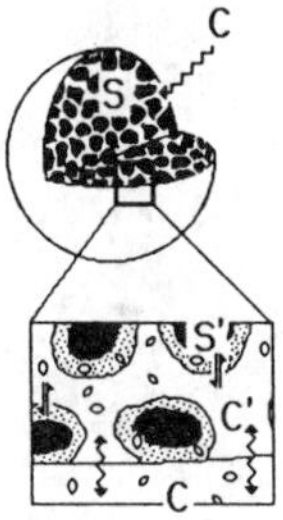	D_{eff}

Figure 11.18. Comparison of three sorption kinetics models. Reprinted with permission from Wu and Gschwend, 1986, Copyright American Chemical Society. K_p is the distribution coefficient, equal to K_d used in this text.

$\partial s / \partial t = 0$, we obtain with $q = \rho_b / \varepsilon \cdot s$, and $q/C = K_d$, that:

$$k_b = k_f / K_d$$

A slow mass-transfer has the effect that the breakthrough of a sorbed chemical is more diffuse than given by simple retardation of the advection-dispersion equation. The two-box model simply adds one additional class of sorption sites with a different mass-transfer rate. The two boxes may be in series, as illustrated in Figure 11.18, or parallel, i.e. both having direct, but different mass-transfer coefficients with the pore solution. When a two-box model is used, it is customary to let one box attain instantaneous equilibrium with the pore solution and in that case there is no different between the parallel, and the series concept. Either the box-models or the diffusion model can be modeled easily as a batch process, and then introduced into the finite difference transport scheme given in Section 9.5.

The mass-transfer rate is related to the effective diffusion coefficient (Bolt, 1982; Van Genuchten, 1985; Rao et al., 1982). Figure 11.19 shows results of a diffusion experiment of bromide out of cubes of chalk, and a diffusion model based on the concept of an equivalent spherical aggregate which replaces the cubes. The radius of the equivalent spheres is chosen such that the volume of the cubes and the spheres is equal. The results are also compared with a one-box model, in which kinetic exchange of the cube and the solution takes place according to:

$$\varepsilon_{im} \partial C_{im} / \partial t = k_{im}(C_m - C_{im})$$

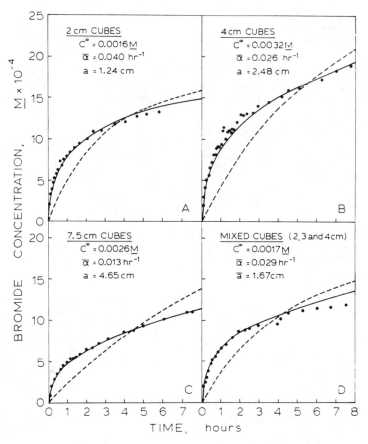

Figure 11.19. Diffusion of bromide out of cubes of chalk; comparison of the diffusion model for a cube as equivalent spheres (solid lines) and a box model (dashed lines). Reprinted with permission from Rao et al., 1982, Copyright Williams & Wilkins.

where k_{im} is the mass-transfer factor (s^{-1}), and subscripts im and m point to immobile and mobile zones.

The mass-transfer factor has been quantified (Bolt, 1982; Van Genuchten, 1985) as:

$$k_{im} = D_f \varepsilon_{im}/ (r^2 f^2)$$

where r is the equivalent spherical radius (m), and f is a shape factor.

The shape factor transforms the actual configuration of the stagnant zone to an equivalent spherical diffusion model. For a layered system, $k_{im} = 3\varepsilon_{im}D_f/L^2$, where L is the vertical thickness of the immobile layer. For a spherical stagnant zone it is $k_{im} = 15\varepsilon_{im}D_f/r^2$ (Bolt, 1982). When additional sorption occurs in the stagnant zone, the diffusion coefficient in free water, D_f, must be divided by the retardation.

The effect of non-equilibrium in column experiments or aquifer clean-up schemes can be assessed by using the so-called stop-flow or flow-interruption technique (Brusseau et al., 1989). When during the increasing limb in the breakthrough curve, the flow is stopped for some time, a non-equilibrium process becomes manifest by a relatively sharp decrease of

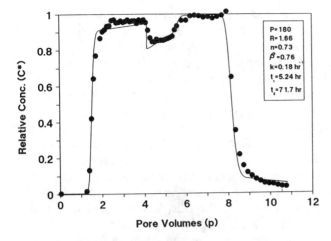

Figure 11.20. Effects of non-equilibrium during flow inter-ruption. The flow is halted when 4 pore volumes have been injected; solid line gives modeled results (Brusseau et al., 1989).

concentrations when the flow is reactivated. Similarly in the decreasing limb, when a chemical is flushed there will be an increase of concentrations when flow is restarted. Observation wells in an aquifer in the course of clean-up would show an increase of concentration due to this effect (Bahr, 1989). The jerks in the concentration are blurred somewhat by diffusion, and the effect in column experiments is most outspoken when concentrations are close to the end-values (Figure 11.20).

11.4 CONCLUSIONS AND A FURTHER OUTLOOK

This chapter together with Chapter 9, presents some calculation methods commonly employed to describe pollution in aquifers, and which might also be profitably used in restoration schemes. Simple formulae were obtained based on few assumptions. The flushing factor which relates number of flushing volumes to the slope of the desorption isotherm is a specifically easy indicator of the time periods which are associated with clean-up. When the slope can be diminished at the 'end' concentration, the flushing time diminishes proportionally. We have tried to indicate some possibilities to decrease the slope and also speculated on desorption of heavy metals which are bound by ion exchange. Geochemical techniques were sought with 'mild' characteristics. A mild treatment may well be necessary to obtain good long-term results. It has been observed that acid leaching of polluted soil, and similar harsh techniques, may lead to higher pollutant concentrations in flora and fauna than found in the untreated polluted soil (in other words, bioconcentration is more pronounced in the treated soil, Versluys et al., 1988). This effect is considered a result of the destruction of the sorptive properties of the soil.

However, it is not easy to assess the validity of assumptions in a complex natural flow system, where flowlines and aquifer properties are largely unknown, and highly variable. In some aquifer cleaning schemes it is found acceptable to continue pumping of contaminated groundwater for decades. Whenever such large time scales are involved, it may become profitable to investigate the possibility of manipulation of the water quality, or to change the physical conditions in the aquifer.

Absorption of non-ionic organic compounds is adequately described as a physical process which would not be much affected by inorganic dissolved constituents. For such pollutants, current research is aiming at reducing the effective solubility in water by adding cosolvents, or colloids which can absorb the pollutant and transport it in the colloidal state. It is of interest to note that transport of colloidal matter is enhanced by relatively high Na^+ concentrations in water, since this causes the colloids (humic acids, or clay particles) to remain peptized, and thus facilitates their transport through the porous medium.

The adsorption of ionic constituents, such as heavy metals or radionuclides, is influenced by water composition and can thus be manipulated by changing the concentrations of other ions. An example has been given for Cd^{2+} desorption with NaCl solutions of different normality. Since adsorption of ionic pollutants is in reality an ion exchange process, their transport requires a complete description of water chemistry, of the type discussed in Chapter 10. Whenever manipulation of the water quality is considered to remedy a pollution of heavy metals or radionuclides, such a model should be used.

A large part of the problems is due to aquifer heterogeneities. Slow diffusion out of large stagnant zones is difficult to accelerate. The diffusion out of the zone with immobile water takes much more time than the timespan of pollution. If the release is so slow that concentrations in the mobile water remain at an acceptable low level, we may consider the aquifer as *effectively* cleaned. This leads to the possibility of trying to *contain* the polluting chemical so that it becomes immobilized. Whatever the possibility, there is considerable scope for creative investigations here.

PROBLEMS

11.1. Draw elution isotherms for a pollutant with $q = 6C$, and another pollutant with $q = 6C^{0.7}$. Compare results with output from Example 9.12.

11.2. The pollutants of Problem 11.1 now enter an aquifer. Calculate arrival time in terms of pore volumes of a concentration $C = 1$ mg/l. The resident concentration is 0.1 mg/l. The isotherms are $q = 6C$, and $q = 6C^{0.7}$.

11.3. Calculate flushing factors for 0.1 μg/l β-HCH from the isotherm relations given in Figure 11.6. Assume $\rho_b/\varepsilon = 6$ kg/l.

REFERENCES

Abdul, A.S., Gibson, Th.L. and Ral, D.N., 1990, Use of humic acids solution to remove organic contaminants from hydrogeologic systems. *Environ. Sci. Technol.* 24, 328-333.

Appelo, C.A.J., Bäer, T., Smit, P.M.H. and Roelofzen, W., 1991, Flushing of column spiked with HCH (in Dutch). Report, Free University.

Appelo, C.A.J., Hendriks, J.A. and Van Veldhuizen, M., 1993. Flushing factors and a sharp front solution for solute transport with multicomponent ion exchange. *J. Hydrol.*, 146, 89-113.

Appelo, C.A.J., Wegener, J.W.M., Van Schaick, M.J.M. and Bosman, B.A., 1990, Adsorption and desorption of HCH's from a sandy aquifer (in Dutch). Report, Free University, 86 pp.

Bahr, J.M. 1989, Analysis of nonequilibrium desorption of volatile organics during field test of aquifer decontamination. *J. Contam. Hydrol.* 4, 205-222.

Bahr, J.M. and Rublin, J., 1987, Direct comparison of kinetic and local equilibrium formulations for solute transport affected by surface reactions. *Water Resour. Res.* 23, 438-452.

Bolt, G.H. (ed),1982, *Soil chemistry, B. Physico-chemical models*. Elsevier, Amsterdam, 527 pp.

Brusseau, M.L. and Rao, P.S.C., 1989, Sorption nonideality during organic contaminant transport in porous media. *Crit. Rev. Env. Control* 19, 33-99.

Brusseau, M.L., Rao, P.S.C., Jessup, R.E. and Davidson, J.M., 1989, Flow interruption: a method for investigating sorption nonequilibrium. *J. Contam. Hydrol.* 4, 223-240.

Charbeneau, R.J., 1982. Calculation of pollutant removal during groundwater restoration with adsorption and ion exchange. *Water Resour. Res.* 18, 1117-1125.

Charbeneau, R.J., 1988, Multicomponent exchange and subsurface transport: characteristics, coherence and the Riemann problem. *Water Resour. Res.* 24, 57-64.

Chiou, C.T., 1989, Theoretical considerations of the partition uptake of nonionic organic compounds by soil organic matter. *Reactions and movement of organic chemicals in soils*. Soil Sci. Soc. Am. Spec. Publ. 22, 1-29.

Chiou, C.T., Porter, P.E. and Shoup, Th.D., 1984, Reply on comments by W.G. MacIntyre and C.L. Smith. *Environ. Sci. Technol.* 18, 295-297.

Domenico, P.A. and Schwartz, F.W., 1990, *Physical and chemical hydrogeology*. Wiley & Sons, New York, 824 pp.

Enfield, C.G., Bengtsson, G. and Lindqvist, R., 1989, Influence of macromolecules on chemical transport. *Environ. Sci. Technol.* 23, 1278-1286.

Freer, R., 1981, Diffusion in silicate minerals and glasses: a data digest and guide to the literature. *Contrib. Mineral. Petrol.* 76, 440-454.

Garbarini, D.R. and Lion, L.W., 1986, Influence of the nature of soil organics on the sorption of toluene and trichloroethylene. *Environ. Sci. Technol.* 20, 1263-1269.

Goltz, M.N. and Roberts, P.V., 1988, Simulation of physical nonequilibrium solute transport models: application to a large scale field experiment. *J. Contam. Hydrol.* 3, 37-63.

Grathwohl, P., 1990, Influence of organic matter from soils and sediments from various origins on the sorption of some chlorinated aliphatic hydrocarbons: implications on K_{oc} correlations. *Environ. Sci. Technol.* 24, 1687-1693.

Griffioen, J., Appelo, C.A.J. and Van Veldhuizen, M., 1992, Practice of chromatography: deriving isotherms from elution curves. *Soil Sci. Soc. Am. J.*, 56, 1429-1436.

Gschwend, P.M. and Wu, S.-C., 1985, On the constancy of sediment-water partition coefficients of hydrophobic organic pollutants. *Environ. Sci. Technol.* 19, 90-96.

Helfferich, F., 1962, *Ion exchange*. McGraw-Hill, New York.

Helfferich, F. and Klein, G., 1970. *Multicomponent chromatography*. Marcel Dekker, New York, 419 pp.

Hornsby, A.G. and Davidson, J.M., 1973, Solution and adsorbed fluometuron concentration distribution in a water-saturated soil: experimental and predicted evaluation. *Soil Sci. Soc. Am. J.* 37, 823-828.

Hwang, Y.-L., Helfferich, F.G. and Leu, R.-J., 1988, Multicomponent equilibrium theory for ion-exchange columns involving reactions, *AIChE J.* 34, 1615-1626.

James, R.V. and Rubin, J., 1979, Applicability of the local equilibrium assumption to transport through soils of solutes affected by ion exchange. In E.A. Jenne (ed.), *Chemical modeling of aqueous systems*. ACS Symp. Ser. 93, 225-235.

Jennings, A., 1987, Critical chemical reaction rates for multicomponent groundwater contamination models. *Water Resour. Res.* 23, 1775-1784.

Kan, A.T. and Tomson, M.B., 1990, Effect of pH concentration on the transport of naphthalene in saturated aquifer media. *J. Contam. Hydrol.* 5, 235-251.

Karickhoff, S.W., 1981, Semi-empirical estimation of sorption of hydrophobic pollutants on natural sediments and soils. *Chemosphere* 10, 833-846.

Karickhoff, S.W., 1984, Organic pollutant sorption in aquatic systems. *J. Hydraul. Eng.* 110, 707-735.

Keay, J. and Wild, A., 1961, The kinetics of cation exchange in vermiculite. *Soil Sci.* 92, 54-60.

Kile, D.E., Chiou, C.T. and Helburn, R.S., 1990, Effect of some petroleum sulfonate surfactants on the apparent water solubility of organic compounds. *Environ. Sci. Technol.* 24, 205-208.

Levenspiel, O., 1972, *Chemical reaction engineering*. Wiley & Sons, New York, 578 pp.

Madhun, Y.A., Young, J.L. and Freed, V.H., 1986, Binding of water soluble organic materials from soil. *J. Env. Qual.* 15, 64-68.

Mehran, M., Olsen, R.L. and Rector, B.M., 1987, Distribution coefficient of trichloroethylene in soil-water systems. *Ground Water* 25, 275-282.

Nkedi-Kizza, P., Biggar, J.W., Selim, H.M., Van Genuchten, M.Th., Wierenga, P.J., Davidson, J.M. and Nielsen, D.R., 1984, On the equivalence of two conceptual models for describing ion exchange during transport through an aggregated oxisol. *Water Resour. Res.* 20, 1123-1130.

Nkedi-Kizza, P., Brusseau, M.L. and Rao, P.S.C., 1989, Nonequilibrium sorption during displacement of hydrophobic organic chemicals and ^{45}Ca through soil columns with aqueous and mixed solvents. *Environ. Sci. Technol.* 23, 814-820.

O'Connor, D.J. and Connolly, J.P., 1980, The effect of concentration of adsorbing solids on the partition coefficient. *Water Res.* 14, 1517-1523.

Perkins, T.K. and Johnston, O.C., 1963, A review of diffusion and dispersion in porous media. *Soc. Petrol. Eng. J.,* March, 70-84.

Pope, G.A., Lake, L.W. and Helfferich, F.G., 1978, Cation exchange in chemical flooding: Part 1 – basic theory without dispersion. *Soc. Petrol. Eng. J.,* Dec., 418-433.

Rainwater, K.A., Wise, W.R. and Charbeneau, R.J., 1987, Parameter estimation through groundwater tracer tests. *Water Resour. Res.* 23, 1901-1910.

Rao, P.S.C., Hornsby, A.G., Kilcrease, D.P. and Nkedi-Kizza, P., 1985, Sorption and transport of hydrophobic organic chemicals in aqueous and mixed solvent systems: model development and preliminary evaluation. *J. Env. Qual.* 14, 376-383.

Rao, P.S.C., Jessup, R.E. and Addiscott, T.M., 1982, Experimental and theoretical aspects of solute diffusion in spherical and nonspherical aggregates. *Soil Sci.* 133, 342-351.

Rhee, H.-K., Aris, R. and Amundson, N.R., 1970, On the theory of multicomponent chromatography. *Trans. Royal Soc., London,* 267 A 419-455.

Rhee, H.-K., Aris, R. and Amundson, N.R., 1989, *First-order partial differential equations*, Vol, II. Prentice Hall, New Yersey, 548 pp.

Rijnaarts, H.H.M., Bachmann, A., Jumelet, J.C. and Zehnder, A.J.B., 1990, Effect of desorption and intraparticle mass transfer on the aerobic biomineralization of α-HCH in a contaminated calcareous soil. *Environ. Sci. Technol.* 24, 1349-1354.

Sillén, L.G., 1951, On filtration through a sorbent layer. IV. The Ψ condition, a simple approach to the theory of sorption columns. *Arkiv Kemi* 2, 477-498.

Sparks, D.L., 1989, *Kinetics of soil chemical processes*. Academic Press, San Diego, Cal., 210 pp.

Urlings, L.G.C.M., Ackermann, V.P., Van Woudenberg, J.C., Van der Pijl, P.P. and Gaastra, J.J., 1988, In situ cadmium removal – full scale remedial action of contaminated soil. TAUW Infra Consult, Deventer.

Valocchi, A.J., 1985, Validity of the local equilibrium assumption for modeling sorbing solute transport through homogeneous soils. *Water Resour. Res.* 21, 808-820.

Van Eijkeren, J.C.H. and Loch, J.P.G., 1984, Transport of cationic solutes in sorbing porous media. *Water Resour. Res.* 20, 714-718.

Van Genuchten, M.Th., 1985, A general approach for modeling solute transport in structured soils. *IAH Mem.* 17, 513-525.

Van Genuchten, M.Th. and Cleary, R.W., 1982, Movement of solutes in soil: computer-simulated and laboratory results. In G.H. Bolt (ed.), *Soil Chemistry, B. Physico-chemical models*. Elsevier, Amsterdam, 349-386.

Van Leeuwen, N.F.M., 1991, Effects of stagnant zones on aquifer restoration (in Dutch). Report, Free University, Amsterdam.

Vermeulen, Th., Le Van, M.D., Hiester, N.K. and Klein, G., 1984, Adsorption and ion exchange. *Perry's chemical engineer's handbook*. 6th ed. McGraw-Hill.

Versluijs, C.W., Aalbers, Th.G., Adema, D.M.M., Assink, J.W., Van Gestel, C.A.M. and Anthonissen, I.H., 1988, Comparison of leaching behaviour of heavy metals in contaminated soils and soils cleaned up with several extractive and thermal methods. In K.Wolf et al. (eds), *Contaminated Soil '88*, 11-21, Kluwer Academic.

Voice, Th.C., Rice, C.P. and Weber, W.J., 1983, Effect of solids concentration on the sorptive partitioning of hydrophobic pollutants in aquatic systems. *Environ. Sci. Technol.* 17, 513-518.

Weber, W.J. and Miller, C.T., 1988, Modeling the sorption of hydrophobic contaminants by aquifer materials. *Water Res.* 22, 457-474.

Wu, S-C. and Gschwend, P.M., 1986, Sorption kinetics of hydrophobic organic compounds to natural sediments and soils. *Environ. Sci. Technol.* 20, 717-725.

Activities, standard states and the law of mass action

The law of mass action and the concepts 'activity' and 'standard state' have been applied routinely in previous chapters without much discussion of the thermodynamic formalisms. However, confusion may arise when concepts are used too loosely, for example when the activity of aqueous solutes is given the dimension 'molality', or the (dimensionless) equilibrium constant K is given a complicated and wrong dimension. Another common error is to assign the dimension 'kg H_2O/mol' to the activity coefficient, with the purpose of making the product of molal concentration and activity coefficient (i.e. the activity) dimensionless. The reader may find a complete treatment in the books on thermodynamics (Guggenheim, 1986; Lewis and Randall, 1961), and a physical interpretation of many thermodynamic relationships is presented by Moore (1972) and Feynman et al. (1989). Here, we present in a nutshell the formal thermodynamic basis of the activity concept.

All thermodynamic properties can be derived from an equation that expresses the Gibbs free energy of a system as a function of thermodynamic variables. These variables are the temperature, pressure, amounts of ions present, the electric charge, ponderable mass, surface area and others, depending on the specific conditions for the system under study. The change of the Gibbs free energy of a reaction with temperature, pressure and masses of the constituents is given by the differential equation:

$$d\Delta G_r = - \Delta S_r \, dT + \Delta V_r \, dP + \sum_i \mu_i \, dn_i \tag{A.1}$$

where ΔG_r is the Gibbs free energy that is liberated with each mole of reaction (J/mol), ΔS_r is the entropy change of the reaction (entropy units, or J/(deg.mol)), T is absolute temperature (Kelvin, or °C + 273.15), ΔV_r is the change in molar volume (cm^3/mol), P is pressure (atm), μ_i is chemical potential of species i (J/mol), n_i is moles of i.

The meaning of molar volume ΔV_r will be clear, but the significance of other terms such as entropy and chemical potential is more difficult to grasp. An understanding can be gained by first considering the pressure dependency of the Gibbs free energy of a reaction. Equation (A.1) implies that at a constant temperature and mass of constituents ($dT = 0$ and $dn_i = 0$), the pressure effect is given by:

$$\left(\frac{\partial \Delta G_r}{\partial P} \right)_{T, n_i} = \Delta V_r \tag{A.2a}$$

If the volume expands during the reaction (the products have a larger molar volume than the reactants) an increase of pressure will increase ΔG_r. The relation between ΔG_r and the equilibrium constant K is $\Delta G_r = - 2.303 \, RT \log K$ (we will derive this relation later on). Combined with Equation (A.2a) it gives

$$\left(\frac{\partial \log K}{\partial P} \right)_{T, n_i} = - \frac{\Delta V_r}{2.3 \, RT}$$

If $\Delta V_r > 0$, a pressure increase will decrease $\log K$ and thus favor the reactants (with a smaller volume than the products). This appears quite logical.

For example for dissolution of calcite:

$$CaCO_3 \leftrightarrow Ca^{2+} + CO_3^{2-} \qquad K = 10^{-8.4}$$

we have the volumina (Berner, 1971):

	$CaCO_3$	Ca^{2+}	CO_3^{2-}	
ΔV^0	36.9	-17.7	-3.7	cm^3/mol

and $\Delta V^0_{r\,(cc)} = (-17.7) + (-3.7) - (36.9) = -58.3 \; cm^3/mol$ (the ΔV may be negative, since these are relative volumina only). From the gas law, $V = RT/P$, we have 1 $cm^3 = 0.1$ J/atm. Hence $\Delta V^0_{r\,(cc)} = -5.9$ J/atm.mol, and at 25°C:

$$\left(\frac{\partial \log K}{\partial P} \right)_{298,\,n_i} = - \frac{-5.9}{2.3 \cdot 8.3 \cdot 298} = 10^{-3}/atm$$

An increase of calcite solubility from $10^{-8.4}$ to $10^{-8.3}$ requires 100 atm pressure increase. Other equilibria display similar small pressure dependence, and in hydrochemistry the pressure effects are normally neglected in thermodynamical calculations.

Likewise, at a constant pressure and with constant amounts of the species ($dP = 0$ and $dn_i = 0$), the change of the free energy with temperature is given by:

$$\left(\frac{\partial \Delta G_r}{\partial T} \right)_{P,\,n_i} = -\Delta S_r \tag{A.2b}$$

A positive reaction entropy denotes that ΔG_r decreases when temperature increases. As a consequence, log K increases with temperature, which implies that the reaction proceeds more easily to the right at a higher temperature. Is it possible to connect a physical concept with entropy, besides the mathematical relationships given by Equation (A.2b)? We expect that the enthalpy of a reaction, which is the difference in heat content of the products and the reactants, has a bearing on the temperature dependence. When heat is consumed during the reaction (this is an endothermal reaction in which the chemical system gains heat, hence $\Delta H_r > 0$), it appears intuitively reasonable that an increase of temperature will be favorable for the reaction (to proceed to the right). Actually, the entropy of a reaction at equilibrium has the following relation with the enthalpy:

$$\Delta S_r = \frac{\Delta H_r}{T}$$

If this relationship is substituted in Equation (A.2b), we see that a positive ΔH_r leads to the expected temperature effect (and also to Equation (3.26) given in Chapter 3).

The chemical potential μ_i, in a similar way, shows how the Gibbs free energy changes at a constant temperature and pressure if the amount of a constituent i varies while all the other constituents remain constant:

$$\left(\frac{\partial \Delta G_r}{\partial n_i} \right)_{T,\,P,\,n_{j \neq i}} = \mu_i \tag{A.2c}$$

Once more, the physical meaning of the chemical potential is not obvious. It must have units of energy per mole to be consistent with Equation (A.2c), and a relationship with concentration can be expected. Such a relationship may be developed from empirical measurements of the energy of a system. However, for perfect gases an equation of state is available which by definition relates pressure and volume (and hence the available energy) to the number of moles of (ideal) gas:

$$P_i V = n_i RT \tag{A.3}$$

where R is the gas constant (8.314 J/deg.mol).

Let us derive the function for μ of an ideal gas, using the gas law (A.3). We can differentiate Equation (A.1) twice, and the result is not affected by the order of the differentiation:

$$\left(\frac{\partial}{\partial P}\left(\frac{\partial \Delta G_r}{\partial n_i}\right)\right)_{T, n_{j \neq i}} = \left(\frac{\partial}{\partial n_i}\left(\frac{\partial \Delta G_r}{\partial P}\right)\right)_{T, n_{j \neq i}} \tag{A.4}$$

This is possible because (A.1) is an exact differential (which means that the path of integration has no influence on the result). Combining Equations (A.2a) and (A.2c) with (A.4) leads to:

$$\left(\frac{\partial V}{\partial n_i}\right)_{T, n_{j \neq i}} = \left(\frac{\partial \mu_i}{\partial P}\right)_{T, n_{j \neq i}} \tag{A.5}$$

which is one of the so-called Maxwell, or reciprocity relations.

From the gas law, Equation (A.3), we obtain the change:

$$\left(\frac{\partial V}{\partial n_i}\right)_{T, n_{j \neq i}} = \left(\frac{\partial\left(\frac{nRT}{P}\right)}{\partial n_i}\right)_{T, n_{j \neq i}} = \left(\frac{RT}{P}\right)_{T, n_{j \neq i}} \tag{A.6}$$

and on combining with Equation (A.5) how the chemical potential of an ideal gas at constant temperature, depends on pressure:

$$\frac{d\mu}{dP} = \frac{RT}{P} \tag{A.7}$$

A reference level can be defined, where $\mu = \mu^0$ when $P = P^0$. Going from this reference pressure to the actual conditions in the container where gas pressure of gas i is P_i means integrating Equation (A.7), or:

$$\int_{\mu_i^0}^{\mu_i} d\mu = \int_{P_i^0}^{P_i} \frac{RT}{P} dP \tag{A.8}$$

which gives:

$$\mu_i = \mu_i^0 + RT \ln (P_i/P_i^0) \tag{A.8}$$

With Equation (A.8) we have obtained a simple relationship between chemical potential and gas pressure under isothermal conditions. The state of the gas at the reference pressure P^0 is called the *standard state*, and the convention is that $P^0 = 1$ atm. It is therefore convenient to use atmosphere as the unit for gas pressure when thermodynamic calculations are performed (despite pascal being the SI-unit for pressure). We observe from Equation (A.8) that the chemical potential of a perfect gas is a logarithmic function of the relative pressure of the gas. This relative pressure is a concentration measure, which is termed *activity:*

$$[i] = P_i/P^0 \tag{A.9}$$

The activity is obviously dimensionless. A perfect gas has activity $[i] = 1$ when the pressure $P_i = 1$ atm, equal to the standard state.

A simple equation of state, as for a perfect gas, is not available for solids or liquids. However, the chemical potential of these substances is assumed to follow a similar relation, and activity is defined likewise from a concentration measure. The conventions for these concentration measures are shown in Table A1. Real substances can perhaps behave ideally in the same way as a perfect gas does, which means that the activity is identical to the concentration measure of Table A1. An ideal substance therefore has activity of one when the concentration is equal to the standard state.

The standard state varies for different substances (cf. Table A1). For a gas it is equal to a gas pressure of 1 atm. For solids, non-aqueous solutions, as well as water itself, it is defined as the pure substance. The standard state for an exchangeable cation is the exchanger covered by one cation only. For solutes in water it is defined as an ideal solution with solute concentration 1 mol/kg $H_2O = 1$ molal.

Table A1. Conventions for standard states and activities.

	Concentration measure	Symbol	Standard state
Gases	Gas pressure (atm)/ 1 atm	P_i/P^0	$P^0 = 1$ atm
Solid-, and liquid mixtures	Mole-fraction	X	$X = 1$
Aqueous solutes	Molality/(1 molal)	m_i/m^0	$m^0 = 1$ mol/kg H_2O
Exchangeable ions	Equiv.-fraction or mole-fraction	β or β^M	$\beta = 1, \beta^M = 1$

Real substances normally exhibit non-ideal behavior. The deviations from ideality are corrected with *activity coefficients*. For example, we have used activity coefficients γ for aqueous solutes, which can now be defined using:

$$[i] = \gamma_i \, m_i \, /m^0 \tag{A.10}$$

The activity coefficents have no dimension, and are obtained most often by empirical measurements. Sometimes a physical theory is available to calculate a-priori activity coefficients, such as the Debije-Hückel theory has provided for aqueous ions. However, when empirical models are used for calculating activity coefficients, one is free to choose the concentration measure. It is quite permissible to use the equivalent fractions for activities of exchangeable cations, because these allow a more convenient calculation of exchanger composition from solution data as explained in Chapter 5.

The Debije-Hückel theory provides a method for calculating individual activity coefficients in aqueous solution. A difficulty is that only the activity of a salt (e.g. NaCl, $CaSO_4$) can be measured since addition of only cations (here Na^+, Ca^{2+}) or anions (Cl^-, SO_4^{2-}) is clearly impossible. It has been agreed, nevertheless, that in an aqueous solution the activity of a single ion can be expressed as in Equation (A.10). The standard state solution has a concentration of 1 molal, and is ideal in the sense that the activity coefficient is 1.

We noted before that the potential of a substance is expressed as a energy per mole. The Gibbs free energy of formation of a substance represents that energy for the standard state:

$$\mu^0 = \Delta G_f^0 \tag{A.11}$$

where ΔG_f^0 is the standard state Gibbs free energy of formation from the elements (in kcal/mol, or kJ/mol).

The ^{0}-sign in ΔG_f^0 points to the standard conditions: 25°C, and 1 atm. The free energy of aqueous Ca^{2+} is, for example, $\Delta G_{f\,Ca^{2+}}^0 = -553.54$ kJ/mol, which indicates that if elemental Ca is thrown in water, a reaction takes place in which Ca transforms into Ca^{2+} with a concomitant loss of 553.54 kJ/mol.

Let us apply the concept of chemical potentials to the equilibrium dissolution of fluorite. Fluorite dissolves according to:

$$CaF_2 \leftrightarrow Ca^{2+} + 2F^- \tag{A.12}$$

Then, at equilibrium, the chemical potentials of reactants and products should be equal:

$$\mu_{Ca^{2+}} + 2\,\mu_{F^-} = \mu_{CaF_2}$$

or:

$$\mu_{Ca^{2+}}^0 + RT \ln [Ca^{2+}] + 2\,\mu_{F^-}^0 + 2\,RT \ln [F^-] - \mu_{CaF_2}^0 - RT \ln [CaF_2] = 0 \tag{A.13}$$

The fluorite is pure solid, and therefore $[CaF_2] = 1$. Collecting the terms yields:

$$RT \ln [Ca^{2+}][F^-]^2 = \mu_{CaF_2}^0 - \mu_{Ca^{2+}}^0 - 2\,\mu_{F^-}^0$$
$$= \Delta G_{f\,CaF_2}^0 - \Delta G_{f\,Ca^{2+}}^0 - 2\,\Delta G_{f\,F^-}^0 = -\Delta G_r^0$$
$$= -1050.8 - (-553.54) - 2(-278.8) = -60.3$$

or

$$[Ca^{2+}][F^-]^2 = \exp(-\Delta G_r^0/RT) = K_{Fluorite} = 10^{-10.57}$$

where ΔG_r^0 is the standard free energy of the reaction.

Thus, we have derived here the law of mass action from Equation (A.1). Thermodynamicists maintain that they can derive any relationship which describes the energy status of a system, using only Equation (A.1): all they need is a little bit of formula manipulation.

REFERENCES

Berner, R.A., 1971, *Principles of chemical sedimentology*. McGraw-Hill, New York, 240 pp.

Feynman, R.P., Leighton, R.B. and Sands, M.L., 1989, *The Feynman lectures on physics*, vol. 1, Chpt. 44, 45, 46. Addison-Wesley, Redwood City, Cal.

Guggenheim, E.A., 1986, *Thermodynamics*. 8th ed. North-Holland Physics Publ., Amsterdam, 390 pp.

Lewis, G.N. and Randall, M., 1961, *Thermodynamics*. 2nd ed. Revised by Pitzer and Brewer. McGraw-Hill, New York, 723 pp.

Moore, W.J., 1972, *Physical chemistry*. 5th ed. Longman, London, 977 pp.

APPENDIX B

Stability constants

The following tables contain stability constants for aqueous species and minerals as well as data concerning the temperature dependency of these constants. Data for major components in groundwater are shown in Table 1 that is reproduced directly from the consistent data set compiled by Nordstrom et al. (1990). The reader is referred to this publication for the sources of constants and further documentation. This data set is consistent with recent versions of PHREEQE (Parkhurst et al., 1990), PHREEQM (Chapter 10) and WATEQ4F (Ball and Nordstrom, 1991).

To provide supplemental information on trace elements, Table 2 lists constants for aqueous species and minerals of some selected trace elements that have been taken from the 1988 version of the MINTEQA2 database (Brown and Allison, 1987). Note that the data between these two tables are not necessarily internally consistent. The temperature dependency of a constant can be calculated either from an analytical expression, or from the van 't Hoff equation and ΔH_r^0 data (see Section 3.3.2). Units of kcal/mol can be converted to kJ/mol by multiplying with 4.184 ($R = 1.987 \cdot 10^{-3}$ kcal/mol.deg or $8.314 \cdot 10^{-3}$ kJ/mol.deg).

REFERENCES

Ball, J.W., and Nordstrom, D.K., 1991, User's manual for WATEQ4F, with revised thermodynamic data base and test cases for calculating speciation of major, trace, and redox elements in natural waters. US Geol. Surv., Open-File Report 91-183.

Brown, D.S., and Allison, J.D., 1987, An equilibrium metal speciation model: Users manual. Env. Res. Lab, Office of Res. and Dev., US Env. Protection Agency, Athens, GA 30613, USA, Rept. No. EPA/600 /3–87/012.

Nordstrom, D.K., Plummer, L.N., Langmuir, D., Busenberg, E., May, H.M., Jones, B.F. and Parkhurst, D.L., 1990, Revised chemical equilibrium data for major water-mineral reactions and their limitation. In Melchior, D.C., and Basset, R.L. (eds), *Chemical modeling of aqueous systems II*. ACS Symp. Ser. 416, 398-413.

Parkhurst, D.L., Thorstenson, D.C., and Plummer, L.N., 1990, PHREEQE–A computer program for geochemical calculations. US Geol. Surv. Water Res. Inv. 80-96 (1990 revision).

Table 1a. Summary of revised thermodynamic data. I: Fluoride and chloride species.

Reaction	ΔH_r^0 (kcal/mol)	log K	Reaction	ΔH_r^0 (kcal/mol)	log K
$H^+ + F^- = HF^0$	3.18	3.18	$Al^{3+} + F^- = AlF^{2+}$	1.06	7.0
$H^+ + 2F^- = HF_2^-$	4.55	3.76	$Al^{3+} + 2F^- = AlF_2^+$	1.98	12.7
$Na^+ + F^- = NaF^0$	----	-0.24	$Al^{3+} + 3F^- = AlF_3^0$	2.16	16.8
$Ca^{2+} + F^- = CaF^+$	4.12	0.94	$Al^{3+} + 4F^- = AlF_4^-$	2.20	19.4
$Mg^{2+} + F^- = MgF^+$	3.2	1.82	$Al^{3+} + 5F^- = AlF_5^{2-}$	1.84	20.6
$Mn^{2+} + F^- = MnF^+$	----	0.84	$Al^{3+} + 6F^- = AlF_6^{3-}$	-1.67	20.6
$Fe^{2+} + F^- = FeF^+$	---	1.0			
$Fe^{3+} + F^- = FeF^{2+}$	2.7	6.2	$Si(OH)_4 + 4H^+ + 6F^-$	-16.26	30.18
$Fe^{3+} + 2F^- = FeF_2^+$	4.8	10.8	$= SiF_6^{2-} + 4H_2O$		
$Fe^{3+} + 3F^- = FeF_3^0$	5.4	14.0	$Fe^{2+} + Cl^- = FeCl^+$	----	0.14
$Mn^{2+} + Cl^- = MnCl^+$	----	0.61	$Fe^{3+} + Cl^- = FeCl^{2+}$	5.6	1.48
$Mn^{2+} + 2Cl^- = MnCl_2^0$	----	0.25	$Fe^{3+} + 2Cl^- = FeCl_2^+$	----	2.13
$Mn^{2+} + 3Cl^- = MnCl_3^-$	----	-0.31	$Fe^{3+} + 3Cl^- = FeCl_3^0$	----	1.13

Mineral	Reaction	ΔH_r^0 (kcal/mol)	log K
Cryolite	$Na_3AlF_6 = 3Na^+ + Al^{3+} + 6F^-$	9.09	-33.84
Fluorite	$CaF_2 = Ca^{2+} + 2F^-$	4.69	-10.6

Redox Potentials	ΔH_r^0 (kcal/mol)	E^0 (volts)	log K
$Fe^{2+} = Fe^{3+} + e^-$	9.68	-0.770	-13.02
$Mn^{2+} = Mn^{3+} + e^-$	25.8	-1.51	-25.51

Reaction	Analytical Expressions for Temperature Dependence
$H^+ + F^- = HF^0$	$logK_{HF} = -2.033 + 0.012645T + 429.01/T$
$CaF_2 = Ca^{2+} + 2F^-$	$logK_{FLUORITE} = 66.348 - 4298.2/T - 25.271 \log T$

Table 1b. Summary of revised thermodynamic data. II: Oxide and hydroxide species.

Reaction	ΔH_r^0 (kcal/mol)	log K	Reaction	ΔH_r^0 (kcal/mol)	log K
$H_2O = H^+ + OH^-$	13.362	-14.000	$Fe^{3+} + H_2O = FeOH^{2+} + H^+$	10.4	-2.19
$Li^+ + H_2O = LiOH^0 + H^+$	0.0	-13.64	$Fe^{3+} + 2H_2O = Fe(OH)_2^+ + 2H^+$	17.1	-5.67
$Na^+ + H_2O = NaOH^0 + H^+$	0.0	-14.18	$Fe^{3+} + 3H_2O = Fe(OH)_3^0 + 3H^+$	24.8	-12.56
$K^+ + H_2O = KOH^0 + H^+$	---	-14.46	$Fe^{3+} + 4H_2O = Fe(OH)_4^- + 4H^+$	31.9	-21.6
$Ca^{2+} + H_2O = CaOH^+ + H^+$	---	-12.78	$2Fe^{3+} + 2H_2O = Fe_2(OH)_2^{4+} + 2H^+$	13.5	-2.95
$Mg^{2+} + H_2O = MgOH^+ + H^+$	---	-11.44	$3Fe^{3+} + 4H_2O = Fe_3(OH)_4^{5+} + 4H^+$	14.3	-6.3
$Sr^{2+} + H_2O = SrOH^+ + H^+$	---	-13.29	$Al^{3+} + H_2O = AlOH^{2+} + H^+$	11.49	-5.00
$Ba^{2+} + H_2O = BaOH^+ + H^+$	---	-13.47	$Al^{3+} + 2H_2O = Al(OH)_2^+ + 2H^+$	26.90	-10.1
$Ra^{2+} + H_2O = RaOH^+ + H^+$	---	-13.49	$Al^{3+} + 3H_2O = Al(OH)_3^0 + 3H^+$	39.89	-16.9
$Fe^{2+} + H_2O = FeOH^+ + H^+$	13.2	-9.5	$Al^{3+} + 4H_2O = Al(OH)_4^- + 4H^+$	42.30	-22.7
$Mn^{2+} + H_2O = MnOH^+ + H^+$	14.4	-10.59			

Mineral	Reaction	ΔH_r^0 (kcal/mol)	log K
Portlandite	$Ca(OH)_2 + 2H^+ = Ca^{2+} + 2H_2O$	-31.0	22.8
Brucite	$Mg(OH)_2 + 2H^+ = Mg^{2+} + 2H_2O$	-27.1	16.84
Pyrolusite	$MnO_2 + 4H^+ + 2e^- = Mn^{2+} + 2H_2O$	-65.11	41.38
Hausmanite	$Mn_3O_4 + 8H^+ + 2e^- = 3Mn^{2+} + 4H_2O$	-100.64	61.03
Manganite	$MnOOH + 3H^+ + e^- = Mn^{2+} + 2H_2O$	----	25.34
Pyrochroite	$Mn(OH)_2 + 2H^+ = Mn^{2+} + 2H_2O$	----	15.2
Gibbsite (crystalline)	$Al(OH)_3 + 3H^+ = Al^{3+} + 3H_2O$	-22.8	8.11
Gibbsite(microcrystalline)	$Al(OH)_3 + 3H^+ = Al^{3+} + 3H_2O$	(-24.5)	9.35
$Al(OH)_3$ (amorphous)	$Al(OH)_3 + 3H^+ = Al^{3+} + 3H_2O$	(-26.5)	10.8
Goethite	$FeOOH + 3H^+ = Fe^{3+} + 2H_2O$	----	-1.0
Ferrihydrite(amorphous to microcrystalline)	$Fe(OH)_3 + 3H^+ = Fe^{3+} + 3H_2O$	----	3.0 to 5.0

Reaction	Analytical Expressions for Temperature Dependence
$H_2O = H^+ + OH^-$	$logK_w = -283.9710 + 13323.00/T - 0.05069842T + 102.24447 \log T - 1119669/T^2$
$Al^{3+} + H_2O = AlOH^{2+} + H^+$	$logK_1 = -38.253 - 656.27/T + 14.327 \log T$
$Al^{3+} + 2H_2O = Al(OH)_2^+ + 2H^+$	$log\beta_2 = 88.500 - 9391.6/T - 27.121 \log T$
$Al^{3+} + 3H_2O = Al(OH)_3^0 + 3H^+$	$log\beta_3 = 226.374 - 18247.8/T - 73.597 \log T$
$Al^{3+} + 4H_2O = Al(OH)_4^- + 4H^+$	$log\beta_4 = 51.578 - 11168.9/T - 14.865 \log T$

Table 1c. Summary of revised thermodynamic data. III: Carbonate species.

Reaction	ΔH_r° (kcal/mol)	log K	Reaction	ΔH_r° (kcal/mol)	log K
$CO_2(g) = CO_2(aq)$	-4.776	-1.468	$Ca^{2+} + CO_3^{2-} = CaCO_3^\circ$	3.545	3.224
$CO_2(aq) + H_2O = H^+ + HCO_3^-$	2.177	-6.352	$Mg^{2+} + CO_3^{2-} = MgCO_3^\circ$	2.713	2.98
$HCO_3^- = H^+ + CO_3^{2-}$	3.561	-10.329	$Sr^{2+} + CO_3^{2-} = SrCO_3^\circ$	5.22	2.81
$Ca^{2+} + HCO_3^- = CaHCO_3^+$	2.69	1.106	$Ba^{2+} + CO_3^{2-} = BaCO_3^\circ$	3.55	2.71
$Mg^{2+} + HCO_3^- = MgHCO_3^+$	0.79	1.07	$Mn^{2+} + CO_3^{2-} = MnCO_3^\circ$	---	4.90
$Sr^{2+} + HCO_3^- = SrHCO_3^+$	6.05	1.18	$Fe^{2+} + CO_3^{2-} = FeCO_3^\circ$	---	4.38
$Ba^{2+} + HCO_3^- = BaHCO_3^+$	5.56	0.982	$Na^+ + CO_3^{2-} = NaCO_3^-$	8.91	1.27
$Mn^{2+} + HCO_3^- = MnHCO_3^+$	---	1.95	$Na^+ + HCO_3^- = NaHCO_3$	---	-0.25
$Fe^{2+} + HCO_3^- = FeHCO_3^+$	---	2.0	$Ra^{2+} + CO_3^{2-} = RaCO_3^\circ$	1.07	2.5

Mineral	Reaction	ΔH_r° (kcal/mol)	log K
Calcite	$CaCO_3 = Ca^{2+} + CO_3^{2-}$	-2.297	-8.480
Aragonite	$CaCO_3 = Ca^{2+} + CO_3^{2-}$	-2.589	-8.336
Dolomite(Ordered)	$CaMg(CO_3)_2 = Ca^{2+} + Mg^{2+} + 2CO_3^{2-}$	-9.436	-17.09
Dolomite(Disordered)	$CaMg(CO_3)_2 = Ca^{2+} + Mg^{2+} + 2CO_3^{2-}$	-11.09	-16.54
Strontianite	$SrCO_3 = Sr^{2+} + CO_3^{2-}$	-0.40	-9.271
Siderite(crystalline)	$FeCO_3 = Fe^{2+} + CO_3^{2-}$	-2.48	-10.89
Siderite(precipitated)	$FeCO_3 = Fe^{2+} + CO_3^{2-}$	---	-10.45
Witherite	$BaCO_3 = Ba^{2+} + CO_3^{2-}$	0.703	-8.562
Rhodocrosite(crystalline)	$MnCO_3 = Mn^{2+} + CO_3^{2-}$	-1.43	-11.13
Rhodocrosite(synthetic)	$MnCO_3 = Mn^{2+} + CO_3^{2-}$	---	-10.39

Reaction	Analytical Expressions for Temperature Dependence
$CO_2(g) = CO_2(aq)$	$\log K_H = 108.3865 + 0.01985076T - 6919.53/T - 40.45154 \log T + 669365/T^2$
$CO_2(aq) + H_2O = H^+ + HCO_3^-$	$\log K_1 = -356.3094 - 0.06091964T + 21834.37/T + 126.8339 \log T - 1684915/T^2$
$HCO_3^- = H^+ + CO_3^{2-}$	$\log K_2 = -107.8871 - 0.03252849T + 5151.79/T + 38.92561 \log T - 563713.9/T^2$
$Ca^{2+} + HCO_3^- = CaHCO_3^+$	$\log K_{CaHCO3}^+ = 1209.120 + 0.31294T - 34765.05/T - 478.782 \log T$
$Mg^{2+} + HCO_3^- = MgHCO_3^+$	$\log K_{MgHCO3}^+ = -59.215 + 2537.455/T + 20.92298 \log T$
$Sr^{2+} + HCO_3^- = SrHCO_3^+$	$\log K_{SrHCO3}^+ = -3.248 + 0.014867T$
$Ba^{2+} + HCO_3^- = BaHCO_3^+$	$\log K_{BaHCO3}^+ = -3.0938 + 0.013669T$
$Ca^{2+} + CO_3^{2-} = CaCO_3^\circ$	$\log K_{CaCO3}^\circ = -1228.732 - 0.299444T + 35512.75/T + 485.818 \log T$
$Mg^{2+} + CO_3^{2-} = MgCO_3^\circ$	$\log K_{MgCO3}^\circ = 0.9910 + 0.00667T$
$Sr^{2+} + CO_3^{2-} = SrCO_3^\circ$	$\log K_{SrCO3}^\circ = -1.019 + 0.012826T$
$Ba^{2+} + CO_3^{2-} = BaCO_3^\circ$	$\log K_{BaCO3}^\circ = 0.113 + 0.008721T$
$CaCO_3 = Ca^{2+} + CO_3^{2-}$	$\log K_{CALCITE} = -171.9065 - 0.077993T + 2839.319/T + 71.595 \log T$
$CaCO_3 = Ca^{2+} + CO_3^{2-}$	$\log K_{ARAGONITE} = -171.9773 - 0.077993T + 2903.293/T + 71.595 \log T$
$SrCO_3 = Sr^{2+} + CO_3^{2-}$	$\log K_{STRONTIANITE} = 155.0305 - 7239.594/T - 56.58638 \log T$
$BaCO_3 = Ba^{2+} + CO_3^{2-}$	$\log K_{WITHERITE} = 607.642 + 0.121098T - 20011.25/T - 236.4948 \log T$

Table 1d. Summary of revised thermodynamic data. IV: Silicate species.

Reaction	ΔH_r^0 (kcal/mol)	log K
$Si(OH)_4^0 = SiO(OH)_3^- + H^+$	6.12	-9.83
$Si(OH)_4^0 = SiO_2(OH)_2^{2-} + 2H^+$	17.6	-23.0

Mineral	Reaction	ΔH_r^0 (kcal/mol)	log K
Kaolinite	$Al_2Si_2O_5(OH)_4 + 6H^+ = 2Al^{3+} + 2Si(OH)_4^0 + H_2O$	-35.3	7.435
Chrysotile	$Mg_3Si_2O_5(OH)_4 + 6H^+ = 3Mg^{2+} + 2Si(OH)_4^0 + H_2O$	-46.8	32.20
Sepiolite	$Mg_2Si_3O_{7.5}(OH)\cdot3H_2O + 4H^+ + 0.5\,H_2O = 2Mg^{2+} + 3Si(OH)_4^0$	-10.7	15.76
Kerolite	$Mg_3Si_4O_{10}(OH)_2\cdot H_2O + 6H^+ + 3H_2O = 3Mg^{2+} + 4Si(OH)_4^0$	---	25.79
Quartz	$SiO_2 + 2H_2O = Si(OH)_4^0$	5.99	-3.98
Chalcedony	$SiO_2 + 2H_2O = Si(OH)_4^0$	4.72	-3.55
Amorphous Silica	$SiO_2 + 2H_2O = Si(OH)_4^0$	3.34	-2.71

Reaction	Analytical Expressions for Temperature Dependence
$Si(OH)_4^0 = SiO(OH)_3^- + H^+$	$logK_1 = -302.3724 - 0.050698T + 15669.69/T + 108.18466\ log\ T - 1119669/T^2$
$Si(OH)_4^0 = SiO_2(OH)_2^{2-} + 2H^+$	$log\beta_2 = -294.0184 - 0.072650T + 11204.49/T + 108.18466\ log\ T - 1119669/T^2$
$Mg_3Si_2O_5(OH)_4 + 6H^+ = 3Mg^{2+} + 2Si(OH)_4^0 + H_2O$	$logK_{CHRYSOTILE} = 13.248 + 10217.1/T - 6.1894\ log\ T$
$SiO_2 + 2H_2O = Si(OH)_4^0$	$logK_{QUARTZ} = 0.41 - 1309/T$
$SiO_2 + 2H_2O = Si(OH)_4^0$	$logK_{CHALCEDONY} = -0.09 - 1032/T$
$SiO_2 + 2H_2O = Si(OH)_4^0$	$logK_{AMORPHOUS\ SILICA} = -0.26 - 731/T$

Table 1e. Summary of revised thermodynamic data. V: Sulfate species.

Reaction	ΔH_r^0 (kcal/mol)	log K	Reaction	ΔH_r^0 (kcal/mol)	log K
$H^+ + SO_4^{2-} = HSO_4^-$	3.85	1.988	$Mn^{2+} + SO_4^{2-} = MnSO_4^0$	3.37	2.25
$Li^+ + SO_4^{2-} = LiSO_4^-$	----	0.64	$Fe^{2+} + SO_4^{2-} = FeSO_4^0$	3.23	2.25
$Na^+ + SO_4^{2-} = NaSO_4^-$	1.12	0.70	$Fe^{2+} + HSO_4^- = FeHSO_4^+$	---	1.08
$K^+ + SO_4^{2-} = KSO_4^-$	2.25	0.85	$Fe^{3+} + SO_4^{2-} = FeSO_4^+$	3.91	4.04
$Ca^{2+} + SO_4^{2-} = CaSO_4^0$	1.65	2.30	$Fe^{3+} + 2SO_4^{2-} = Fe(SO_4)_2^-$	4.60	5.38
$Mg^{2+} + SO_4^{2-} = MgSO_4^0$	4.55	2.37	$Fe^{3+} + HSO_4^- = FeHSO_4^{2+}$	---	2.48
$Sr^{2+} + SO_4^{2-} = SrSO_4^0$	2.08	2.29	$Al^{3+} + SO_4^{2-} = AlSO_4^+$	2.15	3.02
$Ba^{2+} + SO_4^{2-} = BaSO_4^0$	---	2.7	$Al^{3+} + 2SO_4^{2-} = Al(SO_4)_2^-$	2.84	4.92
$Ra^{2+} + SO_4^{2-} = RaSO_4^0$	1.3	2.75	$Al^{3+} + HSO_4^- = AlHSO_4^{2+}$	---	0.46

Mineral	Reaction	ΔH_r^0 (kcal/mol)	log K
Gypsum	$CaSO_4 \cdot 2H_2O = Ca^{2+} + SO_4^{2-} + 2H_2O$	-0.109	-4.58
Anhydrite	$CaSO_4 = Ca^{2+} + SO_4^{2-}$	-1.71	-4.36
Celestite	$SrSO_4 = Sr^{2+} + SO_4^{2-}$	-1.037	-6.63
Barite	$BaSO_4 = Ba^{2+} + SO_4^{2-}$	6.35	-9.97
Radium sulfate	$RaSO_4 = Ra^{2+} + SO_4^{2-}$	9.40	-10.26
Melanterite	$FeSO_4 \cdot 7H_2O = Fe^{2+} + SO_4^{2-} + 7H_2O$	4.91	-2.209
Alunite	$KAl_3(SO_4)_2(OH)_6 + 6H^+ = K^+ + 3Al^{3+} + 2SO_4^{2-} + 6H_2O$	-50.25	-1.4

Reaction	Analytical Expressions for Temperature Dependence
$H^+ + SO_4^{2-} = HSO_4^-$	$\log K_2 = -56.889 + 0.006473T + 2307.9/T + 19.8858 \log T$
$CaSO_4 \cdot 2H_2O = Ca^{2+} + SO_4^{2-} + 2H_2O$	$\log K_{GYPSUM} = 68.2401 - 3221.51/T - 25.0627 \log T$
$CaSO_4 = Ca^{2+} + SO_4^{2-}$	$\log K_{ANHYDRITE} = 197.52 - 8669.8/T - 69.835 \log T$
$SrSO_4 = Sr^{2+} + SO_4^{2-}$	$\log K_{CELESTITE} = -14805.9622 - 2.4660924T + 756968.533/T - 40553604/T^2 + 5436.3588 \log T$
$BaSO_4 = Ba^{2+} + SO_4^{2-}$	$\log K_{BARITE} = 136.035 - 7680.41/T - 48.595 \log T$
$RaSO_4 = Ra^{2+} + SO_4^{2-}$	$\log K_{RaSO_4} = 137.98 - 8346.87/T - 48.595 \log T$
$FeSO_4 \cdot 7H_2O = Fe^{2+} + SO_4^{2-} + 7H_2O$	$\log K_{MELANTERITE} = 1.447 - 0.004153T - 214949/T^2$

Table 2a. Trace elements: Arsenic.

Reaction	ΔH_r° (kcal/mol)	log K
$H_2AsO_3^- + H^+ \rightleftarrows H_3AsO_3$	-6.56	9.228
$HAsO_3^{2-} + 2H^+ \rightleftarrows H_3AsO_3$	-14.199	21.33
$AsO_3^{3-} + 3H^+ \rightleftarrows H_3AsO_3$	-20.25	34.744
$H_4AsO_3^+ \rightleftarrows H_3AsO_3 + H^+$	-	0.305
$H_2AsO_4^- + H^+ \rightleftarrows H_3AsO_4$	1.69	2.243
$HAsO_4^{2-} + 2H^+ \rightleftarrows H_3AsO_4$	0.92	9.1
$AsO_4^{3-} + 3H^+ \rightleftarrows H_3AsO_4$	-3.43	20.597

Mineral	Reaction	ΔH_r° (kcal/mol)	log K
	$AsI_3 + 3H_2O \rightleftarrows H_3AsO_3 + 3I^- + 3H^+$	1.875	4.155
Arsenolite	$As_4O_6 + 6H_2O \rightleftarrows 4H_3AsO_3$	14.33	-2.801
Claudetite	$As_4O_6 + 6H_2O \rightleftarrows 4H_3AsO_3$	13.29	-3.065
Orpiment	$As_2S_3 + 6H_2O \rightleftarrows 2H_3AsO_3 + 3HS^- + 3H^+$	82.89	-60.971
	$As_2O_5 + 3H_2O \rightleftarrows 2H_3AsO_4$	-5.405	6.699
	$AlAsO_4 \cdot 2H_2O + 3H^+ \rightleftarrows H_3AsO_4 + Al^{3+} + 2H_2O$	-	4.8
	$Ca_3(AsO_4)_2 \cdot 6H_2O + 6H^+ \rightleftarrows 2H_3AsO_4 + 3Ca^{2+} + 6H_2O$	-	22.3
	$Cu_3(AsO_4)_2 \cdot 6H_2O + 6H^+ \rightleftarrows 3Cu^{2+} + 2H_3AsO_4 + 6H_2O$	-	6.1
	$FeAsO_4 \cdot 2H_2O + 3H^+ \rightleftarrows H_3AsO_4 + Fe^{3+} + 2H_2O$	-	0.4
	$Mn_3(AsO_4)_2 \cdot 8H_2O + 6H^+ \rightleftarrows 2H_3AsO_4 + 3Mn^{2+} + 8H_2O$	-	12.5
	$Ni_3(AsO_4)_2 \cdot 8H_2O + 6H^+ \rightleftarrows 2H_3AsO_4 + 3Ni^{2+} + 8H_2O$	-	15.7
	$Pb_3(AsO_4)_2 + 6H^+ \rightleftarrows 3Pb^{2+} + 2H_3AsO_4$	-	5.8
	$Zn_3(AsO_4)_2 \cdot 2.5H_2O + 6H^+ \rightleftarrows 2H_3AsO_4 + 3Zn^{2+} + 2.5H_2O$	-	13.65
	$Ba_3(AsO_4)_2 + 6H^+ \rightleftarrows 3Ba^{2+} + 2H_3AsO_4$	2.64	-8.91
Realgar	$AsS + 3H_2O \rightleftarrows H_3AsO_3 + HS^- + 2H^+ + e^-$	30.545	-19.747

Redox Potential	ΔH_r° (kcal/mol)	log K
$H_3AsO_3 + H_2O \rightleftarrows H_3AsO_4 + 2e^- + 2H^+$	30.015	-19.444

Table 2b. Trace elements: Cadmium.

Reaction	ΔH_r^o (kcal/mol)	log K
$CdCl^+ \rightleftarrows Cd^{2+} + Cl^-$	-0.59	-1.98
$CdCl_{2(aq)} \rightleftarrows Cd^{2+} + 2Cl^-$	-1.24	-2.6
$CdCl_3^- \rightleftarrows Cd^{2+} + 3Cl^-$	-3.9	-2.399
$CdF^+ \rightleftarrows Cd^{2+} + F^-$	-	-1.1
$CdF_{2(aq)} \rightleftarrows Cd^{2+} + 2F^-$	-	-1.5
$CdBr^+ \rightleftarrows Cd^{2+} + Br^-$	0.81	-2.17
$CdBr_{2(aq)} \rightleftarrows Cd^{2+} + 2Br^-$	-	-2.899
$CdI^+ \rightleftarrows Cd^{2+} + I^-$	2.37	-2.15
$CdI_{2(aq)} \rightleftarrows Cd^{2+} + 2I^-$	-	-3.59
$Cd(CO_3)_3^{4-} \rightleftarrows Cd^{2+} + 3CO_3^{2-}$	-	-6.22
$CdCO_{3(aq)} \rightleftarrows Cd^{2+} + CO_3^{2-}$	-	-5.399
$CdHCO_3^+ \rightleftarrows Cd^{2+} + CO_3^{2-} + H^+$	-	-12.4
$Cd(OH)^+ + H^+ \rightleftarrows Cd^{2+} + H_2O$	-13.1	10.08
$Cd(OH)_{2(aq)} + 2H^+ \rightleftarrows Cd^{2+} + 2H_2O$	-	20.35
$Cd(OH)_3^- + 3H^+ \rightleftarrows Cd^{2+} + 3H_2O$	-	33.3
$Cd(OH)_4^{2-} + 4H^+ \rightleftarrows Cd^{2+} + 4H_2O$	-	47.35
$Cd(OH)Cl_{(aq)} + H^+ \rightleftarrows Cd^{2+} + H_2O + Cl^-$	-4.355	7.404
$Cd_2(OH)^{3+} + H^+ \rightleftarrows 2Cd^{2+} + H_2O$	-10.899	9.39
$CdNO_3^+ \rightleftarrows Cd^{2+} + NO_3^-$	5.2	-0.399
$CdSO_{4(aq)} \rightleftarrows Cd^{2+} + SO_4^{2-}$	-1.08	-2.46
$Cd(SO_4)_2^{2-} \rightleftarrows Cd^{2+} + 2SO_4^{2-}$	-	-3.5
$Cd(HS)^+ \rightleftarrows Cd^{2+} + HS^-$	-	-10.17
$Cd(HS)_{2(aq)} \rightleftarrows Cd^{2+} + 2HS^-$	-	-16.53
$Cd(HS)_3^- \rightleftarrows Cd^{2+} + 3HS^-$	-	-18.71
$Cd(HS)_4^{2-} \rightleftarrows Cd^{2+} + 4HS^-$	-	-20.9
$Cd(SeO_3)_2^{2-} + 2H^+ \rightleftarrows Cd^{2+} + 2HSeO_3^-$	-	11.189
$CdCN^+ \rightleftarrows Cd^{2+} + CN^-$	-	-5.32
$Cd(CN)_{2(aq)} \rightleftarrows Cd^{2+} + 2CN^-$	13.0	-10.3703
$Cd(CN)_3^- \rightleftarrows Cd^{2+} + 3CN^-$	21.6	-14.8341
$Cd(CN)_4^{2-} \rightleftarrows Cd^{2+} + 4CN^-$	23.56	-18.2938

Table 2c. Trace elements: Cadmium (cont.).

Mineral	Reaction	ΔH_r^o (kcal/mol)	log K
	$CdCl_2 \rightleftarrows Cd^{2+} + 2Cl^-$	-4.47	-0.68
	$CdCl_2 \cdot H_2O \rightleftarrows Cd^{2+} + 2Cl^- + H_2O$	-1.82	-1.71
	$CdCl_2 \cdot 2.5H_2O \rightleftarrows Cd^{2+} + 2Cl^- + 2.5H_2O$	1.71	-1.94
	$CdF_2 \rightleftarrows Cd^{2+} + 2F^-$	-9.72	-2.98
	$CdBr_2 \cdot 4H_2O \rightleftarrows Cd^{2+} + 2Br^- + 4H_2O$	7.23	-2.42
	$CdI_2 \rightleftarrows Cd^{2+} + 2I^-$	4.08	-3.61
Greenockite	$CdS + H^+ \rightleftarrows Cd^{2+} + HS^-$	16.36	-15.93
Monteponite	$CdO + 2H^+ \rightleftarrows Cd^{2+} + H_2O$	-24.76	15.12
	$Cd(OH)_{2(A)} + 2H^+ \rightleftarrows Cd^{2+} + 2H_2O$	-20.77	13.73
	$Cd(OH)_{2(C)} + 2H^+ \rightleftarrows Cd^{2+} + 2H_2O$	-	13.65
	$Cd(OH)Cl + H^+ \rightleftarrows Cd^{2+} + H_2O + Cl^-$	-7.407	3.52
	$Cd_3(OH)_4(SO_4) + 4H^+ \rightleftarrows 3Cd^{2+} + 4H_2O + SO_4^{2-}$	-	22.56
	$Cd_3(OH)_2(SO_4)_2 + 2H^+ \rightleftarrows 3Cd^{2+} + 2H_2O + 2SO_4^{2-}$	-	6.71
	$Cd_4(OH)_6(SO_4) + 6H^+ \rightleftarrows 4Cd^{2+} + 6H_2O + SO_4^{2-}$	-	28.4
Otavite	$CdCO_3 \rightleftarrows Cd^{2+} + CO_3^{2-}$	-0.58	-13.74
	$CdSO_4 \rightleftarrows Cd^{2+} + SO_4^{2-}$	-14.74	-0.1
	$CdSO_4 \cdot H_2O \rightleftarrows Cd^{2+} + SO_4^{2-} + H_2O$	-7.52	-1.657
	$CdSO_4 \cdot 2.67H_2O \rightleftarrows Cd^{2+} + SO_4^{2-} + 2.67H_2O$	-4.3	-1.873
	$CdSe + H^+ \rightleftarrows Cd^{2+} + HSe^-$	18.16	-18.0739
	$CdSeO_4 \rightleftarrows Cd^{2+} + SeO_4^{2-}$	-	-2.2415
	$Cd_3(PO_4)_2 \rightleftarrows 3Cd^{2+} + 2PO_4^{3-}$	-	-32.6
	$CdSiO_3 + H_2O + 2H^+ \rightleftarrows Cd^{2+} + H_4SiO_4$	-16.63	9.06
	$Cd(BO_2)_2 + 2H_2O + 2H^+ \rightleftarrows Cd^{2+} + 2H_3BO_3$	-	9.84
	$Cd_2Fe(CN)_6 \rightleftarrows 2Cd^{2+} + Fe^{2+} + 6CN^-$	-	-28.2243
	$Cd_2Fe(CN)_6 \cdot 7H_2O \rightleftarrows 6CN^- + 2Cd^{2+} + Fe^{2+} + 7H_2O$	-	-62.9824
	$K_2CdFe(CN)_6 \rightleftarrows 2K^+ + Cd^{2+} + Fe^{2+} + 6CN^-$	-	-63.0279
	$K_{12}Cd_8(Fe(CN)_6)_7 \rightleftarrows 12K^+ + 8Cd^{2+} + 7Fe^{2+} + 42CN^-$	-	-441.9853

Redox Potential	ΔH_r^o (kcal/mol)	log K
$Cd_{(s)} \rightleftarrows Cd^{2+} + 2e^-$	-18.0	13.49
$\gamma\text{-}Cd \rightleftarrows Cd^{2+} + 2e^-$	-18.14	13.59

Table 2d. Trace elements: Chromium.

Reaction	ΔH_r° (kcal/mol)	log K
$Cr^{3+} + 2H_2O \rightleftarrows Cr(OH)_2^+ + 2H^+$	20.14	-9.62
$CrCl^{2+} + 2H_2O \rightleftarrows Cr(OH)_2^+ + Cl^- + 2H^+$	13.847	-9.3683
$CrCl_2^+ + 2H_2O \rightleftarrows Cr(OH)_2^+ + 2Cl^- + 2H^+$	9.374	-8.658
$CrF^{2+} + 2H_2O \rightleftarrows Cr(OH)_2^+ + F^- + 2H^+$	16.789	-14.5424
$CrBr^{2+} + 2H_2O \rightleftarrows Cr(OH)_2^+ + Br^- + 2H^+$	11.211	-7.5519
$CrI^{2+} + 2H_2O \rightleftarrows Cr(OH)_2^+ + I^- + 2H^+$	-	-4.8289
$Cr(OH)^{2+} + H_2O \rightleftarrows Cr(OH)_2^+ + H^+$	-	-5.62
$Cr(OH)_{3(aq)} + H^+ \rightleftarrows Cr(OH)_2^+ + H_2O$	-	7.13
$Cr(OH)_4^- + 2H^+ \rightleftarrows Cr(OH)_2^+ + 2H_2O$	-	18.15
$Cr(OH)Cl_{2(aq)} + H_2O \rightleftarrows Cr(OH)_2^+ + 2Cl^- + H^+$	-	-2.9627
$Cr(NH_3)_6^{3+} + 4H^+ + 2H_2O \rightleftarrows Cr(OH)_2^+ + 6NH_4^+$	-	32.5709
$Cr(NH_3)_5(OH)^{2+} + 4H^+ + H_2O \rightleftarrows Cr(OH)_2^+ + 5NH_4^+$	-	30.2759
$CCr(NH_3)_4(OH)_2^+ + 4H^+ \rightleftarrows Cr(OH)_2^+ + 4NH_4^+$	-	29.8574
$TCr(NH_3)_4(OH)_2^+ + 4H^+ \rightleftarrows Cr(OH)_2^+ + 4NH_4^+$	-	30.5537
$Cr(NH_3)_6Cl^{2+} + 2H_2O + 4H^+ \rightleftarrows Cr(OH)_2^+ + 6NH_4^+ + Cl^-$	-	31.7932
$Cr(NH_3)_6Br^{2+} + 4H^+ + 2H_2O \rightleftarrows Cr(OH)_2^+ + 6NH_4^+ + Br^-$	-	31.887
$Cr(NH_3)_6I^{2+} + 4H^+ + 2H_2O \rightleftarrows Cr(OH)_2^+ + 6NH_4^+ + I^-$	-	32.8
$CrNO_3^{2+} + 2H_2O \rightleftarrows Cr(OH)_2^+ + NO_3^- + 2H^+$	15.64	-8.2094
$CrH_2PO_4^{2+} + 2H_2O \rightleftarrows Cr(OH)_2^+ + 4H^+ + PO_4^{3-}$	-	-31.9068
$CrSO_4^+ + 2H_2O \rightleftarrows Cr(OH)_2^+ + SO_4^{2-} + 2H^+$	12.62	-10.9654
$Cr(OH)SO_{4(aq)} + H_2O \rightleftarrows Cr(OH)_2^+ + SO_4^{2-} + H^+$	-	-8.2754
$Cr_2(OH)_2SO_4^{2+} + 2H_2O \rightleftarrows 2Cr(OH)_2^+ + SO_4^{2-} + 2H^+$	-	-16.155
$CrO_2^- + 2H^+ \rightleftarrows Cr(OH)_2^+$	-	17.7456
$CrO_3Cl^- + H_2O \rightleftarrows CrO_4^{2-} + Cl^- + 2H^+$	-	-7.3086
$CrO_3H_2PO_4^- + H_2O \rightleftarrows CrO_4^{2-} + 4H^+ + PO_4^{3-}$	-	-29.3634
$CrO_3HPO_4^{2-} + H_2O \rightleftarrows CrO_4^{2-} + 3H^+ + PO_4^{3-}$	-	-26.6806
$CrO_3SO_4^{2-} + H_2O \rightleftarrows CrO_4^{2-} + SO_4^{2-} + 2H^+$	-	-8.9937
$HCrO_4^- \rightleftarrows CrO_4^{2-} + H^+$	-0.9	-6.5089
$H_2CrO_{4(aq)} \rightleftarrows CrO_4^{2-} + 2H^+$	-	-5.6513
$NaCrO_4^- \rightleftarrows CrO_4^{2-} + Na^+$	-	-0.6963
$KCrO_4^- \rightleftarrows CrO_4^{2-} + K^+$	-	-0.799
$Cr_2O_7^{2-} + H_2O \rightleftarrows 2CrO_4^{2-} + 2H^+$	2.995	-14.5571

Table 2e. Trace elements: Chromium (cont.).

Mineral	Reaction	ΔH_r^o (kcal/mol)	log K
	$CrCl_2 \rightleftarrows Cr^{2+} + 2Cl^-$	-19.666	15.8676
	$CrCl_3 + 2H_2O \rightleftarrows Cr(OH)_2^+ + 3Cl^- + 2H^+$	-27.509	13.5067
	$CrBr_3 + 2H_2O \rightleftarrows Cr(OH)_2^+ + 3Br^- + 2H^+$	-33.777	19.9086
	$CrI_3 + 2H_2O \rightleftarrows Cr(OH)_2^+ + 3I^- + 2H^+$	-32.127	20.4767
	$Cr(OH)_2 + 2H^+ \rightleftarrows Cr^{2+} + 2H_2O$	-8.51	10.8189
	$Cr(OH)_3(A) + H^+ \rightleftarrows Cr(OH)_2^+ + H_2O$	-	-0.75
	$Cr(OH)_3(C) + H^+ \rightleftarrows Cr(OH)_2^+ + H_2O$	-7.115	1.75
	$Cr_2(OH)_2(SO_4)_2 + 2H_2O \rightleftarrows 2Cr(OH)_2^+ + 2SO_4^{2-} + 2H^+$	-	-17.9288
	$CrO_3 + H_2O \rightleftarrows CrO_4^{2-} + 2H^+$	-1.245	-3.2105
	$Ag_2CrO_4 \rightleftarrows CrO_4^{2-} + 2Ag^+$	14.04	-11.5548
	$BaCrO_4 \rightleftarrows CrO_4^{2-} + Ba^{2+}$	6.39	-9.6681
	$Cs_2CrO_4 \rightleftarrows CrO_4^{2-} + 2Cs^+$	7.504	-0.5541
	$CuCrO_4 \rightleftarrows CrO_4^{2-} + Cu^{2+}$	-	-5.4754
	$K_2CrO_4 \rightleftarrows CrO_4^{2-} + 2K^+$	4.25	0.73
	$Li_2CrO_4 \rightleftarrows CrO_4^{2-} + 2Li^+$	-10.822	4.8568
	$MgCrO_4 \rightleftarrows CrO_4^{2-} + Mg^{2+}$	-21.26	5.3801
	$(NH_4)_2CrO_4 \rightleftarrows CrO_4^{2-} + 2NH_4^+$	2.19	0.4046
	$Na_2CrO_4 \rightleftarrows CrO_4^{2-} + 2Na^+$	-4.61	3.2618
	$PbCrO_4 \rightleftarrows CrO_4^{2-} + Pb^{2+}$	10.23	-13.6848
	$Rb_2CrO_4 \rightleftarrows CrO_4^{2-} + 2Rb^+$	5.892	-0.0968
	$SrCrO_4 \rightleftarrows CrO_4^{2-} + Sr^{2+}$	-2.42	-4.8443
	$CaCrO_4 \rightleftarrows Ca^{2+} + CrO_4^{2-}$	-6.44	-2.2657
	$Hg_2CrO_4 \rightleftarrows 2Hg^+ + CrO_4^{2-}$	-	-8.7031
	$Tl_2CrO_4 \rightleftarrows 2Tl^+ + CrO_4^{2-}$	25.31	-12.0136
	$Cr_2O_3 + 2H^+ + H_2O \rightleftarrows 2Cr(OH)_2^+$	-12.125	-3.3937
	$FeCr_2O_4 + 4H^+ \rightleftarrows 2Cr(OH)_2^+ + Fe^{2+}$	-24.86	-0.9016
	$MgCr_2O_4 + 4H^+ \rightleftarrows 2Cr(OH)_2^+ + Mg^{2+}$	-39.86	12.0796
	$Cs_2Cr_2O_7 + H_2O \rightleftarrows 2CrO_4^{2-} + 2Cs^+ + 2H^+$	22.899	-17.7793
	$K_2Cr_2O_7 + H_2O \rightleftarrows 2CrO_4^{2-} + 2K^+ + 2H^+$	18.125	-15.6712
	$Na_2Cr_2O_7 + H_2O \rightleftarrows 2CrO_4^{2-} + 2Na^+ + 2H^+$	5.305	-9.8953

Table 2f. Trace elements: Chromium (cont.) and Copper.

Redox Potential	ΔH_r° (kcal/mol)	log K
$Cr(s) \rightleftarrows Cr^{2+} + 2e^-$	-34.3	32.244
$CrCN \rightleftarrows Cr^{2+} + CN^- + e^-$	-	23.888
$Cr_2CN \rightleftarrows 2Cr^{2+} + CN^- + 3e^-$	-	56.645
$Cr(OH)_2^+ + 2H^+ + e^- \rightleftarrows Cr^{2+} + 2H_2O$	6.36	2.947
$Cr(OH)_2^+ + 2H_2O \rightleftarrows CrO_4^{2-} + 6H^+ + 3e^-$	103.0	-67.376

Cu

Reaction	ΔH_r° (kcal/mol)	log K
$CuCl^+ \rightleftarrows Cu^{2+} + Cl^-$	-8.65	-0.43
$CuCl_2^- \rightleftarrows Cu^+ + 2Cl^-$	0.42	-5.5
$CuCl_{2(aq)} \rightleftarrows Cu^{2+} + 2Cl^-$	-10.56	-0.16
$CuCl_3^{2-} \rightleftarrows Cu^+ + 3Cl^-$	-0.26	-5.7
$CuCl_3^- \rightleftarrows Cu^{2+} + 3Cl^-$	-13.69	2.29
$CuCl_4^{2-} \rightleftarrows Cu^{2+} + 4Cl^-$	-7.78	4.59
$CuF^+ \rightleftarrows Cu^{2+} + F^-$	-1.62	-1.26
$CuCO_{3(aq)} \rightleftarrows Cu^{2+} + CO_3^{2-}$	-	-6.73
$Cu(CO_3)_2^{2-} \rightleftarrows Cu^{2+} + 2CO_3^{2-}$	-	-9.83
$CuHCO_3^+ \rightleftarrows Cu^{2+} + CO_3^{2-} + H^+$	-	-13.0
$Cu(OH)^+ + H^+ \rightleftarrows Cu^{2+} + H_2O$	-	8.0
$Cu(OH)_{2(aq)} + 2H^+ \rightleftarrows Cu^{2+} + 2H_2O$	-	13.68
$Cu(OH)_3^- + 3H^+ \rightleftarrows Cu^{2+} + 3H_2O$	-	26.899
$Cu(OH)_4^{2-} + 4H^+ \rightleftarrows Cu^{2+} + 4H_2O$	-	39.6
$Cu_2(OH)_2^{2+} + 2H^+ \rightleftarrows 2Cu^{2+} + 2H_2O$	-17.539	10.359
$CuSO_{4(aq)} \rightleftarrows Cu^{2+} + SO_4^{2-}$	-1.22	-2.31
$Cu(S_4)_2^{3-} + 2H^+ \rightleftarrows Cu^+ + 2HS^- + 6S^0$	-	-3.39
$CuS_4S_5^{3-} + 2H^+ \rightleftarrows Cu^+ + 2HS^- + 7S^0$	-	-2.66
$Cu(HS)_3^- \rightleftarrows Cu^{2+} + 3HS^-$	-	-25.899
$Cu(CN)_2^- \rightleftarrows Cu^+ + 2CN^-$	29.1	-24.0272
$Cu(CN)_3^{2-} \rightleftarrows Cu^+ + 3CN^-$	40.2	-28.6524
$Cu(CN)_4^{3-} \rightleftarrows Cu^+ + 4CN^-$	51.4	-30.3456

Table 2g. Trace elements: Copper (cont.).

Mineral	Reaction	ΔH_r° (kcal/mol)	log K
Melanothallite	$CuCl_2 \rightleftharpoons Cu^{2+} + 2Cl^-$	-12.32	3.73
	$CuF_2 \rightleftharpoons Cu^{2+} + 2F^-$	-13.32	-0.62
	$CuF_2 \cdot 2H_2O \rightleftharpoons Cu^{2+} + 2F^- + 2H_2O$	-3.65	-4.55
	$CuBr \rightleftharpoons Cu^+ + Br^-$	13.08	-8.21
	$CuI \rightleftharpoons Cu^+ + I^-$	20.14	-11.89
	$Cu(OH)_2 + 2H^+ \rightleftharpoons Cu^{2+} + 2H_2O$	-15.25	8.64
Atacamite	$Cu_2(OH)_3Cl + 3H^+ \rightleftharpoons 2Cu^{2+} + 3H_2O + Cl^-$	-18.69	7.34
	$Cu_2(OH)_3NO_3 + 3H^+ \rightleftharpoons 2Cu^{2+} + 3H_2O + NO_3^-$	-17.35	9.24
Azurite	$Cu_3(OH)_2(CO_3)_2 + 2H^+ \rightleftharpoons 3Cu^{2+} + 2H_2O + 2CO_3^{2-}$	-23.77	-16.92
Malachite	$Cu_2(OH)_2(CO_3) + 2H^+ \rightleftharpoons 2Cu^{2+} + 2H_2O + CO_3^{2-}$	-15.61	-5.18
	$CuCO_3 \rightleftharpoons Cu^{2+} + CO_3^{2-}$	-	-9.63
Covellite	$CuS + H^+ \rightleftharpoons Cu^{2+} + HS^-$	24.01	-23.038
Blaublei I	$Cu_{0.9}^{2+}Cu_{0.2}^{+}S + H^+ \rightleftharpoons 0.9Cu^{2+} + 0.2Cu^+ + HS^-$	-	-24.162
Blaublei II	$Cu_{0.6}^{2+}Cu_{0.8}^{+}S + H^+ \rightleftharpoons 0.6Cu^{2+} + 0.8Cu^+ + HS^-$	-	-27.279
Anilite	$Cu_{0.25}^{2+}Cu_{1.5}^{+}S + H^+ \rightleftharpoons 0.25Cu^{2+} + 1.5Cu^+ + HS^-$	43.535	-31.878
Djurleite	$Cu_{0.066}^{2+}Cu_{1.868}^{+}S + H^+ \rightleftharpoons$		
	$0.066Cu^{2+} + 1.868Cu^+ + HS^-$	47.881	-33.92
Chalcopyrite	$CuFeS_2 + 2H^+ \rightleftharpoons Cu^{2+} + Fe^{2+} + 2HS^-$	35.48	-35.27
	$CuSO_4 \rightleftharpoons Cu^{2+} + SO_4^{2-}$	-18.14	3.01
Chalcanthite	$CuSO_4 \cdot 5H_2O \rightleftharpoons Cu^{2+} + SO_4^{2-} + 5H_2O$	1.44	-2.64
Brochantite	$Cu_4(SO_4)(OH)_6 + 6H^+ \rightleftharpoons 4Cu^{2+} + 6H_2O + SO_4^{2-}$	-	15.34
Langite	$Cu_4(SO_4)(OH)_6 \cdot H_2O + 6H^+ \rightleftharpoons 4Cu^{2+} + 7H_2O + SO_4^{2-}$	-39.61	16.79
Antlerite	$Cu_3(SO_4)(OH)_4 + 4H^+ \rightleftharpoons 3Cu^{2+} + 4H_2O + SO_4^{2-}$	-	8.29
	$Cu_2SO_4 \rightleftharpoons 2Cu^+ + SO_4^{2-}$	-4.56	-1.95
	$CuOCuSO_4 + 2H^+ \rightleftharpoons 2Cu^{2+} + H_2O + SO_4^{2-}$	-35.575	11.53
Tenorite	$CuO + 2H^+ \rightleftharpoons Cu^{2+} + H_2O$	-15.24	7.62
Cupric Ferrite	$Cu(Fe)_2O_4 + 8H^+ \rightleftharpoons Cu^{2+} + 2Fe^{3+} + 4H_2O$	-38.69	5.88
Cuprous Ferrite	$CuFeO_2 + 4H^+ \rightleftharpoons Cu^+ + Fe^{3+} + 2H_2O$	-3.8	-8.92
Dioptase	$CuSiO_3 \cdot H_2O + 2H^+ \rightleftharpoons Cu^{2+} + H_4SiO_4$	-8.96	6.5
	$CuCN \rightleftharpoons Cu^+ + CN^-$	30.2	-19.4974
	$Cu_3(PO_4)_2 \rightleftharpoons 3Cu^{2+} + 2PO_4^{3-}$	-	-36.85
	$Cu_3(PO_4)_2 \cdot 3H_2O \rightleftharpoons 3Cu^{2+} + 2PO_4^{3-} + 3H_2O$	-	-35.12
	$CuSe + H^+ \rightleftharpoons Cu^{2+} + HSe^-$	28.95	-26.5121
	$Cu_2Se_{(\alpha)} + H^+ \rightleftharpoons 2Cu^+ + HSe^-$	51.21	-36.0922

Table 2h. Trace elements: Copper (cont.) and Nickel.

Mineral	Reaction	ΔH_r° (kcal/mol)	log K
	$Cu_3Se_2 + 2H^+ \rightleftarrows 2Cu^+ + Cu^{2+} + 2HSe^-$	81.34	-63.4911
	$CuSeO_3 \cdot 2H_2O + H^+ \rightleftarrows HSeO_3^- + Cu^{2+} + 2H_2O$	-8.81	0.4838
	$Cu_2Fe(CN)_6 \rightleftarrows 2Cu^{2+} + Fe^{2+} + 6CN^-$	-	-61.4168

Redox Potential		ΔH_r° (kcal/mol)	log K
$Cu^+ \rightleftarrows Cu^{2+} + e^-$		-1.65	-2.72
$CuSe_2 + 2H^+ + 2e^- \rightleftarrows Cu^{2+} + 2HSe^-$		33.6	-33.3655
$Cu_2Sb \rightleftarrows Cu^+ + Cu^{2+} + Sb(OH)_3 + 6e^- + 3H^+$		55.745	-34.8827
$Cu_3Sb + 3H_2O \rightleftarrows 3Cu^+ + Sb(OH)_3 + 6e^- + 3H^+$		73.645	-42.5937
$Cu(SbO_3)_2 + 6H^+ + 4e^- \rightleftarrows Cu^{2+} + 2Sb(OH)_3$		-	45.2105

Ni

Reaction	ΔH_r° (kcal/mol)	log K
$NiCl^+ \rightleftarrows Ni^{2+} + Cl^-$	-	-0.399
$NiCl_{2(aq)} \rightleftarrows Ni^{2+} + 2Cl^-$	-	-0.96
$NiF^+ \rightleftarrows Ni^{2+} + F^-$	-	-1.3
$NiBr^+ \rightleftarrows Ni^{2+} + Br^-$	-	-0.5
$Ni(OH)^+ + H^+ \rightleftarrows Ni^{2+} + H_2O$	-12.42	9.86
$Ni(OH)_{2(aq)} + 2H^+ \rightleftarrows Ni^{2+} + 2H_2O$	-	19.0
$Ni(OH)_3^- + 3H^+ \rightleftarrows Ni^{2+} + 3H_2O$	-	30.0
$NiSO_{4(aq)} \rightleftarrows Ni^{2+} + SO_4^{2-}$	-1.52	-2.29
$Ni(SO_4)_2^{2-} \rightleftarrows Ni^{2+} + 2SO_4^{2-}$	-	-1.02
$NiCO_{3(aq)} \rightleftarrows Ni^{2+} + CO_3^{2-}$	-	-6.87
$Ni(CO_3)_2^{2-} \rightleftarrows Ni^{2+} + 2CO_3^{2-}$	-	-10.11
$NiHCO_3^+ \rightleftarrows Ni^{2+} + CO_3^{2-} + H^+$	-	-12.47
$Ni(CN)_{2(aq)} \rightleftarrows Ni^{2+} + 2CN^-$	-	-14.5864
$Ni(CN)_3^- \rightleftarrows Ni^{2+} + 3CN^-$	-	-22.6346
$Ni(CN)_4^{2-} \rightleftarrows Ni^{2+} + 4CN^-$	43.19	-30.1257
$NiH(CN)_4^- \rightleftarrows Ni^{2+} + H^+ + 4CN^-$	-	-36.7482

Table 2i. Trace elements: Nickel (cont.) and Lead.

Reaction	ΔH_r° (kcal/mol)	log K
$NiH_2CN_{4(aq)} \rightleftarrows Ni^{2+} + 2H^+ + 4CN^-$	-	-41.4576
$NiH_3(CN)_4^+ \rightleftarrows Ni^{2+} + 3H^+ + 4CN^-$	-	-43.9498

Mineral	Reaction	ΔH_r° (kcal/mol)	log K
	$Ni(OH)_2 + 2H^+ \rightleftarrows Ni^{2+} + 2H_2O$	30.45	10.8
	$Ni_4(OH)_6SO_4 + 6H^+ \rightleftarrows 4Ni^{2+} + SO_4^{2-} + 6H_2O$	-	32.0
	$NiCO_3 \rightleftarrows Ni^{2+} + CO_3^{2-}$	-9.94	-6.84
Millerite	$NiS + H^+ \rightleftarrows Ni^{2+} + HS^-$	2.5	-8.042
Morenosite	$NiSO_4 \cdot 7H_2O \rightleftarrows Ni^{2+} + SO_4^{2-} + 7H_2O$	2.94	-2.36
Retgersite	$NiSO_4 \cdot 6H_2O \rightleftarrows Ni^{2+} + SO_4^{2-} + 6H_2O$	1.1	-2.04
Bunsenite	$NiO + 2H^+ \rightleftarrows Ni^{2+} + H_2O$	-23.92	12.45
	$NiSeO_4 \rightleftarrows Ni^{2+} + SeO_4^{2-}$	-3.5	-2.6387
	$Ni_3(PO_4)_2 \rightleftarrows 3Ni^{2+} + 2PO_4^{3-}$	-	-31.3
	$Ni_2SiO_4 + 4H^+ \rightleftarrows 2Ni^{2+} + H_4SiO_4$	-33.36	14.54
	$NiSe + H^+ \rightleftarrows Ni^{2+} + HSe^-$	-	-17.7382
	$K_2Ni_3(Fe(CN)_6)_2 \rightleftarrows 2K^+ + 3Ni^{2+} + 2Fe^{2+} + 12CN^-$	-	-123.1267
	$K_4Ni_4(Fe(CN)_6)_3 \rightleftarrows 4K^+ + 4Ni^{2+} + 3Fe^{2+} + 18CN^-$	-	-183.5467

Pb

Reaction	ΔH_r° (kcal/mol)	log K
$PbCl^+ \rightleftarrows Pb^{2+} + Cl^-$	-4.38	-1.6
$PbCl_{2(aq)} \rightleftarrows Pb^{2+} + 2Cl^-$	-1.08	-1.8
$PbCl_3^- \rightleftarrows Pb^{2+} + 3Cl^-$	-2.17	-1.699
$PbCl_4^{2-} \rightleftarrows Pb^{2+} + 4Cl^-$	-3.53	-1.38
$PbF^+ \rightleftarrows Pb^{2+} + F^-$	-	-1.25
$PbF_2 \rightleftarrows Pb^{2+} + 2F^-$	-0.7	-7.44
$PbF_{2(aq)} \rightleftarrows Pb^{2+} + 2F^-$	-	-2.56
$PbF_3^- \rightleftarrows Pb^{2+} + 3F^-$	-	-3.42

Table 2j. Trace elements: Lead (cont.)

Reaction	ΔH_r° (kcal/mol)	log K
$PbF_4^{2-} \rightleftarrows Pb^{2+} + 4F^-$	-	-3.1
$PbBr^+ \rightleftarrows Pb^{2+} + Br^-$	-2.88	-1.77
$PbBr_{2(aq)} \rightleftarrows Pb^{2+} + 2Br^-$	-	-1.44
$PbI^+ \rightleftarrows Pb^{2+} + I^-$	-	-1.94
$PbI_{2(aq)} \rightleftarrows Pb^{2+} + 2I^-$	-	-3.199
$Pb(OH)^+ + H^+ \rightleftarrows Pb^{2+} + H_2O$	-	7.71
$Pb(OH)_{2(aq)} + 2H^+ \rightleftarrows Pb^{2+} + 2H_2O$	-	17.12
$Pb(OH)_3^- + 3H^+ \rightleftarrows Pb^{2+} + 3H_2O$	-	28.06
$Pb(OH)_4^{2-} + 4H^+ \rightleftarrows Pb^{2+} + 4H_2O$	-	39.699
$Pb_2(OH)^{3+} + H^+ \rightleftarrows 2Pb^{2+} + H_2O$	-	6.36
$Pb_3(OH)_4^{2+} + 4H^+ \rightleftarrows 3Pb^{2+} + 4H_2O$	-26.5	23.88
$PbCO_{3(aq)} \rightleftarrows Pb^{2+} + CO_3^{2-}$	-	-7.24
$Pb(CO_3)_2^{2-} \rightleftarrows Pb^{2+} + 2CO_3^{2-}$	-	-10.64
$PbHCO_3^+ \rightleftarrows Pb^{2+} + CO_3^{2-} + H^+$	-	-13.2
$PbSO_{4(aq)} \rightleftarrows Pb^{2+} + SO_4^{2-}$	-	-2.75
$Pb(SO_4)_2^{2-} \rightleftarrows Pb^{2+} + 2SO_4^{2-}$	-	-3.47
$Pb(HS)_{2(aq)} \rightleftarrows Pb^{2+} + 2HS^-$	-	-15.27
$Pb(HS)_3^- \rightleftarrows Pb^{2+} + 3HS^-$	-	-16.57
$PbNO_3^+ \rightleftarrows Pb^{2+} + NO_3^-$	-	-1.17

Mineral	Reaction	ΔH_r° (kcal/mol)	log K
Cotunnite	$PbCl_2 \rightleftarrows Pb^{2+} + 2Cl^-$	5.6	-4.77
Phosgenite	$PbCl\bullet_2PbCO_3 \rightleftarrows 2Pb^{2+} + 2Cl^- + CO_3^{2-}$	-	-19.81
Matlockite	$PbFCl \rightleftarrows Pb^{2+} + Cl^- + F^-$	7.95	-9.43
	$PbBr_2 \rightleftarrows Pb^{2+} + 2Br^-$	8.1	-5.18
	$PbBrF \rightleftarrows Pb^{2+} + Br^- + F^-$	-	-8.49
	$PbI_2 \rightleftarrows Pb^{2+} + 2I^-$	15.16	-8.07
	$Pb(OH)_{2(C)} + 2H^+ \rightleftarrows Pb^{2+} + 2H_2O$	-13.99	8.15
	$Pb_2O(OH)_2 + 4H^+ \rightleftarrows 2Pb^{2+} + 3H_2O$	-	26.2
	$Pb_2(OH)_3Cl + 3H^+ \rightleftarrows 2Pb^{2+} + 3H_2O + Cl^-$	-	8.793
Laurionite	$Pb(OH)Cl + H^+ \rightleftarrows Pb^{2+} + Cl^- + H_2O$	-	0.623
Cerussite	$PbCO_3 \rightleftarrows Pb^{2+} + CO_3^{2-}$	4.86	-13.13

Table 2k. Trace elements: Lead (cont.)

Mineral	Reaction	ΔH_r° (kcal/mol)	log K
Hydrocerussite	$2PbCO_3 \cdot Pb(OH)_2 + 2H^+ \rightleftarrows 3Pb^{2+} + 2CO_3^{2-} + 2H_2O$	-	-17.46
Galena	$PbS + H^+ \rightleftarrows Pb^{2+} + HS^-$	19.4	-15.132
Anglesite	$PbSO_4 \rightleftarrows Pb^{2+} + SO_4^{2-}$	2.15	-7.79
Litharge	$PbO + 2H^+ \rightleftarrows Pb^{2+} + H_2O$	-16.38	12.72
Massicot	$PbO + 2H^+ \rightleftarrows Pb^{2+} + H_2O$	-16.78	12.91
	$PbO \cdot 0.33H_2O + 2H^+ \rightleftarrows Pb^{2+} + 1.33H_2O$	-	12.98
Larnakite	$PbO \cdot PbSO_4 + 2H^+ \rightleftarrows 2Pb^{2+} + SO_4^{2-} + H_2O$	-6.44	-0.28
Plattnerite	$PbO_2 + 4H^+ + 2e^- \rightleftarrows Pb^{2+} + 2H_2O$	-70.73	49.3
Minium	$Pb_3O_4 + 8H^+ + 2e^- \rightleftarrows 3Pb^{2+} + 4H_2O$	-102.76	73.69
	$Pb_2OCO_3 + 2H^+ \rightleftarrows 2Pb^{2+} + H_2O + CO_3^{2-}$	-11.46	-0.5
	$Pb_3O_2SO_4 + 4H^+ \rightleftarrows 3Pb^{2+} + SO_4^{2-} + 2H_2O$	-20.75	10.4
	$Pb_4O_3SO_4 + 6H^+ \rightleftarrows 4Pb^{2+} + SO_4^{2-} + 3H_2O$	-35.07	22.1
	$Pb_3O_2CO_3 + 4H^+ \rightleftarrows 3Pb^{2+} + CO_3^{2-} + 2H_2O$	-26.43	11.02
	$PbSiO_3 + H_2O + 2H^+ \rightleftarrows Pb^{2+} + H_4SiO_4$	-9.26	7.32
	$Pb_2SiO_4 + 4H^+ \rightleftarrows 2Pb^{2+} + H_4SiO_4$	-26.0	19.76
	$Pb_4(OH)_6SO_4 + 6H^+ \rightleftarrows 4Pb^{2+} + SO_4^{2-} + 6H_2O$	-	21.1
	$Pb(BO_2)_2 + 2H_2O + 2H^+ \rightleftarrows Pb^{2+} + 2H_3BO_3$	-5.8	7.61
	$PbHPO_4 \rightleftarrows Pb^{2+} + PO_4^{3-} + H^+$	-	-23.9
	$Pb_3(PO_4)_2 \rightleftarrows 3Pb^{2+} + 2PO_4^{3-}$	-	-44.5
Cl-pyromorphite	$Pb_5(PO_4)_3Cl \rightleftarrows 5Pb^{2+} + 3PO_4^{3-} + Cl^-$	-	-84.43
Hxy-pyromorphite	$Pb_5(PO_4)_3(OH) + H^+ \rightleftarrows 5Pb^{2+} + 3PO_4^{3-} + H_2O$	-	-62.79
Plumbogummite	$PbAl_3(PO_4)_2(OH)_5 \cdot H_2O + 5H^+ \rightleftarrows$		
	$Pb^{2+} + 3Al^{3+} + 2PO_4^{3-} + 6H_2O$	-	-32.79
Tsumebite	$Pb_2Cu(PO_4)(OH)_3 \cdot 3H_2O + 3H^+ \rightleftarrows$		
	$2Pb^{2+} + Cu^{2+} + PO_4^{3-} + 6H_2O$	-	-9.79
Hinsdalite	$(Pb,Sr)Al_3(PO_4)(SO_4)(OH)_6 + 6H^+ \rightleftarrows$		
	$Pb^{2+}/Sr^{2+} + 3Al^{3+} + PO_4^{3-} + SO_4^{2-} + 6H_2O$	-	-2.5
	$PbSeO_4 \rightleftarrows SeO_4^{2-} + Pb^{2+}$	3.8	-6.8387
	$Pb_2Fe(CN)_6 \rightleftarrows 2Pb^{2+} + Fe^{2+} + 6CN^-$	-	-27.5895
	$Pb_2Fe(CN)_6 \cdot 3H_2O \rightleftarrows 2Pb^{2+} + Fe^{2+} + 6CN^- + 3H_2O$	-	-63.6011

Table 21. Trace elements: Some redox reactions.

Redox Potential	ΔH_r^o (kcal/mol)	log K
$Pb_{(s)} \rightleftarrows Pb^{2+} + 2e^-$	0.4	4.27
$Pb_2O_3 + 6H^+ + 2e^- \rightleftarrows 2Pb^{2+} + 3H_2O$	-	61.04

Some redox reactions

Redox Potential	ΔH_r^o (kcal/mol)	log K
$Fe^{2+} \rightleftarrows Fe^{3+} + e^-$	10.0	-13.032
$NO_2^- + H_2O \rightleftarrows NO_3^- + 2H^+ + 2e^-$	43.76	-28.57
$NH_4^+ + 3H_2O \rightleftarrows NO_3^- + 10H^+ + 8e^-$	187.055	-119.077
$Mn^{3+} + e^- \rightleftarrows Mn^{2+}$	-25.76	25.507
$HS^- + 4H_2O \rightleftarrows SO_4^{2-} + 9H^+ + 8e^-$	60.14	-33.66
$CH_4(g) + 3H_2O \rightleftarrows CO_3^{2-} + 8e^- + 10H^+$	61.0	-40.1

Answers to problems

CHAPTER 1

1.1a. The difference: $7.49 - 7.74 = -0.25$ meq/l $= -1.7\%$. This is a fairly good analysis.

Na$^+$	K$^+$	Ca^{2+}	Mg^{2+}	$\Sigma+$	Cl$^-$	HCO$_3^-$	SO$_4^{2-}$	$\Sigma-$
2.70	0.002	4.05	0.74	7.49	−4.74	−3.00	−0.0	−7.74 meq/l

1.1b. Total Hardness $= 4.05 + 0.74 = 4.79$ meq/l
expressed as mg/l CaCO$_3$: 240 mg/l
expressed as °d (German Hardness): 13.5 °d.
Alkalinity (3.00 meq/l) equals 150 mg/l CaCO$_3$.
Hence, non-carbonate hardness is 90 mg/l. It derives from other sources than calcite dissolution and will remain in the water when pH increases and calcite precipitates.

1.2.1. $\Sigma+ = 7.4$ meq/l, $\Sigma- = -6.0$ meq/l.
Comparison of these values with the measured EC of 750 μS/cm suggests that there is an anion shortage. This shortage is probably due to a too low HCO$_3^-$ value, since the calculated $P_{CO_2} = 10^{-3.7}$ is below the atmospheric level.

1.2.2. $\Sigma+ = 3.6$ meq/l, $\Sigma- = -1.5$ meq/l.
In this case there is a cation-surplus and if we look at the EC (150 μS/cm), it seems that Ca^{2+} determination must be wrong since the contribution of measured Ca^{2+} concentration to the EC would be ≈ 280 μS/cm. At the measured pH, the Al content is also rather high.

1.2.3. $\Sigma+ = 4.5$ meq/l, (Fe included as Fe^{2+}), $\Sigma- = -4.5$ meq/l.
Seems all right ($EC = 450$ μS/cm), but NO$_3^-$ cannot occur together with Fe^{2+}. If a brown Fe(OH)$_3$ precipitate is present in the non-acidified sample, the contents of anions must be wrong. Remember that the concentration of HCO$_3^-$ decreases when Fe^{2+} becomes oxidized in the non-acidified sample:

$$Fe^{2+} + \tfrac{1}{4}O_2 + 2HCO_3^- + \tfrac{1}{2}H_2O \rightarrow Fe(OH)_3 + 2CO_2$$

Thus, alkalinity is best determined in the field!

1.3b. A = St. Heddinge waterworks, B = Maarum waterworks, C = Haralds mineral water, D = Rainwater Borris

Sample No.		A	B	C	D
NO_3^-	(mM)	0.565	0.032	0.081	0.073
HCO_3^-	(mM)	4.867	17.748	–	–
SO_4^{2-}	(mM)	0.635	–	–	0.047
Cl^-	(mM)	0.818	5.669	0.480	0.338
$\Sigma-$	(meq/l)	7.52	23.450	0.561	0.505
Ca^{2+}	(mM)	3.019	1.048	–	0.035
Mg^{2+}	(mM)	0.579	1.193	–	0.032
Na^+	(mM)	0.783	19.139	0.391	0.313
K^+	(mM)	0.161	0.358	0.026	0.016
NH_4^+	(mM)	0.001	0.194	–	0.068
$\Sigma+$	(meq/l)	8.134	24.173	0.417	0.531
EC	(μS/cm)	750	1900	88	

ad A:
$$E.N. = \frac{8.134 - 7.520}{8.134 + 7.520} \cdot 100\% = 3.9\%$$

The E.N. is reasonable but not perfect. Since $EC/100 = 750/100 = 7.50$ is in best agreement with $\Sigma-$, one of the cation concentrations must be a bit high, possibly Ca^{2+}.

ad B:
$$E.N. = \frac{24.173 - 23.450}{24.173 + 23.450} \cdot 100\% = 1.5\%$$

Based on the E.N., this is a rather good analysis. However, $EC/100 = 19.00$ is not in good agreement with the sum of cations or anions, but we are at the upper limit where the simple relation $\Sigma+ = \Sigma-$ (in meq/l) $= EC/100$ is justified.

ad C:
The analysis is incomplete and the E.N. cannot be calculated. The hardness in german degrees is $1°dH = 17.8$ mg $CaCO_3$/l $= 0.18$ mmol/l $(Ca^{2+} + Mg^{2+}) = 0.36$ meq/l. The $\Sigma+$ becomes then 0.78 meq/l ~ 78 μS/cm EC, which is in reasonable agreement with the measured EC.

ad D:
$$E.N. = \frac{0.532 - 0.505}{0.532 + 0.505} \cdot 100\% = 2.6\%$$

1.3c. Rainwater contributions calculated from D.

		A	B	C
Concentration factor		2.42	16.75	1.42
SO_4^{2-}	mg/l	10.83	75.04	6.35
Cl^-	mg/l	29.00	201.00	17.00
PO_4^{3-}	mg/l	0.00	0.00	0.00
NO_3^-	mg/l	10.92	75.71	6.40
NH_4^+	mg/l	2.97	20.60	1.74
Ca^{2+}	mg/l	3.38	23.45	1.98
Mg^{2+}	mg/l	1.89	13.07	1.11
Na^+	mg/l	17.40	120.60	10.20
K^+	mg/l	1.52	10.55	0.89

ad A:

The contribution of rainwater is small for most components, except for Na^+ which apparently is derived mainly from rainwater.

ad B:

The concentration factor of 16.75 is much too high for Danish conditions and the high Cl^- concentration indicates a marine influence.

ad C:

Most of the dissolved contents in this sample originates from rainwater.

1.3d.

ad A:

The high Ca^{2+} and HCO_3^- contents indicate the presence of $CaCO_3$ in the aquifer and the high NO_3^- content suggests agricultural activity in the recharge area.

ad B:

The high Na^+ content, which is balanced by HCO_3^-, indicates that ion exchange has influenced this water composition.

ad C:

This sample must originate from a reservoir with little reactive material, since most of the dissolved content is derived from rainwater.

CHAPTER 2

2.4a. In the following table, concentrations of cations and anions are expressed in $\mu eq/l$. The estimated parameters are underlined.

	H^+	NH_4^+	Na^+	K^+	Ca^{2+}	Mg^{2+}	$\Sigma+$	Cl^-	NO_3^-	SO_4^{2-}	$\Sigma-$
Vlissingen	100	49	265	8	78	68	568	297	92	170	559
Deelen	112	119	30	3	32	10	306	37	90	<u>179</u>	<u>306</u>
Beek	<u>26</u>	143	23	11	100	10	<u>313</u>	34	87	192	313

Vlissingen: The error in the charge balance is 0.8%, which is a rather good analysis.

2.4b. The seawater contribution to the total concentration of ion i may be calculated from:

$$C_{i\,(rain)} = \frac{C_{i\,(sea)}}{Cl_{(sea)}} \cdot Cl_{(rain)}$$

	Vlissingen			Deelen			Beek		
	Total	Sea	%Sea	Total	Sea	%Sea	Total	Sea	%Sea
H^+	100	0	0	112	0	0	26	0	0
NH_4^+	49	0	0	119	0	0	143	0	0
Na^+	265	254.5	96	30	31.7	>100	23	29.1	>100
K^+	8	5.6	70	3	0.7	23	11	0.6	6
Ca^{2+}	78	10.5	13	32	1.3	4	100	1.2	1
Mg^{2+}	68	57.7	85	10	7.2	72	10	6.6	66
Cl^-	297	297	100	37	37	100	34	34	100
NO_3^-	92	0	0	90	0	0	87	0	0
SO_4^{2-}	170	31	18	179	3.9	2	192	3.5	2

2.4c. – The very low H^+, NO_3^- and NH_4^+ contents in seawater cannot contribute significantly to the rainwater composition. Instead, their sources in rainwater are industrial processes and the evaporation of ammonia.

– The seawater contribution for the Na^+ concentration is negative for both Deelen and Beek. This might be an effect of industrial Cl^- production. In such cases Na^+ is preferred as conservative ion.

– Dust and fertilizer are likely sources for K^+ in rain.

– The seawater contribution to the Ca^{2+} concentration is low and cement industry seems to be the source of high Ca^{2+} concentrations in rain at Beek.

2.5.

Alps.

Ca^{2+} and HCO_3^- are the only ions that are present in appreciable quantities and calcite dissolution is therefore the dominant process.

Salt Marsh.
1. Diluted seawater (seawater $\cdot$ 2/5).
2. As 1. plus sulfate reduction. The HCO_3^- content increases equivalently.
3. As 2. plus cation exchange of Na^+ for Ca^{2+} and some precipitation of calcite.

Brabant.
This sample corresponds to acid rain, concentrated 3 times by evapotranspiration under a forest.

Kenya.
The water has become concentrated by evaporation in an arid climate and calcite has been precipitated. Silicate weathering may have supplied additional Na^+ ($Na^+ > Cl^-$).

CHAPTER 3

3.1a.

A: $SI = \log IAP - \log K = -11.08 + 10.57 = -0.51$
B: $SI = \log IAP - \log K = -10.07 + 10.57 = 0.50$
C: $SI = \log IAP - \log K = -8.78 + 10.57 = 1.79$

3.1b.

	I	γ_{F^-}	$\gamma_{Ca^{2+}}$	IAP	SI
A:	0.026	0.86	0.54	−11.48	−0.91
B:	0.057	0.81	0.44	−10.61	−0.04
C:	0.199	0.75	0.31	− 9.54	1.03

3.2a. 25°C $\log K_{villiaumite} = -0.49$

3.2b. 10°C $\log K_{villiaumite} = -0.50$

3.2c. $SI_{villiaumite}$
  A: −5.41
  B: −4.45
  C: −2.67

3.3a $SI_{gypsum} = \log IAP - \log K = -5.45 + 4.60 = -0.85$

3.3b. 2.56 mmol gypsum dissolves, $[Ca^{2+}] = 3.24$ mmol/l and $[F^-] = 0.091$ mmol/l.

3.4. Analysis A, but it requires that fluorite is present in the rock and at equilibrium with the water.

3.5. $O_{2(g)} \leftrightarrow O_{2(aq)}$
$\Delta G_r^0 = 16.5 - 0 = 16.5$ kJ/mol
$\Delta G_r^0 = -RT \ln K \Rightarrow K = [O_{2(aq)}]/ P_{O_2} = e^{(-16.5/2.48)} = 10^{-2.89}$

At 25°C:

$P_{O_2} = 10^{-0.68} \rightarrow O_{2(aq)} = 0.27$ mmol/l = 8.6 mg/l.

Van't Hoff-equation:

$\log K_T = \log K_{25} + (-10000/19.1) \cdot (1/298 - 1/T)$

At 15°C:

(288 K): $\log K = -2.83 \rightarrow O_{2(aq)} = 0.31$ mmol/l = 9.9 mg/l

At 5°C:

(278 K): $\log K = -2.76 \rightarrow O_{2(aq)} = 0.36$ mmol/l = 11.5 mg/l.

3.7b. $d(FeS_2)/dt = k(FeS_{surf})^2(S_{surf}) P_{H_2S}$

$k = 2 \cdot 10^{-20}$ mol/l.s.cm^6.atm

3.7c. The second, since the rate depends on the dissolved sulfide (H_2S) concentration.

CHAPTER 4

4.1. $\gamma_1 = 0.90$; $\gamma_2 = 0.67$. At pH = 7 is $[HCO_3^-] = 1/0.9$ times smaller, is $10^{-2.75}$. Similarly $[CO_3^{2-}] = 10^{-5.88}$, and $TIC = 10^{-2.68}$. If we include $m_{HCO_3^-}$ in the calculation of I, $\gamma_1 = 0.897$, $\gamma_2 = 0.65$. Note that complexes, such as $NaHCO_3^0$, are insignificant.
    At pH = 10 concentrations of the carbonate species are so high, that I is outside the range of the Debije-Hückel theory.

4.2. pH = 6.81. $[CO_3^{2-}] = 10^{-6.38}$.

4.3. $m_{CO_3^{2-}} = \alpha \cdot TIC$; $\alpha^{-1} = 10^{16.6} \gamma_2 [H^+]^2 + [H^+] \gamma_2 / (\gamma_1 10^{-10.3}) + 1$; $\gamma_1 = 0.616$; $\gamma_2 = 0.144$.
   At pH = 10.5: $\alpha^{-1} = 5.7 \cdot 10^{-6} + 0.147 + 1 = 1.147$; $m_{CO_3^{2-}} = 2.18 \cdot 10^{-3.0}$
      $[CO_3^{2-}] = 3.14 \cdot 10^{-4}$.
   At pH = 6.3: $\alpha^{-1} = 3779$; $m_{CO_3^{2-}} = 6.62 \cdot 10^{-7.0}$; $[CO_3^{2-}] = 9.53 \cdot 10^{-8.0}$.

4.4. $[H_2CO_3] = 10^{-3.5}$; $[H^+] = 10^{-4.9} = [HCO_3^-]$; $[CO_3^{2-}] = 10^{-10.3}$.

4.5. $Ca^{2+} = 1.58$ mmol/l. pH = 7.3.

4.6. $P_{CO_2} = 0.02$ atm.

4.7. $P_{CO_2} = 10^{-2.45}$ atm.

4.8. $k_1 \cdot [H^+] \approx 0$; $k_2 \cdot [H_2CO_3^*] = 10^{-9.8}$, $k_3 \cdot [H_2O] = 10^{-7.0}$; $R_{fw} = 10^{-7.0}$.
$[Ca^{2+}]_{eq} = 0.5 \cdot 10^{-3}$; $R_b = k_4 \cdot 2 \cdot [Ca^{2+}]^2 = R_{fw}$; $k_4 = 0.2$.
95% saturation, or $m_{Ca^{2+}} = 4.75 \cdot 10^{-4}$; $t = 9159$ s = 2.5 hr.

4.9a. $\log P_{CO_2}$ can be calculated from $\log P_{CO_2} = 7.7 + \log [HCO_3^-] - pH$, which gives:
(1) -1; (2) -2.3; (3) -2.7; (4) -1.7.

4.9b. Saturation with respect to calcite is calculated from:

$$H^+ + CaCO_3 \leftrightarrow Ca^{2+} + HCO_3^- \qquad \log K = 2.0$$

	SI	calcite?	open/closed
(1)	−5.9	no	–
(2)	0.3	yes	open
(3)	0.7	yes	partly closed
(4)	−1.7	partly	

4.9c. (1) acid rain; (2) –; (3) SO_4^{2-} high; (4) K^+, NO_3^- high.

4.10a. CO_2 in water is analyzed as H_2CO_3.

$$H_2CO_3 \leftrightarrow H^+ + HCO_3^- \qquad K = 10^{-6.3}$$

$$[H^+] = 10^{-6.3} \cdot 66 \cdot 10^{-3}/6.38 \cdot 10^{-3} = 10^{-5.3}; pH = 5.3$$

4.10b. Alkalinity will not change if P_{CO_2} decreases.
(Alkalinity $= m_{HCO_3^-} + 2m_{CO_3^{2-}} + m_{OH^-} - m_{H^+}$). CO_2 concentration will reach equilibrium with the atmosphere: $H_2CO_3 = 10^{-5}$ mol/l. Hence:

$$[H^+] = 10^{-6.3} \cdot 10^{-5}/6.38 \cdot 10^{-3} = 10^{-9.1} \qquad pH = 9.1$$

4.10c. $\Sigma+ = 4.45$ meq/l and $\Sigma- = -7.78$ meq/l; Probably Fe precipitated in the non-acidified sample.

4.10d. To maintain charge balance an equivalent amount (in meq/l) of HCO_3^- must be removed from the solution:

$$Fe^{2+} + 2 HCO_3^- + \tfrac{1}{4} O_2 + \tfrac{1}{2} H_2O \rightarrow Fe(OH)_3 + 2 CO_2$$

4.10e. Juvenile = first emergence at the earth's surface.

4.11.1. Constant: if H_2CO_3 dissociates ($H_2CO_3 \leftrightarrow H^+ + HCO_3^-$), H^+ and HCO_3^- compensate each other's contribution to alkalinity.

4.11.2. Decrease: HCO_3^- reacts with added H^+: $HCO_3^- + H^+ \leftrightarrow H_2CO_3$

4.11.3. Decrease: Fe^{3+} will take up OH^-.

4.11.4. Constant.

4.11.5. Decrease: Mn^{2+} is oxidized by O_2, and releases H^+:

$$Mn^{2+} + H_2O + \tfrac{1}{2}O_2 \leftrightarrow MnO_2 + 2 H^+.$$

4.11.6. Increase.

4.11.7. Increase.

4.11.8. Increase.

4.11.9. Constant.

4.12. 60 mg/l SiO_2 = 1 mmol/l ΣSi. At pH = 10.0 :

$$[H_3SiO_4^-] / [H_4SiO_4^0] = (10^{-9.1})/ (10^{-10.0}) = 10^{0.9} = 7.94$$

$$[H_3SiO_4^-] = 7.94/8.94 = 0.89 \text{ mmol/l}$$

When alkalinity is titrated, $H_3SiO_4^-$ associates with H^+, to become $H_4SiO_4^0$. The contribution to the alkalinity is equivalent to the $H_3SiO_4^-$ concentration.

CHAPTER 5

5.1. gfw = 389.91 g/mol. One packet of 20 units has 0.7 charge over 2 basal planes, i.e. CEC = 0.7 · 100/389.91 · 1/20 · 1000 = 9.0 meq/100g. Charge is 0.35 q_e/Å^2 = 0.12 C/m^2.

5.2. gfw = 368.22 g/mol. CEC = 217 meq/100g. Charge = 0.14 C/m^2.

5.3. 10 times concentrated water: β_{Na} = 0.0259; β_{Mg} = 0.134; β_{Ca} = 0.840. Seawater: β_{Na} = 0.584; β_{Mg} = 0.319; β_{Ca} = 0.097.

5.4. γ_1 = 0.85; γ_2 = 0.53; $K_{Na\backslash Cd}$ = 0.286 and 0.339.

5.5a. Sample 1, is rainwater (with the Na/Cl-ratio of seawater) in which some calcite has dissolved. Sulfate is partly reduced. Sample 2 is slightly mixed with salt water (higher Cl$^-$). Cation exchange is evident, Ca^{2+} is replaced by Na^+, causing subsaturation with respect to calcite. Renewed dissolution of calcite gives high alkalinity. Sample 3 has a low Na/Cl-ratio, sulfate is partly reduced. Sample 4: seawater.

5.5b. A: $NaHCO_3$ water, interface moves downward. B: $CaCl_2$ water, interface moves upward.

5.5c.

	β_{Na}	β_K	β_{Mg}	β_{Ca}	$\Sigma\beta$
Sample 1	0.0062	0.014	0.0187	0.974	1
Sample 4	0.558	0.061	0.292	0.089	1

(We used $K_{Na\backslash K}$ = 0.2; $K_{Na\backslash Mg}$ = 0.5; $K_{Na\backslash Ca}$ = 0.4; activity corrections in solutions were neglected)

5.5d.
CEC expressed per l pore water is ρ_b/ε · (CEC, meq/1000 g) = 1.8/0.3 · 10 = 60 meq/l. Hence exchangeable concentrations are:

	Na-X	K-X	Mg-X$_2$	Ca-X$_2$	Σ	
Sample 1	0.4	0.09	1.1	58.4	60	meq/l
Sample 4	33.5	3.7	17.5	5.3	60	meq/l

5.6a. Evaporation concentrates Nile-water in the first place. Using as 'conservative' species, Cl^-, we can calculate the composition:

Concentration factor $Cl^-_{groundwater}/Cl^-_{Nile} = 4.5/.5 = 9.0$

	Na^+	K^+	Mg^{2+}	Ca^{2+}	HCO_3^-	Cl^-	SO_4^{2-}	pH
Nile-water	.5	.1	.4	.7	2.2	.5	.1	
9* concentrated	4.5	.9	3.6	6.3	19.8	4.5	.9	
Groundwater	8	.1	.2	.7	2.7	4.5	1.0	8.1
Difference	3.5	−.8	−3.4	−5.9	−17.1	−	0.1	

The decrease in Ca^{2+} and HCO_3^-,(and some Mg^{2+}) is an effect of calcite precipitation. If we compare K and *IAP* of the reaction:

$$CaCO_3 + H^+ \leftrightarrow Ca^{2+} + HCO_3^- \qquad K = 100 \qquad IAP = 136,$$

the water is clearly oversaturated. The increase of Na^+ is produced by cation exchange of Ca^{2+} and Mg^{2+} for Na^+ from the soil. The decrease of K^+ is an effect of uptake by vegetation or illite.

5.6b. The very high Cl^- concentration in water from borehole 121 can only be explained by salt water intrusion. If we mix groundwater and seawater at the correct ratio, (contribution of seawater 0.473), we obtain:

	Na^+	K^+	Mg^{2+}	Ca^{2+}	HCO_3^-	Cl^-	SO_4^{2-}
Calculated	231	5	26	5.4	2.5	270	14
Analyzed	180	2.5	35	15	7.5	270	1.5
Difference	− 51	−2.5	9	10	5	−	−12

The decrease in Na^+ is an effect of cation exchange. Salt water intrudes in a fresh aquifer, Ca^{2+} will be removed from the exchange complex and replaced by Na^+. The decrease of SO_4^{2-} is an effect of reduction.

5.6c. Calculation of the composition in mg/l:

Na^+	K^+	Mg^{2+}	Ca^{2+}	HCO_3^-	Cl^-	SO_4^{2-}
4140	98	850	601	458	9571	144

The contribution of HCO_3^- to *TDS* = 0.5 · HCO_3^- (mg/l), since half of the HCO_3^- originates from $CO_{2(g)}$ (and would escape when the sample is evaporated to dryness). Hence: *TDS* = 15633 mg/l.

5.7a. Na-X = 160.6; Mg-X_2 = 283.18; Ca-X_2 = 306.22 meq/l pore water.

5.7b. Total cations = 14.66 meq/l. Na^+ = 13.28; Mg^{2+} = 0.43; Ca^{2+} = 0.26 mmol/l

5.7c. Exchanger composition in equilibrium with injected water is: Na-X = 56.35; Mg-X_2 = 81.1; Ca-X_2 = 612.54 meq/l pore water. Assume that Na-X is flushed from 160.6 to 56.35 meq/l, which needs (160.6 − 56.35)/(13.28 − 9.4) = 26.9 pore volumes, which is 6985 m^3 water. (Note that a second front appears, with Mg^{2+} being flushed from the exchanger, cf. Chapter 11).

5.8a. Na-X = 2.2, Mg-X_2 = 16.1, Ca-X_2 = 41.7 meq/l.

5.8b. Na^+ = 28.8, Mg^{2+} = 50.2, Ca^{2+} = 52.8 meq/l.

5.8c. The exchanger should have a final composition: Na-X = 25.8, Mg-X$_2$ = 22.7, Ca-X$_2$ = 11.5 meq/l. It gives $V = (41.7 - 11.5)/(105.6 - 8.3) = 0.31$ pore volumes. Note that the concentration of Ca^{2+} only reaches up to ca. 17 mmol/l in the experiment, because dispersion smoothens the peak.

5.9. pK_{int} = 2.8; $\alpha/\ln 10 = 3.1$; $K_{h\text{-}Al} = 3\cdot10^{-3}$ give:

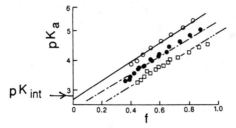

5.11. Smectite with 0.09 C/m^2.

x:	0	10	50	100 Å
ψ_0	−0.393	−0.284	−0.078	−0.015 V
n_+/n_∞	$4.3\cdot10^6$	$6.3\cdot10^4$	20.6	1.8
m_+	43000	63	0.206	0.018 mol/l
n_-/n_∞	$2.3\cdot10^{-7}$	$1.6\cdot10^{-5}$	0.05	0.55

Note that the concentrations of m_+ at 0 and 10 Å are unrealistic.

5.12. *SAR* of irrigation water is 1.7; when 10 times concentrated, *SAR* changes to 5.5.

5.13.

For illite (using activity of Ca^{2+} and K$^+$ in solution obtained with $[i] = \gamma_i m_i$):

	0.1	0.01	0.001	10^{-4} N	
$K^G_{Ca\backslash K}$	0.90	0.76	0.53	0.25	(Gapon)
$K^V_{Ca\backslash K}$	0.62	0.43	0.28	0.13	(Vanselow)
$K_{Ca\backslash K}$	1.09	0.81	0.55	0.25	(Gaines and Thomas)

and similarly for the resin:

	0.1	0.01	0.001	10^{-4} N	
$K^G_{Ca\backslash K}$	2.68	2.41	2.37	1.80	(Gapon)
$K^V_{Ca\backslash K}$	1.53	1.26	1.20	0.90	(Vanselow)
$K_{Ca\backslash K}$	2.87	2.50	2.38	1.81	(Gaines and Thomas)

5.14. $CEC = (\beta^M_I \cdot i + \beta^M_J \cdot j + \beta^M_K \cdot k + ...) \cdot TEC$

$TEC = (\beta_I/i + \beta_J/j + \beta_K/k + ...) \cdot CEC$

$\beta_I \cdot CEC = i \cdot \beta^M_I \cdot TEC$

$\beta_I = i \cdot \beta^M_I \cdot \dfrac{TEC}{CEC} = \dfrac{\beta^M_I \cdot i}{\beta^M_I \cdot i + \beta^M_J \cdot j + \beta^M_K \cdot k + ...}$

$\beta^M_I = \beta_I/i \cdot \dfrac{CEC}{TEC} = \dfrac{\beta_I/i}{\beta_I/i + \beta_J/j + \beta_K/k + ...}$

CHAPTER 6

6.1. K-feldspar, quartz and gibbsite.

6.2. The base production by gibbsite dissolution $(Al(OH)_3)$ is:

$$[OH^-] = 3[Al^{3+}] + 2[AlOH^{2+}] + [Al(OH)_2^+]$$

$$[OH^-] = 3(6 \cdot 10^{-5}) + 2(7.7 \cdot 10^{-6}) + 6 \cdot 10^{-7} = 1.97 \cdot 10^{-4}$$

$$[H^+]_{initial} = 10^{-4.53} + 1.97 \cdot 10^{-4} = 2.26 \cdot 10^{-4} \text{ and } pH_{initial} = 3.65$$

6.3. Combining $K_{gibbsite}$ and $K_{jurbanite}$ gives:

$$2 \log [OH^-] - \log [SO_4^{2-}] = -33.88 + 17.8 = -16.08$$

Combining with the dissociation of water yields:

$$2pH - \log [SO_4^{2-}] = 11.92$$

6.4a. The water chemistry can be explained by:

Dissolving
   0.016 mmol/l NaCl
   0.015 mmol/l gypsum
   0.014 mmol/l biotite
   0.175 mmol/l plagioclase
   0.114 mmol/l calcite

Precipitating
   0.033 mmol/l kaolinite
   0.081 mmol/l Ca-smectite

6.4c. A greater weathering rate of plagioclase than of biotite or calcite is not consistent with commonly noted weathering sequences. However, Feth et al. (1964) stated rather explicitly that calcite is absent in the Sierra Nevada source rocks, and calcite (or gypsum) should in fact not be included in weathering balances for the area. It may be that dry deposition gives a significant contribution to the spring compositions.

CHAPTER 7

7.1a. For example Fe_2O_3 / MnO_2, $Fe_2O_3 / MnCO_3$, $FeCO_3 / MnCO_3$, $FeS_2 / MnCO_3$. For dissolved species Fe^{2+}/Mn^{2+}, Fe^{3+}/Mn^{2+}.

7.1b. For hematite,

$$Fe_2O_3 + 2e^- + 6H^+ \leftrightarrow 2Fe^{2+} + 3H_2O, \text{ and}$$

$$pe = 11.3 - \log [Fe^{2+}] - 3pH$$

7.1c. A gradual decrease in pe or pH or both simultaneously.

7.1d. pH = 7.16

$$[NO_3^-] = 110 \text{ mg/l} = 10^{-2.75} \text{ mol/l}$$

$$[NO_2^-] = 2.1 \text{ mg/l} = 10^{-4.35} \text{ mol/l}$$

$$NO_3^- + 2H^+ + 2e^- \leftrightarrow NO_2^- + H_2O$$

$$\log [NO_2^-] - \log [NO_3^-] + 2pH + 2pe = 28.3 \qquad \rightarrow pe = 7.79$$

$$Mn^{2+} + 2H_2O \leftrightarrow MnO_2 + 4H^+ + 2e^- \qquad \log K = -41.38$$

$$-4pH - 2pe - \log [Mn^{2+}] = -41.38 \qquad \log [Mn^{2+}] = -2.84$$

7.2. The PbO_2/PbO boundary:
Combining reactions 1) and 3) yields:

$$PbO_2 + 2H^+ + 2e^- \leftrightarrow H_2O + PbO, \text{ and}$$

$$\log K = 2pH + 2pe = \log K_1 - \log K_3 = 49.2 - 12.7 = 36.5$$

The PbO/Pb boundary:
Combining reactions 2) and 3) yields:

$$PbO + 2H^+ + 2e^- \leftrightarrow Pb + H_2O, \text{ and}$$

$$\log K = 2pH + 2pe = \log K_3 + \log K_2 = 12.7 - 4.26 = 8.44$$

The PbO_2/Pb^{2+} boundary:
For reaction 1) and $\log [Pb^{2+}] = -6$:

$$\log K_1 = \log [Pb^{2+}] + 4pH + 2pe = 49.2$$

$$4pH + 2pe = 49.2 + 6 = 55.2$$

The PbO/Pb^{2+} boundary:
For reaction 3) and $\log [Pb^{2+}] = -6$:

$$\log K_3 = \log [Pb^{2+}] + 2pH = 12.7$$

$$pH = \frac{1}{2}(12.7 + 6) = 9.35$$

The $Pb(OH)_3^-/Pb^{2+}$ boundary:
For reaction 4):

$$\log K_4 = -3pH + \log [Pb(OH)_3^-] - \log [Pb^{2+}] = -28.1$$

When $\log [Pb(OH)_3^-] = \log [Pb^{2+}]$ then pH = 9.37.

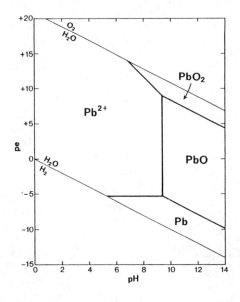

Accordingly, $Pb(OH)_3^-$ never becomes a dominant dissolved species and is therefore not included in the diagram.

The Pb/Pb^{2+} boundary:

For reaction 2) and $\log [Pb^{2+}] = -6$:

$$\log K = -\log [Pb^{2+}] + 2pe = -4.26, \text{ and}$$

$$pe = \tfrac{1}{2}(-4.26 - 6) = -5.13$$

7.3a. Combination of several equations gives:

$$MnCO_3 + CO_2 + H_2O \leftrightarrow Mn^{2+} + 2HCO_3^- \qquad K = 10^{-6.9}$$

In water in contact only with $MnCO_3$:

$$[HCO_3^-] = 2[Mn^{2+}] \Rightarrow 4[Mn^{2+}]^3/P_{CO_2} = 10^{-6.9}$$

$$P_{CO_2} = 10^{-1.5} \rightarrow Mn^{2+} = 1 \text{ mmol/l (i.e. 1 mmol/l } MnCO_3 \text{ will dissolve)}$$

7.3b. Congruent dissolution: The mineral dissolves completely and no secondary phases are produced.

Incongruent dissolution: the stoichiometry of substances appearing in the water does not correspond to their ratio in the dissolving phase; i.e. a secondary precipitate is formed.

7.3c. Rhodochrosite will dissolve according to 7.3a. Mn^{2+} oxidizes in the presence of O_2:

$$Mn^{2+} + H_2O + \tfrac{1}{2}O_2 \rightarrow 2H^+ + MnO_2 \text{ (pyrolusite)}$$

which means that rhodochrosite dissolves incongruently. The H^+ produced reacts with HCO_3^- and thus increases the P_{CO_2}.

7.4. $\quad KMnO_4 + 8H^+ + 5e^- \qquad \leftrightarrow Mn^{2+} + K^+ + 4H_2O$

$\underline{\quad (CH_2O + H_2O \qquad\qquad\quad \leftrightarrow CO_2 + 4H^+ + 4e^-) \cdot 5/4 \qquad\qquad}$ $+$

$\quad KMnO_4 + 3H^+ + 5/4CH_2O \leftrightarrow Mn^{2+} + K^+ + 5/4CO_2 + 11/4H_2O$

7.5.

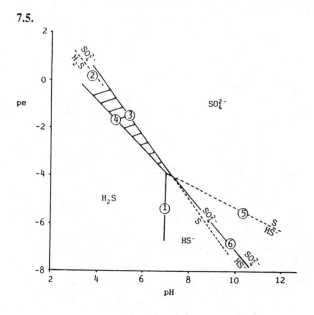

Possible reactions:

1. $H_2S_{(aq)} \leftrightarrow HS^- + H^+$ $K = 10^{-7}$
2. $H_2S_{(aq)} + 4H_2O \leftrightarrow SO_4^{2-} + 10H^+ + 8e^-$ $K = 10^{-40.6}$
3. $S_{(s)} + 4H_2O \leftrightarrow SO_4^{2-} + 8H^+ + 6e^-$ $K = 10^{-36.2}$
4. $(= 3\text{-}2)\ H_2S_{(aq)} \leftrightarrow S_{(s)} + 2H^+ + 2e^-$ $K = 10^{-4.4}$
5. $(= 4\text{-}1)\ HS^- \leftrightarrow S_{(s)} + H^+ + 2e^-$ $K = 10^{2.6}$
6. $(= 2\text{-}1)\ HS^- + 4H_2O \leftrightarrow SO_4^{2-} + 9H^+ + 8e^-$ $K = 10^{-33.6}$

S-precipitation is possible when:
- pH < 6 a 7
- H_2S-concentration $> \pm 5$ mmol/l (at lower levels the stability field of S will disappear)

CHAPTER 8

8.1a. Flowline 1:

$$CaMg(CO_3)_2 + 2CO_2 + 2H_2O \leftrightarrow Ca^{2+} + Mg^{2+} + 4HCO_3^- \qquad K = 10^{-11.6} \qquad (1)$$

From mass balance: $m_{Ca^{2+}} = m_{Mg^{2+}}$, and from charge balance:

$$2m_{Ca^{2+}} + 2m_{Mg^{2+}} + m_{OH^-} = 2m_{CO_3^{2-}} + m_{HCO_3^-} + m_{OH^-}. \qquad (2)$$

Neglect CO_3^{2-}, OH^- and H^+: $4m_{Ca^{2+}} = m_{HCO_3^-}$.
   Insert in (1):

$$m_{Ca^{2+}} \cdot m_{Ca^{2+}} \cdot 256m_{Ca^{2+}}^4 = 10^{-11.6} \cdot (P_{CO_2})^2 \rightarrow m_{Ca^{2+}} = 1.01 \text{ mmol/l}$$

Results:

pH	Ca^{2+}	Mg^{2+}	HCO_3^-	CO_3^{2-}	H_2CO_3	ΣCO_2
7.41	1.01	1.01	4.04	0.005	0.31	4.35 mmol/l

(Neglecting CO_3^{2-}, OH^-, H^+ was correct)

Flowline 2:
   $CO_2 + H_2O \leftrightarrow H_2CO_3$, hence $m_{H_2CO_3} = 10^{-5}$ mol/l. This amount is used for dissolution of dolomite. If CO_3^{2-} etc. is neglected, we obtain from (1): $m_{HCO_3^-} = 2 \cdot m_{H_3CO_{3(start)}} = 2 \cdot 10^{-5}$, and we calculate $m_{Ca^{2+}} = 5 \cdot 10^{-6}$ and pH = 12. This means that the neglect of CO_3^{2-} and even OH^- was not permitted. From mass balance:

$$m_{CO_3^{2-}} + m_{HCO_3^-} = 1 \cdot 10^{-5} + 2m_{Ca^{2+}} \qquad (3)$$

From charge balance:

$$2m_{CO_3^{2-}} + m_{HCO_3^-} + m_{OH^-} = 4m_{Ca^{2+}} \qquad (4)$$

Now express each unknown as a function of H^+.
HCO_3^- in $2 \cdot (3) - (4)$ gives:

$$m_{HCO_3^-} = 2 \cdot 10^{-5} + m_{OH^-} = (10^{-4.7} \cdot m_{H^+} + 10^{-14})/m_{H^+}$$

$$CO_3^{2-}: m_{CO_3^{2-}} = 10^{-10.3} \cdot m_{HCO_3^-}/m_{H^+} = (10^{-15} \cdot m_{H^+} + 10^{-24.3})/(m_{H^+})^2$$

$$Ca^{2+}: m_{Ca^{2+}} = 10^{-8.3}/m_{CO_3^{2-}} = 10^2(m_{H^+})^2/(10^{-4.7} \cdot m_{H^+} + 10^{-14})$$

Insert in (3) and solve by trial and error: $[H^+] = 10^{-10}$

Results:

pH	Ca^{2+}	Mg^{2+}	HCO_3^-	CO_3^{2-}	H_2CO_3	ΣCO_2
10.00	.083	.083	.12	.06	$2 \cdot 10^{-5}$	.18 mmol/l

8.1b. If we compare flowlines 1 and 2, concentrations in 2 are almost negligible. Composition of spring water may be calculated by mixing the cations at the given ratio, calculate HCO_3^- necessary to maintain charge balance and attribute the remaining fraction of ΣCO_2 to H_2CO_3.

Spring water:

pH	Ca^{2+}	Mg^{2+}	HCO_3^-	CO_3^{2-}	H_2CO_3	ΣCO_2
7.8	.55	.55	2.19	.006	.075	2.27 mmol/l

$IAP_{dol} = 10^{-16.9} < 10^{-16.6}$
$IAP_{cc} = 10^{-8.45} < 10^{-8.3}$

8.1c. As an effect of dissolution of gypsum, $[Ca^{2+}]$ will be high. To maintain equilibrium with calcite at $P_{CO_2} = 0.01$, CO_3^{2-} must be low, and is neglected in the charge balance:

$$2m_{Ca^{2+}} = 2m_{SO_4^{2-}} + m_{HCO_3^-} \tag{5}$$

From K_{sp}(gypsum): $m_{SO_4^{2-}} = 10^{-4.7}/m_{Ca^{2+}}$ (6)

From $CO_2 + H_2O + CaCO_3 \leftrightarrow Ca^{2+} + 2HCO_3^-$,

$$m_{HCO_3^-} = 1.58 \cdot 10^{-4}/\sqrt{m_{Ca^{2+}}} \tag{7}$$

Insert (6) and (7) in (5) and obtain by trial and error:

$m_{Ca^{2+}} = 10^{-2.30}$, etc.

Flowline 3:

pH	Ca^{2+}	Mg^{2+}	HCO_3^-	SO_4^{2-}	H_2CO_3	ΣCO_2
7.14	5.0	–	2.23	3.89	.5	2.73 mmol/l

8.1d. Mixing at the ratio (1:9) gives:

pH	Ca^{2+}	Mg^{2+}	HCO_3^-	SO_4^{2-}	H_2CO_3	ΣCO_2
7.68	1.0	.5	2.16	.4	.09	2.25 mmol/l

and: $IAP_{dol} = 10^{-16.9} < 10^{-16.6}$ i.e. undersaturated; $IAP_{cc} = 10^{-8.3}$ i.e. nearly saturated

8.3. $NaAlSi_3O_8 + H_2CO_3 + 4.5H_2O \rightarrow Na^+ + 0.5\ Al_2Si_2O_5(OH)_4 + 2H_4SiO_4 + HCO_3^-$

When: $m_{HCO_3^-} = m_{Na^+}$, and $m_{H_4SiO_4} = 2\ m_{Na^+}$, we have at equilibrium with kaolinite and albite:

$m_{Na^+} = (0.25 \cdot K)^{1/4} \cdot (m_{H_2CO_3})^{1/4}$,

which is a parabolic relationship, that gives subsaturation when waters are mixed which have different P_{CO_2}'s.

8.5a. $m_{SO_4^{2-}} = 1.3$ mmol/l.

8.5b. $P_{CO_2} = 10^{-2.87}$ $SI = 0.43$.

8.5c. We have the equilibrium

$$CaCO_3 + CO_2 + H_2O = Ca^{2+} + 2HCO_3^- \qquad K = 10^{-5.43}$$

With $P_{CO_2} = 10^{-2.87}$ we obtain:

$$[Ca^{2+}] \, [HCO_3^-]^2 = 10^{-8.30}$$

Express this relation in mmol/l (multiply by 10^9), and insert the irrigation water quality 10 times concentrated, and with Δ mmol/l calcite precipitated:

$$(7 - \Delta) \, (27 - 2\Delta)^2 = 5.01$$

which can be solved by trial and error for Δ. A solution can also be obtained by noting that almost all Ca^{2+} will precipitate, hence $\Delta \cong 7$, and $(m_{HCO_3^-})^2 = (27 - 14)^2 = (13)^2 = 169$. Thus, $(7 - \Delta) = 5.01/169$, which gives $\Delta = 6.97$ mmol/l: $m_{Ca^{2+}} = 0.03$, $m_{HCO_3^-} = 13$ mmol/l.

8.5d. Before evaporation is $SAR = 8.4$, afterwards $SAR = 56$. Na-X is 10% of the exchange complex before evaporation, 50% afterwards.

CHAPTER 9

9.2. At 1 km is $v = 20$, $v_D = 6$ m/yr, At 2 km is $v = 40$, $v_D = 12$ m/yr. Travel time is 34.7 yr. Age of water at 10, 20, 30, 40 m depth is 11.2, 25.5, 45.8, 80.5 yr. The formula $t = d \cdot \varepsilon/P$ gives 10, 20, 30 and 40 years.

9.3. 4 years.

9.4. $(h - h_0) = -P/(2kD) \cdot (x^2 - x_0^2)$.

9.6. $D^* = 2/3D$, hence $x_0 = 333$ m. Travel time above clay layer: $2D/3 \cdot \varepsilon \cdot v = 2D/3 \cdot \varepsilon \cdot dx/dt = P \cdot x^* = P \cdot (666 + x)$. This gives $t = (2D \cdot \varepsilon/3P) \cdot (\ln 1666 - \ln 666) = 0.6D\varepsilon/P$.

The flow velocity in the confined part will be equal to the overall velocity at the end of the clay layer, i.e.

$$v_{x = 2000} = 2000 \cdot P/D\varepsilon$$

Hence, the travel time is $t = 0.5 \cdot P/D\varepsilon$, which is 1.2 smaller than in the upper part. (A numerical model gives a ratio of ca. 1.16, which is less than estimated here because more water attempts to flow through the lower aquifer)

9.7. $\rho_b = 2.12$ g/cm^3; $\rho_b/\varepsilon = 10.6$.

9.8. $\varepsilon = 0.26$.

9.9. Naphthalene in peat: $\log K'_d = 3.0 + \log 0.7 = 2.85$; $K'_d = 700$; bulk density of peat is $\rho_b = 1.0 \times 0.4 = 0.4$. Hence $K_d = 700 \times 0.4/0.6 = 467$; velocity is 468 times smaller. Tetracene in sand: $\rho_b = 1.855$; $R = 1 + \rho_b/\varepsilon \cdot 631 = 3904$.

9.11. $\sigma = 25$ and 79 cm.

9.12a. erf(0) = 0;   erfc(0) = 1;   erfc(0.707) = 0.317;   erf(−0.707) = −0.682;
erfc(0.5) = 0.479;   erfc(−0.5) = 1.521.

9.12b. Integrate from 0 to 100 cm:

$$\int_0^{100} C_{x,t}\,dx = \frac{10}{\sqrt{4\pi Dt}} \int_0^{100} \exp\left(-\frac{x^2}{4Dt}\right) dx$$

Substitute $x^2 = 4Dt \cdot s^2$, and simplify:

$$\int_0^{100} C_{x,t}\,dx = \frac{10}{\sqrt{\pi}} \int_0^{\frac{100}{\sqrt{4Dt}}} \exp(-s^2)\,ds = 5\,\text{erf}\left(\frac{100}{\sqrt{4Dt}}\right)$$

We find erf $(100/\sqrt{4Dt})$ = erf (0.89) = 0.792. Hence $5 \cdot 0.792 = 3.96$ is the amount of NaCl between 0 and 100 cm.

9.14. $D_t = D_f \cdot \varepsilon + \alpha\,v$ gives $5 \cdot 10^{-3}$, $5 \cdot 10^{-4}$ and $5.3 \cdot 10^{-5}$ cm^2/s.

9.15. $\alpha_L = 0.02\,x$ and $0.06\,x$.

9.17. Assume concentration in end-cell = B, in adjacent cell = A. Then the concentration on the boundary is: $\frac{1}{4}$A + $\frac{3}{4}$B. We find $B_{t2} = B + f^* (B - 2B + \frac{1}{4}A + \frac{3}{4}B) = B + f^*(\frac{3}{4}B + \frac{1}{4}A - B)$. With $f^* = 4\,mixf$, and $B_{t2} = mixf \cdot A + (1 - mixf) \cdot B$.

9.19a. $x = 550 \cdot \exp(0.3 \cdot t/(50 \cdot 0.36))$;

t years	x m	Flowlength ($=x-x_0$)	Depth m	Thickness m
25	834	284	17.0	3.0
100	2912	2362	40.6	0.9

Note that flowpath increases 8-fold, when time increases 4-fold.

9.19b. Assume CEC = [Ca-X$_2$] = 1 meq per 100g. 100g = 55.6 ml = 20 ml(H$_2$O), hence CEC = 50 meq/l = 25 mmol Ca^{2+}/l.

$$K_d(\text{Cd}) = [\text{Cd-X}_2]/[\text{Cd}^{2+}] = [\text{Ca-X}_2]/[\text{Ca}^{2+}] = 2.5 \cdot 10^{-2}/1 \cdot 10^{-2} = 2.5;\ R(\text{Cd}) = 3.5.$$

$$K_d(\text{TCE}) = q/C = (6 + 6\ln(C))/C = (6 + 6\ln(7.1))/7.1 = 2.5;\ R(\text{TCE}) = 3.5.$$

9.19c. $v_{H_2O} = R \cdot v_i = R \cdot dx_i/dt = Px/D\varepsilon$;　$x = 550 \cdot \exp(0.3 \cdot t/(R \cdot 50 \cdot 0.36))$;

t years	x m	Flowlength ($=x-x_0$)	Depth m	Thickness m
25	620	70	5.6	4.0
100	885	335	18.9	2.8

9.19d. Flowlength after 100 yr = 335 m; α = 10% of 335 m = 33 m. $\sigma^2 = 2Dt = 2 \cdot \alpha \cdot x = 2 \cdot 0.1 \cdot (335)^2 = 22110$; σ = 148 m.

Conc. at 335 + 148 = 483 m: $16/100 \cdot C_0$ = 0.18 mg Cd/l;
335 m: $1/2 \cdot C_0$ = 0.55 mg Cd/l;
335 − 148 = 187 m: $84/100 \cdot C_0$ = 0.94 mg Cd/l.

The TCE-front will be sharper since a convex isotherm gives small concentrations a higher K_d, and hence more retardation.

CHAPTER 11

11.1.

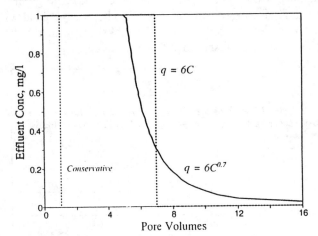

11.2. $R = \Delta q / \Delta C + 1$. Hence $R = 7$ and 6.33.

11.3. At 10°C, in $CaCl_2$: $V^* = 32.8$; in NaCl: $V^* = 17.3$.
At 70°C, in $CaCl_2$: $V^* = 2.9$; in NaCl: $V^* = 1.9$.

PUBLISHER'S NOTE

For your convenience, please use the enclosed postcard to order the computer programs mentioned in this book. If this postcard is no longer in the book, kindly photocopy the blank order form reproduced underneath, complete it, and mail it to the publisher together with your payment. You have permission from the publisher to photocopy this page.

Index

absorption 142, 341, 456
acceleration of clean-up 456
accuracy of analysis 16
acetylene block technique 275
acid mine drainage 264
acid rain 24, 225
   modeling 430, 443
acid volatile sulfide (AVS) 287
acidification
   Al buffering 226-230
   buffering, exchangeable base cations 229, 443-
   445
   exchangeable Al 229-230
   field weathering model 234
   gibbsite 443-445
   groundwater 223-229, 443-445
      buffering 226
      mixing 309
      modeling 443-445
   liming 230
   pyrite oxidation 235, 264, 289
   remediation 230, 445
acidity 5
   of cation exchangers 160, 405-407
activity 49-58, 480-484
   coefficients 49-51, 64, 483
   complexes 51
   dimension 480, 483
   standard state 483
   WATEQP 56-58
   vs. concentration in diffusion eqn 350
adsorbers 148-155
adsorption 142, 181-184, 341
advection in aquifers 331
advection-dispersion, numeric model 380
advective vs. diffusive flux
   of O_2 264
   salinization 174
Al
   acid rain buffer 226-230

acidifying groundwater 443-445
Al hydroxy adsorption 405-408
Al hydroxy complexes 208, 405
Al-F complex adsorption 405-408
amorphous $Al(OH)_3$ 209
bauxite 206, 208
exchange coefficients 405-408
exchangeable cation (Al) 228-229
gibbsite 206, 227
   buffering 443-445
   pH dependence (Al solubility) 208
      solubility 207-210
   groundwater concentrations 209, 226-230
   in goethite 63
   inhibitor in silicate dissolution 222, 233
   jurbanite 210, 229
albite 204, 205
albite dissolution kinetics 222
alkalinity 5
   determination 94
aluminum oxide, acid-base titration 179
ammonia (see also N) 251
amorphous $Al(OH)_3$ 209
amorphous silica 213
amorphous $Fe(OH)_3$ 254
amphibole 202, 223
amphoteric acid/base 175
analysis, chemical 14
   accuracy 16
   concentration units 5
   field analyses 14
   precision 16
analytical solutions, breakthrough curves 451
   diffusive transport 367, 373
   mass transport 365, 367, 371
anion exchange 154, 155
anion exclusion 189
anorthite 204
anoxic 11, 261-271
apparent dissociation constant 177

521

aquiclude 328
aquifer 328
    confined aquifer 328, 335
    depth/time relations 331-336
    depth integrated sampling 336, 361
    flowlines 331-336, 371, 389
    flushing of NO_3^- 337
    freshening 170-172
    heterogeneity 335, 337-339, 388-389
    isochrones 332
    mixed reservoir 337
    phreatic aquifer 331-336
    radial flow 333
    residence time 336
    salinization 173-174
        diffusion vs advection 174
    tracer distribution (in aquifers) 337
    tritium profiles 333, 337
aquifer clean-up 449-476
    acceleration of clean-up 456
    desorption enhancement 456
aquifer thermal energy storage 419, 439-441
aquitard 328
aragonite 88, 129, 136
As 9, 24, 248
    arsenopyrite 248, 265
    black foot disease 248
    redox diagram 249
    release from pyrite 278
    thermodynamic data 491
association constant, geochemical model 401
ATES (aquifer thermal energy storage) 419, 439-441
atmospheric transport 27
atomic weights 7
augite 221
AVS (acid volatile sulfide) 287

background electrolyte 178, 182
backward differences 374
bacteria
    Desulfovibrio 258, 285, 312
    Gallionella 275, 312
    Lepthothrix 312
    Thiobacillus 270, 275
BALANCE 218, 305
basalt 205, 214
bauxite 206, 208
benzene 373
Bermuda 89, 134-136
BET surface area 73, 129
binary exchange 459
biomass cycling 231
biotite 202, 204
black foot disease 248

BOD (biological oxygen demand) 5
Boltzmann energy 185
Borden aquifer (C) 365
bound residue 468
boundary adaptions in numerical models 376
boundary conditions 356
    Dirac delta function 352, 361
    flux type 358
    single shot input 352, 361
    step input 356
Brabant (NL) 147
breakthrough curve
    analytical solution 361-373, 451
    broadening front 451
    chemical non-equilibrium 471
    Damkohler number 472
    desorption isotherm 454
    flushing factor theory 451
    effect of Freundlich exponent 384
    HCH elution 452
    physical non-equilibrium 469
    sharpening fronts 455
    solute profiles along flowline 170, 260, 371, 454
    sorption hysteresis 468, 469
    stop-flow 474
brine composition 299-305
    Ca-brines 304
    dolomitization 299, 305
    salt minerals 302
    salt dissolution 303
    Si-concentrations in brines 305
brine evolution 299
broadening front 349, 451
bronzite 221
buffering in aquifers 226
buffering of acidification, exchangeable cations 443-445
bypass flow 330, 369
C
    alkalinity analysis 94
    carbon dioxide 90-98
    carbonic acid 90-98
    Gran titration 94
    isotopes 288
    methane 259, 288
    organic carbon oxidation 273, 418
    organic matter 286
    pH dependence of carbonate equilibria 92, 93
    redox sequence 257-260
    thermodynamic data 488
    total inorganic carbon (*TIC*) 5, 91
Ca
    Ca/K exchange 158
    Ca/Al exchange 405-408

Ca-brines 304
Ca-silicate stability 212
CaCl$_2$ type water 143-148, 173
calcite solubility 99-103
    cation exchange 435
    column elution 435
    concentration limits 99-103
    dedolomitization 113, 427-430
    New Zealand karst 112
CaF$_2$ 45-49, 78
calcite 79, 88-129, 299
    closed sytem dissolution 102, 107
    decalcification 101, 120
    denudation 101
    deposition in karst stream 119, 129
    dissolution rate 121
        empirical rates 121, 129
        field conditions 123
        inhibitors 129
        laminar flow 124
        saturation length 124
        turbulent flow 124
    dissolution by cation exchange 147
    dissolution kinetics 119
    inhibition 78, 129
    Mg-calcite 88, 130
    mixed carbonate terrain 116
    modeling 398, 419, 439
    open system dissolution 99, 107
    precipitation kinetics 127
    solubility 99-103
        pressure effect 481
    supersaturation 130
carbon dioxide (see also CO$_2$) 90-98
carbonate ions, see C
carbonate aquifers
    Bermuda 89, 134-136
    cave formation 106, 124, 308
    Dolomites 104, 126
    flow mechanisms 110
        conduit/diffuse flow 110, 317
    Madison 113-115
    Mendip Hills 110
    Pennsylvania 107
    Pleistocene 134
    Yucatan 89, 308
carbonate minerals
    aragonite 88, 129, 136
    calcite 88-129
    dolomite 89
    Mg in calcite 88, 130, 134
    otavite 65
    rhodochrosite 63, 290
    siderite 254, 284
    solid solutions 63-66

    solubility 88, 488
    Sr in aragonite 89, 136
carbonic acid 90-98
    influence on silicate weathering 223
catalysis, of nitrate reduction by Cu 279
cation exchange 143-168
    Ca/K exchange 158
    Ca/Al cation exchange 405-408
    CaCl$_2$ type water 143-148, 173
    cation exchangers 148-155
    cation exchange in PHREEQE 422, 426
    cation exchange capacity see *CEC*
    chromatography 165-174, 430, 464-466
    coastal dunes (NL) 144
    elution with Sr 435-437
    fresh/salt displacement 143-148, 170-174
    freshening 170-172
    Gaines and Thomas convention 156, 402-404,
        436
    Gapon convention 156, 404, 436
    geochemical model 403-405, 431, 435
    MgHCO$_3$ type water 166, 170
    NaHCO$_3$ type water 143-148, 168, 170
    salt water upconing 145
    Vanselow convention 156, 405
    well injection and recovery 437, 439
caves 106, 124, 307-308
Cd 65
    exchange vs. Na 164
    flushing with NaCl 462-463
    specific adsorption 183
    thermodynamic data 492-493
CEC (cation exchange capacity) 148
    clay minerals 149
    organic matter 149
    structural charge of clay minerals 152, 153
    surface charge of solids 153
    units recalculation 149
central difference, numerical scheme 374
chalcedony 214, 299
chalcopyrite 265
characteristic velocity, of solutes 167
chemical processes, overview 32
chemical non-equilibrium 471
    Damkohler number 472
chemical potential 480-484
chert 299
chlorite 151
chromatographic separations 166-174, 430, 464-
        466
    concentration effect 173
    field observations 168-174
    salinity front 168
Cl
    Cl profiles, unsaturated zone 330
    conservative ion 3

dry deposition 29
  thermodynamic data 486
clay
  anion exclusion 189
  compaction 189
  electric potential 190
  flocculation and peptization 187
  hyperfiltration 188
  negative adsorption 189
clay minerals 150-153
  c-axis spacing 151
  CEC 149
  chlorite 151
  dissolution kinetics 210
  distribution in soils 203, 207
  gibbsite 206, 218
  illite 152, 202, 207
  interlayer cations 151
  interlayer charge 152-153
  interlayer expansion 152
  interlayer release 437
  kaolinite 151, 202, 207
  layer stacking 151
  montmorillonite 151, 202, 205
    Cd vs. Na exchange 164
  muscovite 152
  octahedral layer 151
  PZC (point of zero charge) 153
  smectite 151
  stability diagram 212
  structural charge 150-153
  tetrahedral layer 151
  vermiculite 152
clogging of wells 312
closed system dissolution 102
CO_2 90
  evapotranspiration and CO_2 99
  equilibria 96
  Iceland 102
  seasonal fluctuations 97
  soil 97
coastal dunes (NL) 144
COD (chemical oxygen demand) 5
colloid stability 457
column elution 431, 435, 461
  chromatographic separation 430, 464
  effluent fronts, numerical modeling 380, 384
  experiment 356, 435-437
  modeling 430-434, 435-437
  sharpening fronts 433-440
column Peclet number 361
common ion effect 47, 309
complexation 51
  complex adsorption 405-408
  complexes in PHREEQE 412

component 399
computer programs
  BALANCE 218, 305
  NETPATH 218, 305
  PIP 409
  PHREEQE 305, 409-445
  PHREEQM 409-445
  PHRQPITZ 56
  WATEQ 56
  WATEQP 56-58
concave isotherm 348
concentration units
  solute ions 5
  exchangeable ions 156
concentration effect in ion exchange 158, 168, 173, 462
conditional constant, geochemical model 404
conduit flow 110, 317
confined aquifer 328, 335
conservation of water samples 14
conservative mixing, fresh and sea water 146, 147
conservative ions (see also Cl) 3
constant capacitance model 178
convergence, geochemical model 399
convex isotherm 348
correlation length, of permeability distribution 388
Coulomb energy 177
Cr
  thermodynamic data 494-496
cristobalite 214
crystal growth 73
  fluorite 78
  inhibition 78, 129
crystal formation 71-80
  nucleation 72
Cs
  multicomponent exchange 464-466
Cu
  catalysis of nitrate reduction 279
  specific adsorption 183
  surface complexation 183
  thermodynamic data 496-498
cumulative diagrams 40
curved isotherms 450
cycling of elements 230-231

Damkohler number 472
Darcy velocity 327
Darcy's law 327
database, geochemical model 401-403, 485-502
Davies equation 50
DDL (diffuse double layer) 184
  Boltzmann energy 185

Debije length 186
dielectric constant 185
eletrical field 185
Gouy Chapman theory 184-186
permittivity 185
Poisson equation 185
practical aspects 187-190
thickness 186, 187
dead-end pores 468
Debije length 186
Debije-Hückel theory 50
decalcification 101, 120
dedolomitization 113, 427-430
    complexation effects 428
    modeling 427
    pH effects 428-430
denitrification 250, 272
denudation 101
depth integrated sampling 10
    effect on dispersivity 336, 361
    mixing effects 311
depth specific samplers 10
depth/time relations in aquifers 331-336
desorption isotherm 451, 454
desorption enhancement 456, 475
Desulfovibrio 258, 285
diagram
    cumulative 40
    Piper 37, 145
    redox 247-256
    silicate stability 212, 214
    Stiff 4, 145
dielectric constant 185
diffuse double layer see DDL
diffuse flow, in carbonate aquifer 110, 317
diffusion 349-354, 424, 434
    activity vs. concentration in diffusion eqn 350
    analytical solution, for diffusive transport 367, 373
    dead-end pores 468
    diffusion coefficient 354, 367, 373, 425, 470
    diffusion coefficient and porosity 369
    Fick's laws 350, 351
    flux through clay liner 373
    model with non-equilibrium flow 473
    numerical model of diffusion 374, 376
    physical non-equilibrium 468
    of seawater 368
    stagnant zones 468
    transport distance 354, 470
    in unsaturated zone 264
diffusive vs. advective transport 354, 470
    of O_2 264
    salinization 174
diopside 221

Dirac delta function 352, 361
discharge estimates 320
discretization, numerical models 422-423
dispersion 350
    analytical solutions 365, 367, 371
    boundary conditions 356
    breakthrough curves 356-358
    column Peclet number 361
    dispersion coefficient 359-361
    dispersivity 360, 361
    macrodispersivity 361
    numerical model 380
    Peclet number 359
dispersion coefficient 359-361
    effect of sampling method 356, 361
    in aquifers 361-365
        macrodispersivity 361
        scale dependence 362
        permeability stratified aquifer 363
    in columns 359-361
        dispersion dominance 360
        low flow velocity 359
dispersion in numerical model 378
    corrections 379, 383
    in PHREEQM 423
dispersivity 360, 361
    macrodispersivity 361
    transversal dispersivity 365
displacement chromatography 165-172, 464-466
dissociation constant 51
    in geochemical model 401
dissolution kinetics 212, 220-235
    pH dependence (silicate weathering) 222, 234
    clay minerals 210
    calcite 121
        field conditions 123
        saturation length 124
    dolomite 124-126
    solubility control 69
dissolved organic carbon see *DOC*
distribution coefficient 64, 162, 339-348, 449
    cations 162
    curved isotherms 450
    hydrophobic organic compounds 341-344
    laboratory determination 341, 346
    lindane 344
    octanol/water 343
    organic carbon 341-343, 457
        H/O ratio 458
    organic carbon/water 343
    organic compounds 341-344
    PCB 344
    pH effect 183
    retardation effects 344
    Sr 163

temperature effect, with desorption 456
trace metals 344-346
    Zn 163
DOC oxidation by O_2 274
dolomite 89, 103, 306
    dedolomitization 113
    dissolution 103
    dissolution kinetics 124-126
    water quality 116
dolomitization 299, 305
Dolomites (N. Italy) 104, 126
double layer see *DDL*
DOWEX resin 405
    Ca/Al exchange 405-408
dry deposition 28, 31
dual porosity medium 369

earth alkaline ions, specific adsorption 181, 183
EC (electrical conductivity) 5, 17
EC routing 320-323
ecosystem balances 33, 232-232
effective permeability 328
effective porosity 328
effective diffusion coefficient 367
effluent fronts from columns, numerical modeling 380, 384
    analytical solutions 452, 461
Egypt 144-145
Eh 241
    measurements 241, 244
    redox classification 259
    relation with pe 246
electric potential 190
electrical balance, neutrality 16
electrical conductivity 5, 17
electrical field, around mineral grains 185
electrochemical cell 240
electron acceptor/donor 239
electron balance
    nitrate vs. organic matter 273
    nitrate vs. pyrite 275
electroneutrality 16
electrostatic models 177
elution of sorbed chemicals 450
    analytical solution for breakthrough 451
    chromatography 165, 450
    elution curve, see breakthrough curve
    elution with Sr 435-437
empirical rates 121, 129
endothermic reactions 62, 481
energy barrier, for nucleation 72
enstatite 221
enthalpy of reaction 61
entropy 481
epidote 234

epm 5
eq/l 5
equation of state (gas law) 481
equivalent fraction 156, 158
    from molar fraction 161, 198
error function 352
    numerical approximation 358, 390
ESR 192
etch patterns in minerals 76, 220
evaporative concentration 297, 302
evaporite minerals 302
evapotranspiration 144
    and CO_2 99
    and acid rain 226-227, 430
    Cl concentration 144, 297, 331
    evapotranspiration factor 2
exchange coefficients 158, 160
    Cd vs. Na 164
    in PHREEQM 405-408
exchange, see ion exchange
exchangeable sodium ratio (*ESR*) 192
exchanger composition 160
    exchanger selectivity 160
    specific adsorption 175, 183
exothermal reaction 62
F 9, 46-49, 81, 310, 405
    Al-F complex adsorption 405-408
    CaF_2 45-49, 78
    concentrations 47
    fluorapatite 46
    fluorite, see CaF_2
    gypsum treatment for lowering F 47, 48
    NaF 81
    villiaumite 81
    thermodynamic data 486
Fe
    am. $Fe(OH)_3$ 254
    cation exchange 442
    column elution 442-443
    Fe bacteria 312
    Fe silicates 279, 281
    Fe^{2+} oxidation 270, 442
        oxidation kinetics 69
    FeS formation, in aquifer 287
    FeS_2 oxidation 55, 264, 289
    goethite 279
        Al content 63
    greigite 287
    hematite 255, 281
        ligand promoted dissolution 282
    iron (oxy)hydroxide 312
    jarosite 264
    lepidocrocite 279, 281
    mackinawite 287
    magnetite 256, 278, 280

marcasite 287
oxide solubilities 283, 487
Pouhon 137
pyrite 256
  formation 82, 256
  oxidation 55, 264, 289
pyrrhotite 256
redox diagram 254-256
redox sequence 259
reductor (Fe^{2+}) 442
siderite 283
solubility control 282-284
sources of solute 279
vivianite 284
FeOOH armoring 269, 280
Fick's diffusion laws 350, 351
field weathering rates 230-233
  model 233-235
field analyses 14, 15
first order reaction 69
flocculation 188
flow in aquifers 331
  mechanisms, in carbonate aquifer 110
  saturated zone 331
  unsaturated zone 329
flowlines in aquifers 331-336, 371, 389
fluorapatite 46
fluorite 46
  crystal growth 78
flushing, of sorbed chemicals 450
  flushing with NaCl 462-463
  flushing of NO_3^- 337
  exchangeable cations 166-170
flushing factor theory 451
  analytical solution for breakthrough 451
  ion exchange 458-466
flux through clay liner 373
flux type boundary condition 358
formation factor 367-369
  and porosity 369
forsterite 223
FORTRAN input formats 409
forward differences, in numerical model 374
fractionation factor 93
fractionation, rainwater vs. seawater 23
fresh/salt water displacement 143-148, 165, 170-174
  well injection 168-170
  model calculations 437-438
Freundlich isotherm 347
  exponent effect on breakthrough curves 384
front dispersion 357
fulvic acids 180

Gaines and Thomas convention 156, 402-404, 436

Gallionella 275
Gapon convention 156, 404, 436
Gårdsjön (S) 231-233
gas pressure, geochemical model 400
gas law 481
Gauss curve 353, 355
geochemical models 56, 217, 396-449, 402
  association constant 401
  cation exchange 403-405, 431, 435
  component 399
  conditional constant 404
  convergence 399
  database 401-403, 485-502
  dissociation constant 401
  fixing gas pressure 400
  governing equations 399
  PHREEQE manual 408-420
  PHREEQM manual 421-426
  species 399
geochemical modeling 56, 217, 397
  acid rainwater 430
  acidification 443
  ATES 439
  Ca/Al, cation exchange 405-408
    complex adsorption 405-408
    hydroxy complexes 405-408
  calcite dissolution 398
  column elution 431, 435, 442
  dedolomitization 427-430
    complexation effects 428
    pH effects 428-430
  evapotranspiration of acid rainwater 430
  pH effects
    evapotranspiration 430-431
    dedolomitization 427-430
    NO reduction 418
    NH oxidation 430
  redox problems 399
  well injection and recovery 437, 439
geochemical balances 33
geometric surface, of grains 73
Gibbs free energy 59, 240, 480
  of crystallization 71
gibbsite 206, 227
  acid rain buffer 443-445
  solubility 207-210
    pH dependence 208
goethite 153, 176, 279-283
  Al content 63
Goldich weathering sequence 203, 221
Gouy Chapman theory 184-186
Gran titration 94
granite 3, 204-216, 231
greigite 287
groundwater (see also aquifer)

acidification 223-229
  contribution to runoff 315
  discharge 327
  mixing 305-313
  groundwater table depth and O_2 262, 274
gypsum 47, 53, 398
  dedolomitization 113
  SO_4^{2-} reduction 70
  solubility 47, 53

H_2S 260, 285
half reactions 239
halite 302
hard water softening 439
hardness 5
HCH elution 452
heating groundwater 439-441
heavy metal adsorption 181
hematite 256, 281-283
  ligand promoted dissolution 282
heterogeneity, effects on flowlines 335, 337-339,
  388-389
heterogeneous nucleation 72
heterovalent ion exchange 156
historic weathering rates 232, 233
homogeneous nucleation 72
hornblende 234
Hubbard Brook (USA) 231, 316-317
humic acids 197
hydroxides, thermodynamic data 487
hydraulic gradient 327
hydraulic conductivity 327
hydrodynamic dispersion coefficient 352, 359
hydrogen electrode 241
hydrogen in groundwater 257
hydrological conditions, and weathering products
  206-207
hydrophobic organic compounds 341-344
hydroxyapatite 75
hyperfiltration 188
hysteresis, with adsorption/desorption 467
  in runoff quantity/quality 319

IAP (ion activity product) 54
Iceland, CO_2 pressure in rivers 102
illite 152, 202, 207
immobile water 369
immobilization of chemicals 468
incongruent dissolution 131, 202
industrial sources of acid rain 25
infiltration area, diagnostic water quality 107
inhibition of reactions 78, 116, 129
integral scale, in aquifers heterogeneity 388
interception, of rain 32

interlayer cations 151
  charge 152-153
  expansion 152
intrinsic dissociation constant 177
ion activity product 54
ion exchange 142-168
  anions 154, 155
  binary exchange 459
  cations 142-168
  chromatography 165
  concentration effect 158, 168, 173, 462
  distribution coefficients 162, 344
  equivalent fraction 156, 158
    from molar fraction 161, 198
  exchange equations 155-158
  exchange conventions 156
  flushing factor theory 458-466
  fresh water intrusion 143, 165
  Gaines and Thomas convention 156
  Gapon convention 156
  heterovalent ion exchange 156
  modeling 431
  molar fraction 156, 158
    from equivalent fraction 198
  seawater intrusion 143, 173
  sharp fronts 463
  trace metals 163, 344
  Vanselow convention 156
ionic strength 30, 50
iron, see Fe
irrigation 191, 302
  evaporative concentration 297, 302
  exchangeable sodium ratio 192
  return flow fraction 194, 302
  salt water 296
  SAR (sodium adsorption ratio) 188-194, 302
  sodium hazard 188-192
  water quality 190-195
isochrones 332
isohypses 328
isotherms, of sorbed over solute concentrations
  347
  concave, convex 348
  curved, effects on retardation 348, 384, 451
  Freundlich 347, 384
  Langmuir 347
  linear 339, 347, 451
isotopes
  H 304, 337-339
  O 304
  S 285
  Sr 232
isotropy, of aquifer properties 328

jarosite 264
jurbanite 210, 229

K
  clay interlayer release 437
  column elution 461
  multicomponent exchange 464
    sharp fronts 464-466
  release from feldspars 440
kaolinite 151, 202-209
karst 106, 112, 307-308
karst stream, calcite deposition 119, 129
K_d (distribution coefficient) 340
kinetics 66-80, 220
  crystallization 71-80
  first order reaction 69
  rate mechanism 68-74
  reaction order 68-70
  silicate weathering 220-223, 233
  solubility control on dissolution rate 69
  specific rate 69
  surface controlled reaction 75
  transport controlled reaction 75
  zero order reaction 70, 71
Kozeny theory 359

lag effects, in quality/discharge 319
laminar flow 124
landfills 261
  diffusive flux through clay liner 373
  leaching 371
  redox zoning 261
Langmuir isotherm 162, 347
law of mass action 45, 484
layer stacking in clay minerals 151
leachate spreading, in aquifer 371
leaching, of rocks (weathering) 205
lepidocrocite 279, 281
Lepthothrix 312
lifetimes of silicate minerals 221
lignite 277
liming 230
lindane 344, 380
linear isotherm 339, 347
linear retardation 339-341, 380, 451
longitudinal dispersion 350, 359
lyotropic series 160

mackinawite 287
macrodispersivity 361
  permeability stratified aquifer 363
  scale dependence 362
  effects of sampling 356, 361
Madison aquifer (USA) 113-115
magnetite 256, 278, 280

marcasite 287
Margules parameters 64
mass balance 215, 230
  calculations 215-220
  BALANCE 218
  Gårdsjön (S) 231-233
  historic weathering rates 232, 233
  Sierra Nevada (USA) 216, 235
  Sr isotopes 232
membrane filtration 188
Mendip Hills (UK) 110
meq/l 5
methane 258, 288
Mg
  cation exchange 435
  column elution 435
  concentration decrease with heating 439
  concentrations in karst water 111, 116
  dedolomitization 113
  dolomite dissolution 103
  in river water 316
  inhibitor 129
  Mg-calcites 88, 130, 134
  $MgHCO_3$ type water 166, 170
  sepiolite precipitation 299
micaschist 2, 119, 205, 214
microbial mediation 258
microcline 202
mineral distribution in soils 203, 207
Mischungskorrosion 307
mixed reservoir 337
mixing cell model 374, 378, 423
  mixing factor 375, 379
mixing of water types 305-323
  acid groundwater 309
  carbonate waters 306
  cave formation 308
  common ion effect 309
  Fe(II)-rich water 311
  groundwater mixing 305-313
  karst 307, 315
  lag effects 319
  Mischungskorrosion 307
  mixing models 315
  modeling with PHREEQE 410
  pH change in mixed water 306
  reduced/oxidized water mixing 311-314
  subsaturation due to mixing 306
  supersaturation due to mixing 308
  surface water mixing 314-323
Mn 255, 258
  cation exchange 442-443
  column elution 442-443
  in redox reactions 239-255
  modeling MnO_2 reduction 442

redox diagram 255
redox sequence 259
solid solution 63
mobile part of pore water 369
  mobile fraction 468
  modeling of stagnant zone transport 472
modeling, see geochemical modeling
molality 5
molar fraction 156, 158
  from equivalent fraction 198
molar weights 7
molarity 5
montmorillonite 152, 202, 205
  Cd vs. Na exchange 164
multicomponent exchange 160, 165, 464
  sharp fronts 464-466
muscovite 151

N
  acetylene block technique 275
  denitrification 250, 272
  NH_4^+ oxidation 430
  nitrate reduction 250, 271-279, 314
  nitrification 250
  nitrite 272, 273
  redox diagram 251
  redox sequence 257-260
Na
  colloid stability 457
  column elution 435-437
  in loess 144
$NaHCO_3$ type water 143-148, 168, 170
  softening hard water 439
NaCl 302
NaF 81
negative adsorption 189
Negev desert 144
Nernst equation 240
NETPATH 217, 305
New Zealand karst 112
Newton iteration 383
NH_4^+ oxidation 430
  clay interlayer release 437
Ni
  thermodynamic data 498-499
Nile delta 144-145
nitrate, see also N
  denitrification 250, 272-275
  injection in reduced aquifer 314
    electron balance 273, 275
  and dissolved organic matter 273
  reduction 271-279, 314, 418
    by Fe^{2+} 278-279
    by Fe-silicates 279
    by lignite 277
    by pyrite 275-278

nitrification 250
nitrite 272, 273
non-equilibrium sorption 472-473
  diffusion model 473
  modeling 472
  two-box model 473
non-linear isotherms 384
normal density distribution 353, 355
normality 5
nucleation of crystals 72
numerical dispersion 378
  corrections 379, 383
  in PHREEQM 433-434, 443
numerical models of transport 373-389
  advection-dispersion 380
  backward differences 374
  boundary adaptions 376
  central differences 374
  diffusion 376
  forward differences 374
  linear sorption isotherm 380
  mixing cell model 374, 378
    mixing factor 375, 379
  non linear isotherms 384
  numerical dispersion 378
    corrections 379, 383
  numerical oscillations 375
  seawater diffusion 376
  stability conditions 375, 376, 379
  Taylor expansion 374
nutritive value, of water 9

O_2
  adsorption on pyrite 269
  advective vs. diffusive flux of O_2 264
  diffusion in unsaturated zone 264
  DOC oxidation by O_2 274
  concentration in groundwater 261
    water table depth and O_2 262, 274
  isotopes 303
  pyrite oxidation by O_2 263, 268
  reduction 261-265, 418
  solubility in water 82, 261
octahedral layer of clay minerals 151
octanol/water distribution coefficient 343
opal 214
open system dissolution 99
organic carbon/water distribution coefficient 343
organic chemicals 456
organic compounds, distribution coefficient 341-344
  H/O ratio 458
organic matter 286
  CEC 149
  DOC (dissolved organic carbon) 5, 273

lignite 277
  oxidation 273, 418
organic pollutants 341-344
oscillations in numerical models 375
otavite 65
overland flow 314
oxidant 239
oxidation, by Fe^{3+} 270
  by NO_3 275-278
  by O_2 263, 268
  kinetics 265-271
  numerical calculation 442-443
oxides, *PZC* 154
  solubilities 283, 487
  thermodynamic data 487
oxygen see O_2

P
  hydroxyapatite 75
  PO_4^{3-} as inhibitor 79, 130, 132
Pb
  redox diagram 290
  specific adsorption 183
  thermodynamic data 499-502
PCB 344
pe 245
  pe calculation in models 416
  relation with *Eh* 246
Peclet number 359
Pennsylvania, carbonate aquifers 107
peptization of colloids 187
percolation rate in unsaturated zone 329
periodic table 7
permeability 327, 328
  variations 387
permeability stratified aquifer 363
permittivity 185
petrographic surface, of mineral grains 73
pH, field analysis 15, 96
  effects
    dedolomitization 427-430
    on distribution coefficient 183
    evapotranspiration 430-431
    NH_4^+ oxidation 430
    NO_3^- reduction 418
  change in mixed water 306
  of soil, in water and 1M KCl 190
  calculated from *TIC* 57, 416
phreatic aquifers 331-336
PHREEQE 305, 409-445
  activity coefficient 413
  applications
    ATES 419, 439-441
    calcite precipitation 419, 439
    dedolomitization 427-430

complexation effects 428
  pH effects 428-430
  mixing 410
  NO_3^- reduction 418
  O_2 reduction 418
  organic matter oxidation 418
  pe calculation 416
  pH from *TIC* 416
  thermal energy storage 419, 439-441
complexes 412
equilibrating 410, 416
PHREEQE 305, 409-445
  keywords and variables 410-419
    ELEMENTS 411
    END 419
    IUNITS 415
    KNOBS 417, 418
    MINERALS 416
    NCOMPS 411
    NEUTRAL 417
    NSTEPS 411
    OPTIONS 410
    PHREEDA 411, 427
    REACTION 417
    SIMIN 414, 416
    SOLUTION 415
    SPECIES 412
    STEPS 417
    SUMS 419
    TEMP 419
    THSP 413
    TITLE 410
    V0 411
  manual 409
  reactions 410
  redox 410
  redox status 413
  saturation index 414, 416
  temperature effects 413
  titrating 410
PHREEQM 409-445
  cation exchange 422, 426, 435-437
  cell length 422-423
  diffusion 424, 434
    diffusion coefficient 425
  discretization 422-423
  dispersivity 423
  equilibration 423, 442
  exchangeable cations 426, 436
  execution time 434
PHREEQM 409-445
  keywords and variables 421-426
    LAYERSOL 422
    DISP 423
    TRANSPRT 424

MEDIUM 425
SIMEX 423
SPECIES 426
PHREEDA 423, 427
NCOL 424
NCELL 422
manual 421-426
mixing cell principle 374, 378, 423
numerical dispersion 433-434, 443
radial flow 423, 437, 439
reactions 424
saturation index 423
spreadsheet files 425
PHRQPITZ 56
physical non-equilibrium 468
piezometers 13
PIP 409
Piper diagram 36, 145
piping 315
piston flow 329
plagioclase 202, 205, 234
platinum electrode 241
Pleistocene carbonate aquifers 134
PO_4^{3-} inhibition 79, 130, 132
point of zero charge see *PZC*
Poisson equation 185
polysulfides 266-267
pore water flow velocity 328, 329, 331
porosity 328
    and formation factor 369
    and diffusion coefficient 369
    effective porosity 328
    water filled porosity 329
potential gradient, for groundwater flow 327
potential determining ions, at charged surface 181-182
Pouhon (spring type) 137
ppb 5
ppm 5
precipitates 299
    calcite 127
    inhibitors 129
precipitation (rain) 22
    surplus 2
precision of analysis 16
preferential flow 330, 369
pressure effect, in reactions 480
pressure filtrate 189
proton adsorption 175-177
Psi condition (of Sillén) 451
pyrite 256, 287
    oxidation 55, 235, 264, 289
        bacterial mediation 275
        by NO_3 275-278
        by O_2 263, 268

    by Fe^{3+} 270
    kinetics 265-271
    FeOOH armoring 269, 280
    formation 82, 287
    O_2 adsorption on pyrite 269
pyroxene 204, 223
pyrrhotite 256
PZC (point of zero charge) 153
    anion exchange 154, 155
    clay minerals 154
    oxides 154

quality changes with discharge 315-323
    lag effects 319
    hysteresis in runoff quantity/quality 319
quartz 202, 213

radial flow 333, 423, 437
rain, evaporative concentration 2, 225-227, 430
rainwater chemistry 22
    acidification 25
    atmospheric transport 27
    dry deposition 28, 31
    fractionation 23
    industrial sources 25
    interception 32
    sampling 22
    seasonal effects 30
    seawater contribution 24
random walk 354
rapid throughflow 315
rate laws 68
rate mechanisms 68-74
ratio principle, with salt precipitation 299
Rb
    multicomponent exchange 464-466
reaction modeling
    PHREEQE 410
    PHREEQM 424
reaction order 68-70
reactions
    endothermal 62, 481
    enthalpy 61, 481
    exothermal 62
    Gibbs free energy 59, 480
    kinetics 66-80
redox cell 241
redox classification 259
redox diagrams 247-256, 290
    As 249
    H_2O 247
    Mn 255
    N 251
    Fe 254-256
    Pb 290
    S 255, 290

redox in PHREEQE 410
  redox status 413
  setting of pe 416
redox potential 242
redox problems, geochemical model 399
redox reactions
  electron donor/acceptor 239
  oxidator/reductor 239
  reaction balancing 240
redox sequence 257-261
redox zoning 257, 261
redoxcline 275
reduced/oxidized water mixing 311-314
reduction 261-265
regular solution 65
release, of ions from feldspars 440
remediation of acidification 230, 445
residence time 336
retardation 339, 344-348, 450
  isotherm effects 348, 349, 450
  linear retardation 339-341, 380
  solute transport 370
retardation factor 341, 450
return flow fraction 194, 302
Rhine sediment 435-437
rhodochrosite 63-64, 255, 290
rhyolite 204
runoff models 314
  groundwater contribution to runoff 315
  mixing models 315
  overland flow 314
  piping 315
  rapid throughflow 315
  runoff contributions 315, 322
  saturated overland flow 315
rutile 181, 183

S
  arsenopyrite 265
  bacterial mediation of redox reactions 275
  chalcopyrite 265
  gypsum solubility 47, 53
  H_2S 260, 285
  isotopes 285
  polysulfides 266-267
  pyrite oxidation 264, 289
  redox sequence 257-261
  redox diagram 255, 290
  reduction by organic matter 286
  sphalerite 265
  sulfate reduction 71, 285-289
  sulfate bacteria 312
  sulfates, thermodynamic data 490
  sulfite 267
  thiosulfate 267, 268

salinity jump 168
salinization 173-174
  diffusion vs advection 174
salt minerals 302
  precipitation 298
  ratio principle 299
  dissolution 303
salt water 296
salt water upconing 145
salt water intrusion 144-148, 173-174
  $CaCl_2$ type water 144-148, 173
  conservative mixing 146, 147
  pH effects 148
  sulfate reduction 148
sampling of water 10
  conservation 14
  depth integrated 10
  depth specific 10
  rainwater 22
  sampling method and dispersion 356, 361
SAR (sodium adsorption ratio) 188, 192, 302
  adjusted for calcite precipitation 194
  irigation water quality 194
saturated overland flow 315
  mixing models 315
saturation index 55
  in PHREEQE 414, 416
  in PHREEQM 423
saturation state 55
scale effect, in dispersivity 362, 363
seasonal fluctuations 30, 97
seawater contribution to rain 24
seawater diffusion, numerical model 376
seawater intrusion 143, 173
self sharpening fronts 349
sepiolite 299
serpentine 119
sharpening fronts 349, 433-440, 454
  with ion exchange 463
Si, see also silicate
  groundwater concentration 204
  in heated groundwater 439-441
  concentration in brines 305
SI (saturation index) 55
siderite 254, 283, 284
Sierra Nevada (USA) 216, 235
silicate minerals, see also clay minerals
  albite 204, 205
  amorphous silica 213
  amphibole 202, 223
  anorthite 204
  augite 221
  biotite 202, 204
  bronzite 221
  Ca-silicate stability 212

chalcedony 214
cristobalite 214
diopside 221
dissolution kinetics 212, 220-235
    pH dependence (silicate weathering) 222, 234
    enstatite 221
    epidote 234
    forsterite 223
    hornblende 234
    lifetimes 221
    microcline 202
    opal 214
    plagioclase 202, 205, 234
    pyroxene 204, 223
    quartz 202, 213
    thermodynamic data 489
    tridymite 213
    weathering sequence 203
silicate rocks
    basalt 205, 214
    granite 3, 205, 214
    groundwater composition 3, 204
    micaschist 205, 214
    rhyolite 205
silicate stability 212, 214
silicate weathering 220-223
single shot input 352, 361
smectite 152
sodium hazard 188-192
sodium adsorption ratio, see SAR
softening of hard water 439
soil, CO_2 pressure 97
    unsaturated zone 98, 105
solid solution 62-66
    Al in goethite 63
    Cd in calcite 65
    Margules parameters 64
    Mg-calcite 89, 130
    regular 65
    rhodochrosite 63-64
    theory 62
solubility 45
    control on dissolution rate 69
solubility product 46
    pressure dependence 480
    temperature dependence 61
solute profiles along flowline 170-174, 258-261, 454
solute transport
    advection in aquifers 331
    analytical solutions for mass transport 365, 367, 371
    broadening front 349, 450-453
    combined physical and chemical effects 370

diffusion 350, 352, 354
dispersion 350
numerical modeling 373-389
retardation 339
sharpening front 349, 433-440, 454
spreading in transport front 357-360
sorption
    bound residue 468
    concave isotherm 348
    convex isotherm 348
    Freundlich isotherm 347
        exponent effect on breakthrough curves 384
    hysteresis 467-469
    immobilization 468
    isotherms 347
        effects on retardation 348
    Langmuir isotherm 347
    linear isotherm 339, 347
species 399
specific adsorption 175, 183
    background electrolyte 182
    Cd, Cu, Pb 183
    earth alkaline ions 181, 183
    protons 175
    on rutile 181, 183
specific discharge 327
sphalerite 265
spreading, in transport front 357, 359, 360
Sr 136, 163, 317
    displacing solution 435-437
    distribution coefficient 163
    in aragonite 89, 136
    in mass balance studies 232
    isotopes 232
    and weathering rate 232
stability constant 51
    pressure dependence 480
    temperature dependence 61
stability of water 247
stability, of numerical calculations 375, 376, 379
stagnant zones 468
standard hydrogen electrode 241
standard potential 241
standard state 482
standards for drinking water 9
step input function 356
Stiff diagram 3, 145
stoichiometric saturation 131
stop-flow, in solute transport 474
structural charge of clay minerals 150-153
subsaturation, due to mixing 306
sulfate, see also S
    reduction 71, 285
        by organic matter 286
        with sea water intrusion 148

bacteria 312
sulfite 267
sulfur see S
supersaturation, for calcite 130
    due to mixing 308
surface area 73
    BET method 73
surface charge, of clay minerals 153
surface complexation 80, 177-180
    background electrolyte 178, 182
    in kinetic mechanisms 80
    metals 182
    protons 177
surface controlled reaction 75
surface etching 73, 129, 221
surface potentials 177, 185
    constant capacitance model 178
surface water, mixing 314-323
swamps 284
swelling of clay minerals 188

talc 116
Taylor expansion 374
TCE 453
TEC (total exchangeable cations) 156
temperature dependence, of *K* 61
    of desorption 456
tetrahedral layer in clay minerals 151
thermal energy storage 419, 439-441
thermodynamics 59, 480-484
Thiobacillus 270, 275
thiosulfate 267, 268
throughput parameter 451
TIC (total inorganic carbon) 5, 91
tidal flat 166
titrating in PHREEQE 410
titrations
    aluminum oxide 179
    fulvic acids 180
    goethite 176
    humic acids 197
    rutile 181, 183
TOC (total organic carbon) 5
toluene 372
tortuosity 359, 367
transmissivity 328
total exchangeable cations 156
trace metals, ion exchange 163
    distribution coefficient 344-346
tracer distribution (in aquifers) 337
transmissivity 328
transport controlled reaction 75
transport, diffusive vs. advective 470
    distance with diffusion 354
    velocity 450

transversal dispersion 350
trichloroethylene 453
tridymite 213
tritium profiles 329, 337
    dating 333, 337
    in aquifers 333, 337
    in unsaturated zone 330
turbulent flow 124
two-box model 473

units recalculation, exchangeable cations 149
    equiv/mol fraction 161, 198
    solute concentrations 6
unsaturated zone 98, 105, 329
    Cl profiles 330
    percolation rate 329
    tritium profiles 330

Van 't Hoff equation 61
Vanselow convention 156, 405
variance, of measurements 355
Veluwe (NL) 2, 108, 137, 224
vermiculite 151
villiaumite 81
vivianite 283

Wadden Sea (NL) 166
waste site (see also landfill) 371
    diffusion through clay liner 373
    leachate spreading 371
WATEQ 56
WATEQP 56-58
water filled porosity 329
water quality routing 320
weathering 202-235
    field weathering rates 230-235
        model 233
    Goldich weathering sequence 203, 221
    historic weathering rates 232
    hydrological conditions and weathering products 205-207
    incongruent dissolution 202
    kinetics of silicate weathering 220-223, 233
    leaching 205
    mass balance 215-220, 231
    rates 230
        rate from Sr isotopes 232
    sequence of silicate weathering 203
well clogging 188, 312-314
well flushing, for sampling 12
well injection 313
    fresh water in brackish water 168-170, 437
    NaNO$_3$ in reduced aquifer 314
well recovery 437, 439

Yucatan 89, 308

Zeeland (NL) 147
zero order reaction 70, 71

Zn
  distribution coefficient 163
  retardation 345
Zuiderzee (NL) 368